Honda CB500 and CBF500
Service and Repair Manual

by Phil Mather

(3753-272-4AH2)

Models covered
CB500. 499cc. 1993 to 2002
CB500S. 499cc. 1998 to 2002
CBF500. 499cc. 2003 to 2007
CBF500A. 499cc. 2003 to 2008

ABCDE
FGHIJ
KLM

© Haynes Publishing 2008

A book in the **Haynes Service and Repair Manual Series**

All rights reserved. No part of this book may be reproduced or transmitted in any form or by any means, electronic or mechanical, including photocopying, recording or by any information storage or retrieval system, without permission in writing from the copyright holder.

ISBN **978 1 84425 770 6**

British Library Cataloguing in Publication Data
A catalogue record for this book is available from the British Library

Printed in USA

Haynes Publishing
Sparkford, Yeovil, Somerset BA22 7JJ, England

Haynes North America, Inc
861 Lawrence Drive, Newbury Park, California 91320, USA

Haynes Publishing Nordiska AB
Box 1504, 751 45 UPPSALA, Sweden

Contents

LIVING WITH YOUR HONDA CB

Introduction
The Birth of a Dream	Page	0•4
Acknowledgements	Page	0•8
About this manual	Page	0•8
Identification numbers	Page	0•9
Buying spare parts	Page	0•9
Bike spec	Page	0•10
Model development	Page	0•11
Safety First!	Page	0•12

Daily (pre-ride) checks
Engine/transmission oil level check	Page	0•13
Brake fluid level checks	Page	0•14
Suspension, steering and final drive checks	Page	0•14
Tyre checks	Page	0•15
Coolant level check	Page	0•16
Legal and safety checks	Page	0•16

MAINTENANCE

Routine maintenance and servicing
Specifications	Page	1•2
Recommended lubricants and fluids	Page	1•2
Maintenance schedule	Page	1•3
Component locations	Page	1•4
Maintenance procedures	Page	1•7

Contents

REPAIRS AND OVERHAUL

Engine, transmission and associated systems

Engine, clutch and transmission	Page	2•1
Cooling system	Page	3•1
Fuel and exhaust systems	Page	4•1
Ignition system	Page	5•1

Chassis and bodywork components

Frame and suspension	Page	6•1
Final drive	Page	6•17
Brakes	Page	7•1
Wheels	Page	7•17
Tyres	Page	7•23
Fairing and bodywork	Page	8•1

Electrical system

	Page	9•1

Wiring diagrams

	Page	9•27

REFERENCE

Tools and Workshop Tips	Page	REF•2
Security	Page	REF•20
Lubricants and fluids	Page	REF•23
Conversion Factors	Page	REF•26
MOT test checks	Page	REF•27
Storage	Page	REF•32
Fault finding	Page	REF•35
Fault finding equipment	Page	REF•45

Index

	Page	REF•49

Introduction

The Birth of a Dream

by Julian Ryder

There is no better example of the Japanese post-war industrial miracle than Honda. Like other companies which have become household names, it started with one man's vision. In this case the man was the 40-year old Soichiro Honda who had sold his piston-ring manufacturing business to Toyota in 1945 and was happily spending the proceeds on prolonged parties for his friends. However, the difficulties of getting around in the chaos of post-war Japan irked Honda, so when he came across a job lot of generator engines he realised that here was a way of getting people mobile again at low cost.

A 12 by 18-foot shack in Hamamatsu became his first bike factory, fitting the generator motors into pushbikes. Before long he'd used up all 500 generator motors and started manufacturing his own engine, known as the 'chimney', either because of the elongated cylinder head or the smoky exhaust or perhaps both. The chimney made all of half a horsepower from its 50 cc engine but it was a major success and became the Honda A-type.

Less than two years after he'd set up in Hamamatsu, Soichiro Honda founded the Honda Motor Company in September 1948. By then, the A-type had been developed into the 90 cc B-type engine, which Mr Honda decided deserved its own chassis not a bicycle frame. Honda was about to become Japan's first post-war manufacturer of complete motorcycles. In August 1949 the first prototype was ready. With an output of three horsepower, the 98 cc D-type was still a simple two-stroke but it had a two-speed transmission and most importantly a pressed steel frame with telescopic forks and hard tail rear end. The frame was almost triangular in profile with the top rail going in a straight line from the massively braced steering head to the rear axle. Legend has it that after the D-type's first tests the entire workforce went for a drink to celebrate and try and think of a name for the bike. One man broke one of those silences you get when people are thinking, exclaiming 'This is like a dream!' 'That's it!' shouted Honda, and so the Honda Dream was christened.

> 'This is like a dream!'
> 'That's it'
> shouted Honda

Mr Honda was a brilliant, intuitive engineer and designer but he did not bother himself with the marketing side of his business. With hindsight, it is possible to see that employing Takeo Fujisawa who would both sort out the home market and plan the eventual expansion into overseas markets was a masterstroke. He arrived in October 1949 and in 1950 was made Sales Director. Another vital new name was Kiyoshi Kawashima, who along with Honda himself, designed the company's first four-stroke after Kawashima had told them that the four-stroke opposition to Honda's two-strokes sounded nicer and therefore sold better. The result of that statement was the overhead-valve 148 cc E-type which first ran in July 1951 just two months after the first drawings were made. Kawashima was made a director of the Honda Company at 34 years old.

The E-type was a massive success, over 32,000 were made in 1953 alone, a feat of mass-production that was astounding by the

Honda C70 and C90 OHV-engined models

Introduction

standards of the day given the relative complexity of the machine. But Honda's lifelong pursuit of technical innovation sometimes distracted him from commercial reality. Fujisawa pointed out that they were in danger of ignoring their core business, the motorised bicycles that still formed Japan's main means of transport. In May 1952 the F-type Cub appeared, another two-stroke despite the top men's reservations. You could buy a complete machine or just the motor to attach to your own bicycle. The result was certainly distinctive, a white fuel tank with a circular profile went just below and behind the saddle on the left of the bike, and the motor with its horizontal cylinder and bright red cover just below the rear axle on the same side of the bike. This was the machine that turned Honda into the biggest bike maker in Japan with 70% of the market for bolt-on bicycle motors, the F-type was also the first Honda to be exported. Next came the machine that would turn Honda into the biggest motorcycle manufacturer in the world.

The C100 Super Cub was a typically audacious piece of Honda engineering and marketing. For the first time, but not the last, Honda invented a completely new type of motorcycle, although the term 'scooterette' was coined to describe the new bike which had many of the characteristics of a scooter but the large wheels, and therefore stability, of a motorcycle. The first one was sold in August 1958, fifteen years later over nine-million of them were on the roads of the world. If ever a machine can be said to have brought mobility to the masses it is the Super Cub. If you add in the electric starter that was added for the C102 model of 1961, the design of the Super Cub has remained substantially unchanged ever since, testament to how right Honda got it first time. The Super Cub made Honda the world's biggest manufacturer after just two years of production.

The CB250N Super Dream became a favorite with UK learner riders of the late seventies and early eighties

Honda's export drive started in earnest in 1957 when Britain and Holland got their first bikes, America got just two bikes the next year. By 1962 Honda had half the American market with 65,000 sales. But Soichiro Honda had already travelled abroad to Europe and the USA, making a special

The GL1000 introduced in 1975, was the first in Honda's line of GoldWings

Introduction

Carl Fogarty in action at the Suzuka 8 Hour on the RC45

An early CB750 Four

point of going to the Isle of Man TT, then the most important race in the GP calendar. He realised that no matter how advanced his products were, only racing success would convince overseas markets for whom 'Made in Japan' still meant cheap and nasty. It took five years from Soichiro Honda's first visit to the Island before his bikes were ready for the TT. In 1959 the factory entered five riders in the 125 class. They did not have a massive impact on the event being benevolently regarded as a curiosity, but sixth, seventh and eighth were good enough for the team prize. The bikes were off the pace but they were well engineered and very reliable.

The TT was the only time the West saw the Hondas in '59, but they came back for more the following year with the first of a generation of bikes which shaped the future of motorcycling – the double-overhead-cam four-cylinder 250. It was fast and reliable – it revved to 14,000 rpm – but didn't handle anywhere near as well as the opposition. However, Honda had now signed up non-Japanese riders to lead their challenge. The first win didn't come until 1962 (Aussie Tom Phillis in the Spanish 125 GP) and was followed up with a world-shaking performance at the TT. Twenty-one year old Mike Hailwood won both 125 and 250 cc TTs and Hondas filled the top five positions in both races. Soichiro Honda's master plan was starting to come to fruition, Hailwood and Honda won the 1961 250 cc World Championship. Next year Honda won three titles. The other Japanese factories fought back and inspired Honda to produce some of the most fascinating racers ever seen: the awesome six-cylinder 250, the five-cylinder 125, and the 500 four with which the immortal Hailwood battled Agostini and the MV Agusta.

When Honda pulled out of racing in '67 they had won sixteen rider's titles, eighteen manufacturer's titles, and 137 GPs, including 18 TTs, and introduced the concept of the modern works team to motorcycle racing. Sales success followed racing victory as Soichiro Honda had predicted, but only because the products advanced as rapidly as the racing machinery. The Hondas that came to Britain in the early '60s were incredibly sophisticated. They had overhead cams where the British bikes had pushrods, they had electric starters when the Brits relied on the kickstart, they had 12V electrics when even the biggest British bike used a 6V system. There seemed no end to the technical wizardry. It wasn't that the technology itself was so amazing but just like that first E-type, it was the fact that Honda could mass-produce it more reliably than the lower-tech competition that was so astonishing.

When in 1968 the first four-cylinder CB750 road bike arrived the world of motorcycling changed for ever, they even had to invent a new word for it, 'Superbike'. Honda raced again with the CB750 at Daytona and won the

Introduction

World Endurance title with a prototype DOHC version that became the CB900 roadster. There was the six-cylinder CBX, the CX500T – the world's first turbocharged production bike, they invented the full-dress tourer with the GoldWing, and came back to GPs with the revolutionary oval-pistoned NR500 four-stroke, a much-misunderstood bike that was more a rolling experimental laboratory than a racer. Just to show their versatility Honda also came up with the weird CX500 shaft-drive V-twin, a rugged workhorse that powered a new industry, the courier companies that oiled the wheels of commerce in London and other big cities.

It was true, though, that Mr Honda was not keen on two-strokes – early motocross engines had to be explained away to him as lawnmower motors! However, in 1982 Honda raced the NS500, an agile three-cylinder lightweight against the big four-cylinder opposition in 500 GPs. The bike won in its first year and in '83 took the world title for Freddie Spencer. In four-stroke racing the V4 layout took over from the straight four, dominating TT, F1 and Endurance championships with the RVF750, the nearest thing ever built to a Formula 1 car on wheels. And when Superbike arrived Honda were ready with the RC30. On the roads the VFR V4 became an instant classic while the CBR600 invented another new class of bike on its way to becoming a best-seller. The V4 road bikes had problems to start with but the VFR750 sold world-wide over its lifetime while the VFR400 became a massive commercial success and cult bike in Japan. The original RC30 won the first two World Superbike Championships is 1988 and '89, but Honda had to wait until 1997 to win it again with the RC45, the last of the V4 roadsters. In Grands Prix, the NSR500 V4 two-stroke superseded the NS triple and became the benchmark racing machine of the '90s. Mick Doohan secured his place in history by winning five World Championships in consecutive years on it.

In yet another example of Honda inventing a new class of motorcycle, they came up with the astounding CBR900RR FireBlade, a bike with the punch of a 1000 cc motor in a package the size and weight of a 750. It became a cult bike as well as a best seller, and with judicious redesigns continues to give much more recent designs a run for their money.

When it became apparent that the high-tech V4 motor of the RC45 was too expensive to produce, Honda looked to a V-twin engine to power its flagship for the first time. Typically, the VTR1000 FireStorm was a much more rideable machine than its opposition and once accepted by the market formed the basis of the next generation of Superbike racer, the VTR-SP-1.

One of Mr Honda's mottos was that technology would solve the customers' problems, and no company has embraced cutting-edge technology more firmly than Honda. In fact Honda often developed new technology, especially in the fields of materials science and metallurgy. The embodiment of that was the NR750, a bike that was misunderstood nearly as much as the original NR500 racer. This limited-edition technological tour-de-force embodied many of Soichiro Honda's ideals. It used the latest techniques and materials in every component, from the oval-piston, 32-valve V4 motor to the titanium coating on the windscreen, it was – as Mr Honda would have wanted – the best it could possibly be. A fitting memorial to the man who has shaped the motorcycle industry and motorcycles as we know them today.

The CX500 – Honda's first V-Twin and a favorite choice of dispatch riders

The VFR400R was a cult bike in Japan and a popular grey import in the UK

Introduction

CB500

CBF500

Jack of all trades

There are some things Honda do better than anyone else and always have done. Giving customers value for money and quality in smaller capacity machines is one of them. Remember the old 400 cc Dream, that workhorse of the 1980s? It took the place in Honda's range previously occupied by the 400/4 of blessed memory. Heresy said the enthusiasts, ignoring the fact that the new twin was in every measurable way an improvement on old multi. It was lighter, faster, better handling and cheaper to name just four areas of improvement. Sure the 250 Dream was underpowered, but that was just a sleeved down 400 for the UK learner market. The 400 Dream, and latterly 400 Super Dream, stayed in the range for what seemed like decades and was the most popular machine around for despatch riders, commuters and newly qualified riders.

The CB500 took on the challenge of doing the same job from 1993 and well into the 21st-Century. Like the old Dream it is a parallel twin in a steel, twin-shock chassis and like the old bike it has remained very largely unchanged through its model life. The addition of a rear disc brake in its third year is the only significant change to the specification made since it arrived late in 1993. Like the Dream, the CB500 is a true all-rounder, unlike the Dream, the CB500 is also a good-looking bike.

In many ways the two bikes show just how far one often significant aspect of motorcycle technology has come despite the outward similarities between the designs. Although the claimed weights of each bike are nearly the same at 170 kg dry, the reality is the new bike is considerably lighter and that is despite the water-cooling it carries to get through modern noise regulations. Its 57 hp gives it a top speed of around 115 mph, the Dream was a good 10 mph slower.

If you judge the CB500 solely by its spec-sheet, you would be forgiven for being underwhelmed. The riding experience is all here, it may have been aimed at the new rider or workaday commuter but riding it is a fun experience. In fact it is such a competent machine in every department that it is impossible to fault, if there is a weakness it is that the Honda is more expensive than its obvious competition in the twin-cylinder ranks. The fact it can handle corners as well as the commuter slog is born out by the all-action CB500 Cup that was run in the UK and France, providing the sort of action not seen since the good old days of the Yamaha 350LC Pro-Am. In the UK, there were classes for novices and national class riders and it was there that we first saw the talent of James Toseland who went on to race in Supersport and Superbike world championships as a works rider.

That bit of race-track cred helps explain why the bike has hardly changed since it first appeared at the end of 1993 ('94 model year), it simply didn't need improving. There were detail changes and a rear disc replaced the drum stopper for the 1997 model year. The Cup model was an option alongside the standard model – it had silver paint and a 'Cup' logo on the tank to go with the race series. The CB500S was introduced in 1998, a half-faired model with square headlight and modified instrumentation. Right through its model life, the other changes to the CB500 have been limited to new paint schemes for each model year.

2004 saw the introduction of the CBF500, an updated and re-styled version using the same parallel twin engine. Changes to the bike though not outwardly remarkable are quite significant. The engine is now hung from a new box-section spine frame and is used as a stressed member. At the rear the design of the swingarm has changed and the bike now has a single shock absorber giving it a modern sleek look that is complemented by twin seats and a new seat cowling and tail section. Electronics of course play their part to an extent with a new instrument cluster that no longer relies on a speedometer cable, and the use of a throttle position sensor to improve fuelling and emissions. Also available is the CBF500A variant which has an anti-lock brake system.

Acknowledgements

Our thanks are due to Bransons Motorcycles of Yeovil who supplied the machines featured in the illustrations throughout this manual. We would also like to thank NGK Spark Plugs (UK) Ltd for supplying the colour spark plug condition photographs, the Avon Rubber Company for supplying information on tyre fitting and Draper Tools Ltd for some of the workshop tools shown.

Thanks are also due to Julian Ryder who wrote 'The Birth of a Dream' and to Honda (UK) Ltd who supplied model photographs.

About this Manual

The aim of this manual is to help you get the best value from your motorcycle. It can do so in several ways. It can help you decide what work must be done, even if you choose to have it done by a dealer; it provides information and procedures for routine maintenance and servicing; and it offers diagnostic and repair procedures to follow when trouble occurs.

We hope you use the manual to tackle the work yourself. For many simpler jobs, doing it yourself may be quicker than arranging an appointment to get the motorcycle into a dealer and making the trips to leave it and pick it up. More importantly, a lot of money can be saved by avoiding the expense the shop must pass on to you to cover its labour and overhead costs. An added benefit is the sense of satisfaction and accomplishment that you feel after doing the job yourself.

References to the left or right side of the motorcycle assume you are sitting on the seat, facing forward.

We take great pride in the accuracy of information given in this manual, but motorcycle manufacturers make alterations and design changes during the production run of a particular motorcycle of which they do not inform us. No liability can be accepted by the authors or publishers for loss, damage or injury caused by any errors in, or omissions from, the information given.

Identification numbers

Frame and engine numbers

The frame serial number is stamped into the right-hand side of the steering head. The engine number is stamped into the top of the crankcase on the right-hand side. Both of these numbers should be recorded and kept in a safe place so they can be furnished to law enforcement officials in the event of a theft. There is also a carburettor identification number on the intake side of each carburettor body, and a colour code label on the top of the rear mudguard, under the seat.

The frame serial number, engine serial number, carburettor identification number and colour code should also be kept in a handy place (such as with your driver's licence) so they are always available when purchasing or ordering parts for your machine.

The procedures in this manual identify models by their code letter (eg X, meaning a 1999 standard model, and SX, meaning a 1999 faired model). The code letter is printed on the colour code label stuck to the top of the rear mudguard. Refer to the colour code label or the table below to identify your exact model.

Buying spare parts

Once you have found all the identification numbers, record them for reference when buying parts. Since the manufacturers change specifications, parts and vendors (companies that manufacture various components on the machine), providing the ID numbers is the only way to be reasonably sure that you are buying the correct parts.

Whenever possible, take the worn part to the dealer so direct comparison with the new component can be made. Along the trail from the manufacturer to the parts shelf, there are numerous places that the part can end up with the wrong number or be listed incorrectly.

The two places to purchase new parts for your motorcycle – the accessory store and the franchised dealer – differ in the type of parts they carry. While dealers can obtain every part for your motorcycle, the accessory dealer is usually limited to normal high wear items such as shock absorbers, tune-up parts, various engine gaskets, cables, chains, brake parts, etc. Rarely will an accessory outlet have major suspension components, camshafts, transmission gears, or cases.

Used parts can be obtained for roughly half the price of new ones, but you can't always be sure of what you're getting. Once again, take your worn part to the breaker for direct comparison.

Whether buying new, used or rebuilt parts, the best course is to deal directly with someone who specialises in parts for your particular make.

Model	Code	Year
CB500	R	1993 to 95
CB500	T	1996
CB500	V	1997
CB500	W	1998
CB500	X	1999
CB500	Y	2000 to 01
CB500	2	2002
CB500S	W	1998
CB500S	X	1999
CB500S	Y	2000 to 01
CB500S	2	2002
CBF500	4	2004 to 05
CBF500A	4	2004 to 05
CBF500	6	2006 to 07
CBF500A	6	2006 to 08

The colour code label is stuck to the rear mudguard

Engine number location

Frame number location

Illegal Copying

It is the policy of Haynes Publishing to actively protect its Copyrights and Trade Marks. Legal action will be taken against anyone who unlawfully copies the cover or contents of this Manual. This includes all forms of unauthorised copying including digital, mechanical, and electronic in any form. Authorisation from Haynes Publishing will only be provided expressly and in writing. Illegal copying will also be reported to the appropriate statutory authorities.

0•10 Bike spec

Bike spec

Weights and dimensions – CB models
Overall length . . . . .2090 mm (82.3 in)
Overall width . . . . 720 mm (28.3 in)
Overall height
  R, T, V, W, X, Y, 2 models . . . . .1050 mm (41.3 in)
  S-W, S-X, S-Y, S-2 models . . . . .1160 mm (45.7 in)
Wheelbase
  R, T, V, W, X, Y, 2 models . . . . .1430 mm (56.3 in)
  S-W, S-X, S-Y, S-2 models . . . . .1435 mm (56.5 in)
Seat height . . . . .775 mm (30.5 in)
Ground clearance . . . . .145 mm (5.7 in)
Engine weight . . . . 56 kg (123 lb)
Dry weight (no fuel and oil)
  R and T models . . . . 170 kg (375 lb)
  V, W, X, Y, 2 models . . . . 173 kg (381 lb)
  S-W, S-X, S-Y, S-2 models . . . . 179 kg (395 lb)
Max. payload (rider, passenger, luggage, accessories) . . 184 kg (406 lb)

Weights and dimensions – CBF models
Overall length . . . . .2170 mm (58.4 in)
Overall width . . . . 765 mm (30.1 in)
Overall height . . . . .1110 mm (43.7 in)
Wheelbase 1480 mm (58.3 in)
Seat height 770 mm (30.3 in)
Ground clearance . . . . .140 mm (5.5 in)
Engine weight . . . . 56 kg (123 lb)
Dry weight (no fluids)
  CBF500 models . . . . 183 kg (403 lb)
  CBF500A models . . . . 186 kg (410 lb)
Dry weight (with fluids)
  CBF500 models . . . . 206 kg (454 lb)
  CBF500A models . . . . 209 kg (461 lb)
Max. payload (rider, passenger, luggage, accessories) . . . . 180 kg (397 lb)

Engine
Type . . . . Liquid-cooled, 8V parallel twin cylinder, 20° inclined from vertical
Capacity . . . . 499 cc
Bore and stroke . . . . 73 x 59.5 mm
Compression ratio . . . . 10.5 :1
Camshafts . . . . DOHC, chain-driven
Carburettors . . . . 2 x 34 mm flat-slide Keihin VP type
Ignition system . . . . Digital transistorised with electronic advance
Clutch . . . . Wet multi-plate, cable-operated
Gearbox . . . . 6-speed constant mesh
Final drive chain . . . . RK 525SMOZ5 or DID 525V8 (108 links – CB; 116 links – CBF)

Bike spec and Model development

Chassis – CB models

Frame type	Double cradle of tubular and box-section steel construction
Fuel tank capacity (inc. reserve)	18 lit (4.0 Imp gal, 4.8 US gal)
Fuel tank reserve	2.5 lit (0.55 Imp gal, 0.66 US gal)
Rake	27°20'
Trail – R, T, V, W, X, Y, 2 models	113 mm (4.45 in)
S-W, S-X, S-Y, S-2 models	108 mm (4.25 in)
Front suspension	
Type	37 mm oil-damped telescopic forks
Travel	115 mm (4.5 in)
Adjustment	non-adjustable
Rear suspension	
Type	Twin shocks acting on a steel, box-section swingarm
Travel	117 mm (4.6 in)
Adjustment	5 position pre-load
Wheels	17 inch 6-spoke alloys
Tyres – Front	110/80-17 57H
Rear	130/80-17 65H
Front brake	
CB500R and T models	single 296 mm disc with Nissin 2-piston sliding caliper
All other models	single 296 mm disc with Brembo 2-piston sliding caliper
Rear brake	
CB500R and T models	160 mm drum
All other models	single 240 mm disc with Brembo single piston sliding caliper

Chassis – CBF models

Frame type	Box-section steel spine
Fuel tank capacity (inc. reserve)	19 lit (4.2 Imp gal, 5.0 US gal)
Fuel tank reserve	3.5 lit (0.77 Imp gal, 0.92 US gal)
Rake	26°
Trail	110 mm (4.3 in)
Front suspension	
Type	41 mm oil-damped telescopic forks
Travel	108 mm (4.3 in)
Adjustment	non-adjustable
Rear suspension	
Type	Single shock absorber acting on a steel, box-section swingarm
Travel	120 mm (4.7 in)
Adjustment	7 position pre-load
Wheels	17 inch 6-spoke alloys
Tyres	
Front	120/70-17 58W
Rear	160/60-17 69W
Front brake	
CB500 models	single 296 mm disc with Nissin 2-piston sliding caliper
CBF500A models	single 296 mm disc with Nissin 3-piston sliding caliper
Rear brake	single 240 mm disc with Nissin single piston sliding caliper

Model development

CB500-R (1994 model year)

The very first CB500 model was the CB500-R, introduced in October 1993.

The CB500 uses a new 180° parallel twin engine, with a balancer shaft to smooth out engine vibration. Drive to the double overhead camshafts is by a centrally-positioned chain for direct operation of the four valves per cylinder. The engine is liquid-cooled, although the cosmetic finning on the cylinders and head give the appearance of an air-cooled engine. The clutch is a conventional wet multi-plate unit and the gearbox is 6-speed. Drive to the rear wheel is by chain and sprockets.

The engine sits in a double cradle frame, with box-section tubing for the upper rails. Front suspension is by conventional oil-damped 37 mm forks. Rear suspension, rather unusually, is by twin shock absorbers. Braking is by a single disc at the front and drum at the rear.

The CB500-R was available in black, red, green and blue colour schemes. It continued unchanged through 1995.

CB500-T (1996 model year)

The CB500-T was introduced in November 1995. There were no significant changes from the R model.

CB500-V (1997 model year)

The CB500-V was introduced in November 1996. The front brake caliper changed to Brembo manufacture, previously Nissin. The rear disc brake, also manufactured by Brembo, replaced the drum brake used on previous models. This year saw the start of CB500 Cup race series. One of the production models was available in silver with cup logo on the fuel tank to celebrate the race series.

CB500-W and S-W (1998)

The CB500-W was introduced in October 1997. There were no significant changes from the V model apart from colours and graphics. The CB500-W was available in red, yellow, black and silver.

The half-faired CB500S Sport model was introduced in February 1998. The S model differed from the standard model in having new instrumentation, a square headlight set in the fairing, and a different handlebar mounting arrangement. Colours were red, yellow and black.

CB500-X and S-X (1999)

The CB500-X and S-X were introduced in December 1998. There were no significant changes from the previous models.

CB500-Y and S-Y (2000)

The CB500-Y and S-Y were introduced in September 1999. There were no changes from previous models apart from graphics. Colours were black, red and metallic blue and all models had gold anodised wheels.

The CB500-Y and S-Y models continued unchanged through 2001.

CB500-2 and S-2 (2002)

There were no changes from previous models apart from graphics.

The CB500-2 and S-2 models continued unchanged through 2003.

CBF500-4 and A-4 (2004)

The first CBF500 models were the CBF500-4 and the CBF500A-4, introduced in November 2003.

The CBF uses the same engine, transmission and final drive as the CB models, but has updated instrumentation, fuelling and emissions systems, and various detail changes to bodywork, running gear and styling.

The engine now hangs below a box-section spine frame, and is used as a stressed member. Front suspension is by conventional oil-damped 41 mm forks. Rear suspension is by a single shock absorber. Braking is by a single disc at the front and rear, with standard models having a twin piston sliding caliper at the front and a single piston sliding caliper at the rear. The A model comes with an anti-lock braking system (ABS) that uses a triple piston sliding front caliper, with the rear using the same single piston caliper as non-ABS models.

The CBF500 was available in black, silver and blue colour schemes, with updated styling and bodywork. It continued unchanged through 2005.

CBF500-6 and A-6 (2006)

There were no changes from previous models, both models continued unchanged through 2008. Colours were black, red and silver, all in metallic finish.

Safety First!

Professional mechanics are trained in safe working procedures. However enthusiastic you may be about getting on with the job at hand, take the time to ensure that your safety is not put at risk. A moment's lack of attention can result in an accident, as can failure to observe simple precautions.

There will always be new ways of having accidents, and the following is not a comprehensive list of all dangers; it is intended rather to make you aware of the risks and to encourage a safe approach to all work you carry out on your bike.

Asbestos

● Certain friction, insulating, sealing and other products - such as brake pads, clutch linings, gaskets, etc. - contain asbestos. Extreme care must be taken to avoid inhalation of dust from such products since it is hazardous to health. If in doubt, assume that they do contain asbestos.

Fire

● Remember at all times that petrol is highly flammable. Never smoke or have any kind of naked flame around, when working on the vehicle. But the risk does not end there - a spark caused by an electrical short-circuit, by two metal surfaces contacting each other, by careless use of tools, or even by static electricity built up in your body under certain conditions, can ignite petrol vapour, which in a confined space is highly explosive. Never use petrol as a cleaning solvent. Use an approved safety solvent.

● Always disconnect the battery earth terminal before working on any part of the fuel or electrical system, and never risk spilling fuel on to a hot engine or exhaust.

● It is recommended that a fire extinguisher of a type suitable for fuel and electrical fires is kept handy in the garage or workplace at all times. Never try to extinguish a fuel or electrical fire with water.

Fumes

● Certain fumes are highly toxic and can quickly cause unconsciousness and even death if inhaled to any extent. Petrol vapour comes into this category, as do the vapours from certain solvents such as trichloro-ethylene. Any draining or pouring of such volatile fluids should be done in a well ventilated area.

● When using cleaning fluids and solvents, read the instructions carefully. Never use materials from unmarked containers - they may give off poisonous vapours.

● Never run the engine of a motor vehicle in an enclosed space such as a garage. Exhaust fumes contain carbon monoxide which is extremely poisonous; if you need to run the engine, always do so in the open air or at least have the rear of the vehicle outside the workplace.

The battery

● Never cause a spark, or allow a naked light near the vehicle's battery. It will normally be giving off a certain amount of hydrogen gas, which is highly explosive.

● Always disconnect the battery ground (earth) terminal before working on the fuel or electrical systems (except where noted).

● If possible, loosen the filler plugs or cover when charging the battery from an external source. Do not charge at an excessive rate or the battery may burst.

● Take care when topping up, cleaning or carrying the battery. The acid electrolyte, evenwhen diluted, is very corrosive and should not be allowed to contact the eyes or skin. Always wear rubber gloves and goggles or a face shield. If you ever need to prepare electrolyte yourself, always add the acid slowly to the water; never add the water to the acid.

Electricity

● When using an electric power tool, inspection light etc., always ensure that the appliance is correctly connected to its plug and that, where necessary, it is properly grounded (earthed). Do not use such appliances in damp conditions and, again, beware of creating a spark or applying excessive heat in the vicinity of fuel or fuel vapour. Also ensure that the appliances meet national safety standards.

● A severe electric shock can result from touching certain parts of the electrical system, such as the spark plug wires (HT leads), when the engine is running or being cranked, particularly if components are damp or the insulation is defective. Where an electronic ignition system is used, the secondary (HT) voltage is much higher and could prove fatal.

Remember...

✗ **Don't** start the engine without first ascertaining that the transmission is in neutral.
✗ **Don't** suddenly remove the pressure cap from a hot cooling system - cover it with a cloth and release the pressure gradually first, or you may get scalded by escaping coolant.
✗ **Don't** attempt to drain oil until you are sure it has cooled sufficiently to avoid scalding you.
✗ **Don't** grasp any part of the engine or exhaust system without first ascertaining that it is cool enough not to burn you.
✗ **Don't** allow brake fluid or antifreeze to contact the machine's paintwork or plastic components.
✗ **Don't** siphon toxic liquids such as fuel, hydraulic fluid or antifreeze by mouth, or allow them to remain on your skin.
✗ **Don't** inhale dust - it may be injurious to health (see Asbestos heading).
✗ **Don't** allow any spilled oil or grease to remain on the floor - wipe it up right away, before someone slips on it.
✗ **Don't** use ill-fitting spanners or other tools which may slip and cause injury.
✗ **Don't** lift a heavy component which may be beyond your capability - get assistance.
✗ **Don't** rush to finish a job or take unverified short cuts.
✗ **Don't** allow children or animals in or around an unattended vehicle.
✗ **Don't** inflate a tyre above the recommended pressure. Apart from over-stressing the carcass, in extreme cases the tyre may blow off forcibly.
✔ **Do** ensure that the machine is supported securely at all times. This is especially important when the machine is blocked up to aid wheel or fork removal.
✔ **Do** take care when attempting to loosen a stubborn nut or bolt. It is generally better to pull on a spanner, rather than push, so that if you slip, you fall away from the machine rather than onto it.
✔ **Do** wear eye protection when using power tools such as drill, sander, bench grinder etc.
✔ **Do** use a barrier cream on your hands prior to undertaking dirty jobs - it wlll protect your skin from infection as well as making the dirt easier to remove afterwards; but make sure your hands aren't left slippery. Note that long-term contact with used engine oil can be a health hazard.
✔ **Do** keep loose clothing (cuffs, ties etc. and long hair) well out of the way of moving mechanical parts.
✔ **Do** remove rings, wristwatch etc., before working on the vehicle - especially the electrical system.
✔ **Do** keep your work area tidy - it is only too easy to fall over articles left lying around.
✔ **Do** exercise caution when compressing springs for removal or installation. Ensure that the tension is applied and released in a controlled manner, using suitable tools which preclude the possibility of the spring escaping violently.
✔ **Do** ensure that any lifting tackle used has a safe working load rating adequate for the job.
✔ **Do** get someone to check periodically that all is well, when working alone on the vehicle.
✔ **Do** carry out work in a logical sequence and check that everything is correctly assembled and tightened afterwards.
✔ **Do** remember that your vehicle's safety affects that of yourself and others. If in doubt on any point, get professional advice.
● If in spite of following these precautions, you are unfortunate enough to injure yourself, seek medical attention as soon as possible.

Daily (pre-ride) checks

Note: *The daily (pre-ride) checks outlined in the owner's manual covers those items which should be inspected on a daily basis.*

Engine/transmission oil level check

The correct oil
- Modern, high-revving engines place great demands on their oil. It is very important that the correct oil for your bike is used.
- Always top up with a good quality oil of the specified type and viscosity and do not overfill the engine.

Oil type	API grade SE, SF or SG
Oil viscosity	SAE 10W30

Before you start:
✔ Start the engine and allow it to reach normal operating temperature.
Caution: Do not run the engine in an enclosed space such as a garage or workshop.
✔ Stop the engine and place the motorcycle on its centre stand on level ground. Allow it to stand undisturbed for a few minutes to allow the oil level to stabilise.

Bike care:
- If you have to add oil frequently, you should check whether you have any oil leaks. If there is no sign of oil leakage from the joints and gaskets the engine could be burning oil (see *Fault Finding*).

1 Unscrew the oil filler cap (arrowed) from the right-hand crankcase cover. The dipstick is integral with the oil filler cap, and is used to check the engine oil level.

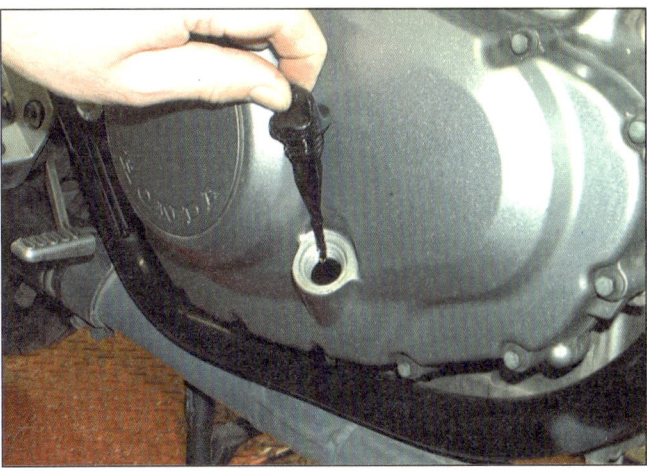

2 Using a clean rag or paper towel, wipe off all the oil from the dipstick. Insert the clean dipstick back into the engine, but **do not** screw it in.

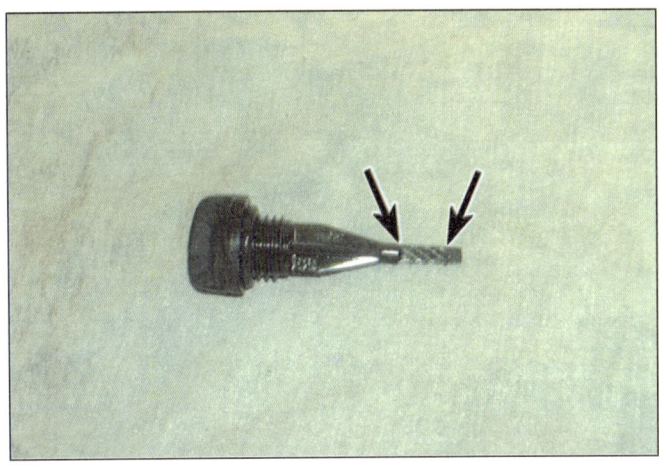

3 Remove the dipstick and observe the level of the oil, which should lie somewhere between the upper and lower level lines (arrowed).

4 If the level is below the lower line, top the engine up with the recommended grade and type of oil, to bring the level up to the upper line on the dipstick. Do not overfill.

0•14 Daily (pre-ride) checks

Brake fluid level checks

⚠️ **Warning:** Brake hydraulic fluid can harm your eyes and damage painted surfaces, so use extreme caution when handling and pouring it and cover surrounding surfaces with rag. Do not use fluid that has been standing open for some time, as it absorbs moisture from the air which can cause a dangerous loss of braking effectiveness.

Bike care:
● The fluid in the front and rear brake master cylinder reservoirs will drop slightly as the brake pads wear down.
● If any fluid reservoir requires repeated topping-up this is an indication of an hydraulic leak somewhere in the system, which should be investigated immediately.
● Check for signs of fluid leakage from the hydraulic hoses and components – if found, rectify immediately.
● Check the operation of both brakes before taking the machine on the road; if there is evidence of air in the system (spongy feel to lever or pedal), it must be bled as described in Chapter 7.

Before you start:
✔ Support the motorcycle on its centre stand on level ground, and turn the handlebars until the top of the front master cylinder is as level as possible. The rear master cylinder reservoir is located below the seat cowling on the right-hand side of the machine.
✔ Make sure you have the correct hydraulic fluid – DOT 4 is recommended.
✔ Wrap a rag around the reservoir being worked on to ensure that any spillage does not come into contact with painted surfaces.
✔ Access to the rear reservoir cap is restricted by the seat cowling. Unscrew the reservoir mounting bolt to gain access to the cap if topping up is required.

1 The front brake fluid level is visible through the sightglass in the reservoir body – it must be above the LOWER level line (arrowed).

2 If the level is below the LOWER level line, remove the two reservoir cap screws and remove the cap, the diaphragm plate and the diaphragm.

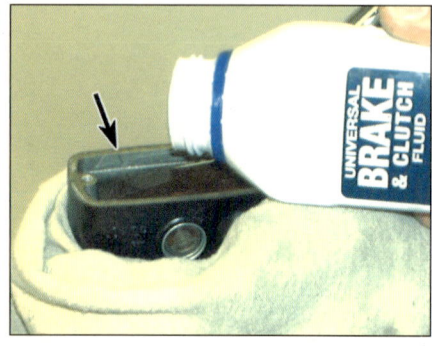
3 Top up with new DOT 4 hydraulic fluid, until the level is just below the UPPER level line cast on the inside of the reservoir (arrowed). Do not overfill, and take care to avoid spills (see **Warning** above).

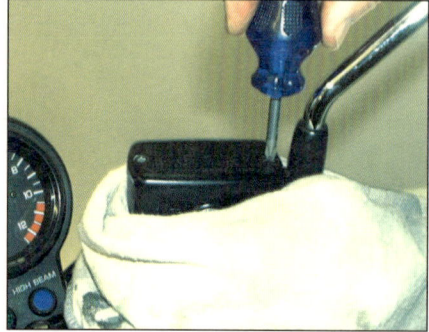
4 Ensure that the diaphragm is correctly seated before installing the plate and cap. Tighten the cap screws securely.

5 The rear brake fluid level is visible through the reservoir body – the fluid level must be between the UPPER and LOWER level lines (arrowed). Remove the reservoir mounting bolt and carefully manoeuvre the reservoir out to access the cap.

Suspension, steering and final drive checks

Suspension and steering:
● Check that the front and rear suspension operate smoothly without binding.
● Check that the rear suspension is adjusted as required.
● Check that the steering moves smoothly from lock-to-lock.

Final drive:
● Check that the drive chain slack isn't excessive, and adjust if necessary (see Chapter 1).
● If the chain looks dry, lubricate it (see Chapter 1).

Daily (pre-ride) checks 0•15

Tyre checks

The correct pressures:
- The tyres must be checked when **cold**, not immediately after riding. Note that low tyre pressures may cause the tyre to slip on the rim or come off. High tyre pressures will cause abnormal tread wear and unsafe handling.
- Use an accurate pressure gauge.
- Proper air pressure will increase tyre life and provide maximum stability and ride comfort.

Loading	Front	Rear
Rider only	29 psi (2.00 Bar)	33 psi (2.25 Bar)
Rider and passenger	29 psi (2.00 Bar)	36 psi (2.50 Bar)

Tyre care:
- Check the tyres carefully for cuts, tears, embedded nails or other sharp objects and excessive wear. Operation of the motorcycle with excessively worn tyres is extremely hazardous, as traction and handling are directly affected.
- Check the condition of the tyre valve and ensure the dust cap is in place.
- Pick out any stones or nails which may have become embedded in the tyre tread. If left, they will eventually penetrate through the casing and cause a puncture.
- If tyre damage is apparent, or unexplained loss of pressure is experienced, seek the advice of a motorcycle tyre fitting specialist without delay.

Tyre tread depth:
- At the time of writing UK law requires that tread depth must be at least 1 mm over $^3/_4$ of the tread breadth all the way around the tyre, with no bald patches. Many riders, however, consider 2 mm tread depth minimum to be a safer limit. Honda recommend a minimum of 2 mm on both tyres.
- Many tyres now incorporate wear indicators in the tread. Identify the triangular pointer or 'TWI' mark on the tyre sidewall to locate the indicator bar and renew the tyre if the tread has worn down to the bar.

1 Check the tyre pressures when the tyres are **cold** and keep them properly inflated.

2 Measure tread depth at the centre of the tyre using a tread depth gauge.

3 Tyre tread wear indicator bar and its location marking (usually either an arrow, a triangle or the letters TWI) on the sidewall (arrowed).

4 Look for the tyre information label on the chainguard.

Daily (pre-ride) checks

Coolant level check

> ⚠️ **Warning:** *DO NOT remove the radiator pressure cap to add coolant. Topping up is done via the coolant reservoir tank filler.* **DO NOT leave open containers of coolant about, as it is poisonous.**

Before you start:
✔ Make sure you have a supply of coolant available (a mixture of 50% distilled water and 50% corrosion inhibited ethylene glycol anti-freeze is needed).
✔ Always check the coolant level when the engine is at normal working temperature. Start the engine and allow it to reach normal temperature, then stop the engine.
Caution: Do not run the engine in an enclosed space such as a garage or workshop.
✔ Place the motorcycle on its centre stand on level ground.

Bike care:
● Use only the specified coolant mixture. It is important that anti-freeze is used in the system all year round, and not just in the winter. Do not top the system up using only water, as the system will become too diluted.
● Do not overfill the reservoir tank. If the coolant is significantly above the UPPER level line at any time, the surplus should be siphoned or drained off to prevent the possibility of it being expelled out of the overflow hose.
● If the coolant level falls steadily, check the system for leaks (see Chapter 1). If no leaks are found and the level continues to fall, it is recommended that the machine is taken to a Honda dealer for a pressure test.

CB models

1 The coolant reservoir is located behind the engine unit on the right-hand side. The coolant UPPER and LOWER level lines (arrowed) are on the front of the reservoir.

2 If the coolant level is not in between the UPPER and LOWER markings, remove the reservoir filler cap and top the coolant level up with the recommended coolant mixture. Fit the cap securely.

CBF models

3 The coolant reservoir is located behind the engine unit on the left-hand side. The coolant UPPER and LOWER level lines (arrowed) are on the side of the reservoir.

4 If the coolant level is not in between the UPPER and LOWER markings, remove the reservoir filler cap and top the coolant level up with the recommended coolant mixture. Fit the cap securely.

Legal and safety checks

Lighting and signalling:
● Take a minute to check that the headlight, tail light, brake light, instrument lights and turn signals all work correctly.
● Check that the horn sounds when the switch is operated.
● A working speedometer graduated in mph is a statutory requirement in the UK.

Safety:
● Check that the throttle grip rotates smoothly and snaps shut when released, in all steering positions. Also check for the correct amount of freeplay (see Chapter 1).
● Check that the engine shuts off when the kill switch is operated.
● Check that sidestand return spring holds the stand securely up when it is retracted.
● Check that the clutch lever operates smoothly and with the correct amount of freeplay (see Chapter 1).
● On R and T models, check the operation of the rear brake drum. If brake stopping power is poor or the brake does not free-off when the pedal is released, investigate the problem immediately.

Fuel:
● This may seem obvious, but check that you have enough fuel to complete your journey. If you notice signs of fuel leakage – rectify the cause immediately.
● Ensure you use the correct grade unleaded fuel – see Chapter 4 Specifications.

Chapter 1
Routine maintenance and servicing

Contents

Air filter and sub-air cleaner – renewal 25	Engine oil pressure – check 33
Battery – charging . . . . see Chapter 9	Engine/transmission oil and filter change 8
Battery – check 12	Exhaust system – check 37
Battery – removal, installation and inspection . . . . see Chapter 9	Front forks – oil change 38
Brake caliper and master cylinder seals – renewal 29	Fuel hoses – renewal 31
Brake fluid – change 24	Fuel system – check 11
Brake hoses – renewal 30	Headlight aim – check and adjustment 16
Brake pads and shoe linings – wear check 3	Idle speed – check and adjustment 2
Rear brake pedal position and freeplay (R and T models) –	Levers, stand pivots and cables – lubrication 6
check and adjustment 4	Nuts and bolts – tightness check 23
Brake system – check 9	Rear brake cam (R and T models) – lubrication 36
Carburettors – synchronisation 15	Spark plugs – check and adjustment 13
Centre and sidestand – check 17	Spark plugs – renewal 27
Clutch – check and adjustment 5	Steering head bearings – check and adjustment 19
Cooling system – check 10	Steering head bearings – lubrication 34
Cooling system – draining, flushing and refilling 28	Suspension – check 18
Crankcase breather – clean 7	Swingarm bearings – lubrication 35
Cylinder compression – check 32	Throttle and choke cables – check and adjustment 14
Drive chain and sprockets – check, adjustment and	Valve clearances – check and adjustment 26
lubrication 1	Wheels and tyres – general check 21
Drive chain slider – wear check 22	Wheel bearings – check 20

Degrees of difficulty

| **Easy,** suitable for novice with little experience | **Fairly easy,** suitable for beginner with some experience | **Fairly difficult,** suitable for competent DIY mechanic | **Difficult,** suitable for experienced DIY mechanic  | **Very difficult,** suitable for expert DIY or professional  |

Specifications

Engine
Cylinder numbering (from left-hand to right-hand side of the motorcycle)................................... 1-2
Spark plugs
   CB models
      Standard.. NGK CR8EH-9 or Nippondenso U24FER-9
      For extended high speed riding .... NGK CR9EH-9 or Nippondenso U27FER-9
   CBF models... NGK CR8EH-9S or Nippondenso U24FER-9S
   Electrode gap....................................... 0.8 to 0.9 mm
Engine idle speed.................................... 1300 rpm (± 100 rpm)
Carburettor synchronisation – max difference between carburettors .. 40 mmHg
Valve clearances (COLD engine)
   Inlet valves .... 0.14 to 0.18 mm
   Exhaust valves .... 0.23 to 0.27 mm
Cylinder compression
   CB models .... 199 psi (14.0 bar) @ 400 rpm
   CBF models .... 213 psi (15.0 bar) @ 400 rpm
Oil pressure (with engine warm)........................ 34 psi (2.4 bar) @ 2000 rpm, oil temp 80°C

Miscellaneous
Drive chain slack .... 30 to 40 mm
Clutch cable freeplay .... 10 to 20 mm
Throttle cable freeplay
   CB models .... 4.5 to 6.5 mm
   CBF models .... 2.0 to 6.0 mm
Rear brake pedal freeplay .... 20 to 30 mm
Tyre pressures and tyre tread depth.................... see Daily (pre-ride) checks

Recommended lubricants and fluids
Engine and transmission oil type .... API grade SE, SF or SG motor oil
Engine and transmission oil viscosity................. SAE 10W30
Engine and transmission oil capacity
   Oil change.. 2.9 litres
   Oil and filter change .... 3.1 litres
   Following engine overhaul – dry engine, new filter... 3.5 litres
Coolant type.. 50% distilled water, 50% corrosion inhibited ethylene glycol anti-freeze
Coolant capacity – CB models
   Radiator and engine................................ 2.0 litres
   Reservoir.. 0.7 litre
Coolant capacity – CBF models
   Radiator and engine................................ 1.95 litres
   Reservoir.. 0.6 litre
Front fork oil.. see Chapter 6
Brake fluid .... DOT 4
Drive chain .... SAE 80 or 90 gear oil or aerosol chain lubricant for O-ring chains

Miscellaneous
Steering head bearings .... Lithium-based multi-purpose grease
Wheel bearings (unsealed) .... Lithium-based multi-purpose grease
Swingarm pivot bearings .... Molybdenum disulphide grease
Bearing seals .... Lithium-based multi-purpose grease
Clutch lever and rear brake pedal pivots .... Molybdenum disulphide grease or dry film lubricant
Front brake lever pivot and piston tip .... Molybdenum disulphide grease or dry film lubricant
Cables .... Cable lubricant or 10W40 motor oil
Centre and sidestand pivots .... Molybdenum disulphide grease
Throttle twistgrip.................................... Multi-purpose grease or dry film lubricant

Torque settings
Rear axle nut
   CB500 models .... 90 Nm
   CBF500 models .... 93 Nm
Steering head bearing adjuster nut .... 25 Nm
Steering stem nut..................................... 105 Nm
Top yoke fork clamp bolts .... 23 Nm
Bottom yoke fork clamp bolts .... 40 Nm
Front brake master cylinder clamp bolts .... 12 Nm
Spark plugs
   CB models .... 12 Nm
   CBF models .... 16 Nm
Oil drain plug .... 35 Nm
Oil filter .... 10 Nm
Water pump drain plug................................. 12 Nm

Maintenance schedule

Note: *The daily (pre-ride) checks outlined in the owner's manual covers those items which should be inspected on a daily basis. Always perform the pre-ride inspection at every maintenance interval (in addition to the procedures listed). The intervals listed below are the intervals recommended by the manufacturer for each particular operation during the model years covered in this manual. Your owner's manual may have different intervals for your model.*

Daily (pre-ride)
- [ ] See 'Daily (pre-ride) checks' at the beginning of this manual.

After the initial 600 miles (1000 km)
Note: *This check is usually performed by a dealer after the first 600 miles (1000 km) from new. Thereafter, maintenance is carried out according to the following intervals of the schedule.*

Every 600 miles (1000 km)
- [ ] Check, adjust and lubricate the drive chain (Section 1)

Every 4000 miles (6000 km) or 6 months (whichever comes sooner)
Carry out all the items under 'Daily (pre-ride) checks' and the 600 mile (1000 km) check, plus the following
- [ ] Check and adjust the idle speed (Section 2)
- [ ] Check the brake pads and shoe linings (Section 3)
- [ ] Check the rear brake pedal position and freeplay – R and T models (Section 4)
- [ ] Check and adjust the clutch (Section 5)
- [ ] Lubricate the clutch lever, brake lever, brake pedal, centre and sidestand pivots and the throttle, choke and clutch cables (Section 6)
- [ ] Clean the crankcase breather (Section 7)

Every 8000 miles (12,000 km) or 12 months (whichever comes sooner)
Carry out all the items under the 4000 mile (6000 km) check, plus the following
- [ ] Change the engine oil and filter (Section 8)
- [ ] Check the brake system and brake light switch operation (Section 9)
- [ ] Check the cooling system (Section 10)
- [ ] Check the fuel system and hoses (Section 11)
- [ ] Check the battery terminals (Section 12)
- [ ] Check the spark plug gaps (Section 13)
- [ ] Check and adjust the throttle and choke cables (Section 14)
- [ ] Check/adjust the carburettor synchronisation (Section 15)
- [ ] Check and adjust the headlight aim (Section 16)
- [ ] Check the centre and sidestand (Section 17)
- [ ] Check the suspension (Section 18)
- [ ] Check and adjust the steering head bearings (Section 19)

Every 8000 miles (12,000 km) or 12 months (whichever comes sooner) (continued)
- [ ] Check the wheel bearings (Section 20)
- [ ] Check the condition of the wheels and tyres (Section 21)
- [ ] Check the drive chain slider (Section 22)
- [ ] Check the tightness of all nuts, bolts and fasteners (Section 23)

Every 12,000 miles (18,000 km) or 18 months (whichever comes sooner)
Carry out all the items under the 4000 mile (6000 km) check, plus the following
- [ ] Change the brake fluid (Section 24)
- [ ] Renew the air filter element and clean the sub-air cleaner (Section 25)

Every 16,000 miles (24,000 km) or two years (whichever comes sooner)
Carry out all the items under the 4000 mile (6000 km) and 8000 mile (12,000 km) checks, plus the following
- [ ] Check and adjust the valve clearances (Section 26)
- [ ] Renew the spark plugs (Section 27)

Every 24,000 miles (36,000 km) or three years (whichever comes sooner)
Carry out all the items under the 4000 mile (6000 km), 8000 mile (12,000 km) and 12,000 mile (18,000 km) checks, plus the following
- [ ] Change the coolant (Section 28)
- [ ] Renew the brake master cylinder and caliper seals (Section 29)

Every four years
- [ ] Renew the brake hoses (Section 30)
- [ ] Renew the fuel hoses (Section 31)

Non-scheduled maintenance
- [ ] Check the cylinder compression (Section 32)
- [ ] Check the engine oil pressure (Section 33)
- [ ] Re-grease the steering head bearings (Section 34)
- [ ] Re-grease the swingarm bearings (Section 35)
- [ ] Re-grease the rear brake cam – R and T models (Section 36)
- [ ] Check the exhaust system (Section 37)
- [ ] Change the front fork oil (Section 38)

1•4 Component locations

Component locations on right-hand side – CB500R

1 Clutch cable lower adjuster
2 Cooling system pressure cap
3 Front brake fluid reservoir
4 Fork seal
5 Oil filter
6 Oil filler/dipstick
7 Coolant reservoir
8 Rear brake pedal height adjuster
9 Rear brake freeplay adjuster
10 Drive chain adjuster

Component location on right-hand side – all other CB models

1 Rear brake fluid reservoir
2 Clutch cable lower adjuster
3 Cooling system pressure cap
4 Front brake fluid reservoir
5 Fork seal
6 Oil filter
7 Oil filler/dipstick
8 Coolant reservoir
9 Drive chain adjuster

Component locations 1•5

Component locations on left-hand side – all CB models

1 Clutch cable upper adjuster
2 Steering head bearing adjuster
3 Idle speed adjuster
4 Air filter
5 Fuel filter
6 Battery
7 Drive chain adjuster
8 Drive chain slider
9 Oil drain bolt
10 Coolant drain bolt
11 Fork seal

Component locations

Component locations on right-hand side – CBF models

1. Rear brake fluid reservoir
2. Steering head bearing adjuster
3. Cooling system pressure cap
4. Front fork oil seals
5. Oil filter
6. Oil filler/dipstick
7. Lower clutch cable adjuster
8. Brake light switch
9. Brake pedal height adjuster
10. Drive chain adjuster

Component locations on left-hand side – CBF models

1. Front fork oil seals
2. Upper clutch cable adjuster
3. Front brake fluid reservoir
4. Spark plugs
5. Fuel filter
6. Air filter
7. Battery
8. Drive chain adjuster
9. Drive chain slider
10. Oil drain bolt
11. Coolant drain bolt
12. Idle speed adjuster

Introduction

1 This Chapter is designed to help the home mechanic maintain his/her motorcycle for safety, economy, long life and peak performance.

2 Deciding where to start or plug into the routine maintenance schedule depends on several factors. If the warranty period on your motorcycle has just expired, and if it has been maintained according to the warranty standards, you may want to pick up routine maintenance as it coincides with the next mileage or calendar interval. If you have owned the machine for some time but have never performed any maintenance on it, then you may want to start at the nearest interval and include some additional procedures to ensure that nothing important is overlooked. If you have just had a major engine overhaul, then you may want to start the maintenance routine from the beginning. If you have a used machine and have no knowledge of its history or maintenance record, you may desire to combine all the checks into one large service initially and then settle into the maintenance schedule prescribed.

3 Before beginning any maintenance or repair, the machine should be cleaned thoroughly, especially around the oil filter, spark plugs, valve cover, oil and coolant drain plugs, carburettors, etc. Cleaning will help ensure that dirt does not contaminate the engine and will allow you to detect wear and damage that could otherwise easily go unnoticed.

4 Certain maintenance information is sometimes printed on decals attached to the motorcycle. If the information on the decals differs from that included here, use the information on the decals.

Every 600 miles (1000 km)

1 Drive chain and sprockets – check, adjustment and lubrication

1 A neglected drive chain won't last long and can quickly damage the sprockets. Routine chain adjustment and lubrication isn't difficult and will ensure maximum chain and sprocket life.

2 To check the chain, place the motorcycle on its centre stand and shift the transmission into neutral. Make sure the ignition switch is OFF. Check the entire length of the chain for damaged rollers, loose links and pins, and missing O-rings and renew it if damage is found. Remove the front sprocket cover (see Chapter 6). Check the teeth on the front and rear sprockets for wear **(see illustration)**.

Note: *Never install a new chain on old sprockets, and never use the old chain if you install new sprockets – renew the chain and sprockets as a set.*

3 To check chain tension, place the motorcycle on its sidestand. Push up on the bottom run of the chain and measure the slack midway between the two sprockets, then compare your measurement to that listed in this Chapter's Specifications **(see illustration)**. As the chain stretches with wear, periodic adjustment will be necessary (see below). Since the chain will rarely wear evenly, rotate the rear wheel so that another section of chain can be checked midway between the sprockets; do this several times to check the entire length of chain.

4 In some cases where, lubrication has been neglected, corrosion and galling may cause the links to bind and kink, which effectively shortens the chain's length. Such links should be thoroughly cleaned and worked free. Mark the tight area with felt pen or paint and check it again after the motorcycle has been ridden. If the chain's still tight in the same area, it may be damaged or worn. Because a tight or kinked chain can damage the transmission output shaft bearing, it's a good idea to renew it.

Caution: *If the machine is ridden with excessive slack in the drive chain, the chain could contact the frame and swingarm, causing severe damage.*

Adjustment

5 Place the motorcycle on its centre stand and rotate the rear wheel until the chain is positioned with the tightest point at the centre of its bottom run.

6 Slacken the rear axle nut and the locknut on each chain adjuster **(see illustration)**.

7 Turn the chain adjusters on both sides of the swingarm a small amount at a time until the specified chain tension is obtained **(see illustration)**. Be sure to turn the adjusters evenly to keep the rear wheel in alignment. If the adjusters reach the end of their travel, the chain is excessively worn and should be renewed (see Chapter 6). The chain wear decals will also indicate the need for chain renewal **(see illustration)**.

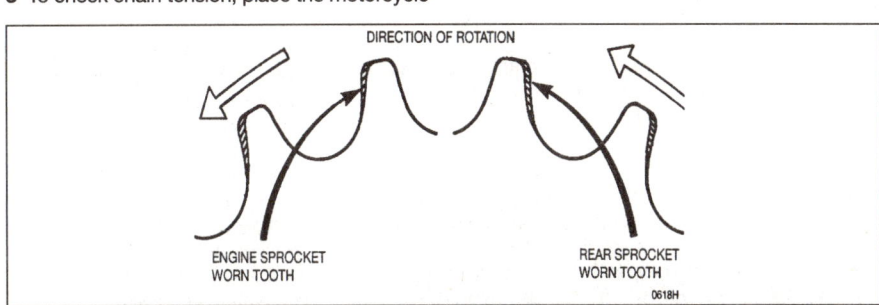

1.2 Check the sprockets in the areas indicated to see if they are worn excessively

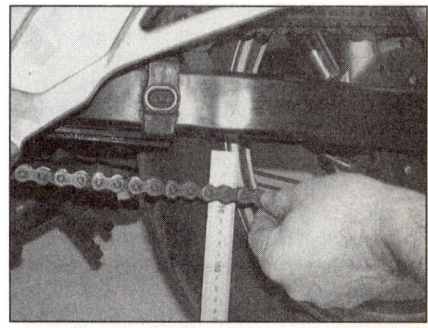

1.3 Push up on the chain and measure the slack

1.6 Rear axle nut (A), locknut (B), adjuster (C)

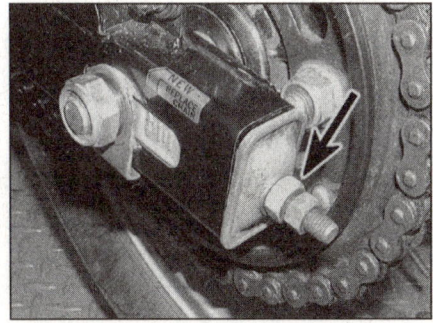

1.7a Turn the adjusters (arrowed) to obtain specified chain tension

1.7b Chain wear decal (A) and alignment marks (B)

1•8 Every 600 miles

8 When the chain has the correct amount of slack, check that the wheel is correctly aligned by making sure the marks on each adjuster are in the same position relative to the slot in the swingarm **(see illustration 1.7b)**. If there is any discrepancy in the chain adjuster positions, adjust one of them so that its position is exactly the same as the other, then recheck the chain freeplay as described above. It is important each adjuster is identically aligned otherwise the rear wheel will be out of alignment with the front.

9 Tighten the axle nut to the torque setting specified at the beginning of this Chapter **(see illustration)**. Tighten the chain adjuster locknuts securely. On R and T models, reset the freeplay on the rear brake pedal by turning the adjuster nut on the brake rod (see Section 4).

Lubrication

10 If the chain is dirty, wash it in paraffin (kerosene), then wipe it off and allow it to dry, using compressed air if available. If the chain is excessively dirty it should be removed from the machine and allowed to soak in the paraffin as described in Chapter 6.
Caution: Don't use petrol, solvent or other cleaning fluids which might damage the internal sealing properties of the chain. Don't use high-pressure water. The entire process shouldn't take longer than ten minutes – if it does, the O-rings in the chain rollers could be damaged.

11 For routine lubrication, the best time to lubricate the chain is after the motorcycle has been ridden. When the chain is warm, the lubricant will penetrate the joints between the side plates better than when cold. **Note:** Honda specifies SAE 80 to SAE 90 gear oil; if you do use aerosol chain lube ensure that it is suitable for O-ring chains. Apply the oil to the area where the side plates overlap – not the middle of the rollers **(see illustration)**.

1.9 Tighten the axle nut to the specified torque

1.11 Apply the lubricant to the overlap in the sideplates

> **HAYNES HiNT** Apply the oil to the top of the lower chain run, so centrifugal force will work the oil into the chain when the motorcycle is moving. After applying the lubricant, let it soak in a few minutes before wiping off any excess.

Every 4000 miles (6000 km) or 6 months

2 Idle speed – check and adjustment

1 The idle speed should be checked and adjusted before and after the carburettors are synchronised (balanced) and when it is obviously too high or too low. Before adjusting the idle speed, make sure the valve clearances and spark plug gaps are correct. Also, turn the handlebars back-and-forth and see if the idle speed changes as this is done. If it does, the throttle cable may not be adjusted or routed correctly, or may be worn out. This is a dangerous condition that can cause loss of control of the motorcycle. Be sure to correct this problem before proceeding.

2 The engine should be at normal operating temperature, which is usually reached after 10 to 15 minutes of stop-and-go riding. Place the motorcycle on its centre stand, and make sure the transmission is in neutral.

3 The idle speed adjuster is located between the carburettors **(see illustration)**. With the engine running and the throttle closed, adjust the idle speed by turning the adjuster screw until the idle speed listed in this Chapter's Specifications is obtained. Turn the screw clockwise to increase idle speed, and anti-clockwise to decrease it.

4 Snap the throttle open and shut a few times, then recheck the idle speed. If necessary, repeat the adjustment procedure.

5 If a smooth, steady idle can't be achieved, the fuel/air mixture may be incorrect (see Chapter 4) or the carburettors may need synchronising (see Section 15).

3 Brake pads and shoe linings – wear check

1 Disc brake pads and rear drum brake shoe linings (R and T models only) all have wear indicators. A quick check of brake wear can be made without dismantling components.

Brake pads

2 The amount of pad wear can be judged by looking at the pads from the underside rear of the caliper (both front and rear) **(see illustration)**. A cutout in the friction material indicates the wear limit **(see illustration)**.

3 If either pad in the caliper has worn down to, or beyond the cutout in the friction material, both pads must be renewed as a set. If the

2.3 Idle speed adjuster screw (arrowed)

3.2a Check pad wear from the underside rear of the caliper

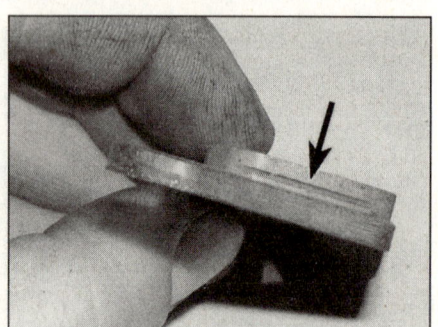

3.2b Brake pad wear limit (arrowed)

Every 4000 miles

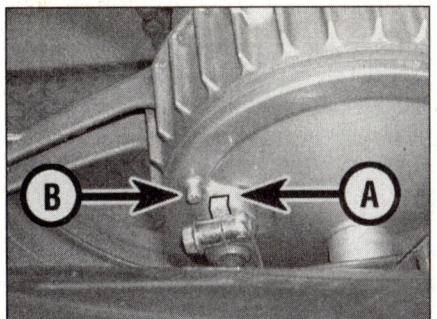

3.5 Rear brake wear indicator (A) and limit reference (B) on R and T models

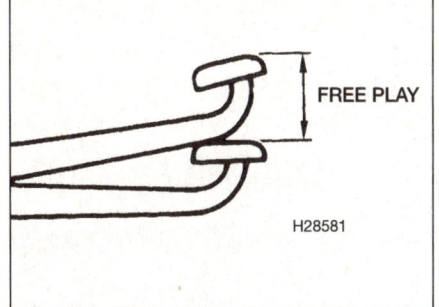

4.1a Measure brake pedal freeplay at the pedal tip

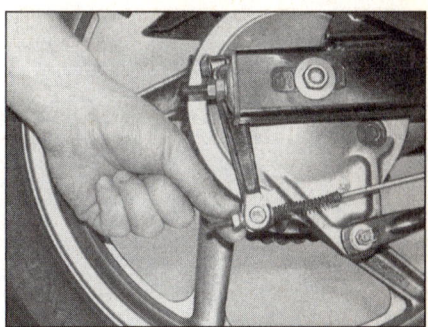

4.1b Adjusting brake pedal freeplay

pads are dirty or if you are in doubt as to the amount of friction material remaining, remove them for inspection (see Chapter 7). **Note:** *Some after-market pads may use different indicators to those on the original equipment as shown.*
4 Refer to Chapter 7 for details of pad renewal.

Rear brake shoe linings – R and T models

5 To check the rear brake linings, press the brake pedal firmly and look at the indicator on the brake backplate **(see illustration)**. If the pointer on the brake arm aligns with the reference mark on the brake backplate the brake shoes must be renewed.
6 Refer to Chapter 7 for details of brake shoe renewal.

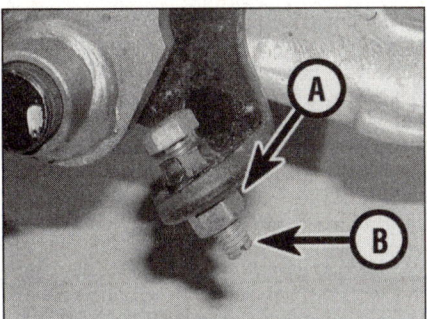

4.6a Slacken locknut (A) and adjust pedal height with screwdriver at (B)

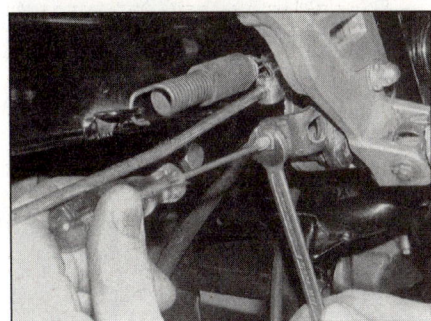

4.6b Ensure locknut is tightened after making adjustment

4 Rear brake pedal position and freeplay (R and T models) – check and adjustment

Pedal freeplay check and adjustment

1 Apply the rear brake and compare the pedal travel with the measurement listed in this Chapter's Specifications **(see illustration)**. If adjustment is necessary, turn the adjuster at the rear end of the brake rod **(see illustration)**.
2 Apply the brake several times and ensure that the wheel turns freely without the brake binding when the brake pedal is released.
3 Make sure that the cut-out on the adjusting nut is properly seated on the brake arm trunnion after making adjustment.
4 If necessary, adjust the brake light switch (see Section 9).

Pedal height adjustment

5 It is important that the brake pedal is positioned correctly in relation to the footrest so that the brake can be engaged quickly and easily without excessive foot movement. Pedal position is largely a matter of personal preference and once set up should not need subsequent alteration.
6 To adjust the position of the pedal, loosen the locknut on the adjusting bolt, turn the slotted end of the bolt with a screwdriver to set the pedal position and tighten the locknut **(see illustrations)**.

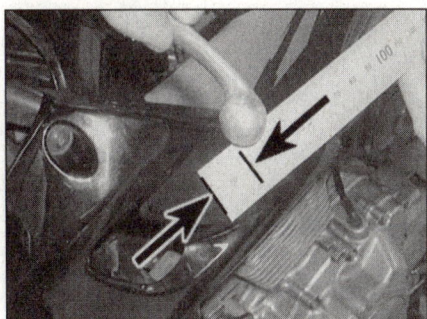

5.3 Measuring clutch cable freeplay

7 If the pedal height has been adjusted, recheck the pedal freeplay as described above.

5 Clutch – check and adjustment

1 Check that the clutch cable operates smoothly and easily.
2 If the clutch lever operation is heavy or stiff, remove the cable (see Chapter 2) and lubricate it (see Section 6). If the cable is still stiff, renew it. Install the lubricated or new cable (see Chapter 2).
3 With the cable operating smoothly, check that the clutch lever is correctly adjusted. Periodic adjustment is necessary to compensate for wear in the clutch plates and stretch in the cable. Check that the amount of cable freeplay at the clutch lever end is within the specifications listed at the beginning of the Chapter **(see illustration)**.
4 If adjustment is required, loosen the adjuster lockring at the lever and turn the adjuster in or out until the required amount of freeplay is obtained **(see illustration)**. To increase freeplay, turn the adjuster clockwise. To reduce freeplay, turn the adjuster anti-clockwise. Tighten the locking ring securely.
5 If all the adjustment has been taken up at the lever, reset the lever adjuster to give the maximum amount of freeplay (ie screw it into the bracket), then set the correct amount of freeplay using the adjuster at the lower end of

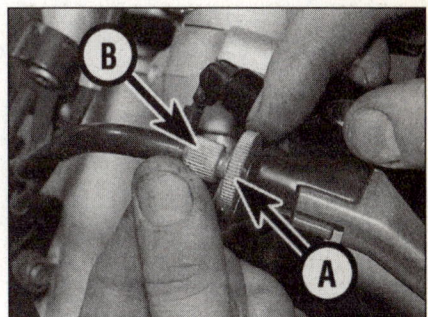

5.4 Slacken the lockring (A) and turn the adjuster (B) in or out as required

1•10 Every 4000 miles

5.5 Clutch cable lower adjuster nuts (arrowed)

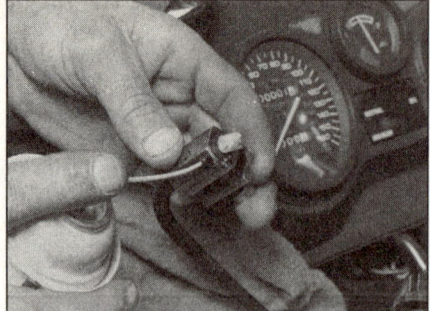

6.3a Lubricating a cable with a cable oiler clamp. Ensure the tool seals around the inner cable

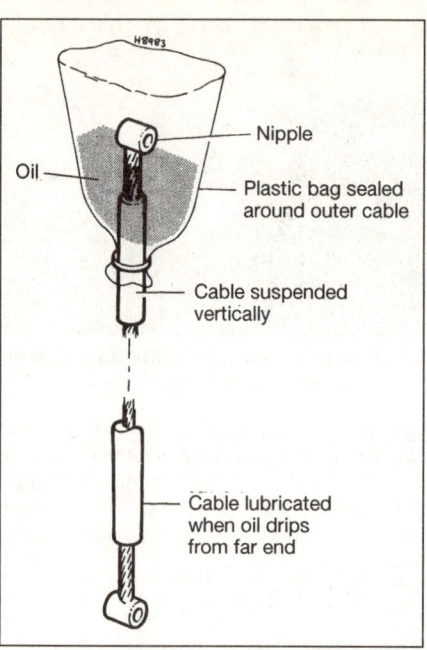

6.3b Lubricating a cable with a makeshift funnel and motor oil

the cable where it passes through the bracket on the right-hand side of the engine (see illustration). To reduce freeplay, slacken the rear nut and tighten the front nut until the freeplay is as specified, then tighten the rear nut against the bracket. To increase freeplay, slacken the front nut and tighten the rear nut until the freeplay is as specified, then tighten the front nut against the bracket. Subsequent adjustments can now be made using the lever adjuster only.

6 Levers, stand pivots and cables – lubrication

Pivot points

1 Since the controls, cables and various other components of a motorcycle are exposed to the elements, they should be lubricated periodically to ensure safe and trouble-free operation.
2 The clutch and brake levers, brake pedal, footrest, centre and sidestand pivots should be lubricated frequently. In order for the lubricant to be applied where it will do the most good, the component should be disassembled and the recommended grease applied (see Specifications). However, if chain and cable lubricant is being used, it can be applied to the pivot joint gaps and will usually work its way into the areas where friction occurs. If motor oil or light grease is being used, apply it sparingly as it may attract dirt (which could cause the controls to bind or wear at an accelerated rate). **Note:** *One of the best lubricants for the control lever pivots is a dry-film lubricant (available from many sources by different names).*

Cables

3 To lubricate the cables, disconnect the relevant cable at its upper end, then lubricate the cable with a cable oiler clamp, or if one is not available, using the set-up shown **(see illustrations)**. See Chapter 4 for the choke and throttle cable removal procedures, and Chapter 2 for the clutch cable.
4 The speedometer cable should be removed (see Chapter 9) and the inner cable withdrawn from the outer cable and lubricated with motor oil or cable lubricant. Do not lubricate the upper few inches of the cable as the lubricant may travel up into the instrument head.

7 Crankcase breather – clean

1 The crankcase breather hose is routed from the valve cover to the air filter housing where vapours are recycled in the combustion process. Sludge from the vapour collects at the bottom of the housing, where it drains into a tube routed to the underside of the motorcycle behind the engine unit. The tube is retained by a bracket on the exhaust system **(see illustration)**.
2 Release the clip that secures the tube plug, remove the plug and drain any deposits into a suitable container **(see illustration)**. Refit the tube plug and ensure the clip is correctly repositioned.

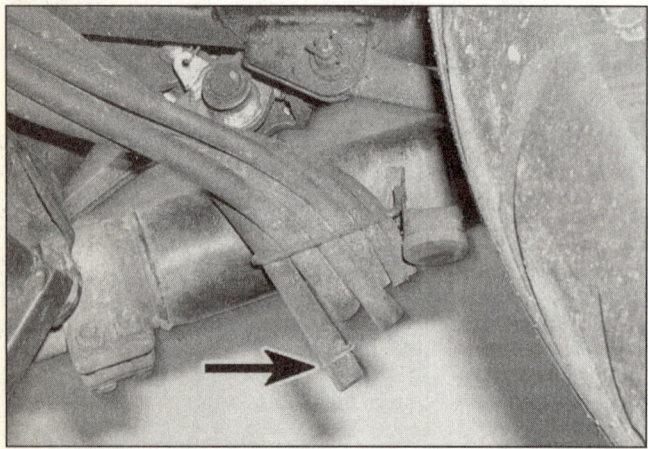

7.1 Crankcase breather tube (arrowed) is located in a bracket on the underside of the silencer

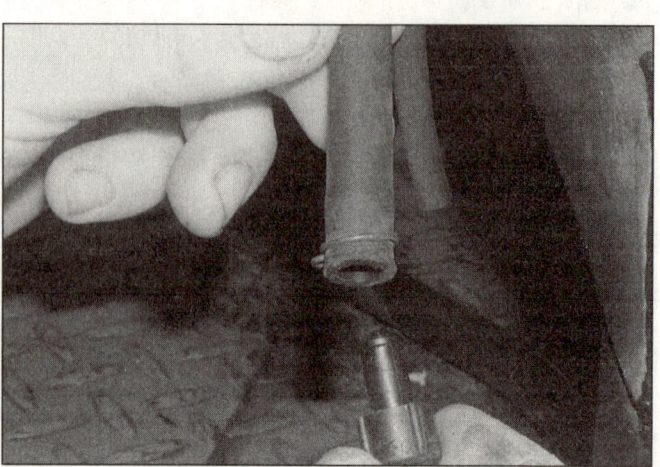

7.2 Remove the plug to drain the breather

Routine maintenance and servicing 1•11

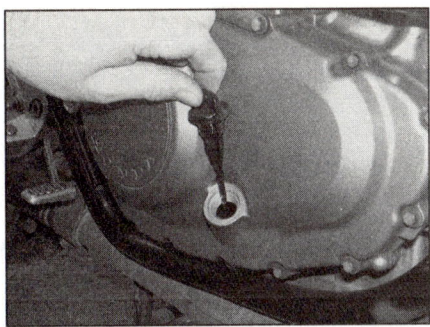

8.3 Unscrew the oil filer cap to vent the crankcase

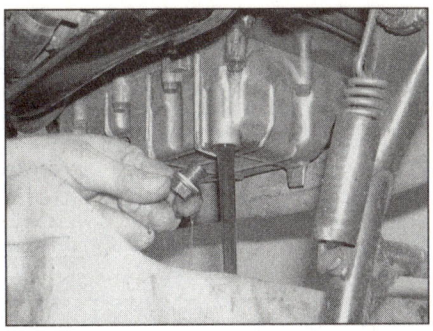

8.4 Remove the drain plug and allow the oil to drain completely

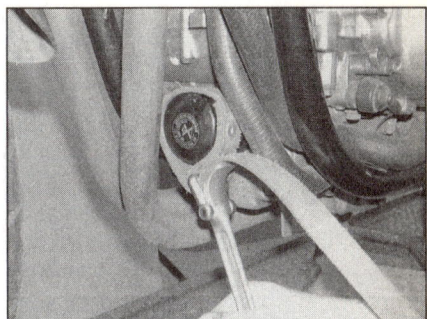

8.6a Unscrew the oil filter . . .

Every 8000 miles (12,000 km) or 12 months

Carry out all the items under the 4000 mile (6000 km) check, plus the following:

8 Engine/transmission – oil and oil filter change

Warning: Be careful when draining the oil, as the exhaust pipes, the engine, and the oil itself can cause severe burns.

1 Consistent routine oil and filter changes are the single most important maintenance procedure you can perform on a motorcycle. The oil not only lubricates the internal parts of the engine, transmission and clutch, but it also acts as a coolant, a cleaner, a sealant, and a protector. Because of these demands, the oil takes a terrific amount of abuse and should be changed often with new oil of the recommended grade and type. Saving a little money on the difference in cost between a good oil and a cheap oil won't pay off if the engine is damaged.
2 Before changing the oil, warm up the engine so the oil will drain easily.
3 Put the motorcycle on its sidestand to ensure complete draining of the sump and position a clean drain tray of approximately 5 litres capacity below the engine. Unscrew the oil filler cap from the clutch cover to vent the crankcase and to act as a reminder that there is no oil in the engine **(see illustration)**.

4 Next, unscrew the oil drain plug from the sump on the bottom of the engine and allow the oil to flow into the drain tray **(see illustration)**. Check the condition of the sealing washer on the drain plug and obtain a new one if it is damaged or worn.

 **HAYNES HiNT** *To help determine whether any abnormal or excessive engine wear is occurring, place a strainer between the engine and the drain tray so that any debris in the oil is filtered out and can be examined.*

5 When the oil has completely drained, fit the plug to the sump, using a new sealing washer if necessary, and tighten it to the specified torque. Avoid overtightening, as damage to the sump will result.
6 Now place the drain tray below the oil filter. Unscrew the oil filter using a filter adapter or a strap wrench and empty the filter into the drain tray **(see illustrations)**. Allow any residual oil to drain from the engine.
7 Smear clean engine oil onto the rubber seal on the new filter, then carefully position the filter on its threads and screw it onto the engine **(see illustration)**. Tighten the filter securely using a filter adapter or a strap wrench **(see illustration)**. To tighten the filter to the specified torque, it is necessary to remove the exhaust system to gain access with a torque wrench.
8 Refill the engine to the proper level using the recommended type and amount of oil see Specifications). With the motorcycle upright on level ground, the oil level should lie between the upper and lower level lines on the dipstick (see *Daily (pre-ride) checks)*. Install the filler cap. Start the engine and let it run for two or three minutes (make sure that the oil pressure light extinguishes after a few seconds). Shut it off, wait a few minutes, then check the oil level. If necessary, add more oil to bring the level up to a mid-point between the upper and lower level lines on the dipstick. Check around the drain plug and the oil filter for leaks.

 **HAYNES HiNT** *Saving a little money on the difference between good and cheap oils won't pay off if the engine is damaged as a result.*

9 The old oil drained from the engine and the filter cannot be re-used and should be disposed of properly. Check with your local refuse disposal company, disposal facility or

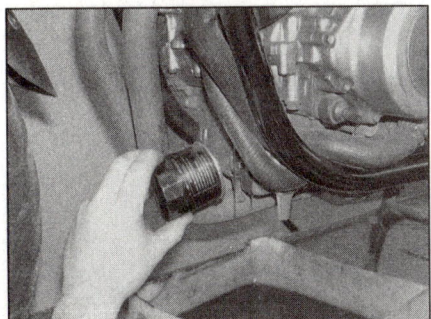

8.6b . . . and allow any residual oil to drain

8.7a Smear the filter seal with clean engine oil . . .

8.7b . . . and tighten the filter securely

1•12 Routine maintenance and servicing

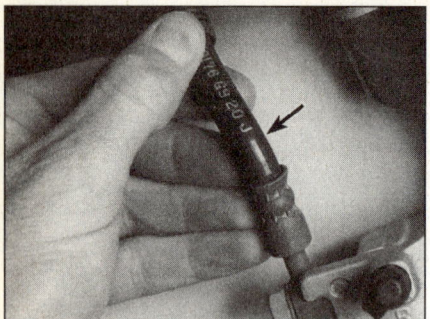

9.3 Flex the brake hose and check for cracks, bulges and leaking fluid

9.5 Rear brake light switch adjuster ring (arrowed)

9.6 Front brake lever span adjuster (arrowed)

environmental agency to see whether they will accept the used oil for recycling. Don't pour used oil into drains or onto the ground.

Note: It is antisocial and illegal to dump oil down the drain. In the UK, call this number free to find the location of your local oil recycling bank. In the USA, note that any oil supplier must accept used oil for recycling.

9 Brake system – check

1 A routine general check of the brake system will ensure that any problems are discovered and remedied before the rider's safety is jeopardised.
2 Check the brake lever and pedal for loose connections, improper or rough action, excessive play, bends and other damage. On R and T models, check the rear brake linkage. Renew any worn or damaged parts (see Chapter 7).
3 Make sure all brake fasteners are tight. On disc brakes, check the brake pads for wear (see Section 3) and make sure the fluid level in the reservoir is correct (see *Daily (pre-ride) checks*). Look for leaks at the hose connections and check for cracks in the hose **(see illustration)**. If the lever or pedal is spongy, bleed the brakes (see Chapter 7). On R and T models, check the rear brake shoes for wear and check for the correct amount of freeplay at the brake pedal tip (see Sections 3 and 4).
4 Make sure the brake light operates when the front brake lever is pulled in. The front brake light switch, mounted on the underside of the master cylinder, is not adjustable. If it fails to operate properly, check it (see Chapter 9).
5 Make sure the brake light comes on just before the rear brake takes effect. If adjustment is necessary, hold the rear brake light switch and turn the adjuster ring on the switch body until the brake light is activated when required **(see illustration)**. The switch is mounted on the inside of the rider's right-hand footrest bracket. If the brake light comes on too late, turn the ring clockwise. If the brake light comes on too soon or is permanently on, turn the ring anti-clockwise. If the switch doesn't operate the brake light, check it (see Chapter 9).
6 On CBF500 models the front brake lever has a span adjuster which alters the distance of the lever from the handlebar **(see illustration)**. Each setting is identified by a notch in the adjuster which aligns with the arrow on the lever. Pull the lever away from the handlebar and turn the adjuster ring until the setting which best suits the rider is obtained. Do not set the adjuster between the notches.

10 Cooling system – check

Warning: The engine must be cool before beginning this procedure.

1 Check the coolant level (see *Daily (pre-ride) checks*).
2 On R, T, V, W, X, Y and 2 models, remove the two radiator side panels; on SW, SX, SY and S2 models, remove the fairing (see Chapter 8). Remove the radiator stone guard (see Chapter 8) and the fuel tank (see Chapter 4). The entire cooling system should be checked for evidence of leakage. Examine each rubber coolant hose along its entire length. Look for cracks, abrasions and other damage. Squeeze each hose at various points. They should feel firm, yet pliable, and return to their original shape when released. If they are cracked or hard, renew them.
3 Check for evidence of leaks at each cooling system joint. Tighten the hose clips carefully to prevent future leaks **(see illustration)**.
4 Check the radiator for leaks and other damage. Leaks in the radiator leave tell-tale scale deposits or coolant stains on the outside of the core below the leak. If leaks are noted, remove the radiator (see Chapter 3) and have it repaired by a specialist.
5 Check the water pump for any sign of leakage from its drain hole (see Chapter 3, Section 8).
Caution: Do not use a liquid leak sealing compound to try to repair leaks.
6 Check the radiator fins for mud, dirt and insects, which may impede the flow of air through the radiator. If the fins are dirty, remove the radiator (see Chapter 3) and clean it using water or low pressure compressed air directed through the fins from the back. If the fins are bent or distorted, straighten them carefully with a screwdriver. Bent or damaged fins will restrict the air flow and impair the efficiency of the radiator causing the engine to overheat. Where there is substantial damage to the radiator's surface area, renew the radiator.
7 Remove the pressure cap from the radiator filler neck by turning it anti-clockwise until it reaches a stop. If you hear a hissing sound when you turn the cap (indicating there is still pressure in the system), wait until it stops before removing the cap. Now press down on the cap and continue turning it until it can be removed **(see illustration)**.

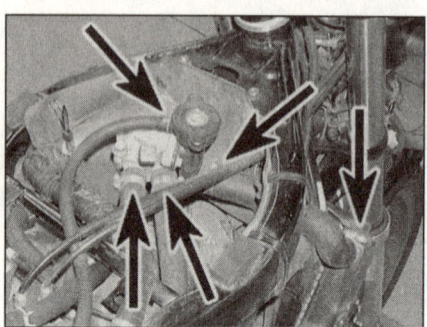

10.3 Check all coolant hoses and joints (arrowed) for leaks

10.7 Turn pressure cap slowly anti-clockwise to remove

Every 8000 miles

8 Check the condition of the coolant in the system. If it is rust-coloured or if accumulations of scale are visible, drain, flush and refill the system with new coolant (see Section 28). Check the cap seal for cracks and other damage. If in doubt about the pressure cap's condition, have it tested by a dealer or renew it.

9 Check the antifreeze content of the coolant with an antifreeze hydrometer. A mixture with less than 40% antifreeze (40/60 antifreeze to distilled water) will not provide proper corrosion protection. Sometimes coolant looks like it's in good condition, but might be too weak to offer adequate protection. If the hydrometer indicates a weak mixture, drain, flush and refill the system (see Section 28). A higher than specified concentration of antifreeze decreases the performance of the cooling system and should only be used when additional protection against freezing is needed.

10 Install the cap by turning it clockwise until it reaches the first stop then push down on the cap and continue turning until it will turn no further.

11 Start the engine and let it reach normal operating temperature, then check for leaks again. As the coolant temperature increases, the fan behind the radiator should come on automatically and the temperature should begin to drop. If it does not, refer to Chapter 3 and check the fan and fan circuit carefully.

12 If the coolant level is consistently low, and no evidence of leaks can be found, have the entire system pressure checked by a Honda dealer.

11 Fuel system – check

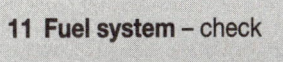

Warning: *Petrol (gasoline) is extremely flammable, so take extra precautions when you work on any part of the fuel system. Don't smoke or allow open flames or bare light bulbs near the work area, and*

11.1a Check the various hoses as described

don't work in a garage where a natural gas-type appliance is present. If you spill any fuel on your skin, rinse it off immediately with soap and water. When you perform any kind of work on the fuel system, wear safety glasses and have a fire extinguisher suitable for a Class B type fire (flammable liquids) on hand.

Check

1 Remove the fuel tank (see Chapter 4) and check the tank, the tank filler cap, the fuel tap, the fuel hose and the fuel tap vacuum hose for signs of leakage, deterioration or damage; in particular check that there is no leakage from the fuel hose. Also check the crankcase breather hose(s) between the air filter housing and the valve cover, and on CBF models the PAIR system air hose between the right-hand side of the air filter housing and the control valve on the front of the engine, and the vacuum hose from each inlet duct to the control valve **(see illustrations)**. Renew any hoses that are cracked or deteriorated.

2 If the fuel tap to tank joint is leaking, tightening the retaining nut may help. Hold the tap to prevent it twisting while tightening the nut. If leakage persists, the tank must be drained of fuel, the tap removed and its O-ring renewed. It is advised that this is done when there is minimal fuel in the tank. In this way the tank can be rested on its right-hand side whilst the tap is removed. Alternatively, on

11.1b On CBF models check the PAIR system air hose (A) and vacuum hoses (B)

CB500 models with the tap in the RES position, apply a vacuum to the end of the vacuum pipe to allow the fuel to drain through the fuel pipe and into a suitable container. Remove the tap by unscrewing the retaining nut. Use a new O-ring **(see illustration)** on refitting and take care not to overtighten the retaining nut.

3 If the carburettor gaskets are leaking, the carburettors should be disassembled and rebuilt using new gaskets and seals (see Chapter 4).

Filter cleaning

4 Cleaning or renewal of the fuel filter is advised after a particularly high mileage has been covered. It is also necessary if fuel starvation is suspected. The fuel filter is mounted in the tank and is integral with the fuel tap.

5 Remove the fuel tank (see Chapter 4) and the fuel tap (see Step 2), and carefully pull the filter off the fuel supply tube **(see illustration)**. Once the filter is dry, clean the gauze with a soft brush or low pressure compressed air to remove all traces of dirt and fuel sediment. Check the gauze for holes. If any are found, a new filter should be fitted (it is available separately). A damaged filter will allow dirt particles to enter the tap body.

6 Check the condition of the tap O-ring and renew it if it is in any way damaged or deteriorated. It is advisable to renew it as a matter of course.

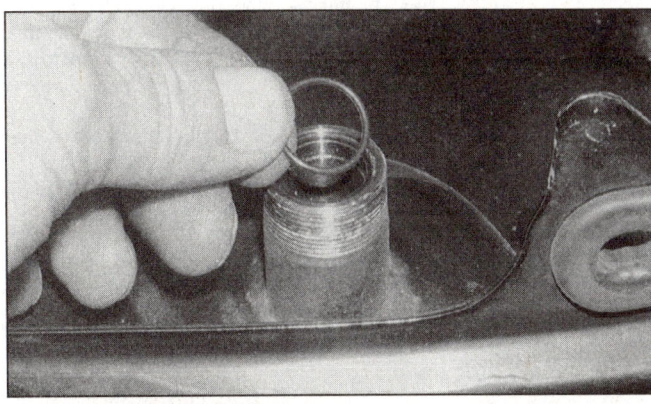

11.2 Renew fuel tap O-ring to cure persistent leaks

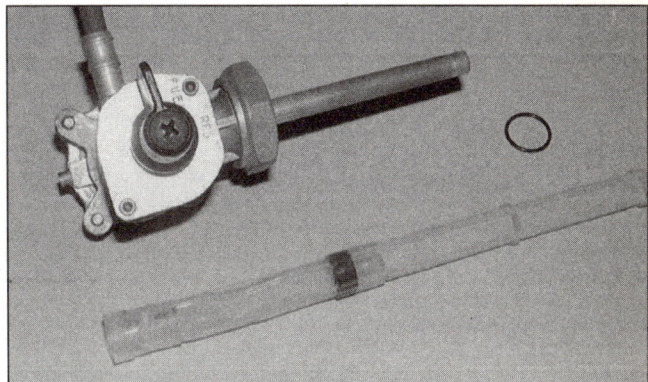

11.5 Fuel tap, O-ring and fuel filter

1•14 Every 8000 miles

13.4 Pull the cap off the spark plug ...

13.5a ... then unscrew the spark plug ...

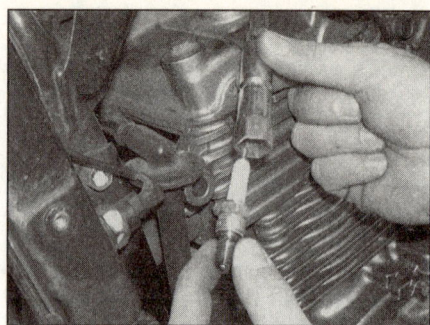
13.5b ... and separate the plug from the tool

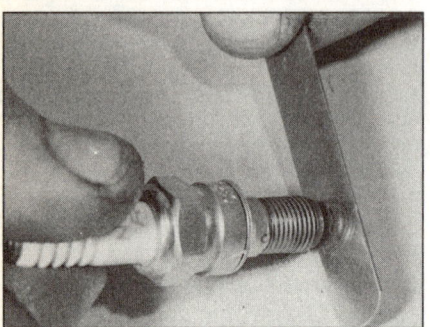
13.9a Using a feeler gauge to measure the spark plug electrode gap

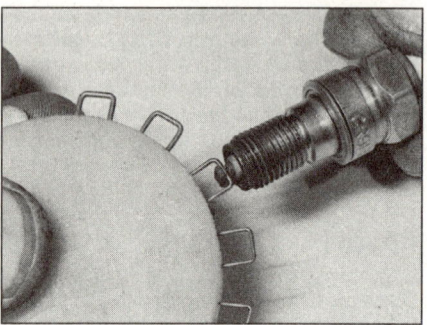
13.9b Using a wire type gauge to measure the spark plug electrode gap

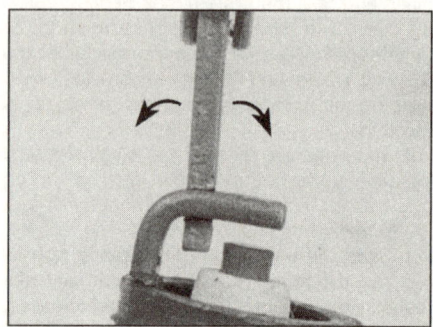
13.9c Adjust the electrode gap by bending the side electrode only

12 Battery – check
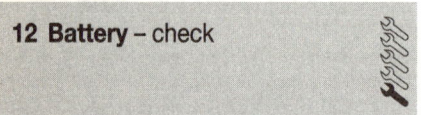

1 All models covered in this manual are fitted with a sealed, maintenance-free battery which requires no topping-up. **Note:** *Do not attempt to remove the battery filler caps, if fitted, to check the electrolyte level or battery specific gravity. Removal will damage the caps, resulting in electrolyte leakage and battery damage.*

2 All that should be done is to check that the battery terminals are clean and tight and that the casing is not damaged or leaking. See Chapter 9 for further details.

3 If the machine is not in regular use, disconnect the battery and give it a refresher charge every month to six weeks, as described in Chapter 9.

13 Spark plugs – check and adjustment
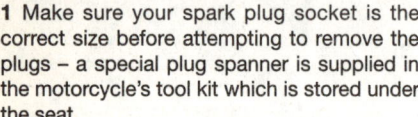

1 Make sure your spark plug socket is the correct size before attempting to remove the plugs – a special plug spanner is supplied in the motorcycle's tool kit which is stored under the seat.
2 On R, T, V, W, X, Y and 2 models, remove the radiator side panels; on SW, SX, SY and S2 models remove the fairing (see Chapter 8).
3 Clean the area around the plug caps to prevent any dirt falling into the spark plug access passages.
4 Check that the cylinder location is marked on each plug lead and mark them accordingly if not. Pull the spark plug cap off each spark plug **(see illustration)**. Clean the area around the base of the plugs with low pressure compressed air to prevent any dirt falling into the engine.
5 Using either the plug spanner supplied in the motorcycle's toolkit or a deep socket type wrench, unscrew the plugs from the cylinder head **(see illustrations)**. Lay each plug out in relation to its cylinder; if either plug shows up a problem it will then be easy to identify the troublesome cylinder.
6 Inspect the electrodes for wear. Both the centre and side electrodes should have square edges and the side electrodes should be of uniform thickness. Look for excessive deposits and evidence of a cracked or chipped insulator around the centre electrode. Compare your spark plugs to the colour spark plug reading chart at the end of this manual. Check the threads, the washer and the ceramic insulator body for cracks and other damage.
7 If the electrodes are not excessively worn, and if the deposits can be easily removed with a wire brush, the plugs can be re-gapped and re-used (if no cracks or chips are visible in the insulator). If in doubt concerning the condition of the plugs, renew them, as the expense is minimal.
8 Cleaning spark plugs by sandblasting is permitted, provided you clean the plugs with a high flash-point solvent afterwards.
9 Before installing the plugs, make sure they are the correct type and heat range and check the gap between the electrodes **(see illustrations)**. Compare the gap to that specified and adjust as necessary. If the gap must be adjusted, bend the side electrode only and be very careful not to chip or crack the insulator nose **(see illustration)**. Make sure the washer is in place before installing each plug.
10 Since the cylinder head is made of aluminium, which is soft and easily damaged, thread the plugs into the head turning the plug tool by hand to start with. Once the plugs are finger-tight, tighten them to the recommended torque setting (see Specifications). Alternatively, using the plug spanner, tighten a re-usable plug 1/8 to 1/4 turn after it seats. Tighten a new plug 1/2 turn after it seats. Take great care not to over-tighten the plugs.

 HAYNES HiNT *As the plugs are quite recessed, slip a short length of hose over the end of the plug to use as a tool to thread it into place. The hose will grip the plug well enough to turn it, but will start to slip if the plug begins to cross-thread in the hole – this will prevent damaged threads.*

Every 8000 miles 1•15

11 Reconnect the spark plug caps, making sure they are securely connected to the correct plug. Install all other components previously removed.

 *Stripped plug threads in the cylinder head can be repaired with a thread insert – see 'Tools and Workshop Tips' in the Reference section.*

14 Throttle and choke cables – check and adjustment

Throttle cables

1 Make sure the throttle twistgrip rotates easily from fully closed to fully open with the handlebars turned at various angles. The twistgrip should return automatically from fully open to fully closed when released.
2 If the twistgrip sticks, this is probably due to a cable fault. Remove the cables (see Chapter 4) and lubricate them (see Section 6). Clean the inside of the twistgrip body and lubricate with light grease. Install the cables, making sure they are correctly routed. If this fails to improve the operation of the twistgrip, the cables must be renewed. Note that in very rare cases the fault could lie in the carburettors rather than the cables, necessitating the removal of the carburettors and inspection of the throttle linkage (see Chapter 4).
3 With the twistgrip operating smoothly, check for a small amount of freeplay in the cables, measured in terms of the amount of twistgrip rotation before the throttle opens, and compare the amount to that listed in this Chapter's Specifications **(see illustration)**. If it's incorrect, adjust the accelerator cable (opening cable) to correct it.
4 Freeplay adjustments can be made at the twistgrip end of the cable. On CBF models pull the rubber boot off the cable adjuster **(see illustration)**. On all models loosen the locknut and turn the adjuster until the specified amount of freeplay is obtained (see this Chapter's Specifications), then retighten the locknut **(see illustrations)**. On CBF models refit the rubber boot.
5 If this adjuster has reached its limit of adjustment, reset it so that the freeplay is at a maximum. On CB models remove the fuel tank (see Chapter 4). Adjust the cable at the carburettor end as follows for your model. On CB models slacken the upper adjuster locknut, then, holding the lower locknut against the bracket, turn the adjuster out, making sure the lower nut remains on the adjuster threads **(see illustration)**. Turn the adjuster until the specified amount of freeplay is obtained, then tighten the upper locknut. On CBF models slacken the upper nut holding the rear cable in the bracket, then turn the lower nut as required to reposition the cable until the specified amount of freeplay is obtained, then hold the cable up and tighten the upper nut onto the bracket **(see illustration)**. Further adjustments can now be made at the twistgrip end. If the cable cannot be adjusted as specified, renew the cable (see Chapter 4).

⚠️ **Warning:** *Turn the handlebars all the way through their travel with the engine idling. Idle speed should not change. If it does, the cable may be routed incorrectly. Correct this condition before riding the motorcycle.*

6 Check that the throttle twistgrip operates smoothly and snaps shut quickly when released.

Choke cable

7 If the choke does not operate smoothly this is probably due to a cable fault. Remove the cable (see Chapter 4) and lubricate it (see Section 6). Install the cable, on CB models making sure it is correctly routed.
8 If this fails to improve the operation of the choke, the cable must be renewed. Note that in very rare cases the fault could lie in the carburettors rather than the cable, necessitating the removal of the carburettors and inspection of the choke plungers (see Chapter 4).
9 Make sure there is a small amount of freeplay in the cable before the plungers move. If there isn't, check that the cable is correctly installed at both ends – on CB models remove the fuel tank to access the carburettor end of the cable (see Chapter 4). To adjust the cable on CB models, slacken the

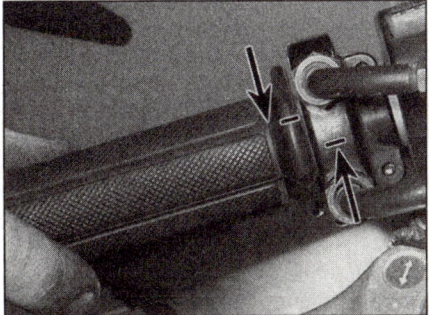
14.3 Throttle cable freeplay is measured in terms of twistgrip rotation

14.4a Pull the rubber boot back to expose the adjuster

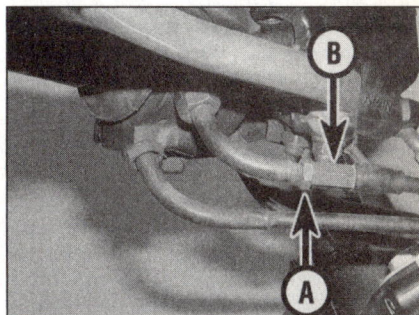
14.4b Throttle cable adjuster locknut (A) and adjuster (B) at twistgrip end

14.4c Tighten locknut against the adjuster when freeplay is satisfactory

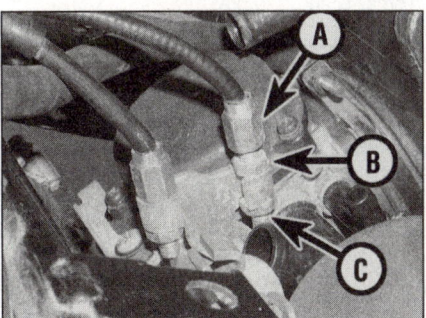
14.5a Throttle cable adjuster (A), upper locknut (B) and lower locknut (C) at carburettor end – CB models

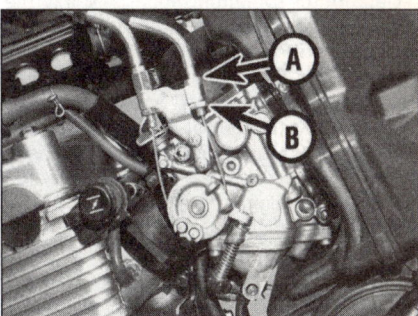
14.5b On CBF models slacken the upper nut (A) then turn the lower nut (B) to adjust the cable

Every 8000 miles

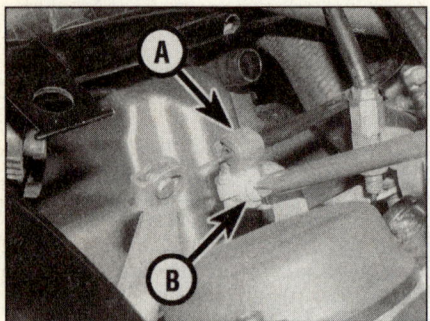

14.9 Choke cable bracket (A) and screw (B)

15.5 Detach the hose (arrowed) from each inlet duct

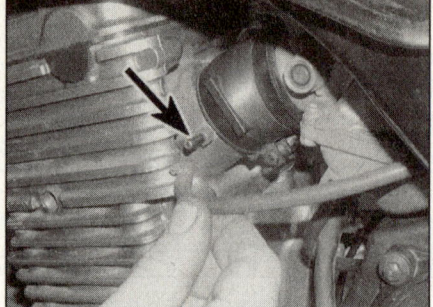

15.6 Fuel tap vacuum hose union on No. 1 cylinder inlet manifold (arrowed)

choke outer cable bracket screw on the carburettor and slide the cable further into the bracket, creating some freeplay (see illustration). On CBF models no adjustment is possible, but shouldn't be required if the cable is correctly installed. Otherwise, renew the cable.

15 Carburettors – synchronisation

⚠ **Warning:** Petrol (gasoline) is extremely flammable, so take extra precautions when you work on any part of the fuel system. Don't smoke or allow open flames or bare light bulbs near the work area, and don't work in a garage where a natural gas-type appliance is present. If you spill any fuel on your skin, rinse it off immediately with soap and water. When you perform any kind of work on the fuel system, wear safety glasses and have a fire extinguisher suitable for a Class B type fire (flammable liquids) on hand.

⚠ **Warning:** Take great care not to burn your hand on the hot engine unit when accessing the gauge take-off points on the inlet manifolds. Do not allow exhaust gases to build up in the work area; either perform the check outside or use an exhaust gas extraction system.

1 Carburettor synchronisation is simply the process of adjusting the carburettors so they pass the same amount of fuel/air mixture to each cylinder. This is done by measuring the vacuum produced in each cylinder. Carburettors that are out of synchronisation will result in decreased fuel mileage, increased engine temperature, less than ideal throttle response and higher vibration levels. Before synchronising the carburettors, make sure the idle speed is correct (see Section 2) and that the valve clearances are properly set (see Section 26).

2 To properly synchronise the carburettors, you will need a set of vacuum gauges or calibrated tubes (manometer) to indicate engine vacuum. The equipment used should be suitable for a two cylinder engine and come complete with the necessary adapter and hoses to fit the take-off points. **Note:** *Because of the nature of the synchronisation procedure and the need for special instruments, most owners leave the task to a Honda dealer.*

3 Start the engine and let it run until it reaches normal operating temperature, then shut it off.

4 On CB models remove the fuel tank (see Chapter 4). Move the coolant hose aside and remove the blanking screw from the vacuum take-off point on the inlet duct of No. 2 cylinder. Install the take-off adapter provided with the vacuum gauges.

5 On CBF models detach the PAIR system vacuum hose from the take-off point on each inlet duct (see illustration).

6 Connect the vacuum gauge hoses to the vacuum take-off points – on CB models the right gauge hose connects to the take-off adapter provided and the left gauge hose connects to the stub on the No. 1 cylinder manifold which the fuel tap vacuum hose normally connects to (see illustration). Make sure the hoses are a good fit because any air leaks will result in false readings.

7 On CB models arrange a temporary fuel supply to the carburettors. Ensure the fuel supply is above the level of the carburettors.

8 Start the engine and make sure the idle speed is correct. If it isn't, adjust it (see Section 2). If the gauges are fitted with damping adjustment, set this so that the needle flutter is just eliminated but so that they can still respond to small changes in pressure.

15.10 Carburettor synchronising screw (arrowed)

9 The vacuum readings for both cylinders should be the same, or at least within the tolerance listed in this Chapter's Specifications. If the vacuum readings vary, adjust the carburettors as follows.

10 Adjust the carburettors by turning the synchronising screw situated in-between the carburettors on the throttle linkage (see illustration). On CBF models use an angled screwdriver, or raise the tank to improve access (see Section 2). **Note:** *Do not press down on the screw whilst adjusting it, otherwise a false reading will be obtained.*

11 When the carburettors are synchronised, open and close the throttle quickly to settle the linkage, and recheck the gauge readings, readjusting if necessary.

12 When the adjustment is complete, check the idle speed and adjust as required by turning the idle adjusting screw (see Section 2) until the idle speed listed in this Chapter's Specifications is obtained. Stop the engine.

13 Remove the vacuum gauges. On CB models remove the adapter from No. 2 cylinder inlet manifold and fit the blanking screw. On CBF models reconnect the PAIR system vacuum hoses to each take-off point.

14 Install the fuel tank (see Chapter 4).

16 Headlight aim – check and adjustment

Note: *An improperly adjusted headlight may cause problems for oncoming traffic or provide poor, unsafe illumination of the road ahead. Before adjusting the headlight aim, be sure to consult with local traffic laws and regulations – refer to MOT Test Checks in the Reference section.*

1 The headlight beam can be adjusted both horizontally and vertically. Before making any adjustment, check that the tyre pressures are correct and the rear suspension is adjusted as required. Make any adjustments to the headlight aim with the machine on level ground, with the fuel tank half full and with an assistant sitting on the seat. If the motorcycle is usually ridden with a passenger on the back, have a second assistant to do this.

Every 8000 miles 1•17

R, T, V, W, X, Y and 2 models

2 Vertical adjustment is made by slackening the headlight mounting bolts and tilting the headlight shell up or down as required. Reference marks for a midway setting are stamped in the headlight brackets and the headlight shell (see illustration). Tighten the bolts securely after the adjustment has been made.

3 Horizontal adjustment is made by turning the adjuster screw on the left-hand side of the headlight rim (see illustration). Turn the screw clockwise to move the beam to the right, and anti-clockwise to move it to the left.

SW, SX, SY and S2 models

4 Vertical adjustment is made by turning the adjuster screw on the lower left-hand side of the headlight unit (see illustration). Turn it clockwise to move the beam up, and anti-clockwise to move it down.

5 Horizontal adjustment is made by turning the adjuster screw on the upper right-hand side of the headlight unit (see illustration 16.4). Turn it clockwise to move the beam to the left, and anti-clockwise to move it to the right.

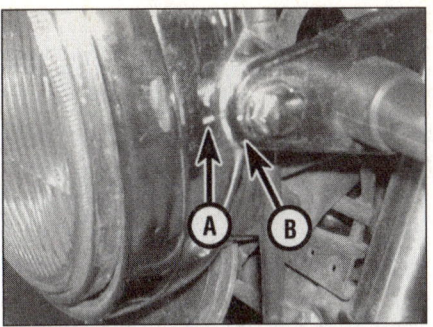

16.2 Headlight alignment reference marks on headlight shell (A) and bracket (B)

16.3 Horizontal beam adjustment screw (arrowed)

17 Centre and sidestand – check

1 The centre and sidestand return springs must be capable of retracting the stands fully and holding them retracted when the motorcycle is in use. If either of the springs is sagged or broken, it must be renewed.

2 Lubricate the stand pivots regularly (see Section 6).

3 The sidestand switch prevents the motorcycle being started with the stand extended unless the transmission is in neutral.

Check its operation by shifting the transmission into neutral, retracting the stand and starting the engine. Pull in the clutch lever and select a gear. Extend the sidestand. The engine should stop as the sidestand is extended. If the sidestand switch does not operate as described, check its circuit (see Chapter 9).

18 Suspension – check

1 The suspension components must be maintained in top operating condition to ensure rider safety. Loose, worn or damaged suspension parts seriously reduce the motorcycle's stability and control.

Front suspension

2 While standing alongside the motorcycle, apply the front brake and push on the handlebars to compress the forks several times. See if they move up-and-down smoothly without binding. If binding is felt, the forks should be disassembled and inspected (see Chapter 6).

3 Inspect the area around the dust seals for signs of oil leakage, then carefully lever up the dust seals using a flat-bladed screwdriver and inspect the area around the fork seal (see illustration). If leakage is evident, the seals must be renewed (see Chapter 6).

4 Check the tightness of all suspension nuts and bolts to be sure none have worked loose.

Rear suspension

5 Inspect the rear shock absorber mounting bolts for tightness. Inspect the shocks for pitting on the damper rods and fluid leakage. If leakage is found, the shocks should be renewed as a pair (see Chapter 6).

6 With the aid of an assistant to support the motorcycle, compress the rear suspension several times. It should move up and down freely without binding. If any binding is felt, the worn or faulty component must be identified and renewed. The problem could be due to either the shock absorber or the swingarm components. The shock absorbers can be removed individually for checking (see Chapter 6).

7 Support the motorcycle on its centre stand

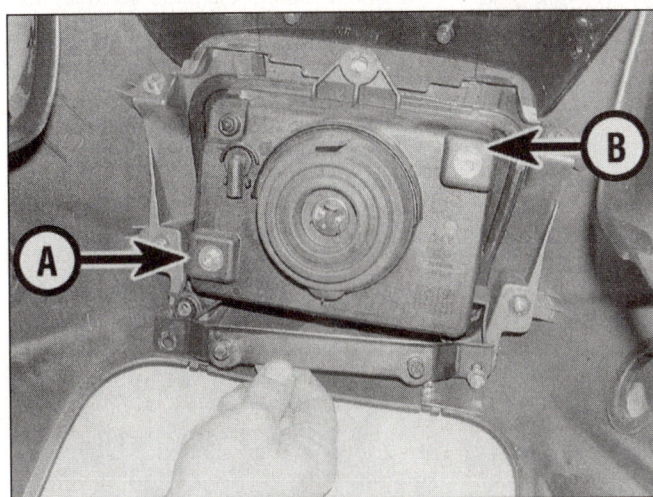

16.4 Vertical adjustment screw (A) and horizontal adjustment screw (B) on faired models

18.3 Prise up the dust seal using a flat-bladed screwdriver

1•18 Every 8000 miles

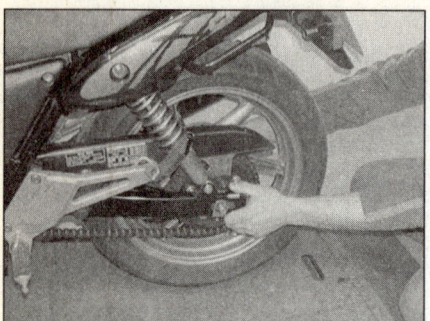

18.7a Checking for play in the swingarm bearings

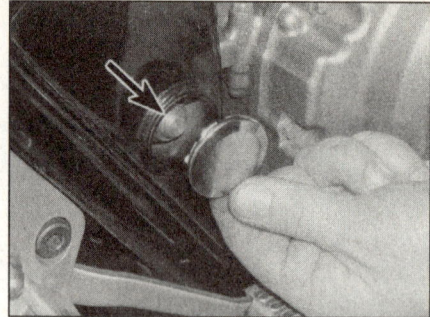

18.7b Check tightness of swingarm nut (arrowed)

18.8 Checking for play in the shock absorber mounts

so that the rear wheel is off the ground. Grab the swingarm and attempt to rock it from side to side – there should be no discernible movement at the rear **(see illustration)**. If there is any movement or a slight clicking can be heard, remove the caps and inspect the tightness of the swingarm nut **(see illustration)**, referring to the torque settings specified at the beginning of Chapter 6, and re-check for movement.

8 Next, grasp the top of the rear wheel and pull it upwards – there should be no discernible freeplay before the shock absorbers begin to compress **(see illustration)**. Any freeplay felt indicates worn shock absorber mountings. The worn components must be renewed (see Chapter 6).

9 To make an accurate assessment of the swingarm bearings, remove the rear wheel (see Chapter 7) and both rear shock absorbers (see Chapter 6). Grasp the rear of the swingarm with one hand and place your other hand at the junction of the swingarm and the frame. Try to move the rear of the swingarm from side-to-side. Any wear (play) in the bearings should be felt as movement between the swingarm and the frame at the front. If there is any play the swingarm will be felt to move forward and backward at the front (not from side-to-side). Next, move the swingarm up and down through its full travel. It should move freely, without any binding or rough spots. If any play in the swingarm is noted or if the swingarm does not move freely, the bearings must be removed for inspection or renewal (see Chapter 6).

19 Steering head bearings – check and adjustment

1 This motorcycle is equipped with caged ball steering head bearings which can become dented, rough or loose during normal use of the machine. In extreme cases, worn or loose steering head bearings can cause steering wobble – a condition that is potentially dangerous.

Check

2 Support the motorcycle on its centre stand. Raise the front wheel off the ground by placing a support under the engine or by having an assistant push down on the rear.

3 Point the front wheel straight-ahead, then slowly move the handlebars from side-to-side. Any dents or roughness in the bearing races will be felt, and if the bearings are too tight the bars will not move smoothly and freely. If the bearings are damaged or the action is rough, they should be renewed (see Chapter 6). If the bearings are too tight they should be adjusted as described below.

> **HAYNES HiNT** *Freeplay in the fork due to worn fork bushes can be misinterpreted for steering head bearing play – do not confuse the two.*

4 Next, grasp the fork sliders and try to move them forward and backward **(see illustration)**. Any looseness in the steering head bearings will be felt as front-to-rear movement of the forks. If play is felt in the bearings, adjust the steering head as follows.

Adjustment

5 Depending on the tools available, access to the steering stem nut may be restricted by the handlebars. If this is the case, remove the handlebar clamps (see Chapter 6) and lay the complete handlebar assembly over the headlight, making sure no strain is placed on the wiring or front brake hydraulic hose. If necessary, secure the handlebar assembly with cable ties and keep the master cylinder reservoir upright to prevent possible fluid leakage.

6 Slacken the fork clamp bolts in the top yoke **(see illustration 19.13)** and unscrew and remove the steering stem nut and washer **(see illustration)**.

7 Gently ease the top yoke upwards off the fork tubes and position it clear of the steering head **(see illustration)**. If necessary, use a rag to protect other components. **Note:** *On R, T, V, W, X, Y and 2 and CBF models the instrument assembly is attached to the top yoke. Ensure no strain is placed on the instrument wiring.*

8 Prise the lockwasher tabs out of the notches in the locknut, unscrew the locknut using either a C-spanner or a suitable drift located in one of the notches and remove the lockwasher.

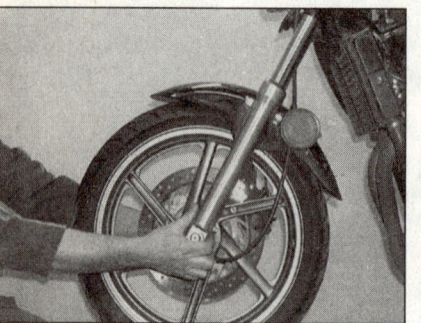

19.4 Checking for play in the steering head bearings

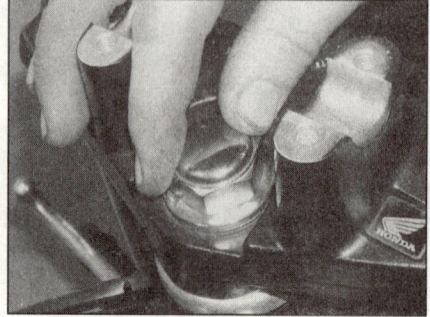

19.6 Remove the steering stem nut and washer

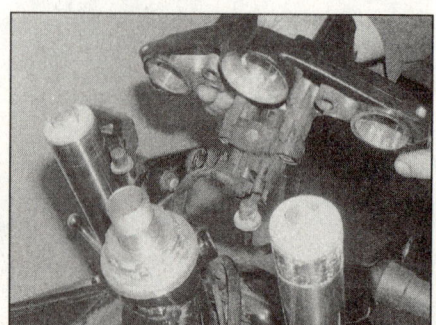

19.7 Lift the top yoke off the fork tubes

Every 8000 miles

19.8 Inspect the lockwasher tabs for cracks and fatigue

19.9 Tighten the adjuster nut carefully to remove freeplay

19.12a Align slots in the locknut with tabs on the lockwasher . . .

Inspect the lockwasher tabs for cracks or signs of fatigue **(see illustration)**. If there are any, discard the lockwasher and use a new one; it is advisable to renew it as a matter of course.

9 Using either the C-spanner or drift, slacken the adjuster nut slightly until pressure on the bearing is just released, then tighten the nut a little at a time until all freeplay is removed **(see illustration)**. Ensure that the steering is able to move smoothly as described in Steps 3 and 4. If the Honda adapter tool (Pt. No. 07946-4300101) is available you can apply the torque setting specified at the beginning of this Chapter.

10 Turn the steering from lock to lock five times to settle the bearings, then recheck the adjustment or the torque setting. The object is to set the adjuster nut so that the bearings are under a very light loading, just enough to remove any freeplay.

Caution: Take great care not to apply excessive pressure because this will cause premature failure of the bearings. If the torque setting is applied and the bearings are still too loose or too tight, set them up according to feel.

11 When the bearings are correctly adjusted, install the lockwasher onto the adjuster nut and bend two of the washer tabs down into the slots in the adjuster nut.

12 Install the locknut and tighten it finger-tight, then tighten it further (to a maximum of 90°) to align the slots in the locknut with the remaining tabs on the lockwasher **(see illustration)**. Hold the adjuster nut to prevent it from moving if necessary. Bend the remaining lockwasher tabs up to secure the locknut **(see illustration)**.

13 Fit the top yoke onto the steering stem and the fork legs. Install the washer and steering stem nut and tighten it and both the fork clamp bolts to the torque settings specified at the beginning of this Chapter **(see illustration)**. If displaced, refit the handlebars (see Chapter 6).

14 Check the bearing adjustment as described in Steps 3 and 4 and re-adjust if necessary.

20 Wheel bearings – check

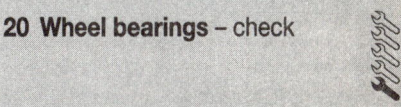

1 Wheel bearings will wear over a period of time and result in handling problems.
2 Support the motorcycle on its centre stand. Check for any play in the bearings by pushing and pulling the wheel against the hub (see

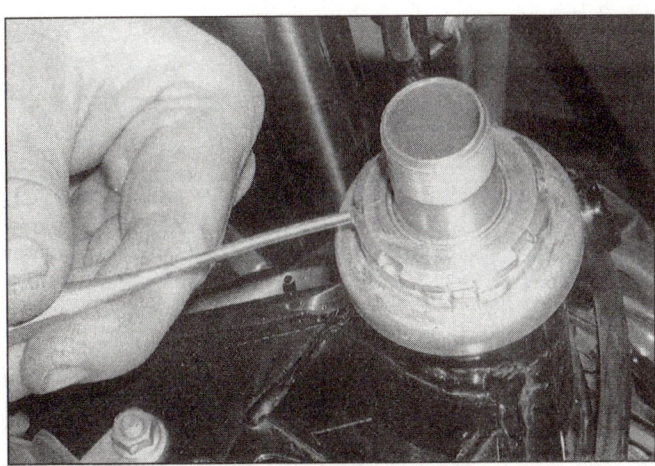

19.12b . . . then bend tabs up to secure locknut

19.13 Tighten the fork clamp bolts to the specified torque

1•20 Every 8000 miles

20.2 Checking for play in the wheel bearings

22.1 Renew the chain slider if wear reaches the indicators (arrowed)

illustration). Also rotate the wheel and check that it rotates smoothly.

3 If any play is detected in the hub, or if the wheel does not rotate smoothly (and this is not due to brake or transmission drag), the wheel bearings must be removed and inspected for wear or damage (see Chapter 7).

21 Wheels and tyres – general check

Tyres

1 Check the tyre condition and tread depth thoroughly – see *Daily (pre-ride) checks*.

Wheels

2 Cast wheels are virtually maintenance free, but they should be kept clean and checked periodically for cracks and other damage. Also check the wheel runout and alignment (see Chapter 7). Never attempt to repair cast wheels; they must be renewed if damaged.
3 Check the valve rubber for signs of damage or deterioration and have it renewed if necessary. Also, make sure the valve stem cap is in place and tight. Check that the wheel balance weights are fixed firmly to the wheel rim. If the weights have fallen off, have the wheel rebalanced by a motorcycle tyre specialist.

22 Drive chain slider – wear check

1 Remove the front sprocket cover (see Chapter 6). Inspect the drive chain slider on the swingarm for wear. If the slider is worn to either limit line indicated by two arrows **(see illustration)** it must be renewed (see Chapter 6).

23 Nuts and bolts – tightness check

1 Since vibration of the machine tends to loosen fasteners, all nuts, bolts, screws, etc. should be periodically checked for proper tightness.
2 Pay particular attention to the following:
   Spark plugs
   Engine oil drain plug
   Engine mounting bolts
   Exhaust system bolts/nuts
   Handlebar clamp bolts
   Headlamp mounting bolts (unfaired models)
   Gearchange pedal bolt
   Footrest and stand bolts
   Front axle nut and axle clamp bolt
   Front fork clamp bolts (top and bottom yoke)
   Shock absorber and swingarm bolts/nuts
   Rear axle nut
   Brake caliper mounting bolts
   Brake hose banjo bolts and caliper bleed valves
   Brake disc bolts
   Rear drum brake torque arm bolts (R and T models)
3 If a torque wrench is available, use it along with the torque specifications at the beginning of this and other Chapters.

Every 12,000 miles (18,000 km) or 18 months

Carry out all the items under the 4000 mile (6000 km) check:

24 Brake fluid – change

1 The brake fluid should be renewed at the prescribed interval or whenever a master cylinder or caliper overhaul is carried out. Refer to the brake bleeding section in Chapter 7.

25 Air filter and sub-air cleaner – renewal

Caution: *If the machine is continually ridden in wet or dusty conditions, the filters should be renewed more frequently.*
1 On CBF models remove the left-hand frame side panel (see Chapter 8). Undo the screws securing the air filter cover (CB models) or air intake duct (CBF models) to the filter housing and remove the cover or duct **(see illustrations)**.

2 Remove the filter element from the housing **(see illustration)**. Tap the element gently to dislodge any dirt, then use compressed air directed in the opposite direction of normal air flow to clean the element. If the element is excessively dirty, and always at the prescribed service interval, replace the element with a new one.
3 Inspect the inside of the filter housing and clean out any dust or dirt particles. On CB models check the air intake passage on the right-hand rear of the housing for obstructions. On CBF models check the air intake duct.

Every 12,000 miles 1•21

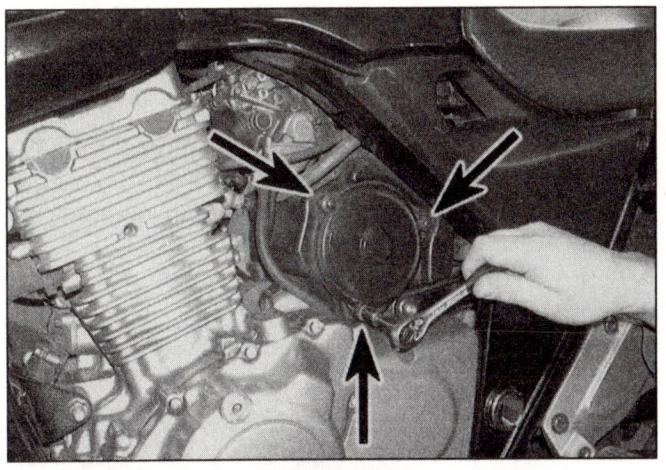

25.1a Remove the air filter housing cover screws...

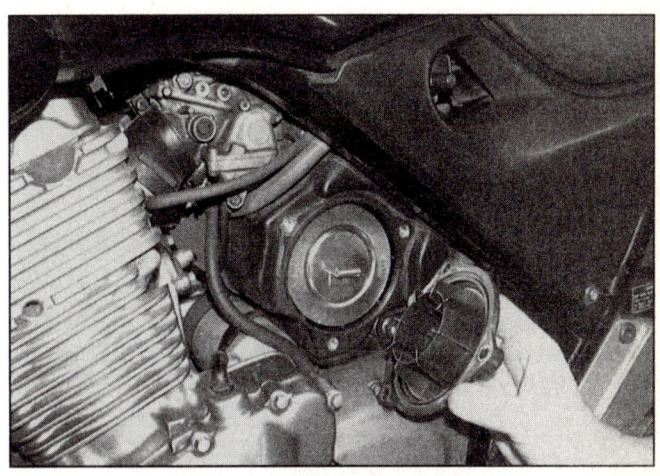

25.1b ...and remove the cover

25.2 Pull the filter element out of the housing

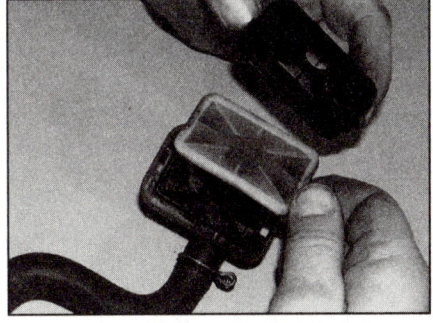

25.5 Check the condition of the sub-air cleaner element

25.7 Remove the plug (arrowed) and drain the hose

4 Install the new filter element, then install the cover and tighten the cover screws carefully to avoid damaging the screw threads.
5 Remove the fuel tank (see Chapter 4). Disconnect the sub-air cleaner hose from the air filter housing and unclip the cover of the sub-air filter **(see illustration)**.

Remove the sub-air cleaner element from its housing and wash the element in warm, soapy water. Rinse the element in clean water and dry before refitting. If the element is damaged or shows signs of deterioration, fit a new one.
6 Install the fuel tank (see Chapter 4).

7 Remove the blanking plug from the bottom of the air filter housing drain hose and allow any sludge to drain out **(see illustration)**. If necessary detach the hose from the housing and clean it through. Check the condition of the hose and replace it with a new one if it is damaged or perished.

Every 16,000 miles (24,000 km) or two years

Carry out all the items under the 4000 mile (6000 km) and 8000 mile (12,000 km) checks:

26 Valve clearances – check and adjustment

1 The engine must be completely cool for this maintenance procedure, so let the machine sit overnight before beginning.
2 Remove the fuel tank (see Chapter 4).
3 Remove the thermostat assembly (see Chapter 3) and sub-air cleaner (see Section 25).
4 Remove the valve cover (see Chapter 2) and the spark plugs (see Section 13).
5 The cylinders are numbered 1 and 2 from left to right. Make a chart or sketch of all valve positions so that a note of each clearance can be made against the relevant valve **(see illustration)**.

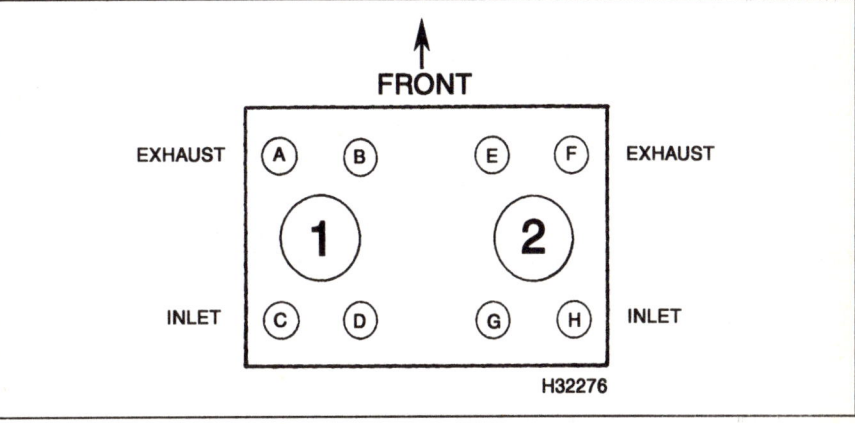

26.5 Note valve clearances on a simple sketch

1•22 Every 16,000 miles

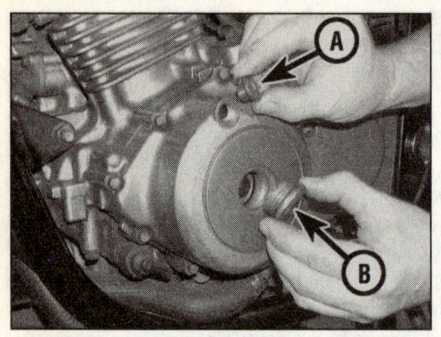

26.6a Remove the timing inspection plug (A) and the centre plug (B)

26.6b Turn the crankshaft anti-clockwise using a socket on the alternator bolt

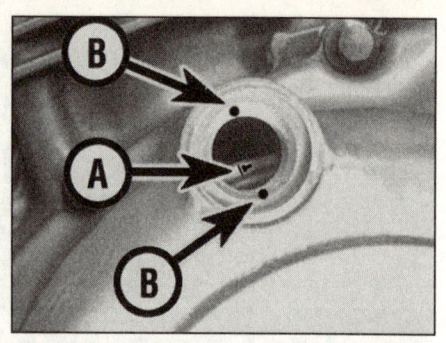

26.7a Turn the engine until the T mark (A) aligns with the static marks (B)

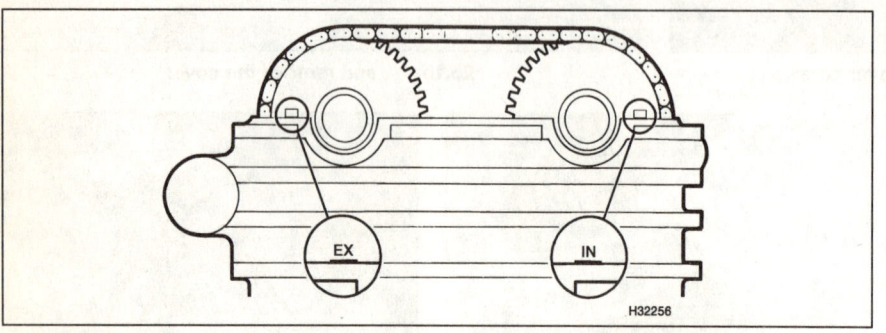

26.7b Check clearances on No. 1 cylinder valves with cam sprockets in this position

26.8 Measure the valve clearance using a feeler gauge

Check

6 Unscrew the timing inspection plug and the centre plug from the alternator cover on the left-hand side of the engine **(see illustration)**. Discard the plug O-rings as new ones should be used. The engine can be turned using a socket on the alternator rotor bolt and turning it in an anti-clockwise direction **(see illustration)**. Alternatively, place the motorcycle on its centre stand so that the rear wheel is off the ground, select a high gear and rotate the rear wheel by hand in its normal direction of rotation.

7 Turn the engine until the line next to the 'T' mark on the alternator rotor aligns with the notches in the timing inspection hole **(see illustration)**. **Note:** *Turn the engine in the normal direction of rotation (anti-clockwise), viewed from the left-hand end of the engine*. Now, check the position of the camshaft sprockets. The 'IN' timing mark on the inlet camshaft sprocket and the 'EX' timing mark on the exhaust camshaft sprocket should align with the cylinder head surface with each mark on the outside of its respective sprocket **(see illustration)**. If the timing marks are on the inside of their respective sprockets, turn the engine one full turn (360°) anti-clockwise until the 'T' mark on the alternator rotor again aligns with the notches in the timing inspection hole. The marks on the camshaft sprockets should now align correctly with the cylinder head surface.

8 Insert a feeler gauge of the same thickness as the correct valve clearance (see Specifications) between the cam base and follower of each valve of the No. 1 cylinder and check that it is a firm sliding fit – you should feel a slight drag when the you pull the gauge out **(see illustration)**. If not, use the feeler gauges to obtain the exact clearance. **Note:** *The inlet and exhaust valve clearances are different*. Record the measured clearance on your chart.

9 Rotate the crankshaft anti-clockwise through 180° (half a revolution). The 'EX' timing mark on the exhaust camshaft should now be facing straight up from the cylinder head surface **(see illustration)**. Measure the valve clearances of the No. 2 cylinder using the method described in Step 8.

Adjustment

10 When all clearances have been measured and noted, identify whether the clearance on any valve falls outside that specified. If it does, the shim between the follower and the valve must be replaced with one of a thickness which will restore the correct clearance.

11 Shim replacement requires removal of the camshafts (see Chapter 2). There is no need to remove both camshafts if shims from only one side of the engine need replacing.

12 With the camshaft removed, remove the cam follower of the valve in question using either a magnet or a pair of pliers **(see illustration)**. Retrieve the shim from either the

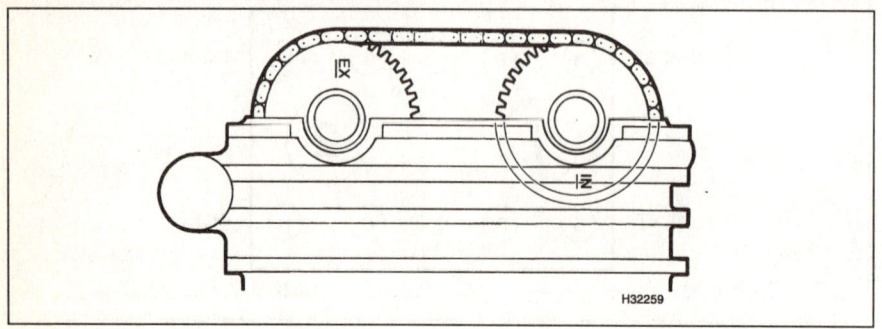

26.9 Check clearances on No. 2 cylinder valves with cam sprockets in this position

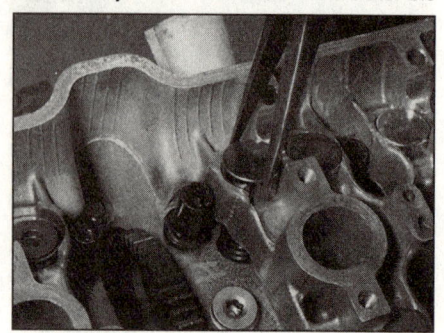

26.12a Remove the follower . . .

Every 16,000 miles 1•23

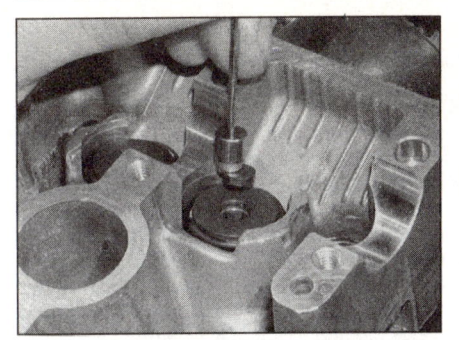

26.12b ... then retrieve the shim from the top of the valve ...

26.12c ... or inside the follower

26.13a The shim size should be marked on its face ...

inside of the follower or pick it out of the top of the valve, using either a magnet or a small screwdriver with a dab of grease on it (the shim will stick to the grease) **(see illustrations)**. Do not allow the shim to fall into the engine.

13 The shim size should be stamped on its face. However, it is recommended that the shim is measured to check that it has not worn **(see illustrations)**. The size marking is in the form of a three figure number, eg 205 indicating that the shim is 2.05 mm thick. The new shim thickness required can then be calculated as follows. *Note: Always aim to get the clearance at the mid-point of the specified range.*

14 If the valve clearance is less than specified, subtract the measured clearance from the specified clearance then deduct the result from the original shim thickness. For example:

Sample calculation – inlet valve clearance too small

Measured clearance: 0.08 mm
Specified clearance:
    0.17 mm (0.16 to 0.18 mm)
Difference: 0.09 mm
Shim thickness fitted: 2.475 mm
Correct shim thickness required is
    2.475 – 0.09 = 2.385 mm

15 If the valve clearance is greater than specified, subtract the specified clearance from the measured clearance, and add the result to the thickness of the original shim. For example:

Sample calculation – exhaust valve clearance too large

Measured clearance: 0.35 mm
Specified clearance:
    0.26 mm (0.25 to 0.27 mm)
Difference: 0.09 mm
Shim thickness fitted: 1.975 mm
Correct shim thickness required is
    1.975 + 0.09 = 2.065 mm

16 Obtain the correct thickness shims from a Honda dealer. Where the required thickness is not equal to the available shim thickness, round off the measurement to the nearest available size. Shims are available in 0.025 mm increments from 1.200 mm to 2.900 mm. *Note: If the required replacement shim is greater than 2.900 mm (the largest available), the valve is probably not seating correctly due to a build-up of carbon deposits and should be checked and cleaned or resurfaced as required (see Chapter 2).*

17 When replacing a shim, lubricate it with molybdenum disulphide oil (a 50/50 mixture of molybdenum disulphide grease and engine oil) and fit it into its recess on the top of the valve, with the size marking facing up **(see illustration)**. Check that the shim is correctly seated, then lubricate the follower with molybdenum disulphide oil and install it onto the valve **(see illustration)**. Repeat the

26.13b ... but measure it anyway

process for any other valves until all the clearances are correct.

18 Install the camshafts (see Chapter 2). Rotate the engine several turns to seat the new shim(s), then check the clearances again.

19 Install all disturbed components in a reverse of the removal sequence. Use new O-rings on the alternator cover centre plug and the timing inspection plug. Smear the O-rings lightly with grease before fitting **(see illustration)**.

27 Spark plugs – renewal

1 Remove the old spark plugs as described in Section 13 and install new ones.

26.17a Fit the valve shim with its size marking facing up ...

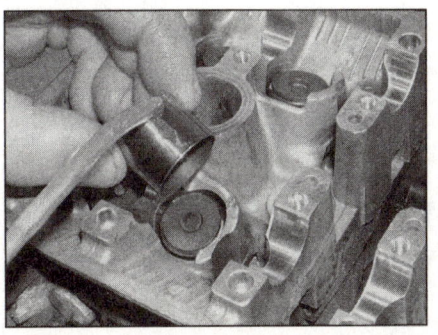

26.17b ... then lubricate and install the follower

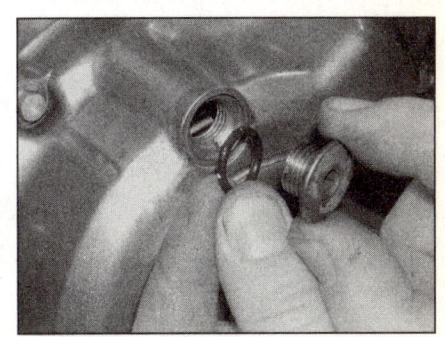

26.19 Fit new O-rings onto the plugs

Routine maintenance and servicing

Every 24,000 miles (36,000 km) or three years

Carry out all the items under the 4000 mile (6000 km), 8000 mile (12,000 km) and 12,000 mile (18,000 km) checks, plus the following:

28 Cooling system – draining, flushing and refilling

Warning: Allow the engine to cool completely before performing this maintenance operation. Also, don't allow antifreeze to come into contact with your skin or the painted surfaces of the motorcycle. Rinse off spills immediately with plenty of water. Antifreeze is highly toxic if ingested. Never leave antifreeze lying around in an open container or in puddles on the floor; children and pets are attracted by its sweet smell and may drink it. Check with local authorities (councils) about disposing of antifreeze. Many communities have collection centres which will see that antifreeze is disposed of safely. Antifreeze is also combustible, so don't store it near open flames.

Draining

1 Remove the fuel tank (see Chapter 4). Remove the pressure cap by turning it anti-clockwise until it reaches a stop. If you hear a hissing sound (indicating there is still pressure in the system), wait until it stops. Now press down on the cap and continue turning until it can be removed **(see illustration 10.7)**.

2 Position a suitable large container beneath the water pump on the left-hand side of the engine. Remove the coolant drain plug and its sealing washer and allow the coolant to completely drain from the system **(see illustrations)**. Retain the old sealing washer for use during flushing.

3 Position the container beneath the reservoir tank. Open the reservoir tank filler cap (see *Daily (pre-ride) checks*), detach the overflow hose from the bottom of the tank and drain the tank **(see illustration)**.

Flushing

4 Flush the cooling system with clean tap water by inserting a garden hose in the pressure cap filler neck. Allow the water to run through the system until it is clear and flows cleanly out of the drain hole. If the drain hole appears to be clogged with sediment, remove the water pump cover and clean the pump (see Chapter 3). If the radiator is extremely corroded, remove it (see Chapter 3) and have it cleaned at a radiator shop. Flush the reservoir tank with clean water, then connect the overflow hose to the tank and fill the tank with clean water.

5 Install the coolant drain plug using the old sealing washer.

6 Fill the cooling system with clean water mixed with a flushing compound. Make sure the flushing compound is compatible with aluminium components and follow the manufacturer's instructions carefully. Fit the pressure cap and install the fuel tank (see Chapter 4).

7 Start the engine and allow it to reach normal operating temperature. Let it run for about ten minutes.

8 Stop the engine and let it cool. Remove the fuel tank, then cover the pressure cap with a heavy rag and turn it anti-clockwise to the first stop, releasing any pressure that may be present in the system. Once the hissing stops, push down on the cap and remove it completely.

9 Remove the coolant drain plug and drain the system once again.

10 Fill the system with clean water and repeat the procedure in Steps 7 to 9.

Refilling

11 Fit a new sealing washer onto the drain plug and tighten it securely. Drain the reservoir tank and ensure that the hose clip is correctly installed when the overflow hose is reconnected.

12 Fill the system with the proper coolant mixture (see this Chapter's Specifications) **(see illustration)**. Note: *Pour the coolant in slowly to minimise the amount of air entering the system.*

13 When the system is full (all the way up to the top of the pressure cap filler neck), install the pressure cap. Also top up the reservoir tank to the UPPER level mark (see *Daily (pre-ride) checks*). Install the fuel tank (see Chapter 4).

14 Start the engine and allow it to idle for 2 or 3 minutes. Flick the throttle twistgrip part open 3 or 4 times, so that the engine speed rises to approximately 4000 – 5000 rpm, then stop the engine.

15 Let the engine cool, then remove the pressure cap as described in Step 8. Check that the coolant level is still up to the pressure cap filler neck. If it's low, add the specified mixture until it reaches the top of the filler neck. Refit the cap.

16 Check the coolant level in the reservoir tank and top up if necessary.

17 Check the system for leaks.

18 Do not dispose of the old coolant by pouring it down the drain. Instead pour it into a heavy plastic container, cap it tightly and take it into an authorised disposal site or service station – see **Warning** at the beginning of this Section.

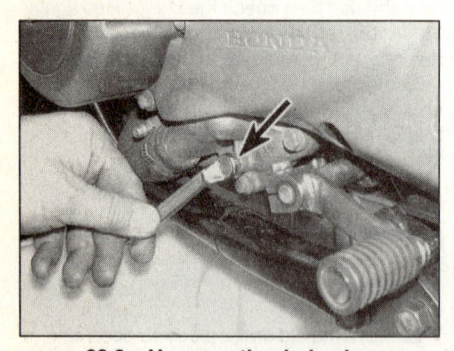

28.2a Unscrew the drain plug (arrowed) . . .

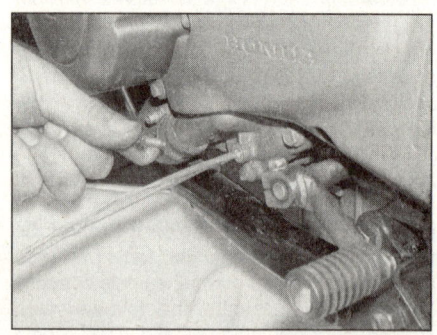

28.2b . . . and allow the coolant to drain completely

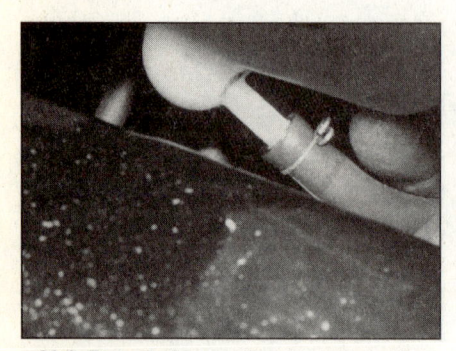

28.3 Detach the overflow hose from the bottom of the reservoir tank

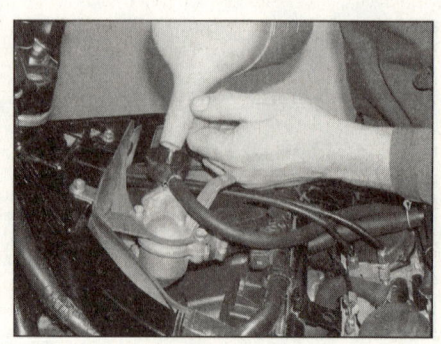

28.12 Fill the system with the specified mixture

29 Brake caliper seals and master cylinder seals – renewal

1 Brake seals will deteriorate over a period of time and lose their effectiveness, leading to sticking operation or fluid loss, or allowing the ingress of air and dirt. Refer to Chapter 7 and dismantle the components for seal renewal.

Routine maintenance and servicing 1•25

Every four years

30 Brake hoses – renewal

1 The brake hoses will in time deteriorate and should be renewed regardless of their apparent condition.
2 Refer to Chapter 7 before disconnecting the brake hoses from the master cylinders and calipers. Always renew the banjo union sealing washers.

31 Fuel hoses – renewal

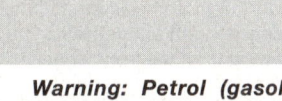

Warning: Petrol (gasoline) is extremely flammable, so take extra precautions when you work on any part of the fuel system. Don't smoke or allow open flames or bare light bulbs near the work area, and don't work in a garage where a natural gas-type appliance is present. If you spill any fuel on your skin, rinse it off immediately with soap and water. When you perform any kind of work on the fuel system, wear safety glasses and have a fire extinguisher suitable for a Class B type fire (flammable liquids) on hand.

1 The fuel delivery and vacuum hoses should be renewed regardless of their condition.
2 Ensure that the ignition is turned OFF and turn the fuel tap to the OFF position as a safety measure. Disconnect the fuel hose from the fuel tap and from the carburettor joint pipe and the vacuum hose from the tap and the No. 1 cylinder inlet manifold, noting the routing of each hose (see Chapter 4 if required).
3 Secure each new hose to its unions using new clips. Run the engine and check that there are no leaks before taking the machine out on the road.

Non-scheduled maintenance

32 Cylinder compression – check

1 Among other things, poor engine performance may be caused by leaking valves, incorrect valve clearances, a leaking head gasket, or worn pistons, rings and/or cylinder walls. A cylinder compression check will help pinpoint these conditions and can also indicate the presence of excessive carbon deposits in the combustion chambers.
2 The only tools required are a compression gauge, with a threaded adapter to fit the spark plug hole, and a spark plug wrench. Depending on the outcome of the initial test, a squirt-type oil can may also be needed.
3 For anything more than an initial check, make sure the valve clearances are correctly set (see Section 26) and that the cylinder head bolts are tightened to the correct torque setting (see Chapter 2).
4 Refer to *Fault Finding Equipment* in the Reference section for details of the compression test.

33 Engine oil pressure – check

1 The oil pressure warning light should come on when the ignition (main) switch is turned ON – this serves as a check that the warning light bulb is sound. If the oil pressure warning light does not come on when the ignition is turned on, check the bulb, the fuse for the tail/signal/brake/horn circuit and the oil pressure switch (see Chapter 9). The light should extinguish a few seconds after the engine is started.
2 If the oil pressure warning light fails to extinguish, or comes on whilst the engine is running, low oil pressure is indicated – stop the engine immediately and carry out an oil level check (see *Daily (pre-ride) checks*). If the level is correct, test the oil pressure switch (see Chapter 9). If the switch is good, carry out an oil pressure check.
3 To check the oil pressure, a suitable pressure gauge (which screws into the crankcase) will be needed. Honda provide a gauge (Pt. No. 07506-3000000) and gauge adapter (Pt. No. 07510-4220100) for this purpose.
4 Warm the engine up to normal operating temperature then stop it.
5 Remove the oil pressure switch (see Chapter 9) and quickly screw the adapter into the crankcase threads. Connect the pressure gauge to the adapter.

Warning: Take great care not to burn your hand on the hot engine unit, exhaust pipe or with engine oil when accessing the switch take-off point on the crankcase. Do not allow exhaust gases to build up in the work area; either perform the check outside or use an exhaust gas extraction system.

6 Start the engine and increase the engine speed to 2000 rpm whilst watching the pressure gauge reading. The oil pressure should be similar to that given in the Specifications at the start of this Chapter.
7 If the pressure is significantly lower than the standard, either the pressure regulator is stuck open, the oil pump is faulty, the oil strainer or filter is blocked, or there is other engine damage. Begin diagnosis by checking the oil filter, strainer and regulator, then the oil pump (see Chapter 2). If those items check out okay, chances are the engine bearing oil clearances are excessive and the engine needs to be overhauled.
8 If the pressure is too high, either an oil passage is clogged, the regulator is stuck closed or the wrong grade of oil is being used.
9 Stop the engine and unscrew the gauge and adapter from the crankcase.
10 Install the oil pressure switch (see Chapter 9). Check the oil level (see *Daily (pre-ride) checks*).

34 Steering head bearings – lubrication

1 Over a period of time the grease in the bearings will harden or may be washed out by incorrect use of jet washes resulting in rapid wear of the bearing surfaces.
2 Disassemble the steering head for re-greasing of the bearings. Refer to Chapter 6 for details.

35 Swingarm bearings – lubrication

1 Over a period of time the grease will harden or dirt will penetrate the bearings due to failed dust seals resulting in rapid wear of the bearing surfaces.
2 Remove the swingarm as described in Chapter 6 for greasing of the bearings.

36 Rear brake cam (R and T models) – lubrication

1 Refer to Chapter 7, and remove the brake shoes and brake cam from the brake backplate as an assembly. If all components are in serviceable condition there is no need to separate the brake shoes. Clean all traces of old grease from the cam spindle and the pivot hole in the brake backplate, and the brake shoe pivot posts.
2 Apply a smear of copper-based grease to the cam spindle and the pivot posts, and to the spring hooks where they engage the holes

Non-scheduled maintenance

in the brake shoes. Install the assembly in the backplate (see Chapter 7).

3 Using the brake arm as a lever, turn the brake cam and clean all traces of old grease from both sides of the cam, then apply a fresh smear of copper-based grease to the two bearing surfaces.

4 Assemble the brake cam and shoes on the brakeplate, install the backplate in the brake drum and install the rear wheel (see Chapter 7).

Caution: Do not apply too much grease otherwise there is a risk of it contaminating the brake drum and shoe linings.

37 Exhaust system – check

1 Periodically check the exhaust system mounting bolts for tightness. Check the cylinder head joints and silencer union for leaks. If tightening the exhaust flange nuts or silencer clamp fails to stop any leaks, renew the gaskets (see Chapter 4).

2 Clean the system and check for corrosion. Rub down any rust spots and treat with a suitable heat resistant paint. The exhaust pipe flange nuts are especially prone to corrosion. Clean the nuts and studs regularly and apply a smear of copper based grease to their threads.

38 Front forks – oil change

1 Fork oil degrades over a period of time and loses its damping qualities. R models are equipped with drain screws in the fork sliders and therefore changing the fork oil is a relatively straightforward task. With all other models the fork legs have to be removed for draining and refilling.

R models

2 Support the motorcycle on its centre stand and place a support under the engine sump to take the weight off the front wheel.

3 Slacken the top yoke clamp bolts **(see illustration 19.13)**, then remove the fork top bolts, spacers, spring seats and springs **(see illustration 7.1 in Chapter 6)**. Use rag to prevent fork oil spillage.

⚠ *Warning: The fork spring is pressing on the fork top bolt with considerable pressure. Unscrew the bolt very carefully, keeping a downward pressure on it and release it slowly as it is likely to spring clear. It is advisable to wear some form of eye and face protection when carrying out this operation.*

4 Remove the drain screw from each fork slider in turn and allow the oil to drain into a suitable container **(see illustration)**.

⚠ *Warning: Do not allow the fork oil to contact the brake disc, pads or tyre. If it does, clean the disc with brake system cleaner, wipe off the tyre and renew the pads before riding the motorcycle. Make up a cardboard chute to direct oil away from the disc and tyre.*

5 Remove the support from under the engine and allow the forks to compress slowly. Gently pump the forks to expel all of the oil then refit the drain screws with new sealing washers.

6 Slowly pour the correct quantity of the specified grade of fork oil into each leg and pump the forks gently at least ten times to distribute the oil evenly. Fully compress the fork tubes into the sliders and measure the fork oil level from the top of the tubes; if necessary, adjust the oil level by adding or subtracting oil until it is at the level specified at the beginning of Chapter 6.

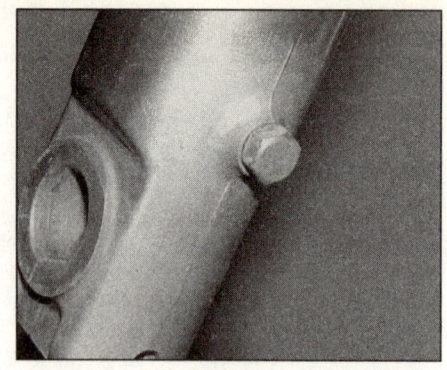

38.4 Fork oil drain screw on R models

7 Replace the support under the engine and fully extend the sliders on the fork tubes. Wipe any surplus oil off the springs and spacers, then install the springs with the widely spaced coils at the bottom, spring seats and spacers. Refer to Chapter 6, Section 7 for order of assembly.

8 Fit new O-rings on the fork top bolts and lubricate the rings with fork oil before installing the top bolts. Tighten the top yoke clamp bolts to the torque setting specified at the beginning of this Chapter.

All other models

9 Remove the forks from the motorcycle as described in Chapter 6, Section 6.

10 Follow Steps 3 to 6 of Chapter 6, Section 7 to drain the fork oil, then Step 25 onwards to refill with fresh oil.

11 Refer to the Specifications at the beginning of Chapter 6 for fork oil type, quantity and level.

Chapter 2
Engine, clutch and transmission

Contents

Alternator – removal and installation . . . . see Chapter 9	Oil and filter – change. . . . . see Chapter 1
Balancer shaft and bearings. 26	Oil level – check . . . . see Daily (pre-ride) checks
Camchain and guides 27	Oil pressure – check. . . . . .see Chapter 1
Camchain tensioner 8	Oil pressure switch – check, removal and installation . . . see Chapter 9
Camshafts – removal, inspection and installation 9	Oil pump – removal, inspection and installation. 17
Clutch cable – replacement 14	Oil sump, oil strainer and pressure relief valve –
Clutch – check . . . . see Chapter 1	removal, inspection and installation 18
Clutch – removal, inspection and installation. 15	Operations possible with the engine in the frame 2
Connecting rods and bearings – removal, inspection	Operations requiring engine removal 3
and installation. 22	Pistons – removal, inspection and installation 23
Crankcase halves and cylinder bores – inspection and servicing . . . 31	Piston rings – inspection and installation. 24
Crankcase halves – separation and reassembly 20	Pulse generator coil assembly –
Crankshaft and main bearings – removal, inspection	removal and installation . . . . see Chapter 5
and installation 25	Recommended running-in procedure 33
Cylinder head – removal and installation 10	Selector drum and forks – removal, inspection and installation 30
Cylinder head and valves – disassembly, inspection	Spark plug gap – check and adjustment . . . . see Chapter 1
and reassembly 12	Starter clutch and idle gear – removal,
Engine – compression check . . . . see Chapter 1	inspection and installation 13
Engine – removal and installation. 5	Starter motor – removal and installation . . . . see Chapter 9
Engine disassembly and reassembly – general information 6	Timing rotor, primary drive gear and balancer gear – removal,
Gearchange mechanism – removal, inspection and installation 19	inspection and installation 16
General information 1	Transmission shafts and bearings – removal and installation 28
Idle speed – check and adjustment . . . . see Chapter 1	Transmission shafts – disassembly, inspection and reassembly 29
Initial start-up after overhaul 32	Valve clearances – check and adjustment. . . . . see Chapter 1
Main and connecting rod bearings – general information 21	Valve cover – removal and installation 7
Major engine repair – general note. 4	Valves/valve seats/valve guides – servicing. 11
Neutral switch – check, removal and installation. . . . . . . see Chapter 9	

Degrees of difficulty

Easy, suitable for novice with little experience	Fairly easy, suitable for beginner with some experience	Fairly difficult, suitable for competent DIY mechanic	Difficult, suitable for experienced DIY mechanic	Very difficult, suitable for expert DIY or professional

Specifications

General
Type	Four-stroke, dohc, in-line twin cylinder
Capacity	499 cc
Bore	73 mm
Stroke	59.6 mm
Compression ratio	10.5 to 1
Cylinder numbering	1 – 2 from left to right
Cooling system	Liquid cooled
Clutch	Wet multi-plate
Transmission	Six-speed constant mesh
Final drive	Chain

Cylinder head
Warpage (max)	0.10 mm

2•2 Engine, clutch and transmission

	Standard	Service limit (min)
Camshafts		
Inlet lobe height	36.280 to 36.360 mm	36.250 mm
Exhaust lobe height	36.370 to 36.450 mm	36.340 mm
Journal diameter	24.949 to 24.970 mm	24.940 mm
Journal oil clearance	0.030 to 0.072 mm	0.100 mm
Runout (max)	0.05 mm	

Valves, guides and springs
Valve clearances... see Chapter 1
Inlet valve
  Stem diameter
    Standard... 4.475 to 4.490 mm
    Service limit (min).. 4.465 mm
  Guide bore diameter
    Standard... 4.500 to 4.412 mm
    Service limit (max)... 4.562 mm
  Stem-to-guide clearance
    Standard... 0.010 to 0.037 mm
    Service limit (max)... 0.075 mm
  Seat width
    Standard... 1.0 mm
    Service limit... 1.5 mm
  Valve guide height above cylinder head........................ 15.30 to 15.50 mm
Exhaust valve
  Stem diameter
    Standard... 4.465 to 4.480 mm
    Service limit (min).. 4.455 mm
  Guide bore diameter
    Standard... 4.500 to 4.512 mm
    Service limit (max)... 4.612 mm
  Stem-to-guide clearance
    Standard... 0.020 to 0.047 mm
    Service limit (max)... 0.085 mm
  Seat width
    Standard... 1.0 mm
    Service limit... 1.5 mm
  Valve guide height above cylinder head........................ 15.30 to 15.50 mm
Valve springs free length (inlet and exhaust)
  Standard... 37.86 mm
  Service limit (min).. 36.1 mm

Clutch
No. of friction plates... 7
No. of plain plates... 6
Friction plate thickness
  Standard... 2.92 to 3.08 mm
  Service limit (min).. 2.60 mm
Plain plate warpage (max)....................................... 0.3 mm
Spring free height
  Standard... 43.2 mm
  Service limit (min).. 42.0 mm
Clutch housing sleeve outside diameter
  CB models
    Standard... 29.993 to 30.007 mm
    Service limit (min).. 29.96 mm
  CBF models
    Standard... 29.987 to 30.000 mm
    Service limit (min).. 29.977 mm
Clutch housing sleeve inside diameter
  CB models
    Standard... 21.980 to 21.993 mm
    Service limit (max)... 22.03 mm
  CBF models
    Standard... 21.991 to 22.016 mm
    Service limit (max)... 22.026 mm
Transmission input shaft outside diameter at sleeve point
  Standard... 21.967 to 21.980 mm
  Service limit (max).. 21.950 mm

Lubrication system
Oil pressure .... see Chapter 1
Oil pump
   Inner rotor tip-to-outer rotor clearance
     Standard .... 0.15 mm
     Service limit (max) .... 0.20 mm
   Outer rotor-to-body clearance
     Standard .... 0.15 to 0.21 mm
     Service limit (max) .... 0.35 mm
   Rotor end-float
     Standard .... 0.02 to 0.09 mm
     Service limit (max) .... 0.12 mm

Cylinder bores
Bore
   Standard .... 73.000 to 73.015 mm
   Service limit (max) .... 73.10 mm
Warpage (max) .... 0.10 mm
Ovality (out-of-round) (max) .... 0.10 mm
Taper (max) .... 0.10 mm
Cylinder compression .... see Chapter 1

Connecting rods
Small-end inside diameter
   Standard .... 17.016 to 17.034 mm
   Service limit (max)
     CB models .... 17.07 mm
     CBF models .... 17.04 mm
Small-end-to-piston pin clearance
   Standard .... 0.016 to 0.040 mm
   Service limit (max) .... 0.06 mm
Big-end side clearance
   Standard .... 0.06 to 0.10 mm
   Service limit (max) .... 0.30 mm
Big-end oil clearance
   Standard .... 0.030 to 0.052 mm
   Service limit (max) .... 0.06 mm

Pistons
Piston diameter (measured at 90° to piston pin axis)
   Standard .... 72.970 to 72.990 mm
   Service limit (min) .... 72.90 mm
Piston-to-bore clearance
   Standard .... 0.010 to 0.045 mm
   Service limit (max) .... 0.10 mm
Piston pin diameter
   Standard .... 16.994 to 17.000 mm
   Service limit (min) .... 16.98 mm
Piston pin bore diameter in piston
   Standard .... 17.002 to 17.008 mm
   Service limit (max) .... 17.02 mm
Piston pin-to-piston pin bore clearance
   Standard .... 0.002 to 0.014 mm
   Service limit (max) .... 0.04 mm

Piston rings
Ring end gap (installed)
   Top ring
     Standard .... 0.15 to 0.30 mm
     Service limit (max) .... 0.50 mm
   2nd ring
     Standard .... 0.30 to 0.45 mm
     Service limit (max) .... 0.60 mm
   Oil ring side-rail
     Standard .... 0.20 to 0.70 mm
     Service limit (max) .... 1.00 mm
Ring-to-groove clearance
   Top ring
     Standard .... 0.025 to 0.060 mm
     Service limit (max) .... 0.08 mm
   2nd ring
     Standard .... 0.015 to 0.050 mm
     Service limit (max) .... 0.08 mm

Crankshaft and bearings
Main bearing oil clearance
    Standard... 0.026 to 0.044 mm
    Service limit (max).................................... 0.05 mm
Runout (max)... 0.05 mm

Transmission
Gear ratios (No. of teeth)
    Primary reduction..................................... 1.947 to 1 (74/38T)
    1st gear.. 3.461 to 1 (45/13T)
    2nd gear.. 2.235 to 1 (38/17T)
    3rd gear.. 1.750 to 1 (35/20T)
    4th gear.. 1.478 to 1 (34/23T)
    5th gear.. 1.280 to 1 (32/25T)
    6th gear.. 1.130 to 1 (26/23T)
Final reduction
    CB models.. 2.666 to 1 (40/15T)
    CBF models... 2.733 to 1 (41/15T)
Input shaft 5th and 6th gears inside diameter
    Standard... 28.000 to 28.021 mm
    Service limit (max).................................... 28.04 mm
Input shaft 5th and 6th gears, bush outside diameter
    Standard... 27.959 to 27.980 mm
    Service limit (min).................................... 27.94 mm
Input shaft 5th and 6th gears gear-to-bush clearance
    Standard... 0.020 to 0.062 mm
    Service limit (max).................................... 0.10 mm
Input shaft 5th gear bush inside diameter
    Standard... 24.985 to 25.006 mm
    Service limit (max).................................... 25.016 mm
Input shaft outside diameter at 5th gear bush point
    Standard... 24.967 to 24.980 mm
    Service limit (min).................................... 24.960 mm
Input shaft-to-bush clearance at 5th gear bush point
    Standard... 0.005 to 0.039 mm
    Service limit (max).................................... 0.06 mm
Output shaft 1st gear inside diameter
    Standard... 24.000 to 24.021 mm
    Service limit (max).................................... 24.04 mm
Output shaft 3rd and 4th gears, gear-to-bush clearance
    Standard... 0.025 to 0.075 mm
    Service limit (max).................................... 0.11 mm
Output shaft 2nd, 3rd and 4th gears inside diameter
    Standard... 31.000 to 31.025 mm
    Service limit (max).................................... 31.04 mm
Output shaft 2nd gear bush outside diameter
    Standard... 30.955 to 30.980 mm
    Service limit (min).................................... 30.94 mm
Output shaft 3rd and 4th gears bush outside diameter
    Standard... 30.950 to 30.975 mm
    Service limit (min).................................... 30.93 mm
Output shaft 2nd gear, gear-to-bush clearance
    Standard... 0.020 to 0.070 mm
    Service limit (max).................................... 0.10 mm
Output shaft 2nd gear bush inside diameter
    Standard... 27.985 to 28.006 mm
    Service limit (max).................................... 28.021 mm
Output shaft outside diameter at 2nd gear bush point
    Standard... 27.967 to 27.980 mm
    Service limit (min).................................... 27.960 mm
Output shaft-to-bush clearance at 2nd gear bush point
    Standard... 0.005 to 0.039 mm
    Service limit (max).................................... 0.06 mm

Selector drum and forks
Selector fork end thickness
    Standard... 5.93 to 6.00 mm
    Service limit (min).................................... 5.90 mm
Selector fork bore inside diameter
    Standard... 12.000 to 12.021 mm
    Service limit (max).................................... 12.03 mm
Selector fork shaft outside diameter
    Standard... 11.969 to 11.980 mm
    Service limit (min).................................... 11.96 mm

Engine, clutch and transmission 2•5

Torque settings

Engine mountings	
CB models	
Mounting bolt nuts	45 Nm
Sub frame bolts and nuts	45 Nm
CBF models	
Mounting bolt nuts	54 Nm
Valve cover bolts	10 Nm
Camchain tensioner bolts	10 Nm
Camshaft holder bolts	12 Nm
Timing inspection plug	10 Nm
Centre plug (in alternator cover)	7 Nm
9 mm cylinder head bolts	48 Nm
6 mm cylinder head bolts	12 Nm
Alternator rotor bolt	95 Nm
Primary drive gear bolt	95 Nm
Balancer gear nut	85 Nm
Clutch nut	
CB500	85 Nm
CBF500	83 Nm
Clutch release plate bolts	12 Nm
Oil pump sprocket bolt	15 Nm
Oil pump assembly bolt	8 Nm
Gearchange lever pinch bolt	23 Nm
Neutral switch	12 Nm
Oil pressure switch	12 Nm
Selector cam bolt	23 Nm
Gearchange mechanism return spring locating pin	23 Nm
Crankcase 10 mm bolts	40 Nm
Crankcase 8 mm bolts	25 Nm
Crankcase 6 mm bolts	12 Nm
Connecting rod big-end cap nuts	34 Nm

1 General information

The engine/transmission unit is a liquid-cooled parallel twin cylinder design, fitted transversely across the frame. The eight valves are operated by double overhead camshafts which are chain driven off the crankshaft. The engine/transmission assembly is constructed from aluminium alloy. The crankcase is divided horizontally and the cylinder block is cast integrally with the top half of the crankcase. A gear driven balancer shaft runs in the crankcases, forward of the crankshaft.

The crankcase incorporates a wet sump, pressure-fed lubrication system for the engine and gearbox which uses a chain-driven oil pump, an oil filter and pressure relief valve assembly, and an oil pressure switch.

The alternator is on the left-hand end of the crankshaft and the primary drive gear, which transmits power to the clutch, is on the right-hand end. The clutch is cable operated and is of the wet, multi-plate type.

The transmission is a six-speed, constant-mesh unit. Final drive to the rear wheel is by chain and sprockets.

2 Operations possible with the engine in the frame

The components and assemblies listed below can be removed without having to take the engine/transmission out of the frame. If however, a number of areas require attention at the same time, removal of the engine is recommended.

Valve cover
Camshafts
Carburettors
Cylinder head
Clutch
Starter clutch and idle gear
Timing rotor, primary drive gear and balancer gear
Gearchange mechanism
Alternator
Oil filter
Oil sump, oil pump, oil strainer and oil pressure relief valve
Starter motor
Water pump

3 Operations requiring engine removal

It is necessary to remove the engine/transmission from the frame to gain access to the following components:

Pistons
Cylinders
Connecting rods and bearings
Balancer shaft
Camchain
Crankshaft and bearings
Transmission shafts and gears
Selector drum and forks

4 Major engine repair – general note

1 It is not always easy to determine when, or if, an engine should be completely overhauled, as a number of factors must be considered.

2 High mileage is not necessarily an indication that an overhaul is needed, while low mileage, on the other hand, does not preclude the need for an overhaul. Frequency of servicing is probably the single most important consideration. An engine that has regular and frequent oil and filter changes, as well as other required maintenance, will most likely give many miles of reliable service. Conversely, a neglected engine, or one which has not been run in properly, may require an overhaul very early in its life.

3 Exhaust smoke and excessive oil consumption are both indications that piston rings and/or valve guides are in need of attention, although make sure that the fault is not due to oil leakage or over filling.

4 If the engine is making obvious knocking or rumbling noises, the connecting rod and/or main bearings are probably the cause.

5 Loss of power, rough running, excessive valve train noise and high fuel consumption also point to the need for urgent attention, especially if they are all present at the same time. If a complete tune-up does not remedy the situation, major mechanical work is the only solution.

6 An engine overhaul involves restoring worn internal parts to the specifications of a new engine. The piston rings and main and

2•6 Engine, clutch and transmission

connecting rod bearings are usually renewed and the cylinder walls honed during a major overhaul. Valve seats are re-ground, since they are usually in less than perfect condition at this point, and valve springs and camchain renewed. The end result should be a like-new engine that will give as many trouble-free miles as the original.

7 Before beginning the engine overhaul, read through the related procedures to familiarise yourself with the scope and requirements of the job. Overhauling an engine is not all that difficult, but it is time consuming. Plan on the motorcycle being tied up for a minimum of two weeks. Check on the availability of parts and gaskets and make sure that any necessary special tools, equipment and materials are obtained in advance.

8 Most work can be done with typical workshop hand tools, although precision measuring tools are required for inspecting parts to determine if they must be renewed. Often a dealer will handle the inspection of parts and offer advice concerning reconditioning and renewal. As a general rule, time is the primary cost of an overhaul so it does not pay to install worn or substandard parts.

9 As a final note, to ensure maximum life and minimum trouble from a rebuilt engine, everything must be assembled with care in a clean and organised environment.

5 Engine – removal and installation

Caution: The engine is very heavy. Engine removal and installation should be carried out with the aid of at least one assistant; personal injury or damage could occur if the engine falls or is dropped. An hydraulic or mechanical floor jack should be used to support and lower or raise the engine if possible.

Removal

1 If the engine is dirty, particularly around the cylinder head and sump, wash and degrease it thoroughly before starting any major dismantling work. This will make the job much easier and rule out the possibility of dirt falling inside.

2 Support the bike on its centre stand. Work can be made easier by raising the machine to a suitable height on an hydraulic ramp or a suitable platform. Make sure the bike is secure and will not topple over (see *Tools and Workshop Tips* in the Reference section).

3 Drain the engine oil (see Chapter 1), also drain the cooling system (see Chapter 1).

4 Remove the seat(s) (see Chapter 8). Disconnect the battery leads, negative (-) first (see Chapter 9).

5 On SW, SX, SY and S2 models, remove the fairing; on R, T, V, W, X, Y and 2 models, remove the radiator side panels (see Chapter 8). Remove the fuel tank (see Chapter 4).

6 Remove the radiator (see Chapter 3).

7 Release the clips and detach the hoses from the thermostat housing to the engine **(see illustration)**.

8 Detach the breather hose from the valve cover and remove the sub-air cleaner **(see illustrations)**.

9 Remove the exhaust system (see Chapter 4). On CBF models remove the PAIR system control valve with its air feed pipes (see Chapter 4). Release the air supply hose from its clips on the frame and secure it out of the way.

10 Remove the carburettors (see Chapter 4). Plug the engine inlet manifolds with clean rag. Remove the air filter housing (see Chapter 4).

11 On CB models disconnect the HT leads from the spark plugs, noting which fits where, and secure the leads clear of the engine. On CBF models remove the ignition coils (see Chapter 5).

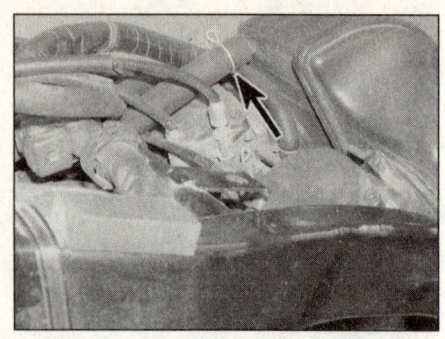

12 Remove the starter motor (see Chapter 9). On CBF models remove the horn (see Chapter 9).

13 Slacken the clutch cable adjuster at the bracket on the right-hand engine cover and slip the cable end out of the release lever **(see illustration)**. Secure the cable clear of the engine.

14 Unscrew the gearchange lever pinch bolt and remove the lever from the shaft, noting the alignment punchmark **(see illustration 19.2)**.

15 Unscrew the bolts securing the front sprocket cover and remove the front sprocket (see Chapter 6).

16 Unscrew the bolt securing the guide and remove the guide **(see illustration)**.

17 Disconnect the neutral switch wiring connector from the switch and secure it clear of the engine **(see illustration)**.

5.7 Release the clips (arrowed) and detach the hoses from the thermostat housing

5.8a Detach the hose from the valve cover . . .

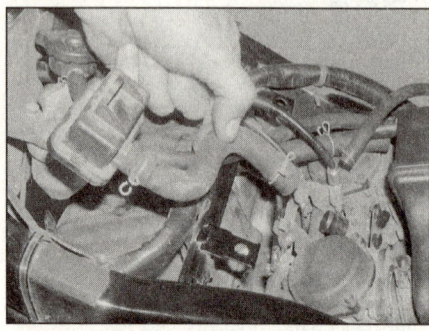

5.8b . . . and remove the sub-air cleaner

5.13 Detach the clutch cable

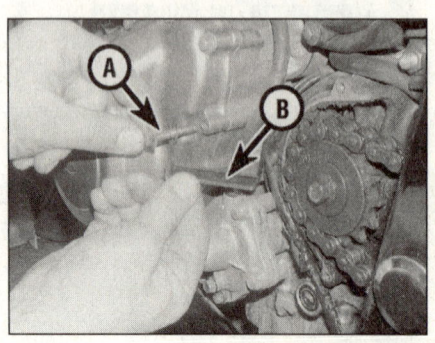

5.16 Remove the bolt (A) and guide (B)

5.17 Detach the neutral switch wiring connector (arrowed)

Engine, clutch and transmission 2•7

5.18a Pull back the boot and detach the oil pressure switch wiring connector

5.18b Disconnect the sidestand switch wiring connector (arrowed)

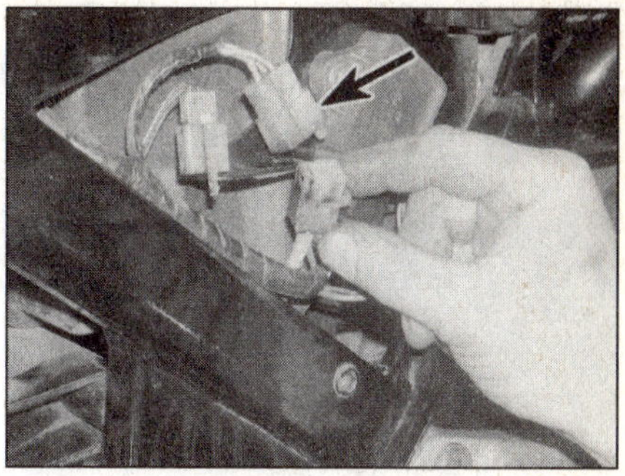

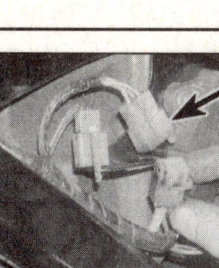

5.18c Disconnect the alternator wiring connector (arrowed)

18 On CB models detach the oil pressure switch wire; it is retained to the switch body by a single screw **(see illustration)**. Trace the sidestand switch wiring from the switch and disconnect it at the green 3-pin connector on the frame left-hand side **(see illustration)**. Also trace the alternator wiring from the alternator cover on the left-hand side of the engine and disconnect it at the white 3-pin connector on the frame left-hand side **(see illustration)**. Next trace the pulse generator coil wiring back from the right-hand engine cover and disconnect it at the white 4-pin connector inside the rubber boot on the right-hand side of the frame **(see illustration 4.2a in Chapter 5)**. Release the wiring from any clips or ties, noting its routing, and coil it so that it does not impede engine removal.

19 On CBF models release the cable-tie on the top of the crankcase to free the wiring. Pull the rubber boot off the wiring connectors on the left-hand side and disconnect the alternator (white 3-pin), speed sensor (black 3-pin), ignition pulse generator (blue 2-pin) and neutral switch/oil pressure switch (black 2-pin) wiring connectors **(see illustration)**.

20 At this point, position an hydraulic or mechanical jack under the engine with a block of wood between the jack head and sump. Make sure the jack is centrally positioned so the engine will not topple when the last mounting bolt is removed. Take the weight of the engine on the jack so it is just supported rather than lifted, otherwise you won't be able to get the bolts out. You may have to adjust the jack slightly to take any weight from the bolts as they are removed.

21 Unscrew the nut on the left-hand end of the engine front mounting bolt on CB models, and on the right-hand end on CBF models **(see illustrations)**. Withdraw the bolt and remove the spacers (fitted to R, T, V, W, X, Y and 2 models and all CBF models) between the engine and the frame on the left and right-hand sides **(see illustration)**.

22 On SW, SX, SY and S2 models, unscrew the nuts on the four bolts securing the front frame cross-member to the frame and remove the cross-member **(see illustration)**.

23 On CB models unscrew the nuts and bolts securing the left-hand sub-frame and remove the sub-frame. The lower left-hand bolt has a captive nut **(see illustration)**. Unscrew the nut

5.19 Pull back the boot (arrowed) and disconnect the relevant wiring connectors

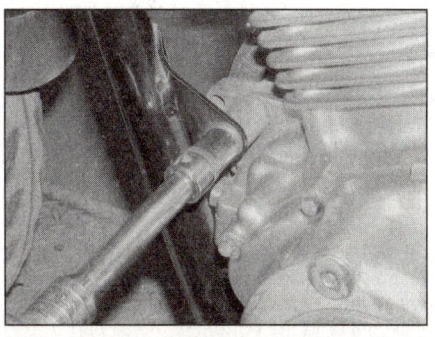

5.21a Unscrew the front mounting bolt nut – CB models

5.21b Front mounting bolt nut (arrowed) – CBF models

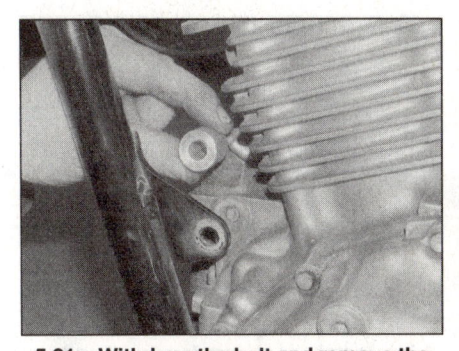

5.21c Withdraw the bolt and remove the spacers

5.22 On SW, SX, SY and S2 models remove frame cross-member

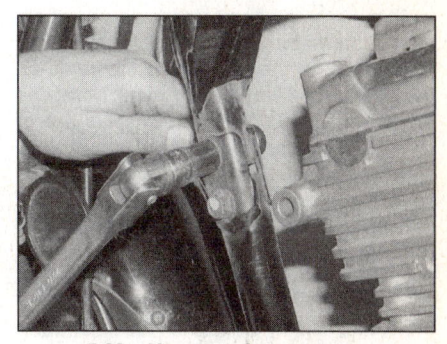

5.23a Unscrew the upper . . .

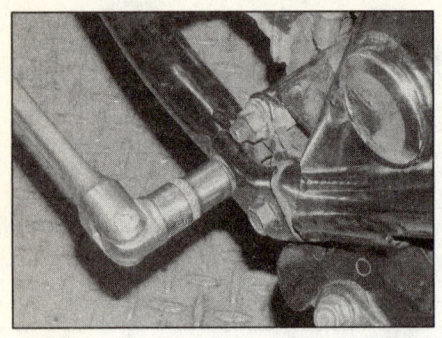

5.23b ... and lower left-hand sub-frame bolts

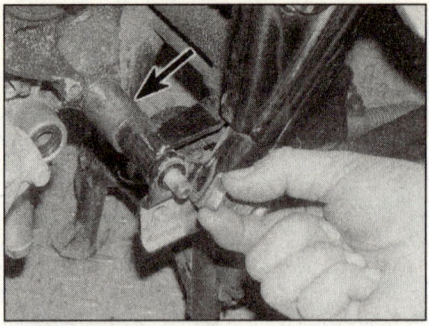

5.23c Remove the lower rear mounting nut, bolt and spacer (arrowed)

5.24 Upper and lower rear mounting bolt nuts (arrowed) – CBF models

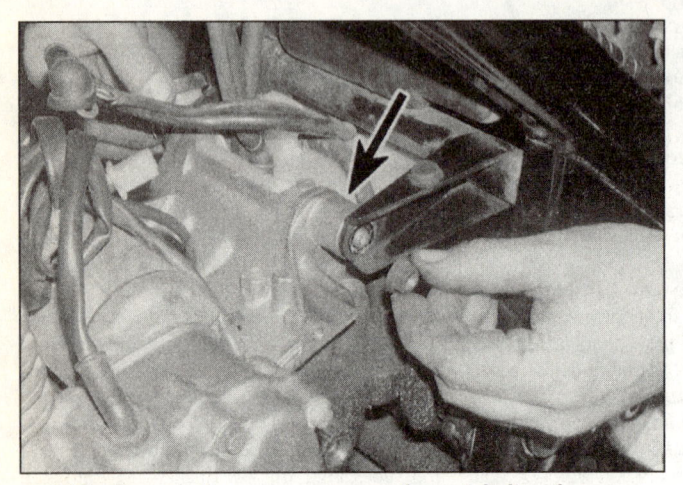

5.26a Remove the upper rear mounting nut, bolt and spacer (arrowed)

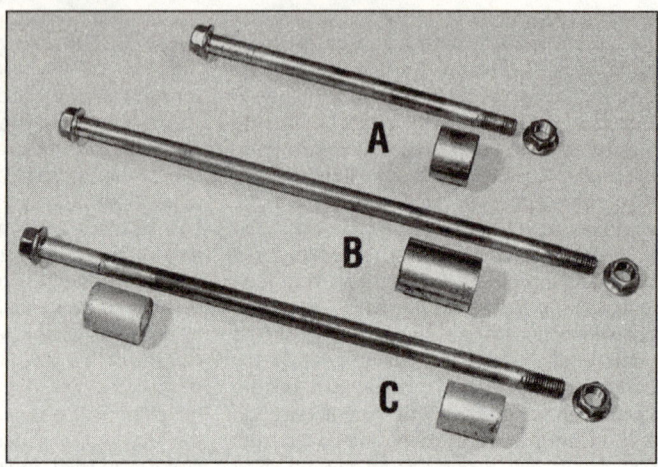

5.26b Bolts and spacer lengths differ. Upper rear (A), lower rear (B) and front (C)

on the left-hand end of the lower rear mounting bolt, then withdraw the bolt and remove the spacer fitted between the engine and the frame on the left-hand side (see illustration).
24 On CBF models unscrew the nut on the right-hand end of the lower rear mounting bolt, then withdraw the bolt and remove the spacer fitted between the engine and the frame on the right-hand side (see illustration).
25 Make sure the engine is properly supported on the jack, and have an assistant support it as well.
26 On CB models unscrew the nut on the left-hand end of the upper rear mounting bolt, then withdraw the bolt and remove the spacer fitted between the engine and the frame on the left-hand side (see illustration). Note: *The engine mounting bolts and spacers are of different lengths and must not be interchanged. Keep the spacers on their respective bolts ready for reassembly* (see illustration).
27 On CBF models unscrew the nut on the right-hand end of the upper rear mounting bolt, then withdraw the bolt and remove the spacer fitted between the engine and the frame on the right-hand side (see illustration 5.24). Note: *The engine mounting bolts and spacers are of different lengths and must not be interchanged. Keep the spacers on their respective bolts ready for reassembly.*
28 The engine can now be removed from the frame. Check that all relevant wiring, cables and hoses are disconnected and secured well clear, then carefully lower the engine and manoeuvre it out of the side of the frame (see *Caution* above).
29 With the aid of an assistant lift the engine unit off the jack and lower it carefully onto the work surface, taking care not to break the long fins cast onto the bottom of the oil pan. These fins are there specifically for the engine to stand on and keep it in an upright position.

Installation

30 With the aid of an assistant, place the engine unit on top of the jack and block of wood and carefully raise the engine unit into position in the frame (see illustration). Make sure no wires, cables or hoses become trapped between the engine and the frame.
31 Position the spacer for the upper rear mounting bolt between the frame and the engine on the left-hand side on CB models and the right-hand side on CBF models, and slide the bolt through from the right on CB models and from the left on CBF models (see illustration). Fit the nut, tightening it hand-tight

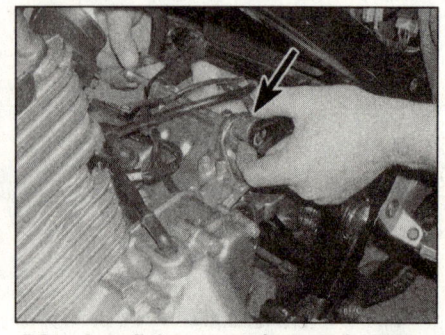

5.30 Raise engine unit into position with the help of a jack

5.31a Install the spacer (arrowed) and the upper rear mounting bolt ...

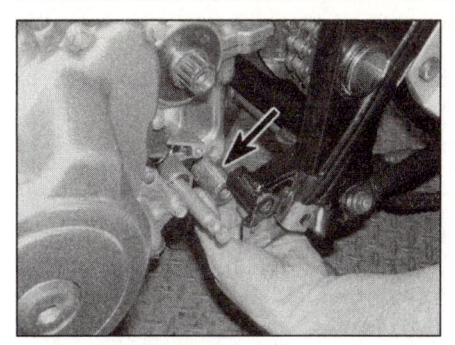

Engine, clutch and transmission 2•9

5.31b . . . then the spacer (arrowed) and the lower rear mounting bolt – CB model shown

5.32a Loosely assemble the left-hand sub-frame . . .

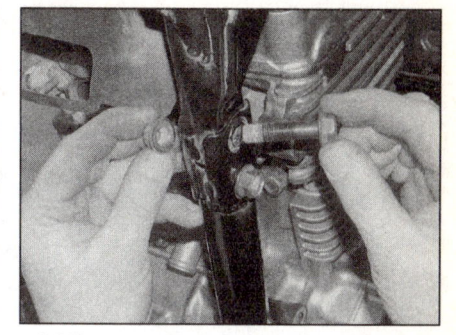

5.32b . . . and sub-frame nuts and bolts

only at this stage. Position the spacer for the lower rear mounting bolt between the frame and the engine on the left-hand side on CB models and the right-hand side on CBF models, and slide the bolt through from the right on CB models and from the left on CBF models **(see illustration)**. Fit the nut, tightening it hand-tight only at this stage.

32 On CB models position the left-hand sub-frame and install the nuts and bolts and tighten them hand-tight only **(see illustrations)**.

33 On SW, SX, SY and S2 models, position the front frame cross-member and install the nuts and bolts securing it to the frame, tightening hand-tight only.

34 Position the left and right-hand spacers (R, T, V, W, X, Y and 2 models and all CBF models only) for the front mounting bolt between the frame and the engine and slide the bolt through from the right on CB models and from the left on CBF models **(see illustration)**. Fit the nut, tightening it hand-tight only at this stage.

35 Ensure all parts are correctly aligned, then tighten all the nuts and bolts to the torque settings specified at the beginning of the Chapter starting with the lower rear mounting bolt. Next tighten the upper rear mounting bolt, on CB models the four sub-frame bolts, and on SW, SX, SY and S2 models the bolts securing the frame cross-member to the frame. Finally tighten the front mounting bolt nut **(see illustrations)**.

36 The remainder of the installation procedure is the reverse of removal, noting the following points.

a) Make sure all wires, cables and hoses are correctly routed and connected, and secured by the relevant clips or ties.

b) Adjust the throttle and clutch cable freeplay (see Chapter 1).
c) Adjust the drive chain (see Chapter 1).
d) Refill the engine with oil and coolant (see Chapter 1).
e) Before riding the bike, start the engine and check that there are no signs of coolant/oil leakage.

6 Engine disassembly and reassembly – general information

Disassembly

1 Before disassembling the engine, the external surfaces of the unit should be thoroughly cleaned and degreased. This will prevent contamination of the engine internals,

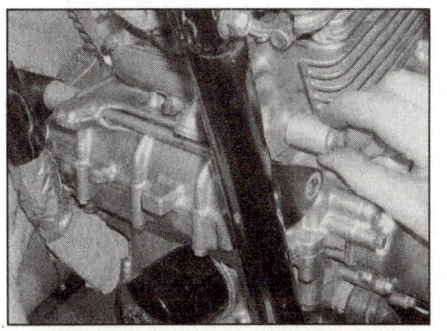

5.34 Position the front bolt spacers on each side (CB500 and CBF500 only)

5.35a Tighten the lower rear mounting bolt . . .

5.35b . . . followed by the upper rear mounting bolt . . .

5.35c . . . the lower sub-frame bolts . . .

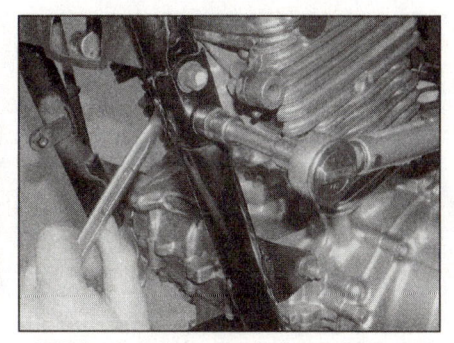

5.35d . . . the upper sub-frame bolts . . .

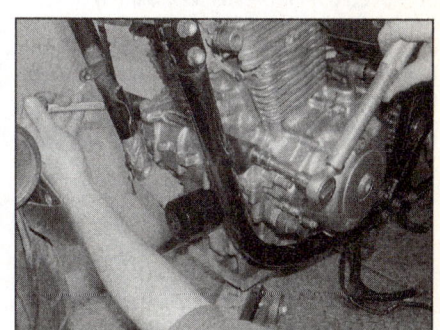

5.35e . . . and the front mounting bolt – CB500 model shown

2•10 Engine, clutch and transmission

7.5 Unscrew the bolts and lift off the valve cover

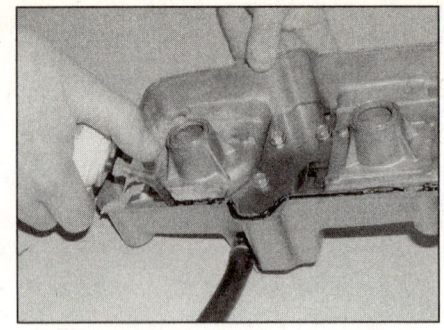

7.8a Apply sealant to the valve cover...

7.8b ...and the cut-outs in the cylinder head

and will also make working a lot easier and cleaner. A high flash-point solvent, such as paraffin (kerosene) can be used, or better still, a proprietary engine degreaser such as Gunk. Use old paintbrushes and toothbrushes to work the solvent into the various recesses of the engine casings. Take care to exclude solvent or water from the electrical components and inlet and exhaust ports.

⚠ **Warning: The use of petrol (gasoline) as a cleaning agent should be avoided because of the risk of fire.**

2 When clean and dry, arrange the unit on the workbench, leaving suitable clear area for working. Gather a selection of small containers and plastic bags so that parts can be grouped together in an easily identifiable manner. Some paper and a pen should be on hand so that notes can be made and labels attached where necessary. A supply of clean rag is also required.

3 Before commencing work, read through the appropriate section so that some idea of the necessary procedure can be gained. When removing components it should be noted that great force is seldom required, unless specified. In many cases, a component's reluctance to be removed is indicative of an incorrect approach or removal method – if in any doubt, re-check with the text.

4 When disassembling the engine, keep 'mated' parts together (including gears, pistons, connecting rods, valves, etc. that have been in contact with each other during engine operation). These 'mated' parts must be reused or renewed as an assembly.

5 A complete engine/transmission disassembly should be done in the following general order with reference to the appropriate Sections.

   Remove the valve cover
   Remove the camchain tensioner
   Remove the camshafts
   Remove the cylinder head
   Remove the starter motor (see Chapter 9)
   Remove the alternator (see Chapter 9)
   Remove the starter clutch
   Remove the timing rotor, primary drive gear and balancer gear
   Remove the clutch
   Remove the oil pump
   Remove the water pump (see Chapter 3)
   Remove the gearchange mechanism external components
   Remove the oil sump
   Separate the crankcase halves
   Remove the connecting rods and pistons
   Remove the crankshaft
   Remove the balancer shaft
   Remove the camchain and guides
   Remove the transmission shafts
   Remove the selector drum and forks

Reassembly

6 Reassembly is accomplished by reversing the general disassembly sequence.

7 Valve cover – removal and installation

Note: *This procedure can be carried out with the engine in the frame. If the engine has been removed, ignore the steps that do not apply.*

Removal

1 On SW, SX, SY and S2 models, remove the fairing; on R, T, V, W, X, Y and 2 models remove the radiator side panels (see Chapter 8). Remove the fuel tank (see Chapter 4).
2 Remove the thermostat assembly (see Chapter 3) and detach the sub-air cleaner **(see illustration 5.8b)**.

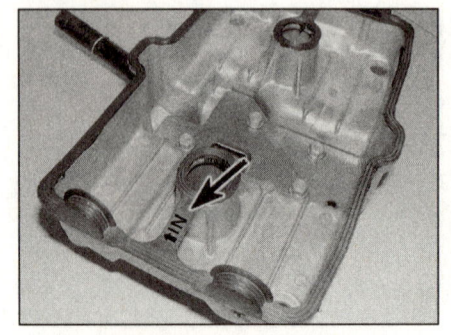

7.8c Install the cover gasket with the IN mark (arrowed) pointing to the rear

3 Disconnect the valve cover breather hose from the air filter housing **(see illustration 5.8a)**.
4 Disconnect the HT leads from the spark plugs and secure them clear of the engine, noting which fits where.
5 Unscrew the four bolts securing the valve cover. Remove the bolts and their sealing washers fitted in the cover. Lift the cover off the cylinder head **(see illustration)**. If it is stuck, do not try to lever it off with a screwdriver. Tap it gently around the sides with a rubber hammer or block of wood to dislodge it. Remove the gasket.

Installation

6 Examine the valve cover gasket and the rubber washers for signs of damage or deterioration and renew them if necessary.
7 Clean the mating surfaces of the cylinder head and the valve cover with lacquer thinner, acetone or brake system cleaner.
8 Apply a smear of a suitable sealant to the valve cover and into the cutouts in the cylinder head **(see illustrations)**. Install the gasket onto the valve cover. The gasket is marked 'IN' which should be on the inlet side (rear) of the cover. Making sure the gasket is fitted correctly **(see illustration)**.
9 The valve cover must be installed with the breather pointing to the rear. Position the valve cover on the cylinder head, making sure the gasket stays in place. If removed, fit the sealing washers into the cover, using new ones if required, and making sure they are installed with the 'UP' mark facing up **(see illustration)**. Install the cover bolts and tighten them to the

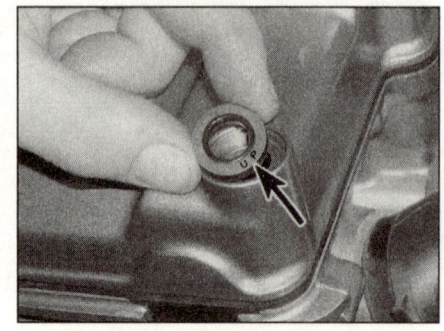

7.9a Install cover bolt sealing washers with UP showing...

Engine, clutch and transmission 2•11

torque setting specified at the beginning of the Chapter **(see illustration)**.
Install the remaining components in the reverse order of removal.

8 Camchain tensioner – removal and installation

Note: *This procedure can be carried out with the engine in the frame. If the engine has been removed, ignore the steps that do not apply.*

Removal

1 Unscrew the camchain tensioner centre bolt and sealing washer **(see illustration)**. If the Honda locking key or a home-made equivalent is available, insert it into the centre bolt hole so that it engages the slotted plunger. Turn the key fully clockwise then push it into the end of the tensioner body so that the key shoulders lock it in this position **(see illustrations)**. Unscrew the camchain tensioner fixing bolts and withdraw the tensioner from the engine.
2 If the locking key is not available, use a small flat-bladed screwdriver to turn the plunger. Hold the screwdriver in place while unscrewing the camchain tensioner fixing bolts, then withdraw the tensioner from the engine **(see illustration)**. The plunger will spring back out once the tension on the screwdriver is released, but can be easily reset on installation **(see illustration)**.
3 Discard the gasket as a new one must be used on reassembly. Do not dismantle the tensioner

Installation

4 Ensure the tensioner and engine faces are clean and dry and fit a new gasket to the tensioner. If the locking key is available, insert it in the tensioner body and turn it fully clockwise to retract the plunger into the body, then engage the key shoulders to lock the plunger. Install the tensioner on the engine and tighten its bolts to the torque setting specified at the beginning of this Chapter.

7.9b . . . then install the bolts and tighten to specified torque

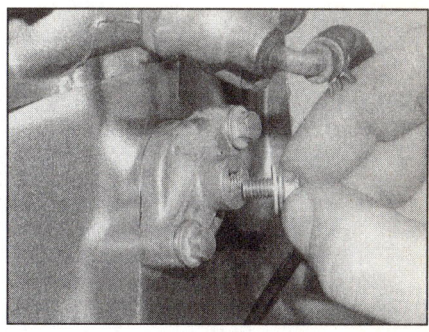

8.1a Remove camchain tensioner bolt and washer

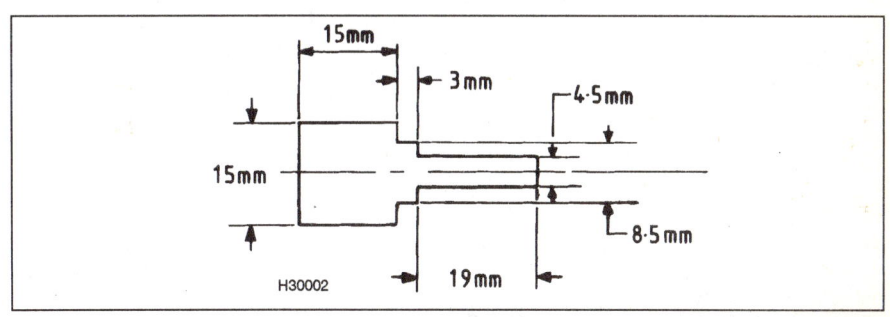

8.1b A copy of Honda's tensioner locking key can be made from a piece of 1 mm mild steel

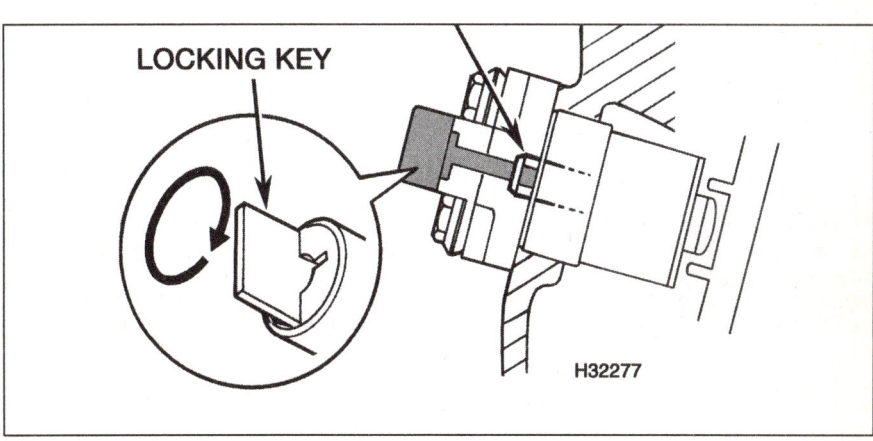

8.1c Locking key in position, tensioner plunger shown retracted

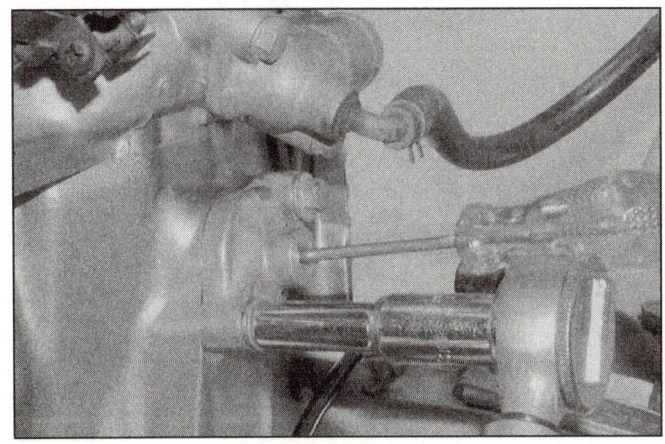

8.2a Tensioner plunger can be retracted using a small flat-bladed screwdriver

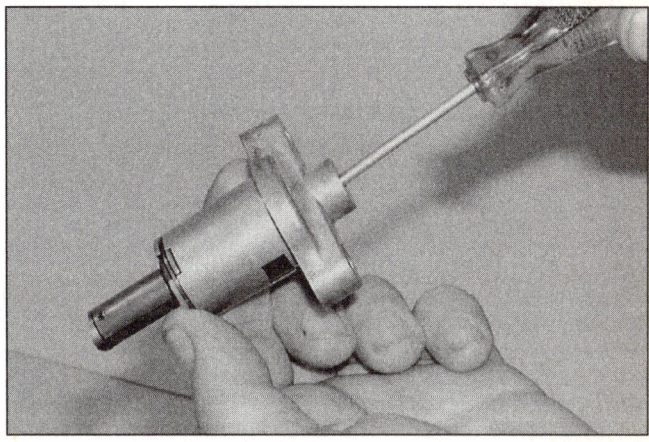

8.2b Tensioner plunger fully extended with tension released

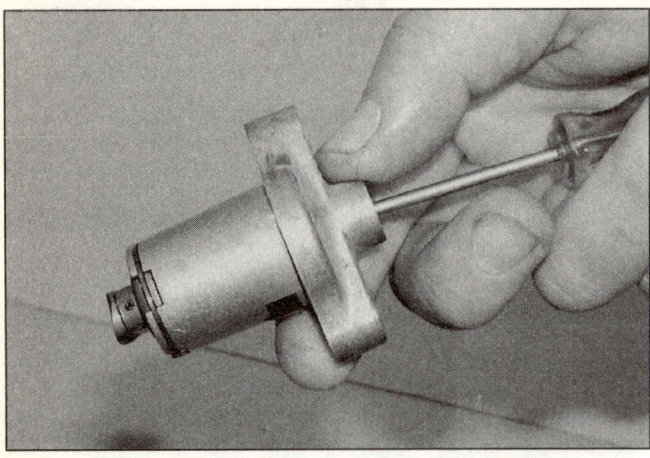

2•12 Engine, clutch and transmission

8.5a Retract tensioner plunger prior to installation

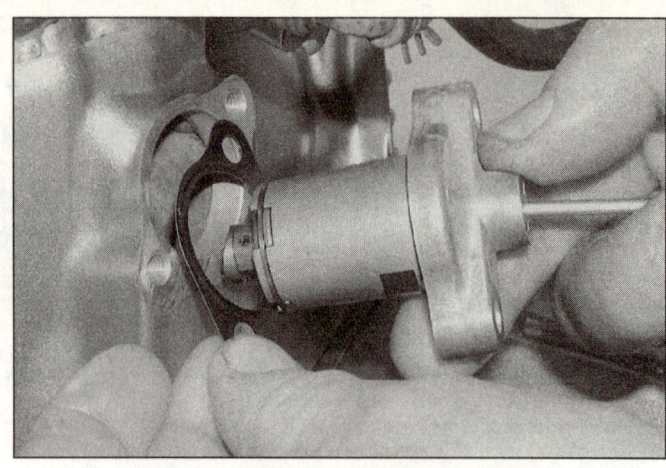

8.5b Install tensioner with a new gasket

Remove the key and install the tensioner centre bolt with a new sealing washer.

5 If the key is not available, turn the plunger fully clockwise with a flat-bladed screwdriver and hold it in this position while the tensioner is installed on the engine and the fixing bolts are tightened to the specified torque setting **(see illustrations)**. Release the tension on the screwdriver and remove it. Install the tensioner centre bolt with a new sealing washer.

9 Camshafts – removal, inspection and installation

Note: *This procedure can be carried out with the engine in the frame. If the engine has been removed, ignore the steps that do not apply. Place rags over the spark plug holes and the cam chain hole to prevent any component from dropping into the engine on removal.*

Removal

1 Remove the valve cover (see Section 7).

2 Unscrew the timing inspection plug and the centre plug from the left-hand engine cover. Discard the plug O-rings as new ones should be used. The engine can be turned using a socket on the alternator rotor bolt and turning it in an anti-clockwise direction **(see illustration)**. Alternatively, place the motorcycle on its centre stand so that the rear wheel is off the ground, select a high gear and rotate the rear wheel by hand in its normal direction of rotation.

3 Turn the engine until the line next to the 'O' mark on the alternator rotor aligns with the notches in the timing inspection hole, and the 'O' marks on the inlet and exhaust camshaft sprockets are aligned with the cylinder head surface with each mark on the outside of its respective sprocket **(see illustrations)**. If the

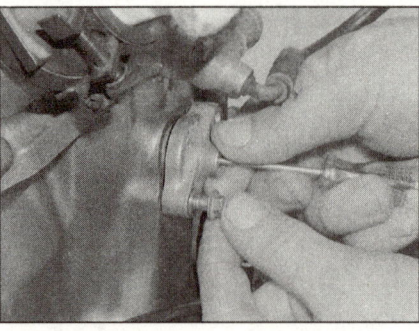

8.5c Hold tensioner retracted while the fixing bolts are tightened

9.2 Use a socket on the rotor bolt to turn the engine anti-clockwise

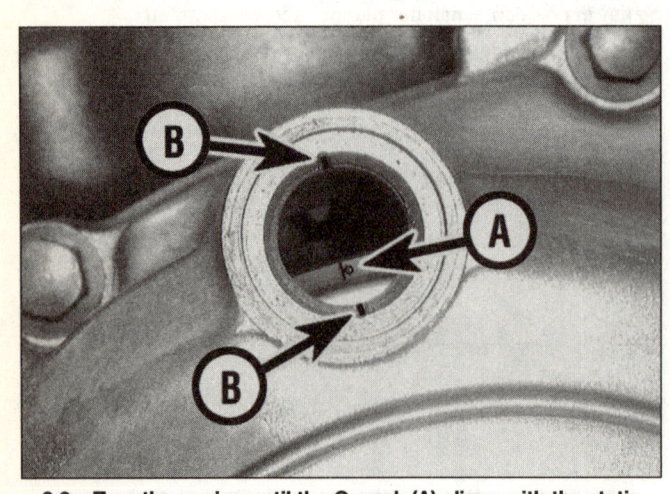

9.3a Turn the engine until the O mark (A) aligns with the static marks (B) . . .

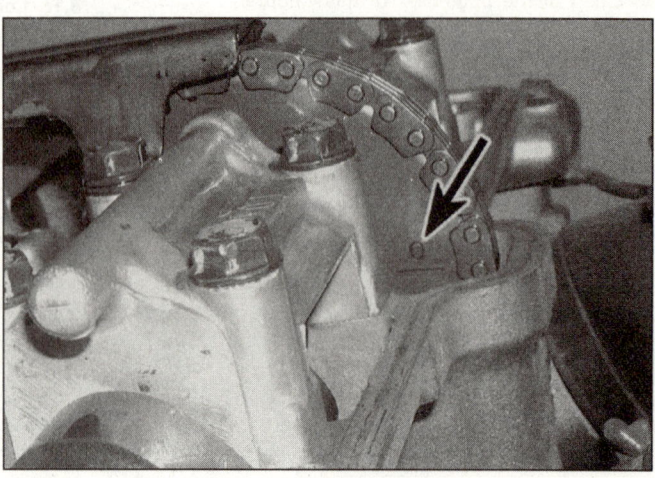

9.3b . . . and the O marks on the camshaft gears (arrowed) align with the cylinder head surface

Engine, clutch and transmission 2•13

9.5 Marks on camshaft holders correspond with marks on cylinder head

9.6a Unscrew the camshaft holder bolts evenly

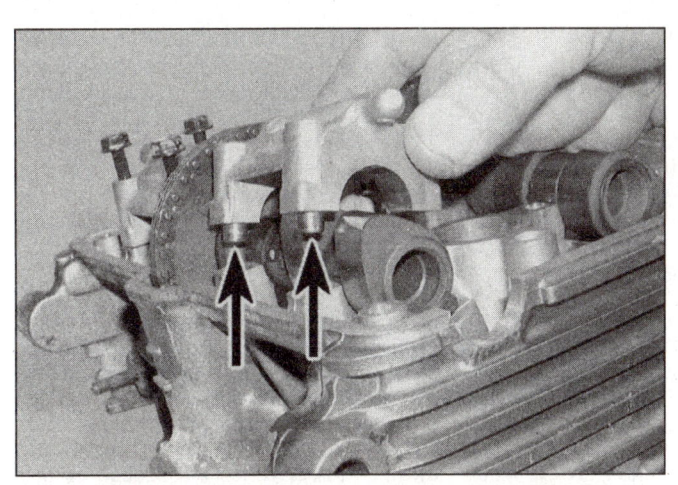

9.6b Retrieve the dowels (arrowed) for safekeeping

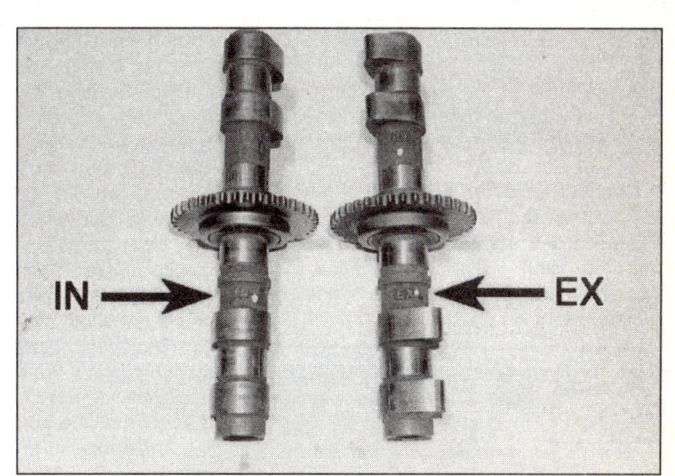

9.6c Note the camshaft identification marks

'O' marks are on the inside of their respective sprockets, turn the engine one full turn (360°) until the 'O' mark on the alternator rotor again aligns with the notches in the inspection hole. The camshaft marks will now be correctly aligned.

4 Remove the camchain tensioner (see Section 8).

5 Mark the front edge of the upper camchain guide to aid reassembly. The four camshaft holders are marked A, B, C and D, with corresponding marks on the inside of the cylinder head **(see illustration)**. These markings ensure that the holders can be matched up to their original locations on installation. If no markings are visible, make your own using a felt pen. Before disturbing the camshaft holders, make a sketch of the layout as a further aid for installation.

6 Unscrew the camshaft holder bolts for the camshaft being worked on, evenly and a little at a time in a criss-cross pattern, until they are all loose **(see illustration)**. While slackening the bolts make sure that each holder is lifting squarely away from the cylinder head and is not sticking on its locating dowels.

Caution: If the bolts are carelessly loosened and the holder does not come squarely away from the head, the holder is likely to break. If this happens the complete cylinder head assembly must be renewed as the holders are matched to the cylinder head and cannot be renewed separately. Also, a camshaft could be damaged if the holder bolts are not slackened evenly and the pressure from a depressed valve causes the shaft to bend.

Remove the bolts, noting their positions, and the upper camchain guide, then lift off the camshaft holders. Retrieve the dowels from either the holder or the cylinder head if they are loose **(see illustration)**. Disengage each camshaft sprocket from the camchain and lift each camshaft out of the head. The camshafts are marked for identification. The inlet camshaft is marked 'IN' and the exhaust camshaft is marked 'EX' **(see illustration)**. Secure the camchain with a length of wire to prevent it dropping into the crankcase.

Inspection

7 Inspect the bearing surfaces of the camshaft holders and the corresponding journals on the camshaft. Look for score marks, deep scratches and evidence of spalling (a pitted appearance) **(see illustration)**.

9.7 Check the journal surfaces of the camshaft for scratches or wear

2•14 Engine, clutch and transmission

9.8a Check the lobes of the camshaft for wear – here's an example of damage requiring repair or renewal

9.8b Measure the height of the camshaft lobes with a micrometer

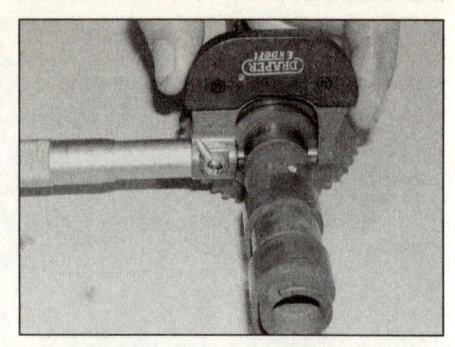

9.11 Measure the cam bearing journals with a micrometer

8 Check the camshaft lobes for heat discoloration (blue appearance) score marks, chipped areas, flat spots and spalling **(see illustration)**. Measure the height of each lobe with a micrometer **(see illustration)** and compare the results to the minimum lobe height listed in this Chapter's Specifications. If damage is noted or wear is excessive, the camshaft must be renewed. Also check the condition of the cam followers in the cylinder head (see Section 12).

9 Check the amount of camshaft runout by supporting each end of the camshaft on V-blocks, and measuring any runout at the journals using a dial gauge. If the runout exceeds the specified limit the camshaft must be renewed.

> **HAYNES HiNT** *Refer to Tools and Workshop Tips in the Reference section for details of how to read a micrometer and dial gauge.*

10 The camshaft bearing oil clearance should now be checked. There are two possible ways of doing this, either by direct measurement (see Steps 11 to 13) or by the use of a product known as Plastigauge (see Steps 14 to 19).
11 If the direct measurement method is to be used, make sure the camshaft holder dowels are in position then fit the holders, making sure they are in their correct location (see Step 5). Tighten the holder bolts evenly and a little at a time in a criss-cross pattern to the torque setting specified at the beginning of the Chapter. Using telescoping gauges and a micrometer (see Tools and Workshop Tips), measure the diameter of the journal holders. Now measure the diameter of the corresponding camshaft journals with a micrometer **(see illustration)**. To determine the journal oil clearance, subtract the journal diameter from the holder diameter and compare the result to the clearance specified. If any clearance is greater than specified, it is an indication of wear on the camshaft, the holder, or both.
12 First check to see if the camshaft journals are worn below the service limit. If they are, a new camshaft must be fitted. However, since it is likely that the holder is also worn, ensure that the specified journal diameter for a new camshaft will restore the oil clearance to within specification before buying a new camshaft.
13 If the camshaft journals are good, or if fitting a new camshaft will not restore the oil clearance to within specification, the holders and cylinder head will have to be renewed.
14 If the Plastigauge method is to be used, clean the camshafts, the bearing surfaces in the cylinder head and camshaft holders with a clean lint-free cloth, then lay the camshafts in place in the cylinder head.
15 Cut strips of Plastigauge and lay one piece on each bearing journal, parallel with the camshaft centreline **(see illustration)**. Make sure the camshaft holder dowels are in position then fit the holders, making sure they are in their correct location (see Step 5). Tighten the holder bolts evenly and a little at a time in a criss-cross pattern to the torque setting specified at the beginning of the Chapter. While doing this, don't let the camshafts rotate.
16 Now unscrew the bolts evenly and a little at a time in a criss-cross pattern, and carefully lift off the camshaft holders.
17 To determine the oil clearance, compare the crushed Plastigauge (at its widest point) on each journal to the scale printed on the Plastigauge container **(see illustration)**. Compare the results to this Chapter's Specifications. Carefully clean away all traces of the Plastigauge using a fingernail or other object which will not score the bearing surfaces.
18 If any clearance is greater than specified, it is an indication of wear on the camshaft, the holder, or both. First check to see if the camshaft journals are worn below the service limit. If they are, a new camshaft must be

9.15 Lay a strip of Plastigauge across each bearing journal, parallel with the camshaft centreline

9.17 Compare the width of the crushed Plastigauge to the scale printed on the Plastigauge container

Engine, clutch and transmission 2•15

9.25a Locate the camchain on the camshaft sprockets, keeping the front run of the chain tight

9.25b Check that all the marks are in alignment

fitted. However, since it is likely that the holders are also worn, use the journal diameter specification for a new camshaft to calculate whether fitting a new camshaft will restore the oil clearance to within specification.

19 If the camshaft journals are good, or if fitting a new camshaft will not restore the oil clearance to within specification, the holders and cylinder head must be renewed.

HAYNES HINT *Before renewing camshafts or holders because of damage, check with local machine shops specialising in motorcycle engine work. In the case of the camshafts, it may be possible for cam lobes to be welded, reground and hardened, at a cost far lower than that of a new camshaft. If the bearing surfaces in the holders are damaged, it may be possible for them to be bored out to accept bearing inserts. Due to the cost of new components it is recommended that all options be explored.*

20 Except in cases of oil starvation, the camchain wears very little. Honda specify no service limit for the camchain, but if the chain has stretched excessively, which makes it difficult to maintain proper tension, replace it with a new one (see Section 27).

21 Check the camchain upper guide for wear or damage. If it is worn or damaged, the condition of the camchain will be doubtful. Refer to Section 27 for camchain renewal.

22 Check the sprockets on each camshaft for wear, cracks and other damage. If the sprockets are worn or damaged, the condition of the camchain will be doubtful. In these circumstances, it is also important to check the condition of the camchain sprocket on the crankshaft (see Section 27). The camchain sprockets are integral with the camshafts and the crankshaft and cannot be renewed separately. If severe wear is apparent, the entire engine should be disassembled for inspection.

Installation

23 Make sure the bearing surfaces on the camshafts and in the holders are clean, then apply a 50/50 mixture of molybdenum disulphide grease and engine oil to each of them. Also apply it to the camshaft lobes and the followers.

24 The camshafts and holders must be installed in their correct locations according to their identification marks (see Steps 5 and 6).

25 Verify that the line next to the 'O' mark on the alternator rotor is still aligned with the notches in the timing inspection hole **(see illustration 9.3a)**. Ensure that the camchain is properly located on the crankshaft sprocket before engaging it with the camshaft sprockets as the camshafts are installed – keep the front run of the camchain tight **(see illustration)**. Make sure all the marks are in exact alignment (see Step 3) **(see illustration)**. Take extra care at this stage as it is easy to be one tooth out on the timing without it appearing as a drastic misalignment of the timing marks.

26 Make sure the camshaft holder dowels are in position then fit the holders, making sure they are in their correct location (see Step 5) **(see illustration)**. Position the upper camchain guide, making sure that it is the right way round (see Step 5) **(see illustration)**. Apply clean engine oil to the threads and seating surfaces of the camshaft holder bolts and install the bolts finger-tight. Tighten the bolts evenly and a little at a time in a criss-cross pattern to the torque setting specified at the beginning of the Chapter. Whilst tightening the bolts, make sure the holders are being pulled squarely down and are not binding on the dowels. When you have finished, check that each camshaft is not pinched by turning the engine a few degrees in each direction.

Caution: The holders are likely to break if they are not tightened down evenly and squarely.

27 Install the camchain tensioner (see Section 8).

28 Turn the engine until the line next to the 'T' mark on the alternator rotor aligns with the notches in the timing inspection hole and the

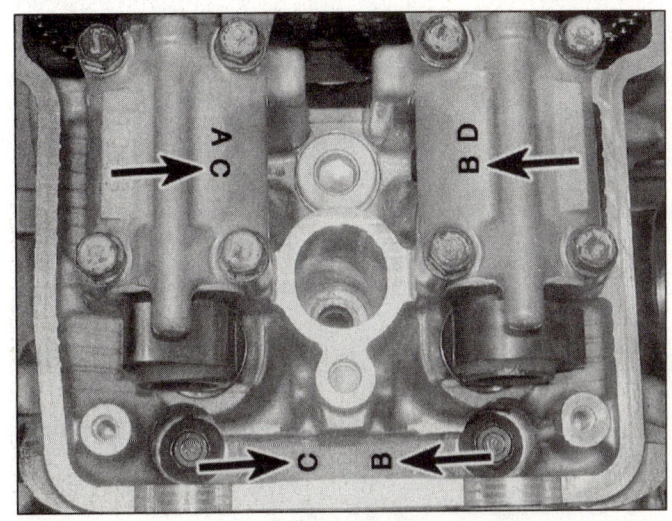

9.26a Check the camshaft holder location with marks on holders and cylinder head

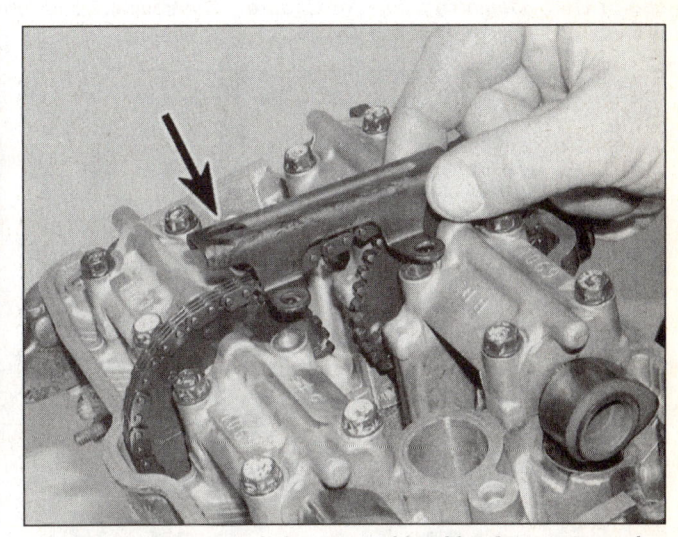

9.26b Install the camchain upper guide with reference to mark made on disassembly (arrowed)

2•16 Engine, clutch and transmission

9.30 Install the plugs using new O-rings

10.4a Remove the follower . . .

10.4b . . . and the valve shim

'IN' timing mark on the inlet camshaft and the 'EX' timing mark on the exhaust camshaft align with the cylinder head surface, with each mark on the outside of its respective sprocket.

Caution: If the marks are not aligned exactly as described, the valve timing will be incorrect and the valves may strike the pistons, causing extensive damage to the engine.

29 Check the valve clearances and adjust them if necessary (see Chapter 1). This is essential if new components have been installed or if the camshafts were removed to change the shims.

30 Use new O-rings on the timing inspection plug and centre plug and smear them and the plug threads with a 50/50 mixture of molybdenum disulphide grease and engine oil **(see illustration)**. Tighten the plugs to the torque setting specified at the beginning of the Chapter.

31 Install the valve cover (see Section 7).

10 Cylinder head – removal and installation

Caution: The engine must be completely cool before beginning this procedure or the cylinder head may become warped.

Note: *This procedure can be carried out with the engine in the frame. If the engine has been removed, ignore the steps that do not apply.*

Removal

1 Drain the cooling system (see Chapter 1).
2 Remove the exhaust system and the carburettors (see Chapter 4). On CBF models also remove the PAIR system control valve with its air feed pipes (see Chapter 4).
3 Remove the camshafts (see Section 9).
4 Obtain a container which is divided into eight compartments, and label each compartment with the location of a corresponding valve in the cylinder head. Remember, the left-hand cylinder is the No. 1 cylinder, the right-hand cylinder is the No. 2 cylinder. If a container is not available, use labelled plastic bags. Lift each cam follower out of the cylinder head using either a magnet or a pair of pliers and store it in its corresponding compartment in the container **(see illustration)**. Retrieve the valve shim from either the inside of the follower or pick it off the top of the valve, using either a magnet, a small screwdriver with a dab of grease on it (the shim will stick to the grease), or a screwdriver and a pair of pliers **(see illustration)**. Do not allow the shim to fall into the engine.

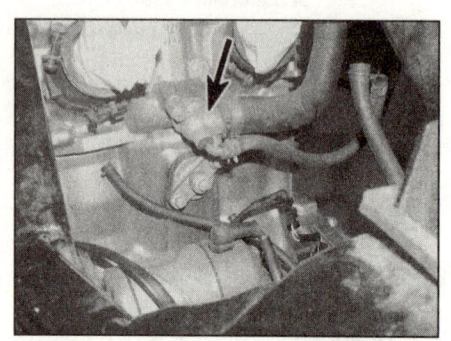

10.5 Unscrew the bolts securing the coolant pipe union (arrowed)

5 Unscrew the bolts securing the coolant pipe union to the rear of the cylinder head **(see illustration)**. Discard the O-ring as a new one must be used.
6 The cylinder head is secured by eight 9 mm bolts and two 6 mm bolts **(see illustration)**. First slacken and remove the two 6 mm bolts located in the cam chain tunnel. Note their positions as they are of different lengths. Now slacken the remaining bolts evenly and a little at a time in a criss-cross sequence, working from the outside towards the centre, until they are all slack and can be removed with their washers **(see illustration)**.

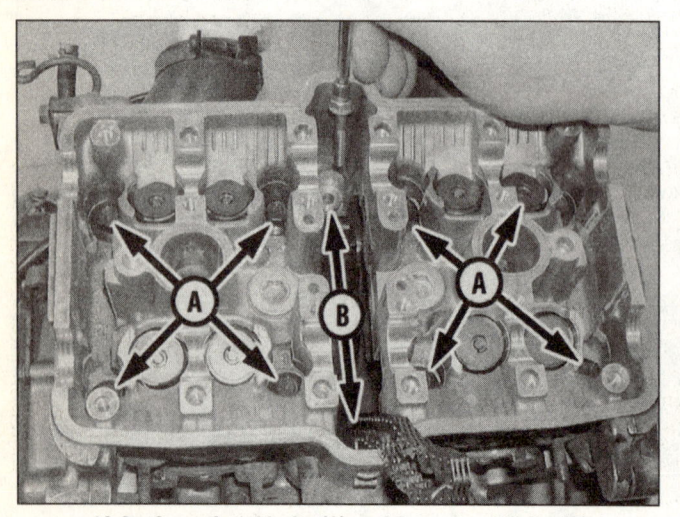

10.6a 9 mm head bolts (A) and 6 mm head bolts (B)

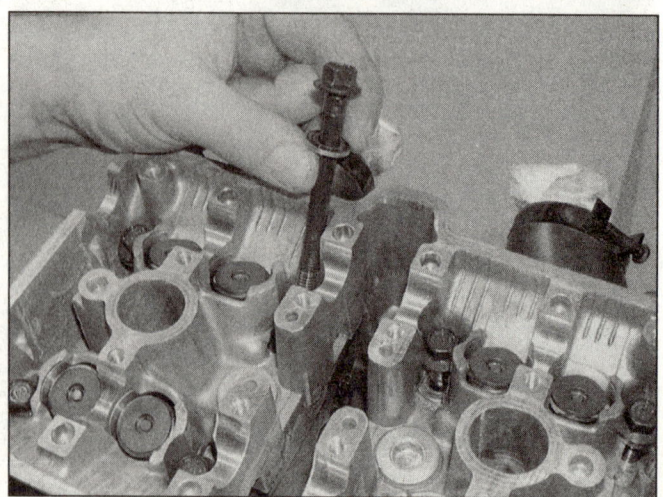

10.6b Remove the head bolts with their washers

Engine, clutch and transmission 2•17

10.7 Secure camchain with length of wire (arrowed)

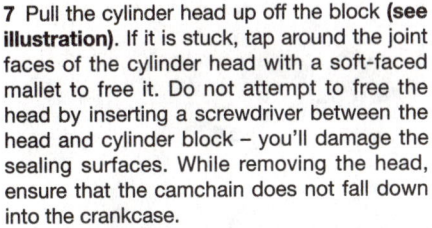

10.10 Clean jointing faces of cylinder head and block thoroughly

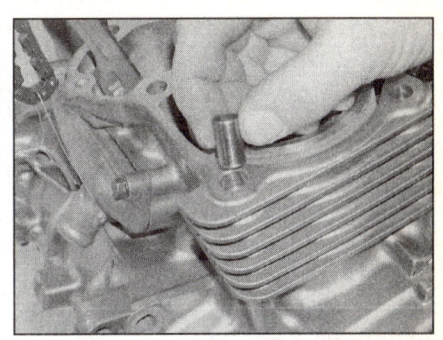

10.11 Install dowels in the cylinder block . . .

7 Pull the cylinder head up off the block **(see illustration)**. If it is stuck, tap around the joint faces of the cylinder head with a soft-faced mallet to free it. Do not attempt to free the head by inserting a screwdriver between the head and cylinder block – you'll damage the sealing surfaces. While removing the head, ensure that the camchain does not fall down into the crankcase.

8 Remove the old cylinder head gasket and discard it as a new one must be used. If they are loose, remove the two dowels from the front outer bolt holes in the cylinder block. If they appear to be missing they are probably stuck in the underside of the cylinder head.

9 Check the cylinder head gasket and the mating surfaces on the cylinder head and block for signs of leakage, which could indicate warpage. Refer to Section 12 and check the flatness of the cylinder head.

10 Clean all traces of old gasket material from the cylinder head and block **(see illustration)**. If a scraper is used, take care not to scratch or gouge the soft aluminium. Be careful not to let any of the gasket material fall into the crankcase, the cylinder bores or the oil and coolant passages.

Installation

11 Lubricate the cylinder bores with engine oil. If removed, install the dowels into the front outer bolt holes in the cylinder block **(see illustration)**.

12 Ensure both cylinder head and block mating surfaces are clean, then lay the new head gasket in place on the cylinder block. Make sure all the holes in the gasket are correctly aligned and that it locates correctly onto the dowels. Check that the 'YUP' lettering stamped out of the gasket is positioned along the rear edge and is the correct way up **(see illustration)**. Never re-use the old gasket.

13 Carefully lower the cylinder head onto the block while feeding the camchain up through the head. Make sure the head locates correctly onto the dowels and secure the camchain with a length of wire to prevent it dropping into the crankcase.

14 Apply clean engine oil to the threads and seating surfaces of the cylinder head bolts and install all the bolts and washers and tighten them finger-tight. First tighten the eight 9 mm bolts evenly and a little at a time in a criss-cross pattern, starting from the centre and working outwards, to the torque setting specified at the beginning of this Chapter **(see illustration)**. Then tighten the two 6 mm bolts located in the camchain tunnel to the specified torque.

15 Fit a new O-ring onto the coolant pipe union and secure it to the rear of the head with the bolts **(see illustration)**.

16 Lubricate each valve shim with a 50/50 mixture of molybdenum disulphide grease and engine oil and fit it into its recess on the top of the valve, with the size marking on each shim facing up **(see illustration)**. Check that the shim is correctly seated, then lubricate the follower with molybdenum disulphide oil and install it onto the valve **(see illustration)**. *Note: It is most important that the shims and followers are returned to their original valves otherwise the valve clearances will be inaccurate.*

17 Install the camshafts (see Section 9).

18 Install the exhaust system and the carburettors (see Chapter 4).

19 Fill the cooling system (see Chapter 1).

10.12 . . . then install gasket with YUP (arrowed) on rear edge

10.14 Tighten the cylinder head bolts as described to the specified torque settings

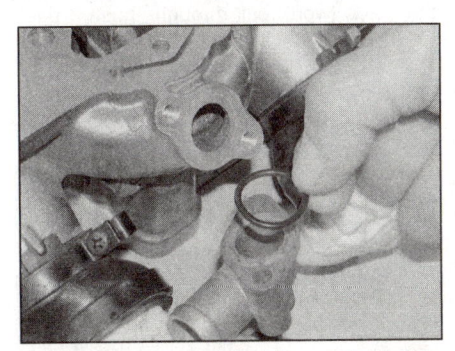

10.15 Fit a new O-ring to the coolant pipe union

10.16a Fit the valve shim with its size marking facing up . . .

10.16b . . . then lubricate and install the follower

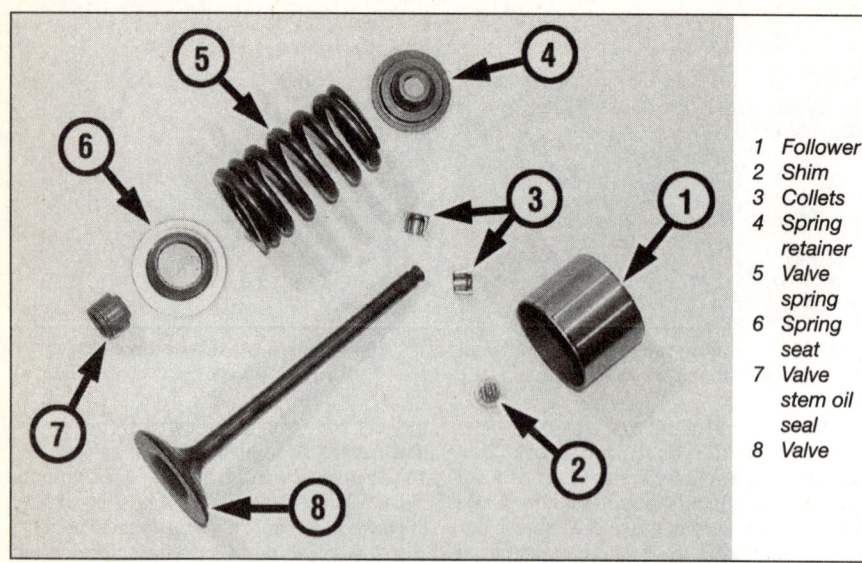

2•18 Engine, clutch and transmission

12.3 Valve components

1 Follower
2 Shim
3 Collets
4 Spring retainer
5 Valve spring
6 Spring seat
7 Valve stem oil seal
8 Valve

12.5a Compress the valve spring . . .

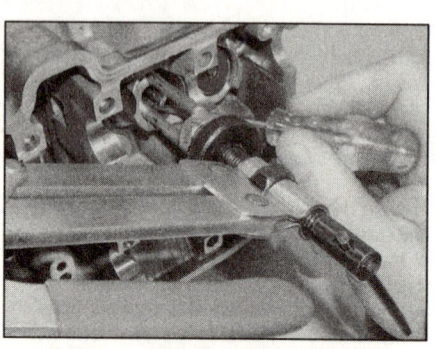

12.5b . . . and remove the collets as described

> **HAYNES HiNT** *Refer to Tools and Workshop Tips for details of gasket removal methods.*

11 Valves/valve seats/valve guides – servicing

1 Because of the complex nature of this job and the special tools and equipment required, most owners leave servicing of the valves, valve seats and valve guides to a professional.

2 The home mechanic can, however, remove the valves from the cylinder head, clean and check the components for wear and assess the extent of the work needed, and, unless a valve service is required, grind in the valves (see Section 12).

3 A Honda dealer or motorcycle engineer will remove the valves and springs, renew the valves and guides, recut the valve seats, check and renew the valve springs, spring retainers and collets (as necessary), renew the valve seals and reassemble the valve components.

4 After the valve service has been performed, the head will be in like-new condition. When the head is returned, be sure to clean it again very thoroughly before installation on the engine to remove any metal particles or abrasive grit that may still be present from the valve service operations. Use compressed air, if available, to blow out all the holes and passages.

12 Cylinder head and valves – disassembly, inspection and reassembly

1 As mentioned in the previous section, valve servicing, valve seat re-cutting and valve guide renewal should be left to a Honda dealer or motorcycle engineer. However, disassembly, cleaning and inspection of the valves and related components can be done (if the necessary special tools are available) by the home mechanic. This way no expense is incurred if the inspection reveals that overhaul is not required at this time.

2 To disassemble the valve components without the risk of damaging them, a valve spring compressor is absolutely necessary.

Disassembly

3 Before proceeding, arrange to label and store the valves along with their related components in such a way that they can be returned to their original locations without getting mixed up **(see illustration)**. A good way to do this is to use the same container as the followers and shims are stored in (see Section 10), or obtain a separate container which is divided into eight compartments, and label each compartment with the identity of the valve which will be stored in it (ie number of cylinder, inlet or exhaust valve, inner or outer valve). Alternatively, labelled plastic bags will do just as well.

4 If not already done, clean all traces of old gasket material from the cylinder head. If a scraper is used, take care not to scratch or gouge the soft aluminium.

5 Compress the valve spring on the first valve with a spring compressor, making sure it is correctly located onto each end of the valve assembly **(see illustration)**. Do not compress the springs any more than is absolutely necessary and take care not to mark the surface of the cam follower hole with the spring compressor. Remove the collets, using either needle-nose pliers, tweezers, a magnet or a screwdriver with a dab of grease on it **(see illustration)**. Carefully release the valve spring compressor and remove the spring retainer, noting which way up it fits, the spring, the spring seat, and the valve from the head **(see illustrations)**. If the valve binds in the guide (won't pull through), push it back

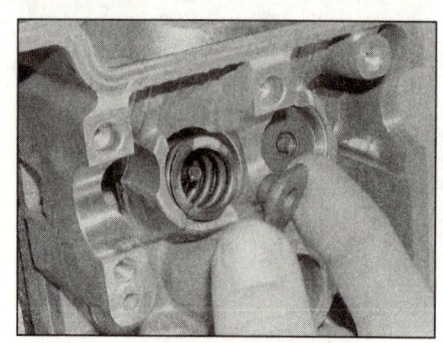

12.5c Remove the spring retainer . . .

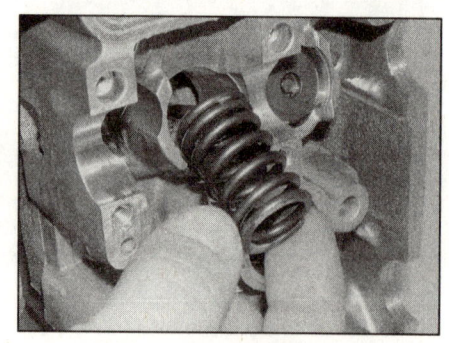

12.5d . . . the valve spring . . .

Engine, clutch and transmission 2•19

12.5e ... and the valve spring seat

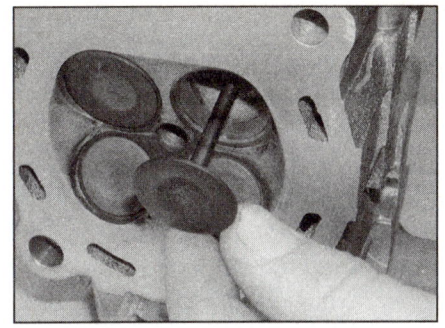

12.5f The valve should slide out of the combustion chamber, but if it sticks ...

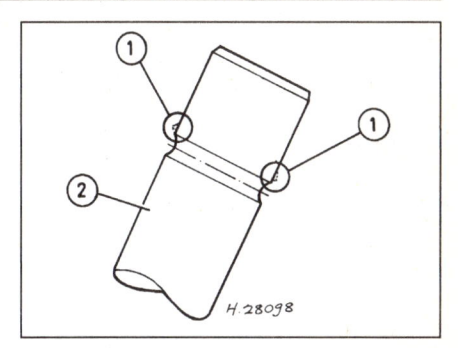

12.5g ... check the valve stem (2) above the collet groove (1) and remove any burrs

into the head and deburr the area around the collet groove with a very fine file or whetstone **(see illustration)**.

6 Repeat the procedure for the remaining valves. Remember to keep the parts for each valve together and separate from the other valves so they can be reinstalled in the same location.

7 Once the valves have been removed and labelled, pull the valve stem seals off the top of the valve guides with pliers and discard them **(see illustration)**. The old seals should not be reused.

8 Next, clean the cylinder head with solvent and dry it thoroughly. Compressed air will speed the drying process and ensure that all holes and recessed areas are clean.

9 Clean all the valve springs, collets, retainers and spring seats with solvent and dry them thoroughly. Clean the parts from one valve at a time so that no mixing of parts between valves occurs.

10 Scrape off any deposits that may have formed on the valve, then use a motorised wire brush to remove deposits from the valve heads and stems. Again, make sure the valves do not get mixed up.

Inspection

11 Inspect the head very carefully for cracks and other damage. If cracks are found, a new head will be required. Check the cam bearing surfaces for wear and evidence of seizure. Check the camshafts for wear as well (see Section 9).

12 Inspect the outer surfaces of the cam followers for evidence of scoring or other damage. If a follower is in poor condition, it is probable that the bore in which it works is also damaged. Check for clearance between the followers and their bores. Whilst no specifications are given, if slack is excessive, renew the followers. If the bores are seriously out-of-round or tapered, the cylinder head and the followers must be renewed.

13 Using a precision straight-edge and a feeler gauge set to the warpage limit listed in the specifications at the beginning of this Chapter, check the head gasket mating surface for warpage. Refer to *Tools and Workshop Tips* in the Reference section for details of how to use the straight-edge. If the head is warped beyond the limit, consult a Honda dealer or take the head to an engineer for rectification.

14 Examine the valve seats in the combustion chamber. If they are pitted, cracked or burned, the head will require work beyond the scope of the home mechanic. Measure the valve seat width and compare it to this Chapter's Specifications **(see illustration)**. If it exceeds the service limit, or if it varies around its circumference consult your Honda dealer or take the head to a specialist repair shop.

15 Measure the valve stem diameter **(see illustration)**. Clean the valve guides to remove any carbon build-up, then measure the inside diameters of the guides (at both ends and the centre of the guide) with a small-hole gauge and micrometer **(see illustrations)**. The guides are measured at the ends and at the centre to determine if they are worn in a bell-mouth pattern (more wear at the ends). Subtract the stem diameter from the valve guide diameter

12.7 Discard the old valve stem seals (arrowed)

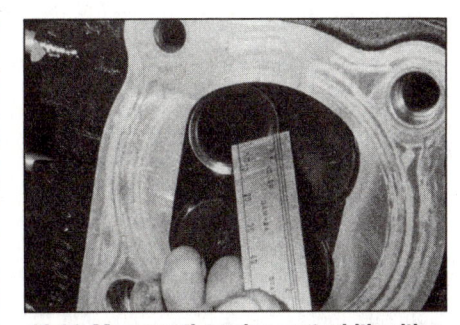

12.14 Measure the valve seat width with a ruler (or for greater precision use a vernier caliper)

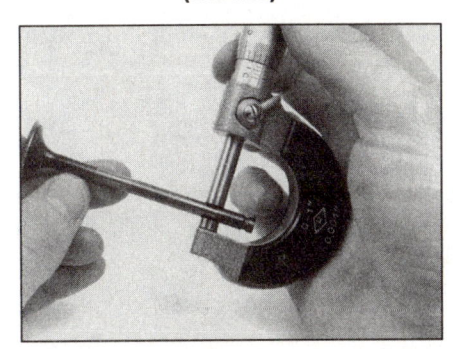

12.15a Measure the valve stem diameter with a micrometer

12.15b Insert a small-hole gauge into the valve guide and expand it so there's a slight drag when it's pulled out

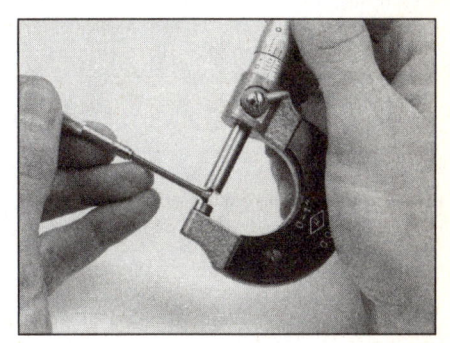

12.15c Measure the small-hole gauge with a micrometer

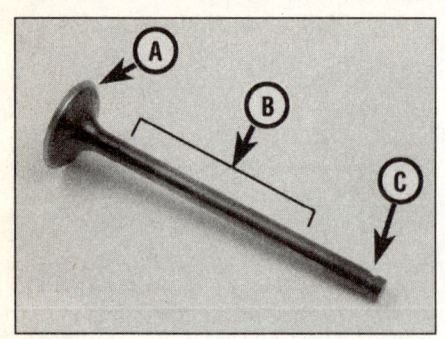

12.16 Check the valve face (A), stem (B) and collet grove (C) for signs of wear and damage

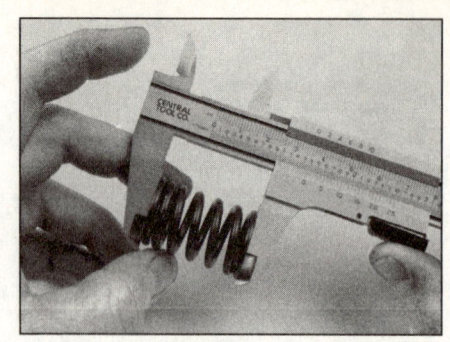

12.17a Measure the free length of the valve springs

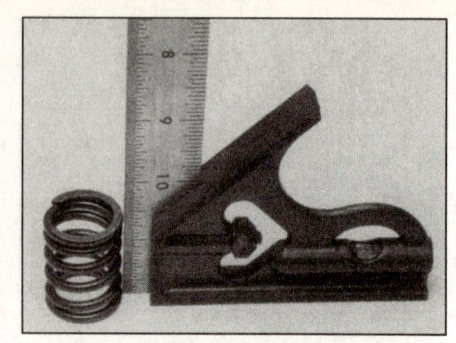

12.17b Check the valve springs for squareness

to obtain the valve stem-to-guide clearance. If the stem-to-guide clearance is greater than listed in this Chapter's Specifications, the guides and valves will have to be renewed. If the valve stem or guide is worn beyond its limit, or if the guide is worn unevenly, it must be renewed.

16 Carefully inspect each valve face for cracks, pits and burned spots. Check the valve stem and the collet groove area for scuffing and cracks **(see illustration)**. Rotate the valve and check for any obvious indication that it is bent. Check the end of the stem for pitting and excessive wear. The presence of any of the above conditions indicates the need for valve servicing.

17 Check the end of each valve spring for wear and pitting. Measure the spring free length and compare it to that listed in the specifications **(see illustration)**. If any spring is shorter than specified it must be renewed. Also place the spring upright on a flat surface and check it for bend by placing a ruler against it **(see illustration)**. If the bend in any spring is excessive, it must be renewed.

18 Check the spring retainers and collets for obvious wear and cracks. Any questionable parts should not be reused, as extensive damage will occur in the event of failure during engine operation.

19 If the inspection indicates that no overhaul work is required, the valve components can be reinstalled in the head.

Reassembly

20 Unless a valve service has been performed, before installing the valves in the head they should be ground in (lapped) to ensure a positive seal between the valves and seats. This procedure requires coarse and fine valve grinding compound and a valve grinding tool. If a grinding tool is not available, a piece of rubber or plastic hose can be slipped over the valve stem (after the valve has been installed in the guide) and used to turn the valve.

21 Apply a small amount of coarse grinding compound to the valve face, then slip the valve into the guide **(see illustration)**. **Note:** *Make sure each valve is installed in its correct guide and be careful not to get any grinding compound on the valve stem.*

22 Attach the grinding tool (or hose) to the valve and rotate the tool between the palms of your hands. Use a back-and-forth motion (as though rubbing your hands together) rather than a continuous circular motion (ie so that the valve rotates alternately clockwise and anti-clockwise rather than in one direction only) **(see illustration)**. Lift the valve off the seat and turn it at regular intervals to distribute the grinding compound properly. Continue the grinding procedure until the valve face and seat contact area is of uniform width and unbroken around the entire circumference of the valve face and seat **(see illustrations)**.

23 Carefully remove the valve from the guide and wipe off all traces of grinding compound. Use solvent to clean the valve and wipe the seat area thoroughly with a solvent soaked cloth.

24 Repeat the procedure with fine valve grinding compound, then repeat the entire procedure for the remaining valves.

25 Lay the spring seat for the first valve in place in the cylinder head, with its narrower shouldered side up so that it fits into the base of the spring, then install a new valve stem

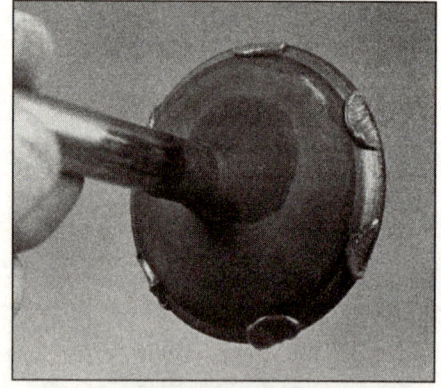

12.21 Apply grinding compound very sparingly, in small dabs, to the valve face only

12.22a Rotate the valve grinding tool back and forth between the palms of your hands

12.22b The valve face and seat should show a uniform unbroken ring . . .

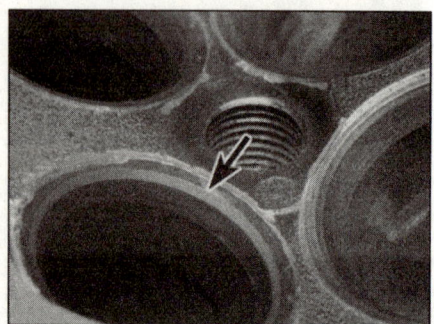

12.22c . . . and the seat (arrowed) should be the specified width all the way round

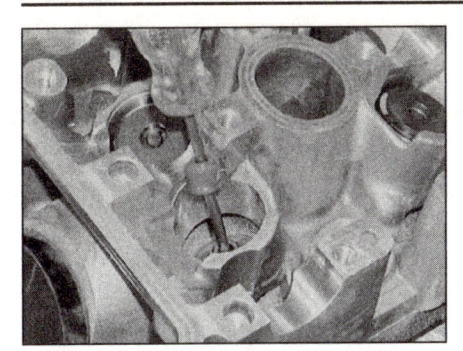

Engine, clutch and transmission 2•21

12.25a Use a small screwdriver to locate a new valve stem seal onto the guide . . .

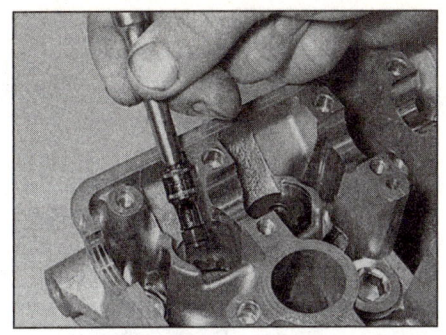

12.25b . . . then press the seal into position with a suitable deep socket

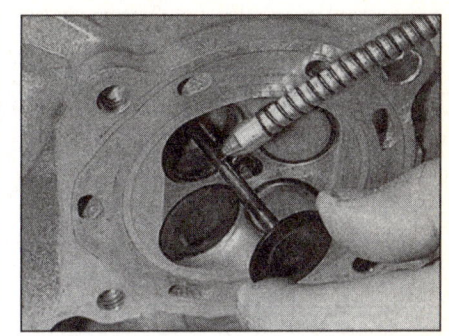

12.26 Lubricate the stem and slide the valve into its correct location

12.27a A small dab of grease will keep the collets in place on the valve during reassembly

12.27b Compress the springs and install the collets, making sure they locate in the groove

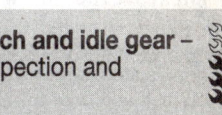

seal onto the guide **(see illustration)**. Use an appropriate size deep socket to push the seal over the end of the valve guide until it is felt to clip into place **(see illustration)**. Don't twist or cock it, or it will not seal properly against the valve stem. Also, don't remove it again or it will be damaged.

26 Coat the valve stem with molybdenum disulphide grease, then install it into its guide, rotating it slowly to avoid damaging the seal **(see illustration)**. Check that the valve moves up and down freely in the guide. Next, fit the spring onto the spring seat with its closer-wound coils facing down into the cylinder head, then fit the spring retainer, with its shouldered side facing down so that it fits into the top of the spring.

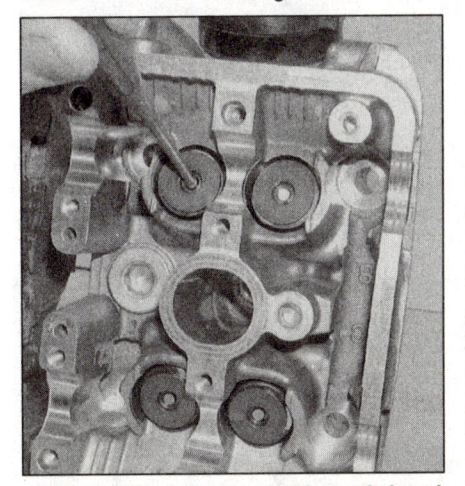

27 Apply a small amount of grease to the collets to help hold them in place **(see illustration)**. Compress the spring with the valve spring compressor and install the collets **(see illustration)**. When compressing the spring, depress it only as far as is absolutely necessary to slip the collets into place. Make certain that the collets are securely locked in the retaining groove.

28 Repeat the procedure for the remaining valves. Remember to keep the parts for each valve together and separate from the other valves so they can be reinstalled in the same location.

29 Support the cylinder head on blocks so the valves can't contact the workbench top, then very gently tap each of the valve stems with a soft-faced punch **(see illustration)**. This will help seat the collets in their grooves.

> **HAYNES HiNT** *Check for proper sealing of the valves by pouring a small amount of solvent into each of the valve ports. If the solvent leaks past any valve into the combustion chamber area the valve grinding operation on that valve should be repeated.*

13 Starter clutch and idle gear – removal, inspection and installation

Note: *This procedure can be carried out with the engine in the frame. If the engine has been removed, ignore the steps which do not apply.*

Removal

1 Drain the engine oil (see Chapter 1).
2 Unscrew the two bolts securing the front sprocket cover and remove the cover and the chain guide **(see illustration)**.
3 Remove the left-hand frame panel (see Chapter 8). Trace the alternator wiring from the left-hand side of the engine and disconnect it at the white 3-pin connector **(see illustration)**. Release the wiring from any clips or ties, and feed it through to the left-hand engine cover, noting its routing.

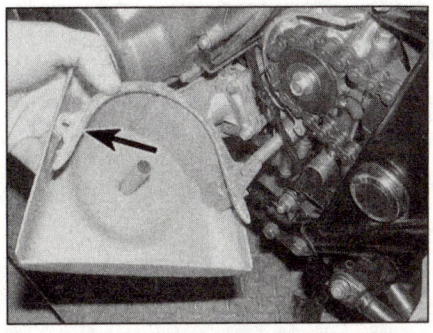

12.29 Tap the valve stem with a soft-faced punch to seat the collets

13.2 Remove the sprocket cover and chain guide (arrowed)

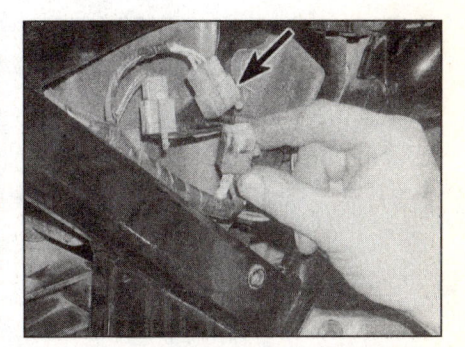

13.3 Disconnect the white 3-pin connector for the alternator wiring

2•22 Engine, clutch and transmission

13.4 Remove the left-hand engine cover

13.5 Check the operation of the starter clutch by rotating the idle gear (arrowed)

13.6 Withdraw the shaft and remove the gear

4 Unscrew the bolts securing the left-hand engine cover and wiring clamp and remove the cover and clamp, being prepared to catch any residual oil **(see illustration)**. If the cover will not lift away easily, break the gasket seal by tapping gently around the sides with a rubber hammer or block of wood and levering carefully on the leverage points around the edge of the cover. Protect the frame with a piece of wood or rag while doing this. Discard the gasket as a new one must be used. Remove the dowel from either the cover or the crankcase if it is loose.

5 The operation of the starter clutch can be checked while it is in place. Remove the starter motor (see Chapter 9). Check that the idle gear is able to rotate freely anti-clockwise as you look at it from the left-hand side of the bike, but locks when rotated clockwise **(see illustration)**. If not, the starter clutch is faulty and should be removed for inspection.

6 Withdraw the idle gear shaft from the crankcase and remove the gear **(see illustration)**.

7 Remove the alternator rotor – the starter clutch is mounted on the back of it (see Chapter 9). **Note:** *Before removing the alternator rotor, slacken the six starter clutch bolts while holding the rotor centre bolt* **(see illustration)**. *If the rotor has already been removed from the bike, hold the rotor with a strap wrench to slacken the bolts.*

8 Withdraw the starter driven gear from the starter clutch **(see illustration)**. If the gear appears stuck, rotate it anti-clockwise as you withdraw it to free it from the starter clutch. If the starter driven gear does not come away with the alternator rotor, slide it off the crankshaft.

Inspection

9 Inspect the starter idle gear and driven gear teeth and renew them as a pair if any teeth are chipped or missing. Check the idler shaft and gear bearing surfaces for signs of wear or damage, and renew if necessary.

10 With the alternator rotor face-down, check that the starter driven gear rotates freely in an anti-clockwise direction and locks against the rotor in a clockwise direction **(see illustration)**. If it doesn't, the starter clutch should be dismantled.

11 Unscrew the six bolts and remove the clutch housing and the clutch from the back of the alternator rotor **(see illustration)**.

12 Inspect the condition of the rollers and the roller cage inside the clutch assembly **(see illustration)**. If they are damaged or worn at any point, the starter clutch should be renewed – individual components are not available.

13 Inspect the driven gear contact surfaces for signs of wear and scoring.

Installation

14 Clean the starter clutch bolts and apply a drop of locking compound to their threads.

15 Fit the clutch assembly and the clutch housing to the rear of the rotor and tighten the

13.7 Slacken the starter clutch bolts before removing the alternator rotor

13.8 Withdraw the gear and inspect for wear or damage

13.10 Gear should rotate anti-clockwise (A) and lock clockwise (B)

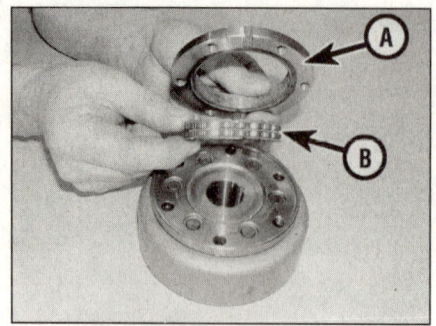

13.11 Remove clutch housing (A) and clutch (B) from back of alternator rotor

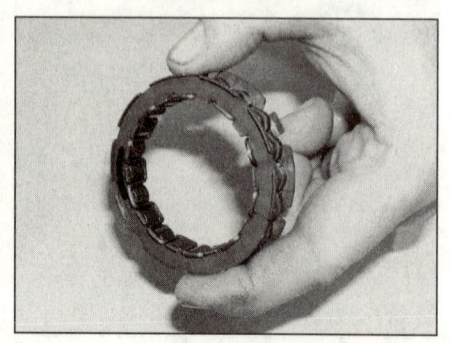

13.12 Inspect the rollers and roller cage

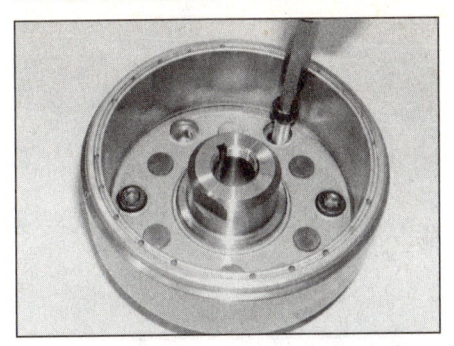

Engine, clutch and transmission 2•23

13.15 Tighten the clutch housing bolts securely

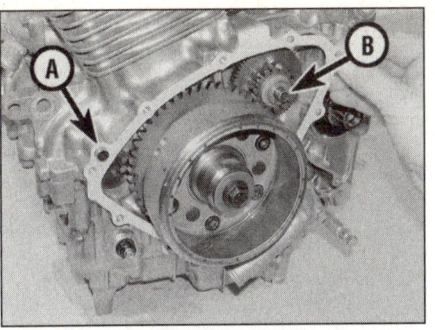

13.20a Ensure the engine cover locates on the dowel (A) and gear shaft (B)

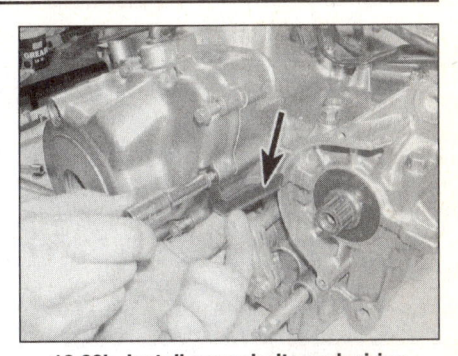

13.20b Install cover bolts and wiring clamp (arrowed)

bolts securely **(see illustration)**. Note: If a strap wrench is not available to hold the alternator rotor, final tightening of the bolts can take place once the rotor has been fitted to the crankshaft.

16 Lubricate the hub of the starter driven gear with clean engine oil, then install the starter driven gear into the clutch, rotating it anti-clockwise as you do so to spread the rollers and allow the hub of the gear to enter. With the rotor face down, ensure that the driven gear rotates freely anti-clockwise and locks against the rotor in a clockwise direction.

17 Install the alternator (see Chapter 9).

18 Lubricate the idle gear shaft with clean engine oil. Position the idle gear, making sure the teeth of the smaller pinion mesh correctly with the teeth of the starter driven gear, then slide the shaft through the gear into the crankcase.

19 Install the starter motor (see Chapter 9).

20 If removed, insert the dowel in the crankcase, then install the left-hand engine cover using a new gasket, making sure the cover locates correctly onto the dowel and the idle gear shaft **(see illustration)**. Install the cover bolts and wiring clamp and tighten the bolts evenly in a criss-cross sequence **(see illustration)**.

21 Feed the alternator wiring back to its connector, making sure it is correctly routed, and reconnect it. Refit the left-hand frame side panel.

22 Refit the chain guide and front sprocket cover.

23 Refill the engine with oil (see Chapter 1).

14 Clutch cable – replacement

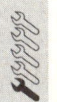

1 Fully slacken the lockring on the adjuster at the handlebar end of the cable then screw the adjuster fully in to create slack in the cable **(see illustration)**.

2 Fully slacken the adjuster nuts on the threaded section in the cable bracket on the right-hand engine cover and disconnect the inner cable end from the clutch release lever, noting how it fits **(see illustration)**. Thread the end nut off the inner cable and draw the cable through the bracket **(see illustration)**.

3 Align the slots in the adjuster and lockring at the handlebar end of the cable with that in the lever bracket, then pull the outer cable end from the socket in the adjuster and release the inner cable from the lever **(see illustrations)**. Remove the cable from the machine, noting its routing through the top yoke and guide on the right-hand side of he frame.

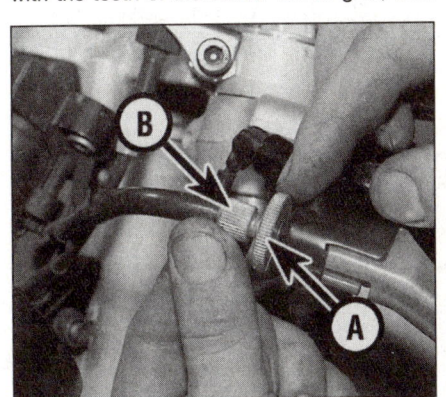

14.1 Fully slacken the lockring (A) and screw the adjuster (B) into the bracket

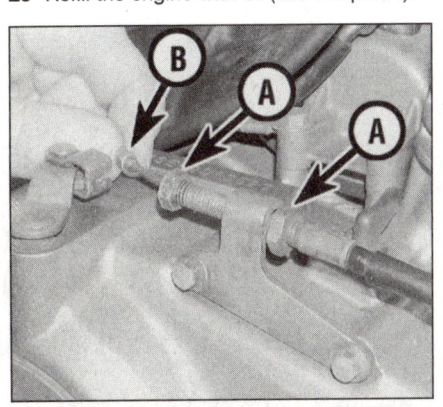

14.2a Slacken adjuster nuts (A) and disconnect the inner cable (B) . . .

> **HAYNES HiNT**
> *Before removing the cable from the bike, tape the lower end of the new cable to the upper end of the old cable. Slowly pull the lower end of the old cable out, guiding the new cable down into position. Using this method will ensure the cable is routed correctly.*

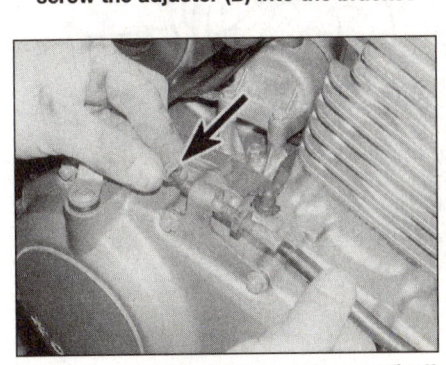

14.2b . . . then thread the nut (arrowed) off the inner cable

14.3a Align the slots as shown . . .

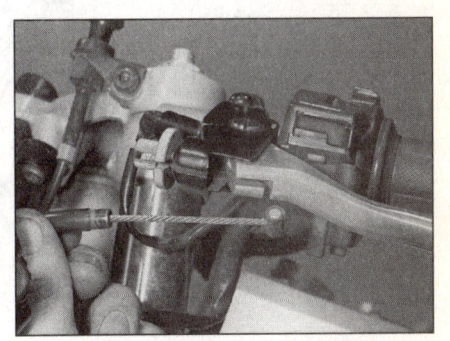

14.3b . . . and detach the inner cable

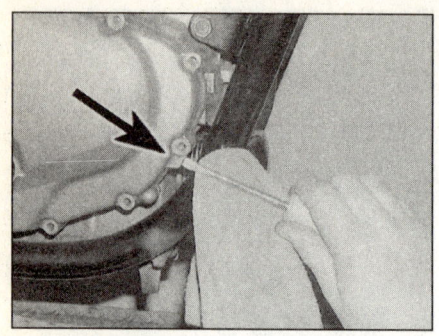

2•24 Engine, clutch and transmission

15.3a Protect the frame with a rag and use a flat-bladed tool . . .

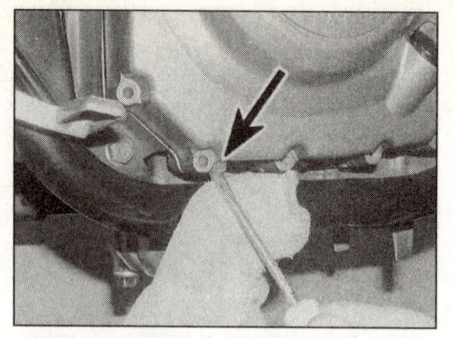

15.3b . . . on the leverage points around the cover (arrowed)

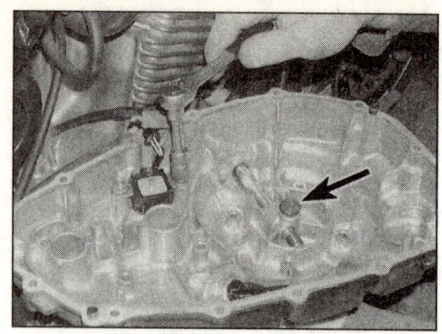

15.4 Remove the ignition pulse generator and the clutch pushrod (arrowed)

4 Installation is the reverse of removal. Apply grease to the cable ends. Make sure the cable is correctly routed. Adjust the amount of clutch lever freeplay (see Chapter 1).

15 Clutch – removal, inspection and installation

Note: *This procedure can be carried out with the engine in the frame. If the engine has been removed, ignore the steps which do not apply.*

Removal

1 Drain the engine oil (see Chapter 1).

2 Slacken the clutch cable adjuster at the bracket on the right-hand engine cover and slip the cable end out of the release lever and the bracket (see Section 14). Secure the cable clear of the engine.

3 Unscrew the bolts securing the engine right-hand cover, noting that two secure the clutch cable bracket. Be prepared to catch any residual oil as you remove the cover. If the cover will not lift away easily, break the gasket seal by tapping gently around the sides with a rubber hammer or block of wood and levering carefully on the leverage points around the edge of the cover. Protect the frame with a piece of wood or rag while doing this **(see illustrations)**. Discard the gasket as a new one must be used. Remove the dowels from either the cover or the crankcase if they are loose.

4 Unscrew the two bolts securing the ignition pulse generator coil to the engine cover and secure the pulse generator clear of the engine. Note the pushrod fitted in the bottom of the shaft in the cover and remove it for safekeeping **(see illustration)**. It is possible that it is stuck to the release bearing in the centre of the clutch release plate.

5 Working evenly in a criss-cross pattern, gradually slacken the clutch release plate bolts until spring pressure is released, then

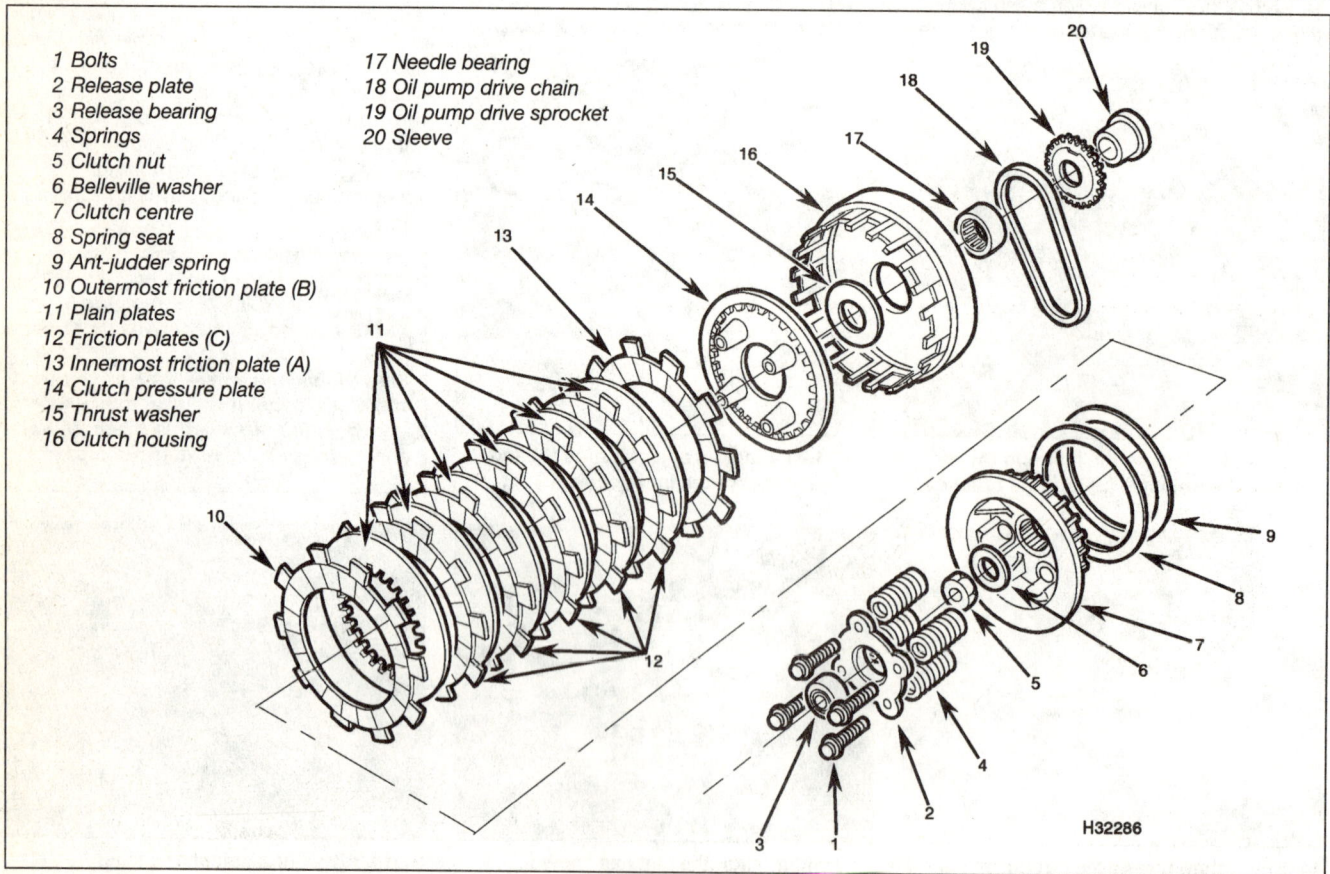

1 Bolts
2 Release plate
3 Release bearing
4 Springs
5 Clutch nut
6 Belleville washer
7 Clutch centre
8 Spring seat
9 Ant-judder spring
10 Outermost friction plate (B)
11 Plain plates
12 Friction plates (C)
13 Innermost friction plate (A)
14 Clutch pressure plate
15 Thrust washer
16 Clutch housing
17 Needle bearing
18 Oil pump drive chain
19 Oil pump drive sprocket
20 Sleeve

15.5a Clutch assembly

Engine, clutch and transmission 2•25

15.5b Unscrew the bolts and remove the plate and springs

15.6a Unstake the clutch nut (arrowed) with a small punch ...

15.6b ... and slacken it as described

remove the bolts, plate and springs (see illustration).

6 The clutch nut is staked against a groove in the transmission input shaft (see illustration 15.33d). Unstake the nut using a small punch (see illustration). To slacken the clutch nut the input shaft must be locked. This can be done in several ways:
 a) If the engine is in the frame, engage 1st gear and have an assistant hold the rear brake on hard with the rear tyre in firm contact with the ground.
 b) Use the service tool, available from a Honda dealer (Pt. No. 07GMB-KT80100).

TOOL TIP

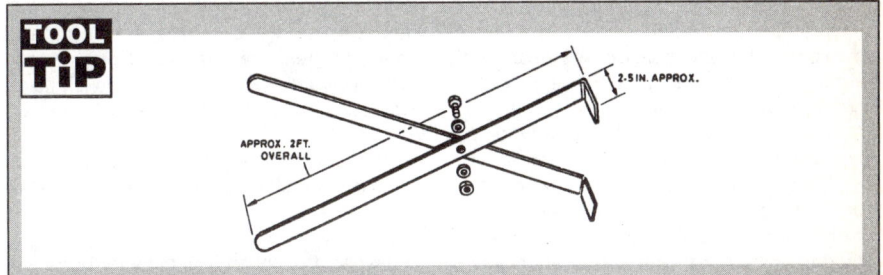

A clutch centre holding tool can easily be made using two strips of steel bent over at the ends and bolted together in the middle. Note that proprietary versions of this tool are available which use self-locking grips

 c) A home-made tool made from two strips of steel bent at the ends and bolted together in the middle (see **Tool Tip**), can be used to stop the clutch centre from turning whilst the nut is slackened (see illustration).

7 Unscrew the clutch nut and remove the Belleville washer, and on CBF models the plain washer. Discard the nut as a new one must be used.

8 Screw a clutch release plate bolt halfway into its thread in the clutch pressure plate and use it to pull the clutch centre, the clutch plates and the pressure plate off the input shaft as an assembly (see illustration). Unless the plates are being renewed, keep them in their original order. Note that there are three types of friction plate – the innermost plate having a different internal diameter to fit over tabs on the clutch pressure plate and the outermost to fit over the anti-judder spring and spring seat (see illustration). Take care not to mix them up.

9 Remove the thrust washer from the input shaft (see illustration).

10 The primary drive gear is in two parts, spring loaded to eliminate backlash in the gear teeth. Before removing the clutch housing, screw a 6 mm thread diameter bolt, 12 mm in length, into the threaded hole in the inner primary drive gear to lock the two halves together and keep the teeth of both halves aligned. Precise alignment

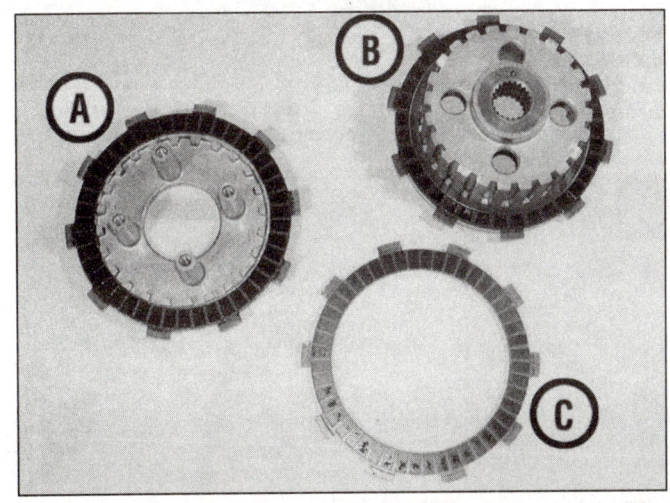

15.8a Pull the clutch off the shaft as an assembly

15.8b Friction plates – innermost plate (A), outermost plate (B) and standard plate (C)

15.9 Remove the thrust washer

2•26 Engine, clutch and transmission

15.10a Align the teeth of the primary drive gear (note hole for 6 mm bolt) . . .

15.10b . . . and remove the clutch housing

15.11a Unscrew the oil pump sprocket bolt . . .

of the teeth can be achieved by inserting a screwdriver into the holes in the gears and twisting it **(see illustration)**. Remove the clutch housing **(see illustration)**.

11 Use a steel rod and a block of wood to lock the oil pump sprocket against the crankcase **(see illustration)**. Unscrew the sprocket bolt and pull the pump sprocket, chain, drive sprocket and clutch housing sleeve off the input shaft **(see illustration)**.

Inspection

12 Check the teeth of the primary driven gear on the back of the clutch housing and the corresponding teeth of the primary drive gear on the end of the crankshaft. Renew the clutch housing and/or drive gear if any teeth are worn or chipped (see Section 16 for removal and disassembly of the primary drive gear).

13 After an extended period of service the clutch friction plates will wear and promote clutch slip. Measure the thickness of each plate using a vernier caliper **(see illustration)**. If the thickness is less than the service limit given in the Specifications at the beginning of the Chapter, the friction plates must be renewed as a set. Also, if any of the plates smell burnt or are glazed, they must be renewed.

14 The plain plates should not show any signs of excess heating (bluing). Check for warpage using a surface plate and feeler gauges **(see illustration)**. If any plate appears warped, or shows signs of bluing, all plain plates must be renewed as a set.

15 Measure the free length of each clutch spring **(see illustration)**. If any spring is below the service limit specified, renew all the springs as a set.

16 Inspect the clutch assembly for burrs and indentations on the edges of the protruding tabs of the friction plates and/or slots in the edge of the housing with which they engage. Similarly check for wear between the inner teeth of the plain plates and the slots in the clutch centre. Wear of this nature will cause clutch drag and slow disengagement during gear changes, since the plates will snag when the pressure plate is lifted. With care, a small amount of wear can be corrected by dressing with a fine file, but if this is excessive the worn components should be renewed.

17 Check the release plate, release bearing, and pushrod for signs of roughness, wear or damage, and renew any parts as necessary. Check that the bearing outer race is a tight fit in the centre of the plate, and that the inner race rotates freely without any rough spots.

18 Measure the clutch housing sleeve inner and outer diameters, and the diameter of the transmission input shaft where the bush fits. Compare the measurements to the specifications at the beginning of this Chapter and renew the bush and/or the shaft if they are worn beyond their service limits.

19 Check the needle bearing in the clutch housing for signs of roughness or wear, and renew if necessary.

20 Check the pressure plate for signs of roughness, wear or damage, and renew if necessary.

21 Check the anti-judder spring and spring seat for distortion, wear or damage, and renew them if necessary.

22 Check that the clutch release shaft rotates smoothly in the engine cover and that the return spring operates properly. With the pushrod in place, the spring should be under slight tension. If the action is rough or stiff, draw the shaft out of the cover and inspect and clean the bearing surfaces **(see**

15.11b . . . and pull the sprockets and chain off as an assembly

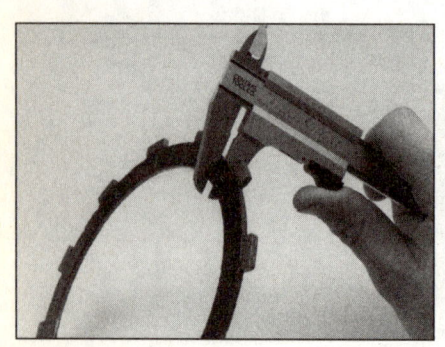

15.13 Measuring clutch friction plate thickness

15.14 Checking the plain plates for warpage

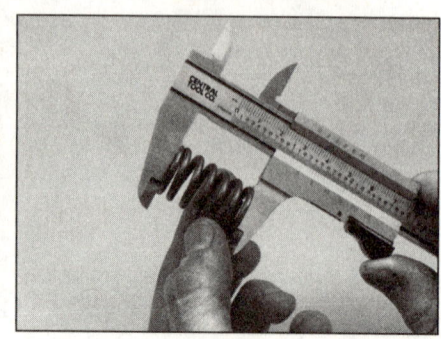

15.15 Measuring clutch spring free length

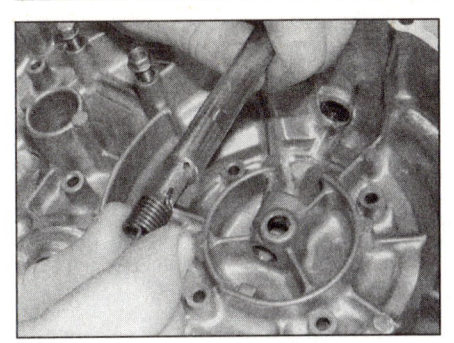

Engine, clutch and transmission 2•27

15.22 Inspect the clutch release shaft and return spring

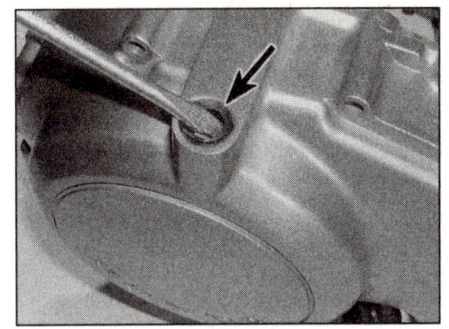

15.23a Prise oil seal (arrowed) out of case with a flat-bladed screwdriver . . .

15.23b . . . and check the condition of the needle bearing

illustration). Check the condition of the spring – if it has lost its tension, renew it.

23 Renew the release shaft oil seal if it is damaged or deteriorated, or shows any signs of leakage (see illustration). Check the needle bearing in the engine cover for signs of roughness or wear and renew it if necessary (see illustration).

Installation

24 Remove all traces of old gasket from the crankcase and engine cover surfaces.
25 Smear the clutch housing sleeve with clean engine oil and slide the sleeve onto the transmission input shaft, followed by the oil pump drive sprocket, with its pins facing out, the chain and the pump sprocket as one assembly (see illustrations).
26 Locate the pump sprocket onto the pump shaft. Apply a drop of locking compound to the sprocket bolt and install the bolt. Lock the sprocket (see Step 11) and tighten the bolt to the torque setting specified at the beginning of this Chapter.
27 Slide the clutch housing onto the sleeve on the input shaft, making sure that the teeth of the primary driven gear on the back of the housing engage fully with those of the primary drive gear, and that the pins on the oil pump drive sprocket engage with the slots in the back of the housing (see illustrations).
28 Slide the thrust washer onto the shaft (see illustration).
29 Fit the spring seat onto the clutch centre, followed by the anti-judder spring, making sure its outer edge is raised off the spring seat – if it is touching the seat and the inner edge is raised, it is the wrong way round (see illustrations).
30 Coat each clutch plate with engine oil, then build up the plates on the clutch centre, making sure the outermost friction plate and the innermost friction plate are correctly identified and fitted (see Step 8). Start with the

15.25a Install the clutch housing sleeve . . .

15.25b . . . and the oil pump drive chain and sprockets

15.27a Engage pins on sprocket (A) in slots (B) . . .

15.27b . . . by turning oil pump sprocket (arrowed)

15.28 Install the thrust washer

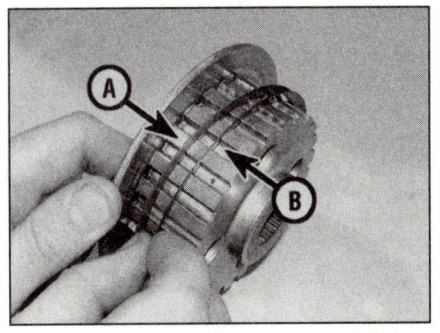

15.29a Fit the spring seat (A) and the spring (B) . . .

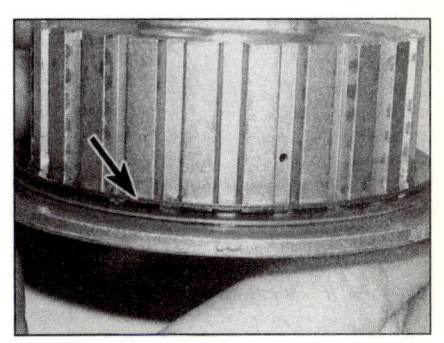

15.29b . . . so that the outer edge of the spring (arrowed) is off the seat

2•28 Engine, clutch and transmission

15.30a Build up the clutch starting with the outermost friction plate . . .

15.30b . . . followed by a plain plate as described

15.31 Fit the pressure plate onto the back of the clutch

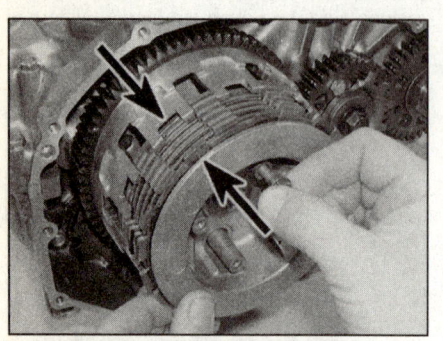

15.32 Align tabs on outer friction plate with short slots in clutch housing

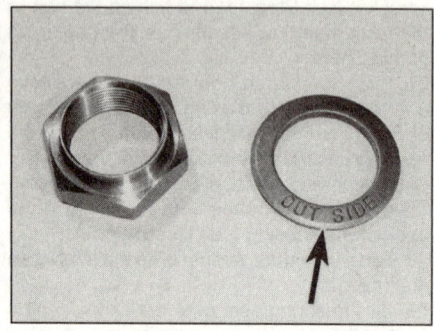

15.33a Fit a new clutch nut with the washer marking (arrowed) facing out

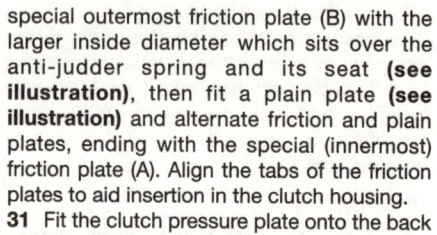

special outermost friction plate (B) with the larger inside diameter which sits over the anti-judder spring and its seat (see illustration), then fit a plain plate (see illustration) and alternate friction and plain plates, ending with the special (innermost) friction plate (A). Align the tabs of the friction plates to aid insertion in the clutch housing.

31 Fit the clutch pressure plate onto the back of the clutch centre (see illustration).

32 Turn the assembly over carefully and slide it into the clutch housing, making sure that the tabs on all except the outer friction plate align with the long slots in the housing and the tabs on the outer friction plate align with the short slots in the clutch housing (see illustration).

33 On CBF models slide the plain washer onto the shaft. On all models slide the Belleville washer on with its OUTSIDE marking facing out, then fit a new clutch nut (see illustrations). Lock the input shaft as on removal (see Step 6) and tighten the nut to the torque setting specified at the beginning of this Chapter. Check that the clutch centre rotates freely after tightening, then stake the rim of the nut into the indent in the end of the shaft using a suitable punch (see illustrations).

34 Install the clutch springs, release plate and release plate bolts and tighten them evenly in a criss-cross sequence to the specified torque setting (see illustration).

35 Unscrew the 6 mm bolt from the primary drive gear (see illustration).

15.33b Stake the rim (arrowed) of the clutch nut . . .

15.33c . . . into the indent on the input shaft

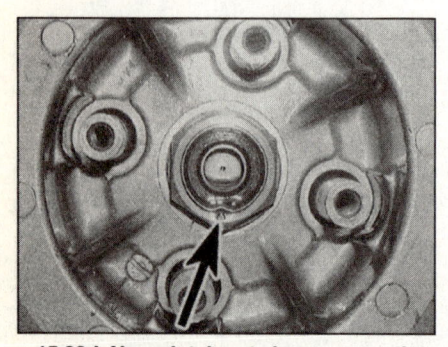

15.33d New clutch nut shown correctly staked in place

15.34 Fit the springs, plate and bolts as described

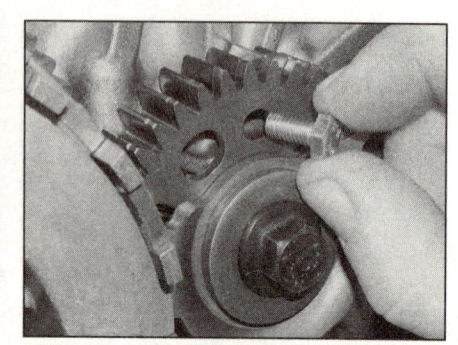

15.35 Remove the bolt locking the primary drive gear halves together

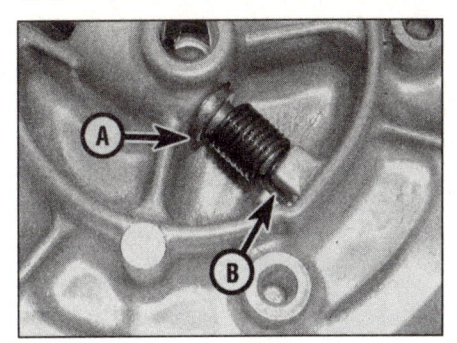

Engine, clutch and transmission 2•29

15.36a Locate one end of the spring in the shaft (A), the other against the case (B) . . .

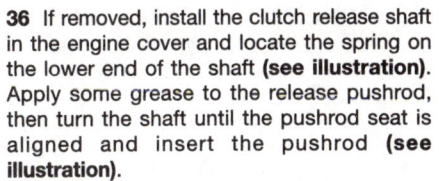

15.36b . . . and turn the shaft to align the pushrod seat (arrowed)

15.37 Install the ignition pulse generator (A) and the wiring grommet (B)

36 If removed, install the clutch release shaft in the engine cover and locate the spring on the lower end of the shaft **(see illustration)**. Apply some grease to the release pushrod, then turn the shaft until the pushrod seat is aligned and insert the pushrod **(see illustration)**.

37 Install the ignition pulse generator into the engine cover and tighten the bolts securely. Apply a smear of suitable sealant into the cutout in the cover and install the wiring grommet **(see illustration)**.

38 If removed, insert the dowels in the crankcase. Install the engine cover using a new gasket, making sure it locates correctly onto the dowels, then install all the cover bolts with the clutch cable bracket **(see illustrations)**. Tighten the bolts evenly in a criss-cross sequence.

39 If the engine is in the frame, fit the clutch cable through the bracket and into the release lever (see Section 14). Adjust the clutch cable (see Chapter 1).

40 Refill the engine with oil (see Chapter 1).

16 Timing rotor, primary drive gear and balancer gear – removal, inspection and installation

Note: *This procedure can be carried out with the engine in the frame. If the engine has been removed, ignore the steps which do not apply.*

Removal

1 Unscrew the timing inspection plug and the centre plug from the engine left-hand cover. Turn the crankshaft anticlockwise until the line next to the 'T' mark on the alternator rotor aligns with notches in the timing inspection hole (see Section 9). If the engine cover has been removed, align the 'T' mark with the arrow on the crankcase

2 Remove the engine right-hand cover (see Section 15). The primary drive gear bolt and the balancer gear nut can be slackened individually by locking the two gears together, either with the Honda service tool (Pt. No. 07724 – 0010100), or a stout piece of cloth jammed between them, without the risk of any damage should either prove tight.

3 Unscrew the primary gear bolt and remove the bolt and its washer and slide the timing rotor off the crankshaft. Unscrew the balancer gear nut and remove the nut and its washer.

4 Align the teeth of the two halves of the primary drive gear and the balancer gear and lock the halves together (see Section 15, Step 10), then slide the gears off their shafts.

Inspection

5 The two halves of the primary drive gear and the balancer gear can be separated by easing them apart with a flat-bladed screwdriver after removing the bolt locking the halves together – disassemble each gear separately to avoid interchanging parts. Note how the tabs on the inside of the outer half locate against the spring ends, how the holes align, and how the spring washer fits with its raised outer edge facing out **(see illustration)**.

6 No specifications are available for the springs, but check that they are of equal length and undamaged. Renew all springs as a set. Check the gear teeth for wear or damage and renew if necessary.

7 Inspect the timing rotor – this is not a wearing part but in the unlikely event that the rotor triggers or splines are damaged, the rotor must be renewed.

Installation

8 If the two halves of the balancer gear have been separated, apply molybdenum disulphide grease to the inside face of the outer gear half before assembly and fit the spring washer between the halves with its raised edge facing out. Align the 1.T mark on the outer gear half with the wide spline on the inner half **(see illustration)**. Align the teeth of the two halves of the gear and lock them together with the bolt used previously.

15.38a Locate a new gasket on the crankcase dowels (arrowed) . . .

15.38b . . . then install the cover and clutch cable bracket

16.5 Note how the tabs and springs locate (arrowed) before disassembly

16.8 1.T mark (A), wide spline (B) and locking bolt (C)

2•30 Engine, clutch and transmission

16.9 Install the balancer gear on its shaft

16.12 Check alignments (A) and (B) as described

16.13a Refit the primary gear, timing rotor, bolt and washer . . .

16.13b . . . and the balancer gear nut and washer . . .

16.13c . . . then remove the locking bolts (arrowed) from both gears

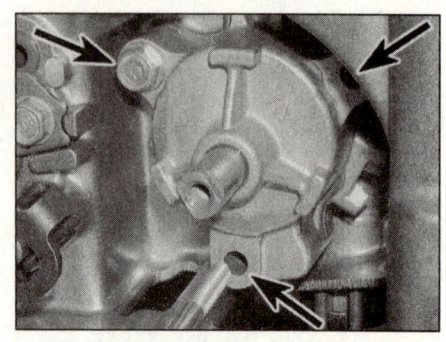

17.2 Unscrew the three bolts (arrowed) and remove the pump

9 Apply molybdenum disulphide grease to the inside face of the balancer gear and slide the gear onto its shaft, aligning the wide splines **(see illustration)**.
10 Check that the line next to the 'T' mark on the alternator rotor aligns with notches in the timing inspection hole (see Step 1).
11 If separated, assemble the components of the primary drive gear (see Step 8), align the teeth of the two halves of the gear and lock the halves together (see Section 15).
12 Apply molybdenum disulphide grease to the inside face of the primary drive gear and slide the gear onto the crankshaft, aligning the wide splines. Ensure that the 1.T mark on the drive gear aligns with the arrow on the crankcase and that the index line on the drive gear aligns with the index line on the balancer gear **(see illustration)**.
13 Install the timing rotor onto the crankshaft, aligning the wide splines. Apply clean engine oil to the threads of the primary drive gear bolt and refit the bolt and washer finger-tight, then apply clean engine oil to the threads of the balancer gear nut and refit the nut and washer finger-tight **(see illustration)**. Unscrew the bolts locking the halves of the primary drive and balancer gears **(see illustration)**.
14 Lock the two gears together (see Step 2) and tighten the primary drive gear bolt and the balancer gear nut to the torque settings specified at the beginning of this Chapter.
15 Refit the right-hand engine cover (see Section 15).
16 Install the timing inspection plug and the centre plug in the left-hand engine cover.

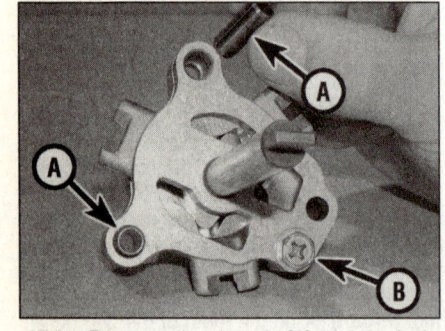

17.3a Remove the dowels (A) and the bolt (B) . . .

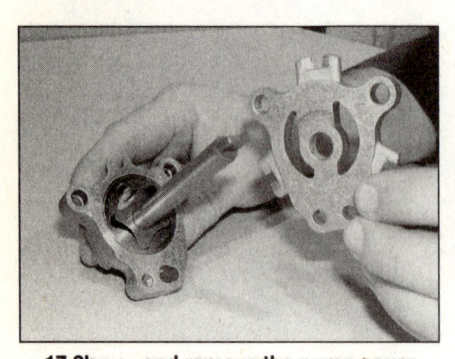

17.3b . . . and remove the pump cover

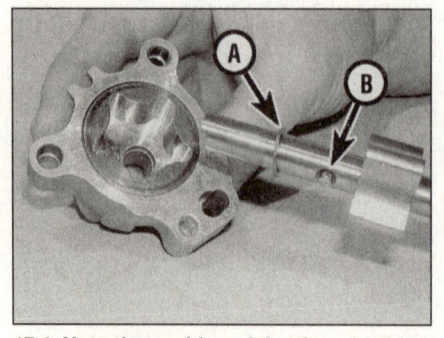

17.4 Note the position of the thrust washer (A) and the pin (B)

17 Oil pump – removal, inspection and installation

Note: *This procedure can be carried out with the engine in the frame. If the engine has been removed, ignore the steps which do not apply.*

Removal

1 Remove the clutch and the oil pump drive sprocket, pump sprocket and chain (see Section 15).
2 Unscrew the three bolts securing the pump to the crankcase, then remove the pump, noting how it fits **(see illustration)**.

Inspection

3 Remove the dowels from the pump body then unscrew the bolt securing the pump cover to the body and remove the cover **(see illustrations)**.
4 Withdraw the pump driveshaft along with the inner rotor. Retrieve the thrust washer from the pump body if it isn't on the shaft. Note how the pin locates through the pump shaft and in the notches in the inner rotor **(see illustration)**.

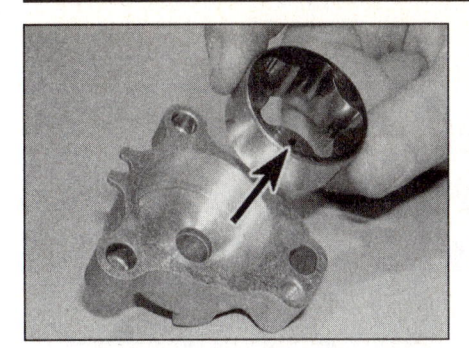

Engine, clutch and transmission 2•31

17.5 Withdraw the outer rotor, noting position of the punchmark (arrowed)

17.7 Look for scoring and wear, as on this outer rotor

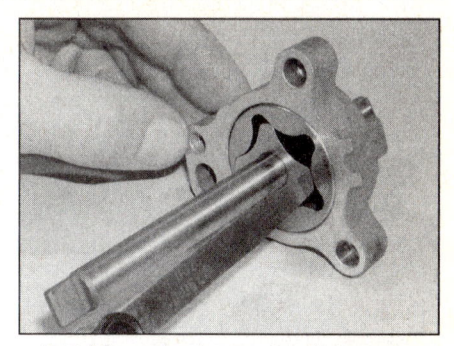

17.8 Measuring inner rotor tip-to-outer rotor clearance

5 Remove the outer rotor from the pump body **(see illustration)**. **Note:** *The outer rotor is punchmarked to ensure that it is fitted the correct way round. Honda specify that the punchmark should face the pump body but on the bike we stripped down the punchmark faced the pump cover. The important thing is to install the rotor the same way as it was when it was removed so that mated surfaces continue to run together.*

6 Clean all the components in solvent.

7 Inspect the pump body and rotors for scoring and wear **(see illustration)**. If any damage, scoring or uneven or excessive wear is evident, renew the pump (individual components are not available). If the engine is being rebuilt, it is advisable to fit a new oil pump as a matter of course.

8 Reassemble the pump components and measure the clearance between the inner rotor tip and the outer rotor with a feeler gauge and compare it that listed in the specifications at the beginning of this Chapter **(see illustration)**. If the clearance measured is greater than that listed, renew the pump.

9 Measure the clearance between the outer rotor and the pump body with a feeler gauge and compare it to that listed in the specifications at the beginning of this Chapter **(see illustration)**. If the clearance measured is greater than that listed, renew the pump.

10 Lay a straight-edge across the rotors and the pump body and, using a feeler gauge, measure the rotor end-float (the gap between the rotors and the straight-edge (see illustration). If the clearance measured is greater than that listed, renew the pump.

11 Check the pump drive chain and sprockets for wear or damage, and renew them as a set if necessary.

12 If the pump is good, make sure all the components are clean, then lubricate them with new engine oil.

13 Install the outer rotor into the pump body with the punchmark facing the same way as noted on removal (see Step 5 and **Note**).

14 Fit the drive pin into the pump driveshaft and slide the inner rotor onto the thin-tabbed end of the shaft, making sure the pin ends fit into the notches in the bottom of the rotor. Slide the thrust washer onto the thick-tabbed end of the shaft and against the bottom of the inner rotor. Fit the shaft and inner rotor into the pump body **(see illustration 17.4)**.

15 Fit the dowels into the pump body, install the cover and tighten the bolt to the specified torque setting **(see illustration)**. Ensure that the pump driveshaft rotates freely.

Installation

16 Slide the pump into position, turning the driveshaft until it engages correctly with the water pump. Ensure the dowels locate correctly in the crankcase. Fit the pump mounting bolts and tighten them securely **(see illustrations)**.

17 Install the oil pump drive sprocket, pump sprocket and chain and the clutch (see Section 15).

17.9 Measuring outer rotor-to-body clearance

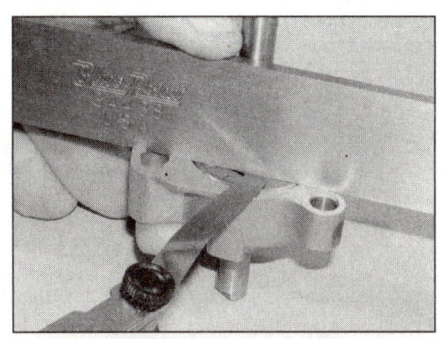

17.10 Measuring rotor end-float

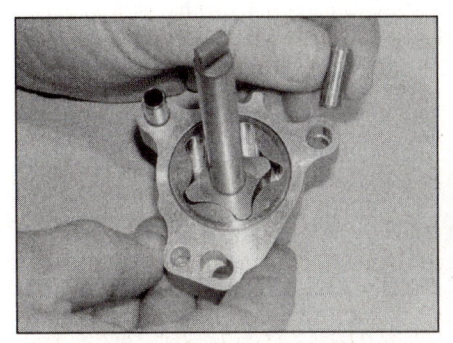

17.15 Install dowels before fitting cover

17.16a Engage flat on driveshaft (arrowed) with water pump . . .

17.16b . . . then install the bolts and tighten them

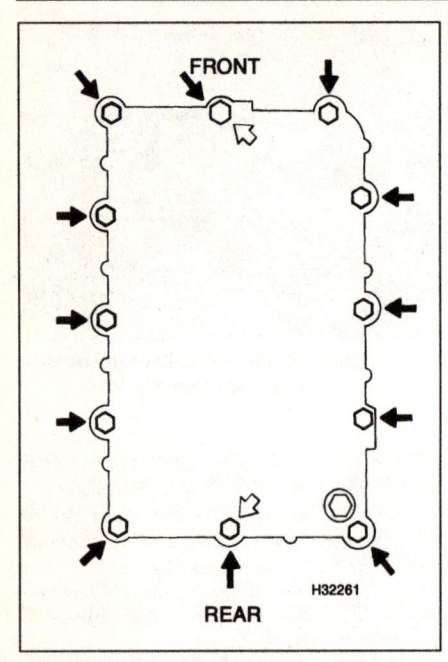

2•32 Engine, clutch and transmission

18.2a Sump bolts (arrowed) – note cast arrows for tightening procedure

18.2b Remove the sump and discard the gasket

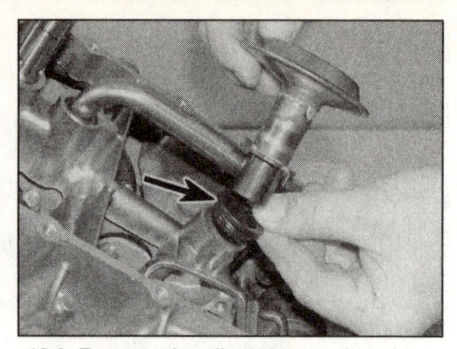

18.3 Remove the oil strainer and packing collar (arrowed)

18 Oil sump, oil strainer and pressure relief valve – removal, inspection and installation

Note: *This procedure can be carried out with the engine in the frame.*

Removal

1 Drain the engine oil (see Chapter 1).
2 Unscrew the sump bolts, slackening them evenly in a criss-cross sequence to prevent distortion, and remove the sump **(see illustrations)**. Discard the gasket as a new one must be used.
3 Pull the oil strainer out of its socket in the crankcase, noting the position of the locating tab in the crankcase casting **(see illustration)**. Remove the packing collar and discard it as a new one must be used.
4 Pull the oil pipe out of its sockets in the crankcase, noting which way round it fits – the short end faces towards the front of the engine, the long end towards the rear **(see illustration)**. Discard the O-rings as new ones must be used.
5 Pull the pressure relief valve out of its socket in the crankcase. Discard the O-ring as a new one must be used **(see illustration)**.

Inspection

6 Remove all traces of gasket from the sump and crankcase mating surfaces. Clean the inside of the sump with solvent and inspect the sockets and sealing surfaces for the strainer, oil pipe and pressure relief valve in the underside of the crankcase.
7 Clean the oil strainer in solvent and remove any debris caught in the mesh. Inspect the strainer for any signs of damage and renew it if necessary.
8 Clean the oil pipe in solvent. Inspect the pipe for any signs of damage and renew it if necessary.

9 Push the relief valve plunger into the valve body and check that it moves smoothly and freely against the spring pressure **(see illustration)**. If the valve operation is rough or sticky, remove the circlip **(see illustration)**, noting that it is under spring pressure, and withdraw the washer, spring and piston. Clean all the parts in solvent and inspect the plunger and bore for wear and damage **(see illustration)**. Apply oil to all parts and reassemble the valve. Check its operation again as above – if it is still rough or sticky it must be renewed – individual components are not available.

Installation

10 Fit a new O-ring onto the relief valve and smear it with clean oil, then push the valve into its socket in the crankcase **(see illustration 18.5)**.

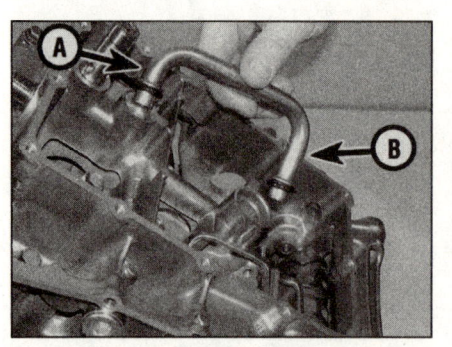

18.4 Oil pipe short end (A) faces front, long end (B) faces rear

18.5 Discard O-ring (arrowed) on relief valve

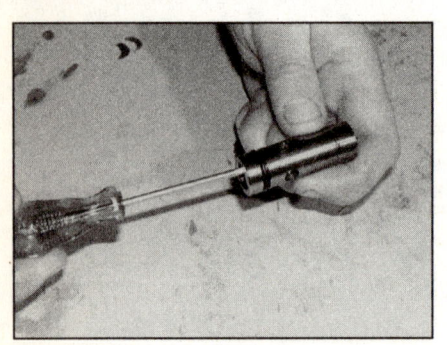

18.9a Check the operation of the valve as described

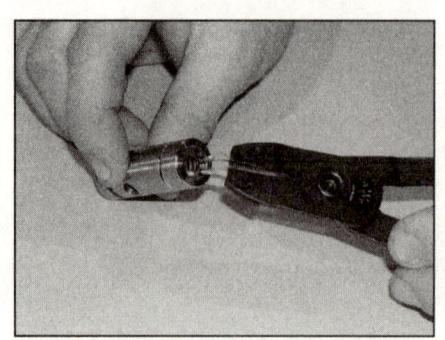

18.9b Remove circlip to dismantle valve

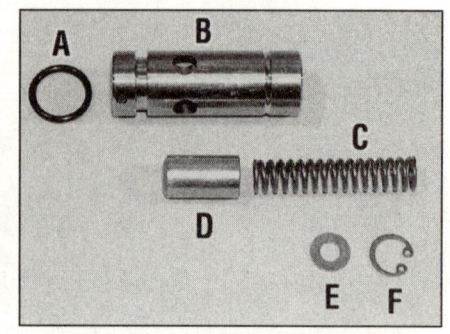

18.9c O-ring (A), body (B), spring (C), plunger (D), washer (E), circlip (F)

Engine, clutch and transmission 2•33

18.12 Align tab on strainer (arrowed) with notch in crankcase

18.13 Fit a new gasket on reassembly

19.2 Note the alignment of the punchmark (arrowed) before removing lever

11 Fit new O-rings onto the oil pipe and smear them with clean oil, then push the pipe into its sockets in the crankcase (see Step 4).
12 Fit a new packing collar onto the oil strainer and smear it with clean oil, then align the tab on the strainer with the notch in the crankcase and push the strainer into its socket **(see illustration)**.
13 Lay a new gasket onto the sump (if the engine is in the frame) or onto the crankcase (if the engine has been removed and is positioned upside down on the work surface) **(see illustration)**. Make sure the holes in the gasket align correctly with the bolt holes.
14 Position the sump onto the crankcase and install the bolts finger-tight. Tighten the bolts evenly in a criss-cross pattern, noting that the centre bolts at the front and rear, next to the cast arrows, should be tightened first **(see illustration 18.2a)**.

15 Install the oil drain plug with a new sealing washer.
16 Refill the engine with oil (see Chapter 1). Start the engine and check that there are no leaks around the sump.

19 Gearchange mechanism – removal, inspection and installation

Note: *The gearchange mechanism (external components) can be removed with the engine in the frame.*

Removal

1 Make sure the transmission is in neutral. Remove the clutch (see Section 15). There is no need to remove the oil pump drive chain and sprockets.

2 Unscrew the gearchange lever pinch bolt and remove the lever from the shaft, noting the alignment punchmark **(see illustration)**. If no mark is visible, make your own before removing the lever so that it can be correctly aligned with the shaft on installation.
3 Wrap some insulating tape around the gearchange shaft splines to avoid damaging the oil seal as it is removed.
4 On the right-hand side of the engine, note how the gearchange shaft return spring ends fit on each side of the locating pin in the casing, and how the shaft selector arm pawls engage with the pins on the selector cam **(see illustration)**. Withdraw the gearchange shaft and its thrust washer from the casing **(see illustration)**.
5 Note where the ends of the gearchange stopper arm return spring locate and how the roller on the arm rests in the neutral detent on the selector cam **(see illustration)**. Unscrew the bolt securing the stopper arm and withdraw the arm, thrust washer and spring from the casing **(see illustration)**.

Inspection

6 Inspect the stopper arm return spring and the gearchange shaft return spring. If they are fatigued, worn or damaged they must be renewed. The shaft return spring is retained

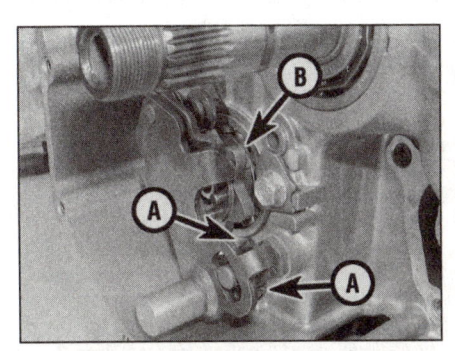

19.4a Ends of the gearchange shaft return spring (A) and gearchange cam (B)

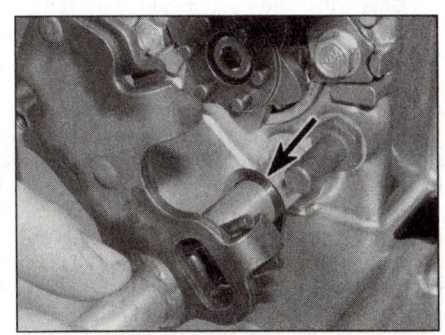

19.4b Gearchange shaft thrust washer (arrowed)

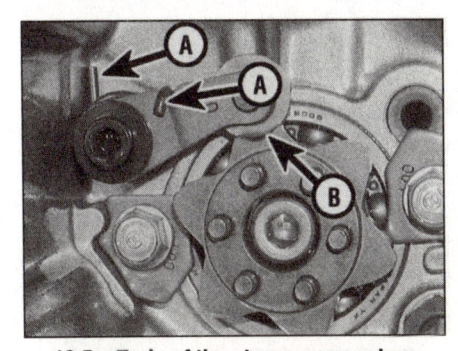

19.5a Ends of the stopper arm return spring (A) and neutral detent on cam (B)

19.5b Remove the gearchange stopper arm assembly . . .

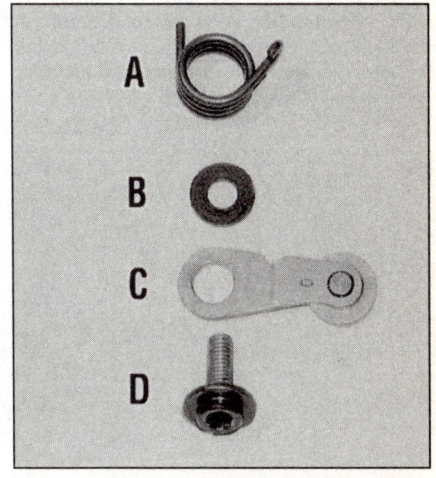

19.5c . . . and inspect the return spring (A), thrust washer (B), arm (C) and bolt (D)

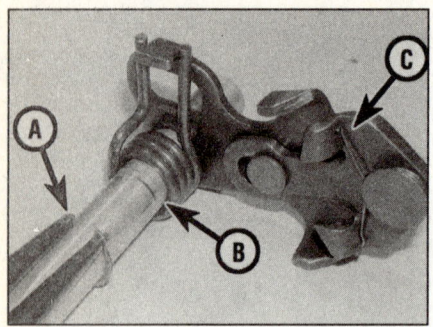

2•34 Engine, clutch and transmission

19.6 Gearchange shaft circlip (A), circlip groove (B) and pawl mechanism (C)

19.8a Prise the old oil seal out with a flat-bladed screwdriver . . .

19.8b . . . press the new seal in with a suitably sized socket

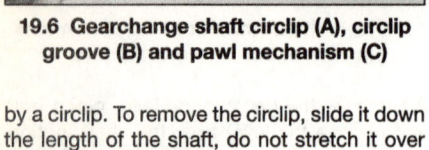

by a circlip. To remove the circlip, slide it down the length of the shaft, do not stretch it over the shaft **(see illustration)**.

7 Also check that the shaft return spring locating pin in the casing is securely tightened. If it is loose, remove it and apply a suitable non-permanent thread-locking compound, then tighten it to the torque specified at the beginning of this Chapter.

8 Check the gearchange shaft for straightness and damage to the splines. If the shaft is bent you can attempt to straighten it, but if the splines are damaged the shaft must be renewed. Also check the condition of the shaft oil seal in the casing and renew it if damaged or deteriorated. Lever the old seal out using a flat-bladed screwdriver and press the new seal in squarely with a suitably sized socket **(see illustrations)**.

9 Inspect the gearchange shaft selector arm for wear where it bears on the return spring. Also inspect the selector arm pawl mechanism and spring for wear or damage **(see illustration 19.6)**. The gearchange shaft and selector arm mechanism must be renewed as one assembly.

10 Check for wear on the selector cam pins and cam lobes, and the stopper arm and stopper arm roller. If they are worn or damaged they must be renewed.

11 Remove the gearchange cam by unscrewing the bolt in its centre. A recess in the back of the cam locates on a pin in the end of the selector drum. Remove the pin for safekeeping **(see illustration)**.

Installation

12 If removed, install the pin in the end of the selector drum. Install the selector cam, locating the pin in the recess in the back of the cam **(see illustration 19.11)**. Clean the threads of the centre bolt, then apply a suitable non-permanent thread-locking compound. Install the bolt and tighten it to the specified torque setting.

13 Fit the stopper arm assembly onto the shoulder on the fixing bolt and screw the bolt into the casing, making sure the ends of the return spring are located correctly over the stopper arm and against the casing **(see illustration 19.5a)**. Position the stopper arm roller in the neutral detent on the selector cam and ensure that the stopper arm is free to move and is returned by the pressure of the spring **(see illustration)**.

14 Check that the gearchange shaft return spring is properly positioned and that the circlip is in its groove. Slide the washer onto the shaft. Lightly grease the inside of the gearchange shaft oil seal in the left-hand side of the engine and slide the shaft into place from the right-hand side **(see illustration)**.

15 Locate the selector arm pawls onto the pins on the selector cam and the ends of the return spring onto each side of the locating pin **(see illustration 19.4a)**.

16 Align the gearchange lever with the punchmark on the gearchange shaft and refit the lever. Tighten the lever pinch bolt to the torque setting specified at the beginning of this Chapter.

17 Install the clutch (see Section 15).

20 Crankcase halves – separation and reassembly

Note 1: *To separate the crankcase halves, the engine must be removed from the frame.*
Note 2: *If the crankcases are being separated to inspect the crankshaft without removing it, or to inspect or remove the transmission shafts, the cylinder head can remain in place. However, if removal of the crankshaft or pistons and connecting rods is intended, full disassembly of the top-end is necessary. The gearchange mechanism external components can remain in place unless the selector drum and forks are being removed.*

Separation

1 To access the pistons, connecting rods, camchain and guides, crankshaft, balancer shaft, bearings, transmission shafts and the selector drum and forks, the crankcase must be split into two parts.

2 Remove the engine from the frame (see Section 5). Before the crankcase can be split the following components must be removed:
a) Valve cover (Section 7) (see **Note 2**).
b) Camshafts (Section 9) (see **Note 2**).
c) Cylinder head (Section 10) (see **Note 2**).
d) Starter motor (Chapter 9).

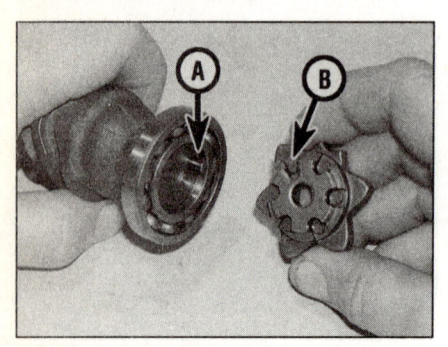

19.11 Cam locating pin (A) and locating recess (B)

19.13 Position the stopper arm roller in the neutral detent

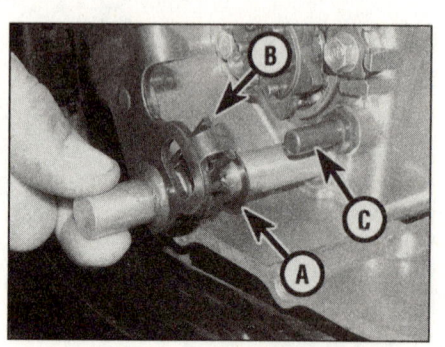

19.14 Install shaft with washer (A) and locate return spring (B) on pin (C)

Engine, clutch and transmission 2•35

e) *Alternator (Chapter 9).*
f) *Clutch (Section 15).*
g) *Oil pump (Section 17).*
h) *Water pump (Chapter 3).*
i) *Primary drive and balancer gear (Section 16).*
j) *Oil sump (Section 18).*
k) *Gearchange mechanism external components (Section 19)* (see **Note 2**).

3 Unscrew the five 6 mm upper crankcase bolts **(see illustration)**. Unscrew the bolts evenly, a little at a time in a criss-cross pattern until they are finger-tight, then remove them, noting their different lengths. **Note:** *As each bolt is removed, store it in its relative position in a cardboard template of the crankcase halves. This will ensure all bolts are installed in their correct locations on reassembly.* Sealing washers are fitted with bolts whose locations are identified by a triangle cast into the crankcase **(see illustration)**. Discard the washers as new ones should be fitted on reassembly.

4 Unscrew the two 8 mm upper crankcase bolts evenly and a little at a time until they are finger-tight, then remove them and place them in the template. Discard the sealing washers as new ones should be fitted on reassembly **(see illustration 20.3a)**.

5 Unscrew the single 10 mm upper crankcase bolt from the rear left-hand corner of the crankcase **(see illustration 20.3a)**.

6 Turn the engine upside down. **Note:** *If the cylinder head has been removed, rest the engine on a block of wood to avoid damaging the camchain and camchain guides.*

7 Unscrew the eight 6 mm lower crankcase bolts evenly, a little at a time in a criss-cross pattern, until they are finger-tight, then remove them **(see illustration)**. **Note:** *As each bolt is removed, store it in its relative position in a cardboard template of the crankcase halves. This will ensure all bolts are installed in their correct locations on reassembly.*

8 Unscrew the twelve 8 mm lower crankcase (crankshaft and balancer shaft journals) bolts **(see illustration 20.7)**. Unscrew the bolts evenly, a little at a time in a criss-cross pattern until they are finger-tight, then remove them, along with the washers and place them in the template. Discard the washers as new ones should be fitted on reassembly.

9 Carefully lift the lower crankcase half off the upper half, using a soft-faced hammer to tap around the joint to initially separate the halves if necessary. **Note:** *If the halves do not

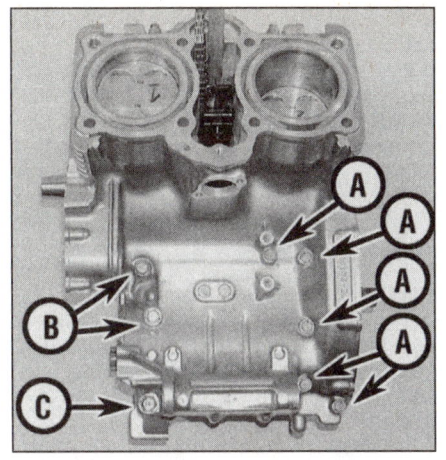

20.3a Upper crankcase 6 mm bolts (A), 8 mm bolts (B) and 10 mm (C) bolt

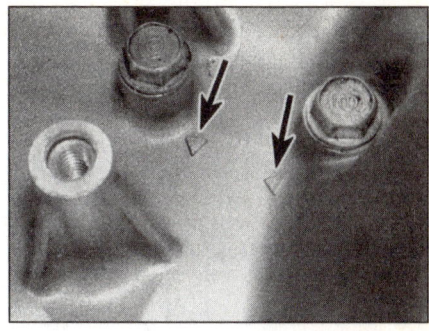

20.3b Triangle (arrowed) denotes use of sealing washer on adjacent bolt

separate easily, make sure all fasteners have been removed. Do not try and separate the halves by levering against the crankcase mating surfaces as they are easily scored and will leak oil on reassembly.* The lower crankcase half will come away with the gearchange mechanism external components (if not already removed) and the selector drum and forks, leaving the crankshaft and transmission shafts in the upper crankcase half. The lower main bearing shells and the lower balancer shaft bearing shells should also come away with the lower crankcase half. Remove them if they are loose, noting their locations (see Sections 25 and 26).

10 If required, remove the three locating dowels from the crankcase if they are loose (they could be in either crankcase half), noting their locations **(see illustration)**. Similarly remove the two oil nozzles, noting which way up they fit.

11 Refer to Sections 21 to 30 for the removal, inspection and installation of the components housed within the crankcases.

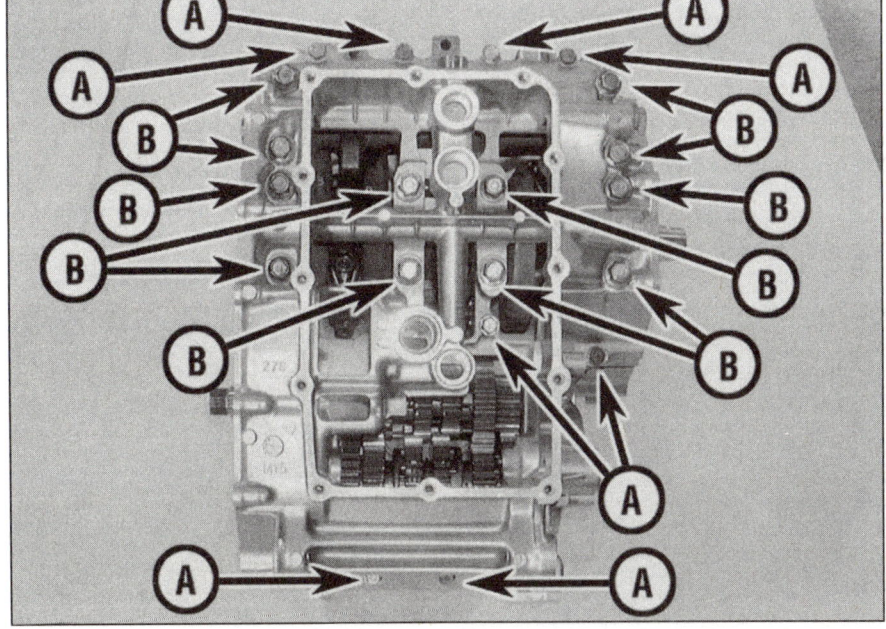

Reassembly

12 Remove all traces of sealant from the crankcase mating surfaces.

13 Ensure that all components and their bearings are in place in the upper and lower crankcase halves. If the transmission shafts have not been removed, check the condition of the output shaft oil seal on the left-hand end of the shaft and renew it if it is damaged or deteriorated (see Section 28); it is sound practice to renew this seal anyway. Apply

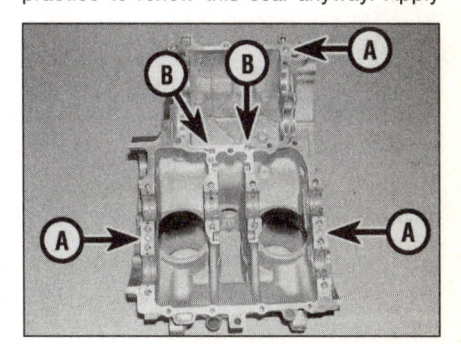

20.7 Lower crankcase 6 mm bolts (A) and 8 mm bolts (B)

20.10 Locating dowels (A) and oil nozzles (B)

2•36 Engine, clutch and transmission

20.13 Renew the oil seal (arrowed) if it is worn or shows signs of leaking

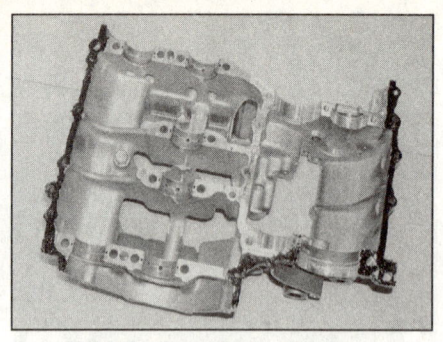

20.16 Apply sealant to the lower crankcase half where illustrated

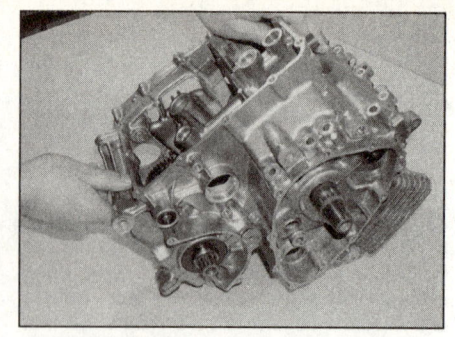

20.17 Carefully lower the lower crankcase half onto the upper crankcase half

some grease to the inside of the new seal on installation **(see illustration)**.
14 Generously lubricate the crankshaft and transmission shafts, particularly around the bearings, with clean engine oil, then use a rag soaked in high flash-point solvent to wipe over the mating surfaces of both crankcase halves to remove all traces of oil.
15 If removed, install the three locating dowels and the two oil nozzles in the upper crankcase half **(see illustration 20.10)**. The oil nozzles must be fitted with the wider aperture facing out, ie towards the crankcase lower half.
16 Apply a small amount of suitable sealant to the outer mating surface of the lower crankcase half **(see illustration)**.
Caution: Do not apply an excessive amount of sealant as it will ooze out when the case halves are assembled and may obstruct oil passages. Do not apply the sealant on or too close to any of the bearing inserts or surfaces.
17 Check again that all components are in position, particularly that the bearing shells are still correctly located in the lower crankcase half and that the camchain guides are correctly located in the upper crankcase half. Carefully install the lower crankcase half down onto the upper crankcase half, making sure each gear selector fork locates correctly in the groove in its pinion, that the guide pins on the forks locate correctly in the grooves in the selector drum and that the crankcase dowels all locate correctly into the lower crankcase half **(see illustration)**.
18 Check that the lower crankcase half is correctly seated. **Note:** *The crankcase halves should fit together without being forced. If the casings are not correctly seated, remove the lower crankcase half and investigate the problem. Do not attempt to pull them together using the crankcase bolts as the casing will crack and be ruined.*
19 Clean the threads of the twelve 8 mm lower crankcase (crankshaft and balancer shaft journals) bolts. Apply clean engine oil to the threads and seating surfaces of the bolts and install them with new washers in their original locations **(see illustration)**. Secure all bolts finger-tight at first, then tighten them evenly and a little at a time in a criss-cross pattern, starting from the centre and working outwards, to the torque setting specified at the beginning of this Chapter.
20 Clean the threads of the eight 6 mm lower crankcase bolts and insert them in their original locations. Secure all bolts finger-tight at first, then tighten them evenly a little at a time in a criss-cross pattern to the torque setting specified at the beginning of this Chapter **(see illustration)**.
21 Turn the engine over. Ensure the camchain is secured to prevent it dropping into the crankcase. Clean the threads of the single 10 mm upper crankcase bolt and insert it into the rear left-hand corner of the crankcase. Tighten the bolt to the specified torque **(see illustration)**.
22 Clean the threads of the two 8 mm upper crankcase bolts and insert them in their original locations with new sealing washers. Secure the bolts finger-tight at first, then tighten them evenly a little at a time to the torque setting specified at the beginning of this Chapter **(see illustration)**.
23 Clean the threads of the five 6 mm upper crankcase bolts and insert them in their original locations. Fit new sealing washers where the bolt locations are identified by a triangle cast into the crankcase. Secure the bolts finger-tight at first, then tighten them evenly a little at a time in a criss-cross pattern to the torque setting specified at

20.19 Install the 8 mm lower crankcase bolts first . . .

20.20 . . . then install the 6 mm bolts

20.21 Install the single 10 mm bolt in the upper crankcase

20.22 Install the 8 mm upper crankcase bolts . . .

Engine, clutch and transmission 2•37

20.23 . . . then install the 6 mm bolts

the beginning of this Chapter (see illustration).
24 With all crankcase fasteners tightened, check that the crankshaft and transmission shafts rotate freely. Check that the transmission shafts rotate independently in neutral, then turn the selector drum by hand and select each gear whilst rotating the input shaft. Check that all gears can be selected and that the shafts rotate freely in every gear. If there are any signs of undue stiffness, tight or rough spots, or of any other problem, the fault must be rectified before proceeding further.
25 Install all the removed assemblies in the reverse of the sequence given in Step 2.

21 Main and connecting rod bearings – general information

1 Even though main and connecting rod bearings are generally renewed during the engine overhaul, the old bearings should be retained for close examination as they may reveal valuable information about the condition of the engine.
2 Bearing failure occurs mainly because of lack of lubrication, the presence of dirt or other foreign particles, overloading the engine and/or corrosion. Regardless of the cause of bearing failure, it must be corrected before the engine is reassembled to prevent it from happening again.
3 When examining the bearings, match them with their corresponding journal on the crankshaft to help identify the cause of any problem.
4 Dirt and other foreign particles get into the engine in a variety of ways. They may be left in the engine during assembly or they may pass through filters or breathers. They may get into the oil and from there into the bearings. Metal chips from machining operations and normal engine wear are often present. Abrasives are sometimes left in engine components after reconditioning operations, especially when parts are not thoroughly cleaned using the proper cleaning methods. Whatever the source, these foreign objects often end up imbedded in the soft bearing material and are easily recognised. Large particles will not imbed in the bearing and will score or gouge the bearing and journal. The best prevention for this cause of bearing failure is to clean all parts thoroughly and keep everything spotlessly clean during engine reassembly. Frequent and regular oil and filter changes are also recommended.
5 Lack of lubrication or lubrication breakdown has a number of interrelated causes. Excessive heat (which thins the oil), overloading (which squeezes the oil from the bearing face) and oil leakage or throw off (from excessive bearing clearances, worn oil pump or high engine speeds) all contribute to lubrication breakdown. Blocked oil passages will starve a bearing and destroy it. When lack of lubrication is the cause of bearing failure, the bearing material is wiped or extruded from the steel backing of the bearing. Temperatures may increase to the point where the steel backing and the journal turn blue from overheating.

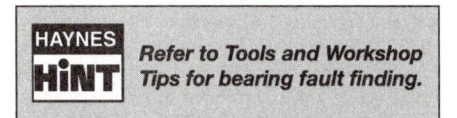

HAYNES HiNT *Refer to Tools and Workshop Tips for bearing fault finding.*

6 Riding habits can have a definite effect on bearing life. Full throttle low speed operation, or labouring the engine, puts very high loads on bearings, which tend to squeeze out the oil film. These loads cause the bearings to flex, which produces fine cracks in the bearing face (fatigue failure). Eventually the bearing material will loosen in pieces and tear away from the steel backing. Short trip riding leads to corrosion of bearings, as insufficient engine heat is produced to drive off the condensed water and corrosive gases produced. These products collect in the engine oil, forming acid and sludge. As the oil is carried to the engine bearings, the acid attacks and corrodes the bearing material.
7 Incorrect bearing installation during engine assembly will lead to bearing failure as well. Tight fitting bearings which leave insufficient bearing oil clearances result in oil starvation.

22.2 Measuring big-end side clearance

Dirt or foreign particles trapped behind a bearing insert result in high spots on the bearing which lead to failure.
8 To avoid bearing problems, clean all parts thoroughly before reassembly, double check all bearing clearance measurements and lubricate the new bearings with clean engine oil during installation.

22 Connecting rods and bearings – removal, inspection and installation

Note: *To remove the connecting rods, the engine must be removed from the frame.*

Removal

1 Remove the engine from the frame (see Section 5). Remove the cylinder head (see Section 10), and separate the crankcase halves (see Section 20). **Note:** *Once the cylinder head has been removed, rest the engine upside-down on a block of wood to avoid damaging the camchain and camchain guides, and to provide sufficient clearance for the pistons and connecting rods to be withdrawn from the cylinders.*
2 Before removing the rods from the crankshaft, measure the big-end side clearance with a feeler gauge (see illustration). If the clearance on any rod is greater than the service limit listed in this Chapter's Specifications, renew that rod.
3 Using paint or a felt marker pen, mark the cylinder identity on each piston, connecting rod and cap. Mark across the cap-to-connecting rod join and note which side of the rod faces the front of the engine to ensure that the cap and rod are fitted the correct way around on reassembly. The oil hole in the big-end of each connecting rod should face the back (inlet side) of the engine. **Note:** *The number already across the rod and cap indicates rod size, not cylinder number* (see illustration).
4 Unscrew the big-end cap nuts and separate the caps, complete with the lower bearing

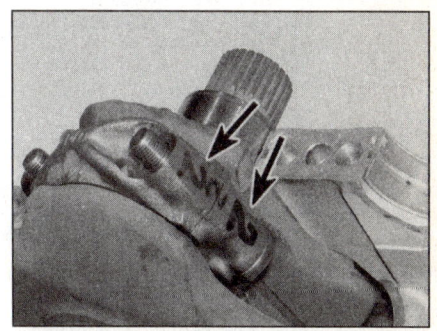

22.3 Mark the connecting rods and big-end caps with the cylinder number

2•38 Engine, clutch and transmission

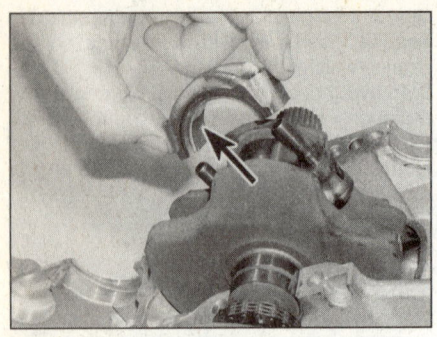

22.4 Detach the big-end caps with the lower bearing shells (arrowed)

22.5a Remove the piston and connecting rod from the top end of the cylinder

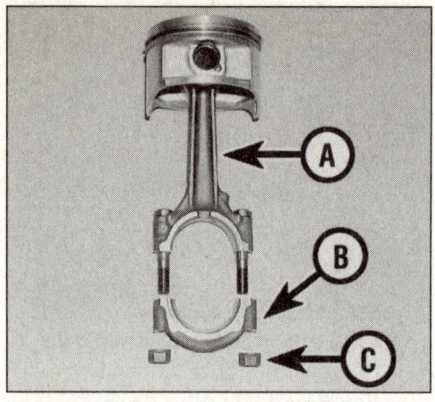

22.5b Keep the piston and rod (A), big-end cap (B) and cap nuts (C) together as an assembly

shells, from the crankpins (see illustration), then detach the connecting rods, complete with the upper bearing shells, from the crankpins.

5 Ensure the camchain is free to be lifted out of the engine and lift the crankshaft with the camchain attached out of the upper crankcase half, taking care not to dislodge the main bearing shells. Push each piston/connecting rod assembly down the cylinder and remove it, making sure the connecting rod does not mark the cylinder bore **(see illustration)**. To ease withdrawal of the piston/connecting rod assembly, remove any ridge of carbon built up on the top of each cylinder bore. If there is a pronounced wear ridge, remove it using a ridge reamer.

Caution: Do not try to remove the piston/ connecting rod assemblies from the bottom of the cylinder bore. The piston will not pass the crankcase main bearing webs. If the piston is pulled right to the bottom of the bore the oil control ring will expand and lock the piston in position. If this happens it is likely the ring will be broken.

Install the relevant bearing shells (if removed), bearing cap and nuts on each connecting rod assembly so that they are all kept together as a matched set **(see illustration)**.

6 Do not remove the bolts from the connecting rods. **Note:** *It is not necessary to renew the big-end bolts when the connecting rods are removed. If, for any reason, the bolts have to be renewed, drive the old bolts out of the rods by tapping gently with a hammer. Under no circumstances twist the bolts while they are in the rods. Because the bolts have knurled shanks and are a tight fit in the rods, take care to align the bolt heads with the machined insets in the rod shoulders before pressing* them into place **(see illustrations)**.

7 Remove the pistons from the connecting rods (see Section 23).

Inspection

8 Check the connecting rods for cracks and other obvious damage. Check that the oil hole in the connecting rod big-end is clear **(see illustration)**.

9 Apply clean engine oil to the piston pin, insert it into the connecting rod small-end and check for any freeplay between the two **(see illustration)**. Measure the pin external diameter and the small-end bore diameter, then calculate the difference to obtain the small-end-to-piston pin clearance **(see illustrations)**. Compare the result to the

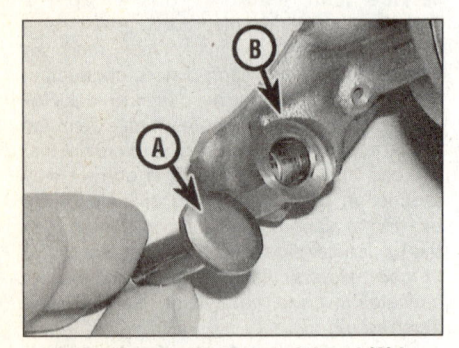

22.6a Ensure the new bolt head (A) is correctly aligned with the inset (B) . . .

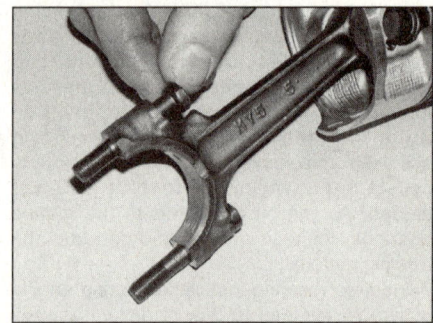

22.6b . . . before pressing the bolt into place

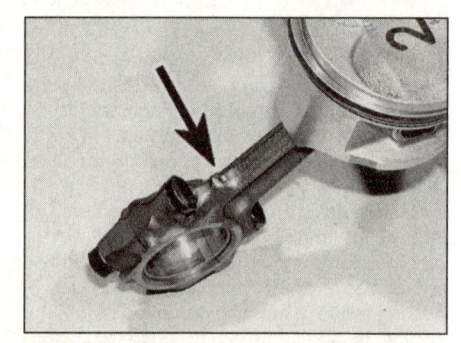

22.8 Ensure oil hole (arrowed) is clear

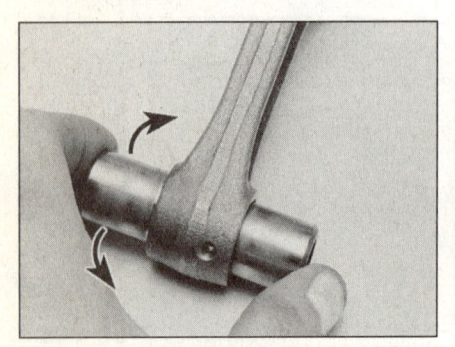

22.9a Rock the piston pin back and forth to check for freeplay

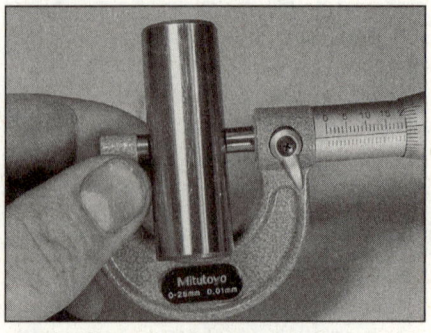

22.9b Measuring the external diameter of the pin . . .

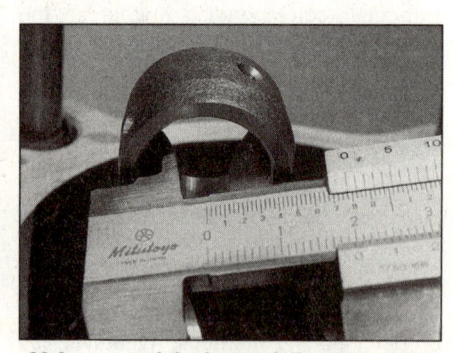

22.9c . . . and the internal diameter of the small-end

Engine, clutch and transmission 2•39

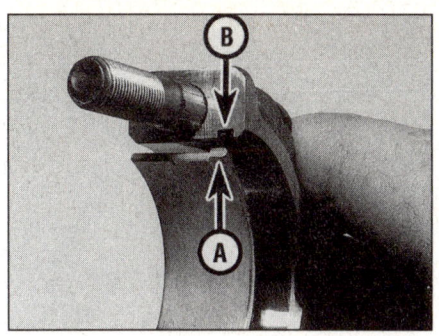

22.13 Make sure the tab (A) locates in the notch (B)

22.20a Crankpin journal size code (arrowed)

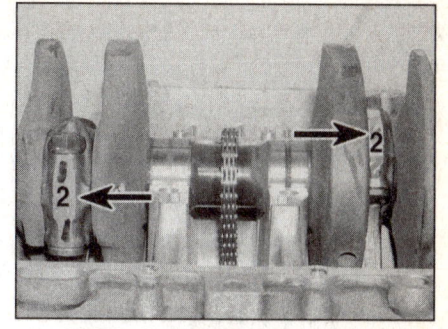

22.20b Connecting rod size code (arrowed)

specifications at the beginning of this Chapter. If the clearance is greater than specified, renew the components that are worn beyond their specified limits Connecting rods and caps are supplied in paired sets. Fit new big-end bolts to a new rod, taking care to align the bolt heads with the machined insets in the rod shoulders (see Step 6).

10 Refer to Section 21 and examine the connecting rod bearing shells. If they are scored, badly scuffed or appear to have seized, new shells must be installed. Always renew the shells in the connecting rods as a set. If they are badly damaged, check the corresponding crankpin. Evidence of extreme heat, such as discoloration, indicates that lubrication failure has occurred. Be sure to thoroughly check the oil pump and pressure relief valve as well as all oil holes and passages before reassembling the engine.

11 Have the rods checked by a Honda dealer if you are in doubt about their straightness.

Oil clearance check

12 The connecting rod bearing oil clearance should be checked before the engine is reassembled, even if the bearing shells and the crankpin journals show no apparent sign of wear. Bearing oil clearance is measured with a product called Plastigauge.

13 Clean the backs of the bearing shells, the bearing locations in both the connecting rod and cap, and the crankpin journals with a suitable solvent or brake system cleaner.

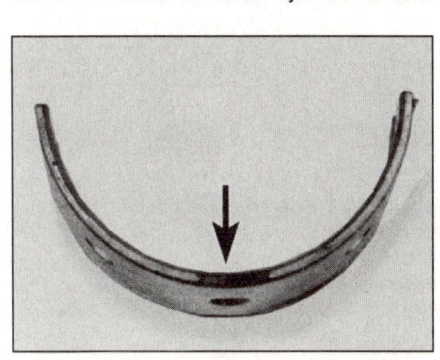

22.21 Bearing shell colour code location (arrowed)

Press the bearing shells into their locations, ensuring that the tab on each shell engages the notch in the connecting rod/cap **(see illustration)**. Clean the bearing surfaces with solvent, taking care not to touch any bearing surface with your fingers.

14 Cut an appropriate size length of Plastigauge (it should be slightly shorter than the width of the crankpin). Place a strand of Plastigauge on a cleaned crankpin journal. Do not place Plastigauge over the oil holes in the crankpins. Fit the cleaned connecting rod, shells and cap. Make sure the cap is fitted the correct way round so that the previously made markings align and that the rod is facing the right way (see Step 3). Tighten the cap nuts evenly, in two or three stages, to the torque setting specified at the beginning of this Chapter, whilst ensuring that the connecting rod does not rotate on the crankshaft. Slacken the cap nuts and remove the connecting rod, again taking great care not to rotate the crankshaft.

15 Compare the width of the crushed Plastigauge on the crankpin journal to the scale printed on the Plastigauge envelope to obtain the connecting rod bearing oil clearance **(see illustration 25.17)**. Compare the reading to the specifications at the beginning of this Chapter. If the oil clearance falls within the specified range and the bearings are in perfect condition, they can be reused.

16 Carefully clean away all traces of the Plastigauge from the crankpin journal and bearing shells using a fingernail or other object which will not score the bearing surfaces.

17 If the clearance is beyond the service limit, measure the diameter of the crankpin journal with a micrometer and compare the result with the appropriate figure in the table for connecting rod bearing shell selection. For example, if the journal being measured is code B, the table indicates that the service limit for that journal is 35.988 mm. If the journal diameter is larger than the service limit, new bearing shells can be fitted (see Steps 20 and 21). If the journal diameter is smaller than the service limit, the crankshaft should be renewed, although seek the advice of a Honda dealer or engineer to explore all alternatives.

18 Repeat the oil clearance check for the remaining connecting rod. Always renew all of the shells (on both connecting rods) at the same time.

19 Install the new shells and check the oil clearance once again.

Bearing shell selection

20 New bearing shells for the connecting rods are supplied on a selected fit basis. Code letters and numbers marked on the crankshaft webs and the connecting rods are used to identify the correct new bearings. The crankpin journal size codes are marked on the inner crankshaft webs and are either an A, B or C **(see illustration)**. The connecting rod size code is marked on the flat face of the connecting rod and cap and will be either a 1, 2 or 3 **(see illustration)**.

21 A corresponding range of bearing shells is available. To select the correct shells for a particular big-end, using the table below cross-refer the crankpin journal code letter with the connecting rod code number to determine the colour code of the shells required. For example, if the connecting rod size is 2, and the crankpin size is A, then the bearing shells required are Green. The colour is marked on the side of the shells **(see illustration)**.

Connecting rod code	Crankpin code		
	A – (35.994 to 36.000 mm)	B – (35.988 to 35.994 mm)	C – (35.982 to 35.988 mm)
1 – (39.000 to 39.006 mm)	Yellow	Green	Brown
2 – (39.006 to 39.012 mm)	Green	Brown	Black
3 – (39.012 to 39.018 mm)	Brown	Black	Blue

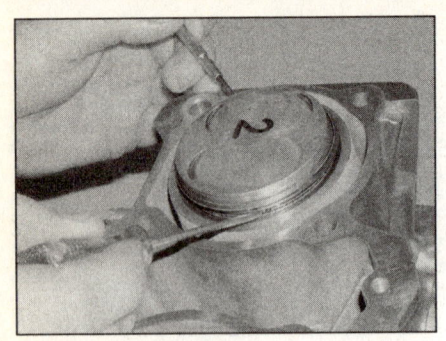

2•40 Engine, clutch and transmission

22.25 Feed the rings into the bore with care

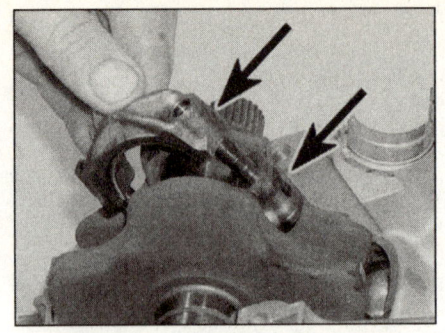

22.27 Fit the cap and shell onto the corresponding rod, aligning the markings (arrowed)

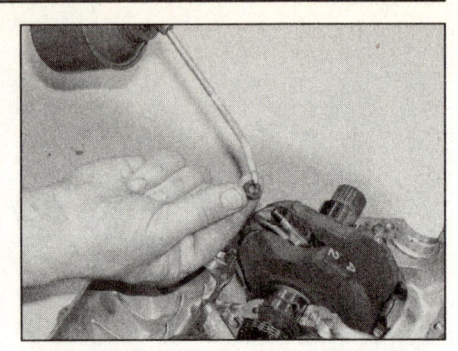

22.28 Lubricate the cap nuts before fitting

Installation

22 Install the pistons onto the connecting rods (see Section 23).

23 Clean the backs of the bearing shells and the bearing seats in the connecting rods and caps. If new shells are being fitted, ensure that all traces of protective grease are cleaned off using paraffin (kerosene). Wipe the shells and the bearing seats dry with a lint-free cloth. Make sure all the oil passages and holes are clear, and blow them through with compressed air if available.

24 Press the bearing shells into their seats, aligning the tab in the bearing with the groove in the rod or cap **(see illustration 22.13)**. Ensure the shells are seated correctly and take care not to touch any shell's bearing surface with your fingers.

25 Lubricate the pistons, rings and cylinder bore with clean engine oil. Stagger the piston ring end gaps (see Section 23) and insert the piston/connecting rod assembly into the top of its bore (from the underside with the engine upside down on the work surface), taking care not to allow the connecting rod to mark the bore. Make sure the 'IN' mark on the piston crown and the oil hole in the big-end are on the inlet side of the bore, then carefully compress and feed each piston ring into the bore until the piston crown is flush with the top of the bore **(see illustration)**. If available, a piston ring compressor makes installation a lot easier.

26 Ensure that the connecting rod bearing shell is still correctly installed. Liberally lubricate the crankpin with a 50/50 mixture of molybdenum disulphide grease and clean engine oil. Fit the connecting rod onto the crankpin.

27 Fit the bearing cap with its shell onto the connecting rod **(see illustration)**. Make sure the cap is fitted the correct way round so the connecting rod and bearing cap markings are aligned.

28 Lubricate the threads and seats of the cap nuts with clean engine oil **(see illustration)**. Fit the nuts to the connecting rod bolts and tighten them evenly, in two or three stages, to the specified torque setting.

29 Check that the crankshaft is free to rotate easily. Check to make sure that all components have been returned to their original locations using the marks made on disassembly.

30 Check that the rods rotate smoothly on the crankpin. If there are any signs of roughness or tightness, remove the rods and re-check the bearing clearance. Sometimes tapping the bottom of the connecting rod cap will relieve tightness, but if in doubt, recheck the clearances.

31 Reassemble the crankcase halves (see Section 20).

23 Pistons – removal, inspection and installation

Note: *To remove the pistons the engine must be removed from the frame.*

Removal

1 Remove the engine from the frame (see Section 5). Remove the cylinder head (see Section 10), and separate the crankcase halves (see Section 20). **Note:** *Once the cylinder head has been removed, rest the engine upside-down on a block of wood to avoid damaging the camchain and camchain guides, and to provide sufficient clearance for the pistons and connecting rods to be withdrawn from the cylinders.*

2 Remove the piston/connecting rod assemblies (see Section 22).

3 Before removing the piston from the connecting rod, use a sharp scriber or felt marker pen to write the cylinder identity on the piston crown, or on the inside of the skirt if the piston is dirty and going to be cleaned. Each piston should also have an 'IN' mark on its crown with corresponding large valve cutaways which should face the inlet side of the engine **(see illustration)**. If the 'IN' mark is not visible, mark the piston accordingly so that it can be installed the correct way round.

4 Carefully prise out the circlip on one side of the piston using needle-nose pliers or a small flat-bladed screwdriver inserted into the notch **(see illustration)**. Push the piston pin out from the other side to free the piston from the connecting rod **(see illustration)**. Remove the other circlip and discard them as new ones must be used. When the piston has been removed, install its pin back into its bore so that related parts do not get mixed up.

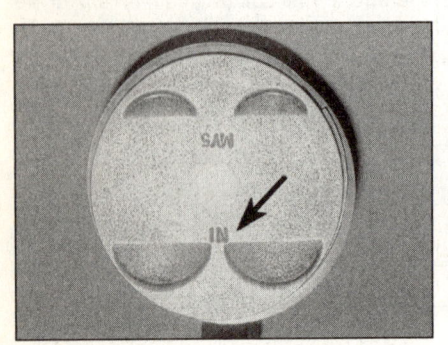

23.3 Note the IN mark which faces the inlet side of the cylinder

23.4a Prise out the circlip . . .

23.4b . . . then push out the pin and remove the piston

Engine, clutch and transmission 2•41

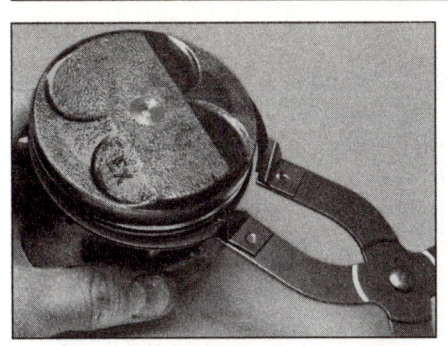

23.6a Using a ring removal and installation tool

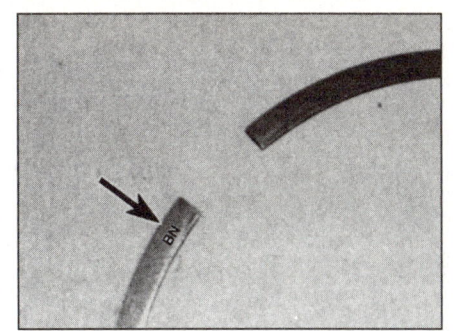

23.6b Piston ring reference (arrowed)

23.12 Measuring the piston ring-to-groove clearance with a feeler gauge

HAYNES HiNT *If a piston pin is a tight fit in the piston bosses, soak a rag in boiling water then wring it out and wrap it around the piston – this will expand the alloy piston sufficiently to release its grip on the pin. If the piston pin is particularly stubborn, extract it using a drawbolt tool, but be careful to protect the piston's working surfaces.*

Inspection

5 Before the inspection process can be carried out, the pistons must be cleaned and the old piston rings removed.

6 Using your thumbs or a piston ring removal and installation tool, carefully remove the rings from the pistons **(see illustration)**. Do not nick or gouge the pistons in the process. Note which way up each ring fits and in which groove. The top and second compression rings have different profiles and they must be installed in their original positions if being re-used. The upper surface of each ring is usually marked at one end. This information will also be required if you are fitting new rings **(see illustration)**.

7 Scrape all traces of carbon from the tops of the pistons. A hand-held wire brush or a piece of fine emery cloth can be used once most of the deposits have been scraped away. Do not, under any circumstances, use a wire brush mounted in a drill motor to remove deposits from the pistons; the piston material is soft and will be eroded away by the wire brush.

8 Use a piston ring groove cleaning tool to remove any carbon deposits from the ring grooves. If a tool is not available, a piece broken off an old ring will do the job. Be very careful to remove only the carbon deposits. Do not remove any metal and do not nick or gouge the sides of the ring grooves.

9 Once the deposits have been removed, clean the pistons with solvent and dry them thoroughly. If the identification previously marked on the piston is cleaned off, be sure to re-mark it with the correct identity. Make sure the oil return holes below the oil ring groove are clear.

10 Carefully inspect each piston for cracks around the skirt, at the pin bosses and at the ring lands. Normal piston wear appears as even, vertical wear on the thrust surfaces of the piston and slight looseness of the top ring in its groove. If the skirt is scored or scuffed, the engine may have been suffering from overheating and/or abnormal combustion, which caused excessively high operating temperatures. The oil pump should be checked thoroughly. Also check that the circlip grooves are not damaged.

11 A hole in the piston crown, an extreme to be sure, is an indication that abnormal combustion (pre-ignition) was occurring. Burned areas at the edge of the piston crown are usually evidence of spark knock (detonation). If any of the above problems exist, the causes must be corrected or the damage will occur again.

12 Measure the piston ring-to-groove clearance by laying each piston ring in its groove and slipping a feeler gauge in beside it **(see illustration)**. Make sure you have the correct ring for the groove. Check the clearance at three or four locations around the groove. If the clearance is greater than specified, renew both the piston and rings as a set. If new rings are being used, measure the clearance using the new rings. If the clearance is greater than that specified, the piston is worn and must be renewed.

13 Check the piston-to-bore clearance by measuring the bore (see Section 31) and the piston diameter. Make sure each piston is matched to its correct cylinder. Measure the piston 15 mm up from the bottom of the skirt and at 90° to the piston pin axis **(see illustration)**. Subtract the piston diameter from the bore diameter to obtain the clearance. If it is greater than the specified figure, the piston must be renewed (assuming the bore itself is within limits). Oversize pistons are available in +0.25mm and +0.50mm sizes. Consult a Honda dealer for details of a reboring service.

14 Apply clean engine oil to the piston pin, insert it into the piston and check for any freeplay between the two **(see illustration)**. Measure the pin external diameter and the pin bore in the piston **(see illustration)**. Calculate the difference to obtain the piston pin-to-piston pin bore clearance. Compare the result to the specifications at the beginning of this Chapter. If the clearance is greater than

23.13 Measuring the piston diameter at the specified distance from the bottom of the skirt

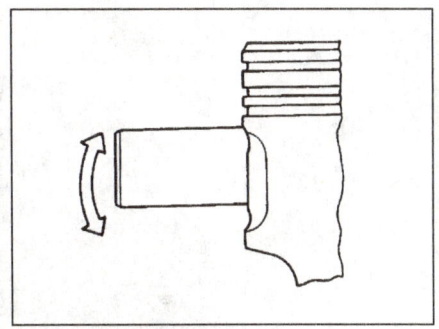

23.14a Rock the piston pin back and forth to check for freeplay

23.14b Measuring the pin bore in the piston

2•42 Engine, clutch and transmission

specified, renew the components that are worn beyond their specified limits. Check the measurements between the pin and the connecting rod small-end (see Section 22).

Installation

15 Inspect and install the piston rings (see Section 24).
16 Lubricate the piston pin, the piston pin bore and the connecting rod small-end bore with a 50/50 mixture of molybdenum disulphide grease and clean engine oil.
17 When installing the pistons onto the connecting rods, make sure that the 'IN' mark is on the same side as the oil hole in the connecting rod. Install a new circlip in one side of the piston (do not re-use old circlips). Line up the piston on its correct connecting rod, and insert the piston pin from the other side. Secure the pin with the other new circlip. When installing the circlips, compress them only just enough to fit them in the piston, and make sure they are properly seated in their grooves with the open end away from the removal notch (see illustration).
18 Install the piston/connecting rod assemblies (see Section 22). Reassemble the crankcase halves (see Section 20), install the cylinder head (see Section 10), then fit engine into the frame (see Section 5).

24 Piston rings – inspection and installation

1 It is good practice to renew the piston rings when an engine is being overhauled. Measure the pistons carefully to ensure you get new rings of the correct size.
2 Before installing the new piston rings, the ring end gaps must be checked with the rings installed in the cylinder. Lay out the pistons and the new ring sets so the rings will be matched with the same piston and cylinder during the end gap measurement procedure and engine assembly. Identify the top and second compression rings (see illustration 24.13).
3 To measure the installed ring end gap, insert the top ring into the cylinder and square it up with the cylinder walls by pushing it in with the top of the piston. The ring should be about 20 mm from the bottom of the cylinder. To measure the end gap, slip a feeler gauge between the ends of the ring and compare the measurement to the specifications at the beginning of this Chapter (see illustration).
4 If the gap is larger or smaller than specified, double check to make sure that you have the correct rings before proceeding.
5 If the gap is too small, it must be enlarged or the ring ends may come in contact with each other during engine operation, which can cause serious damage. The end gap can be increased by filing the ring ends very carefully with a fine file. When performing this operation, file only from the outside in (see illustration).
6 Excess end gap is not critical unless it exceeds the service limit. Again, make sure you have the correct rings for your engine and check that the cylinder is not worn. Cylinder wear will be apparent by measuring the ring gap near the top of the bore and comparing it with the measurement from the bottom of the bore (see Section 31).
7 Repeat the procedure for each ring that will be installed in the cylinder. Remember to keep the rings, pistons and cylinders matched up.
8 Once the ring end gaps have been checked/corrected, the rings can be installed on the pistons.
9 The oil control ring (lowest on the piston) is installed first. It is composed of three separate components, namely the expander and the upper and lower side rails. Slip the expander into the groove, then install the lower side rail. Do not use a piston ring installation tool on the oil ring side rails as they may be damaged. Instead, place one end of the side rail into the groove between the expander and the ring land. Hold it firmly in place and slide a finger or thin blade around the piston while pushing the rail into the groove. Next, install the upper side rail in the same manner (see illustrations). Make sure the ends of the expander do not overlap.

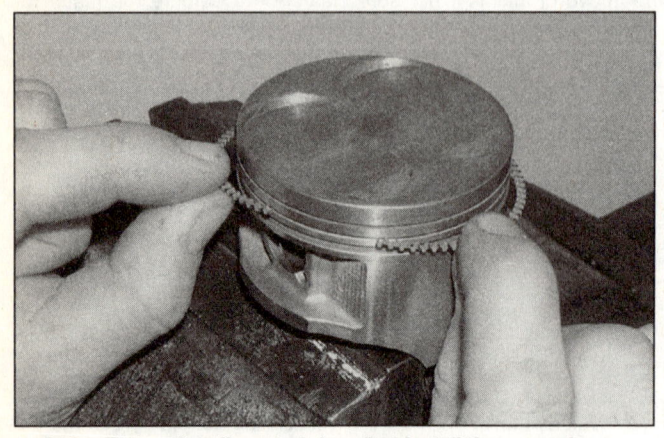

23.17 Ensure the circlips are correctly seated in their grooves

24.3 Measuring installed ring end gap

24.5 Enlarging ring end gap with the file clamped in a vice

24.9a Fitting the oil control ring, first install the expander ...

24.9b ... then the lower and upper rails with the aid of a thin blade

24.11 Check that the compression rings are the right way up before fitting

10 After the three oil ring components have been installed, check to make sure that both the upper and lower side rails can be turned smoothly in the ring groove.

11 The upper surface of each compression ring should be marked at one end near the gap **(see illustration 23.6b)**. Make sure that the identification mark or letter is facing up. Fit the middle ring (2nd compression ring) into the middle groove in the piston. Do not expand the ring any more than is necessary to slide it into place. To avoid breaking the ring, use a piston ring installation tool or the method described in Step 9 **(see illustration)**.

12 Finally, install the top ring in the same manner into the top groove in the piston. Make sure the identification letter near the end gap is facing up.

13 Once the rings are correctly installed, check they move freely without snagging and stagger their end gaps as shown **(see illustration)**.

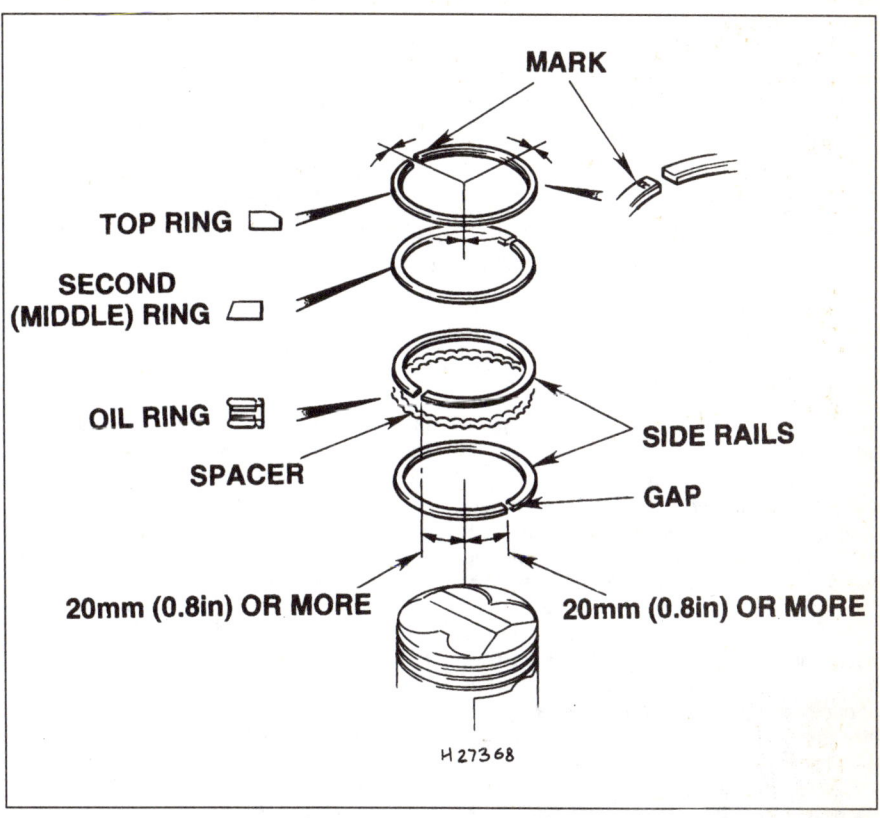

24.13 Stagger the ring end gaps as shown

25 Crankshaft and main bearings – removal, inspection and installation

Note: *To remove the crankshaft the engine must be removed from the frame.*

Removal

1 Remove the engine from the frame (see Section 5). Remove the cylinder head (see Section 10) and separate the crankcase halves (see Section 20).

2 Separate the connecting rods from the crankshaft (see Section 22). **Note:** *If no work is to be carried out on the piston/connecting rod assemblies there is no need to remove them from the cylinders. Remove the connecting rod bearing caps (see Section 22, Steps 3 and 4) and push the pistons up the bores so that the connecting rod ends are clear of the crankshaft. Wrap clean rag around the connecting rods to prevent damage to the cylinder bores.*

3 Ensure the camchain is free to be lifted out of the engine and lift the crankshaft with the camchain attached out of the upper crankcase half, taking care not to dislodge the main bearing shells **(see illustration)**. Remove the camchain from the crankshaft.

Caution: With the crankshaft removed, do not try to remove the piston/connecting rod assemblies from the bottom of the cylinder bore. The piston will not pass the crankcase main bearing webs. If the piston is pulled right to the bottom of the bore the oil control ring will expand and lock the piston in position. If this happens it is likely the ring will be broken.

4 The main bearing shells can be removed from the crankcase halves by pushing their centres to the side, then lifting them out **(see illustration)**. Keep the shells in order.

Inspection

5 Clean the crankshaft with solvent, using a rifle-cleaning brush to scrub out the oil passages. If available, blow the crank dry with compressed air, and also blow through the oil passages. Check the camchain sprocket for wear or damage. If any of the teeth are excessively worn, chipped or broken, the crankshaft must be renewed. If wear or damage is found, also check the camchain, camchain guides and camshaft sprockets (see Section 27).

6 Inspect the splines on the right-hand end of the shaft and ensure that the primary drive gear is a good fit on the splines. Inspect the taper and the slot in the left-hand end of the shaft for the alternator Woodruff key. Damage or wear that prevents either the primary drive gear or the alternator from being fitted securely

25.3 Lift the crankshaft and camchain out together

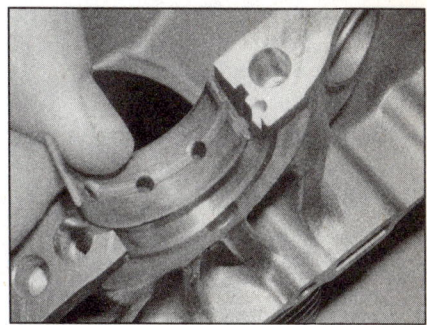

25.4 Push the shells out, don't lever them

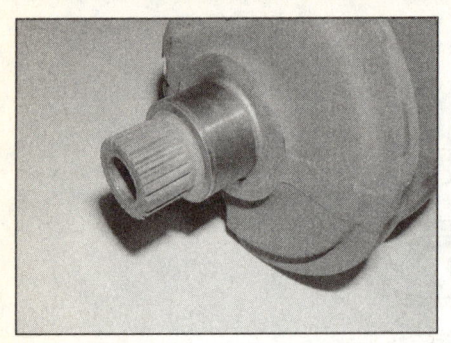

2•44 Engine, clutch and transmission

25.6a Inspect the primary gear splines ...

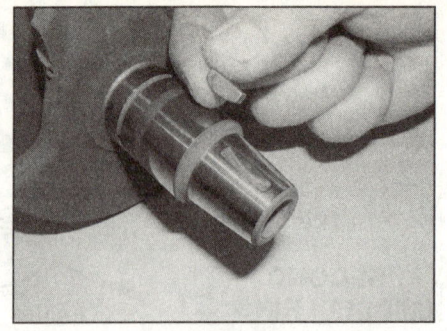

25.6b ... and the alternator taper and Woodruff key

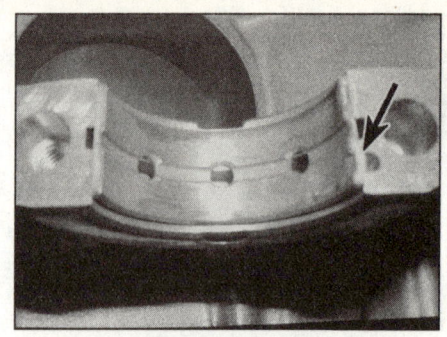

25.12 Fit the shells locating the tab (arrowed) in the notch

will require a new crankshaft **(see illustrations)**.

7 Refer to Section 21 and examine the main bearing shells. If they are scored, badly scuffed or appear to have been seized, new shells must be installed. Always renew the main bearing shells as a set. If they are badly damaged, check the corresponding crankshaft journals. Evidence of extreme heat, such as discoloration, indicates that lubrication failure has occurred. Be sure to thoroughly check the oil pump and pressure relief valve as well as all oil holes and passages before reassembling the engine.

8 Give the crankshaft journals a close visual examination, paying particular attention where damaged bearings have been discovered. If the journals are scored or pitted in any way a new crankshaft will be required. Undersize main bearing shells are not available, precluding the option of re-grinding the crankshaft.

9 Place the crankshaft on V-blocks and check the runout at the main bearing journals using a dial gauge. Compare the reading to the maximum specified at the beginning of this Chapter. If the runout exceeds the limit, the crankshaft must be renewed.

Oil clearance check

10 The crankshaft main bearing oil clearances should be checked before the engine is reassembled even if the bearing shells and the crankshaft journals show no apparent sign of wear. Bearing oil clearance is measured with a product called Plastigauge.

11 Clean the backs of the bearing shells, the bearing housings in both crankcase halves and the crankshaft journals with a suitable solvent or brake system cleaner.

12 Press the bearing shells into their locations, ensuring that the tab on each shell engages in the notch in the crankcase **(see illustration)**. Clean the bearing surfaces with solvent, taking care not to touch any bearing surface with your fingers.

13 Lay the crankshaft in position in the upper crankcase **(see illustration)**.

14 Cut four appropriate size lengths of Plastigauge (they should be slightly shorter than the width of the crankshaft journals). Place a strand of Plastigauge on each cleaned journal **(see illustration)**. Do not place Plastigauge over the oil holes in the crankshaft. Make sure the crankshaft is not rotated.

15 If removed, install the three crankcase locating dowels, then carefully install the lower crankcase half on to the upper half. Check that the lower crankcase half is correctly seated. **Note:** *Do not tighten the crankcase bolts if the casing is not correctly seated.* Clean the threads of the eight 8 mm lower crankcase (crankshaft journal) bolts and insert them with their washers in their original locations **(see illustration)**. Secure all bolts finger-tight at first, then tighten them evenly and a little at a time in a criss-cross pattern to the torque setting specified at the beginning of this Chapter. Make sure that the crankshaft is not rotated as the bolts are tightened.

16 Slacken each bolt evenly a little at a time in a criss-cross pattern until they are all finger-tight, then remove the bolts and washers. Carefully lift off the lower crankcase half, making sure the Plastigauge is not disturbed.

17 Compare the width of the crushed Plastigauge on each crankshaft journal to the scale printed on the Plastigauge envelope to obtain the main bearing oil clearance **(see illustration)**. Compare the reading to the specifications at the beginning of this Chapter. If the oil clearance falls within the specified range and the bearings are in perfect condition, they can be reused.

18 Carefully clean away all traces of the Plastigauge from the crankshaft journal and bearing shells using a fingernail or other object which will not score the bearing surfaces.

19 If the clearance is beyond the service limit, measure the diameter of each crankshaft

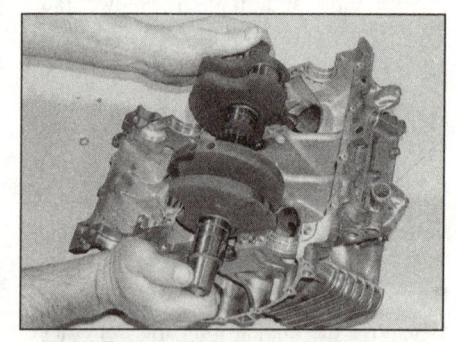

25.13 Ensure the bearing shells are not displaced when positioning the crankshaft

25.14 Lay a strip of Plastigauge on each journal parallel to the crankshaft centreline

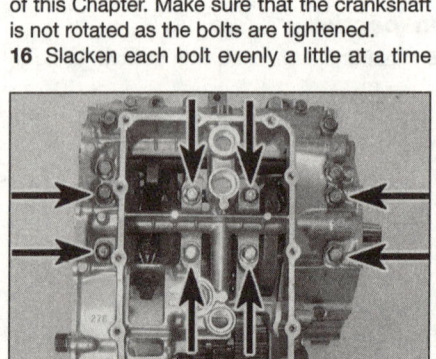

25.15 Main bearing bolts

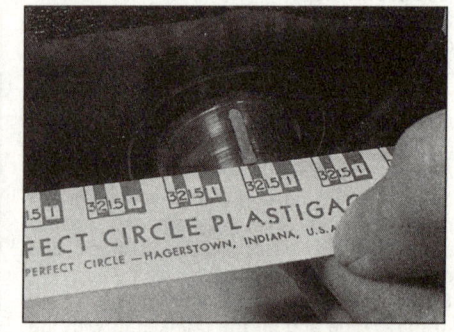

25.17 Measuring the width of the crushed Plastigauge

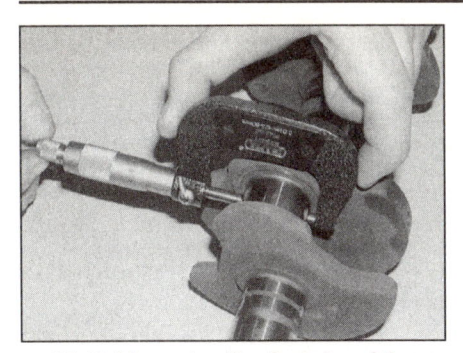

Engine, clutch and transmission 2•45

25.19 Measuring the diameter of the crankshaft journal

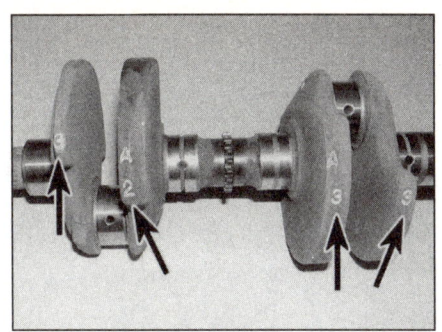

25.21a Crankshaft main bearing journal size codes

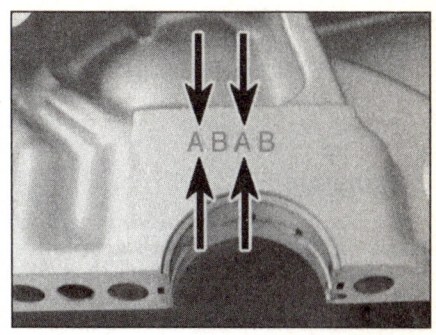

25.21b Main bearing housing size codes

journal with a micrometer **(see illustration)** and compare the result with the appropriate figure in the table for main bearing shell selection. For example, if the journal being measured is code 2, the table indicates that the service limit for that journal is 34.001 mm. If the journal diameter is larger than the service limit, new bearing shells can be fitted (see Steps 21 and 22). If the journal diameter is smaller than the service limit the crankshaft should be renewed, although seek the advice of a Honda dealer or engineer concerning alternatives.

20 Always renew all of the shells at the same time. Install the new shells and check the oil clearance once again.

Main bearing shell selection

21 New bearing shells for the main bearings are supplied on a selected fit basis. Code letters and numbers marked on the crankshaft webs and the left-hand side of the upper crankcase half are used to identify the correct new bearings. The crankshaft main bearing journal size codes are marked on the four crankshaft webs and will be either a 1, 2, or 3 **(see illustration)**. The corresponding main bearing housing size codes are stamped into the upper crankcase half and will be either an A, B or C **(see illustration)**. The left-hand letter corresponds to the left-hand journal, and so on from left to right respectively.

22 A range of bearing shells are available. To select the correct bearing for a particular journal, using the table below cross-refer the main bearing journal size code (marked on the adjacent crank web) with the main bearing housing size code (stamped on the crankcase) to determine the colour code of the bearing required. For example, if the journal code is 1, and the housing code is B, then the bearing required is Green. The colour is marked on the side of the shell **(see illustration 22.21)**.

Installation

23 Clean the backs of the bearing shells and the bearing seats in both crankcase halves. If new shells are being fitted, ensure that all traces of protective grease are cleaned off using paraffin (kerosene). Wipe the shells and crankcase halves dry with a lint-free cloth. Make sure all the oil passages and holes are clear, and blow them through with compressed air if it is available.

24 Press the bearing shells into their seats, aligning the tab on each shell with the groove in the crankcase **(see illustration 25.12)**. Make sure the bearings are seated correctly and take care not to touch any shell's bearing surface with your fingers. Lubricate each shell with a 50/50 mixture of molybdenum disulphide grease and clean engine oil.

25 Fit the camchain over the crankshaft sprocket and lower the crankshaft into position in the upper crankcase, making sure all bearings remain in place.

26 Fit the connecting rods onto the crankshaft (see Section 22).

27 Reassemble the crankcase halves (see Section 20).

26 Balancer shaft and bearings – removal, inspection and installation

Note: *To remove the balancer shaft the engine must be removed from the frame.*

Removal

1 Remove the engine from the frame (see Section 5) and separate the crankcase halves (see Section 20).

2 Turn the balancer shaft so that it can be lifted out of the upper crankcase half without disturbing the crankshaft **(see illustration)**. Take care not to dislodge the bearing shells.

3 The balancer shaft bearing shells can be removed from the crankcase halves by pushing their centres to the side, then lifting them out. Keep the shells in order.

Inspection

4 Clean the balancer shaft with solvent. If available, blow through the oil passages with compressed air. **Note:** *Thorough cleaning of the crankcase oil passages can only be achieved with both the balancer shaft and crankshaft removed (see section 25).*

5 Inspect the splines on the right-hand end of the shaft and ensure that the drive gear is a good fit on the splines. Damage or wear that prevents the drive gear from being fitted securely will require a new balancer shaft.

6 Refer to Section 21 and examine the balancer shaft bearing shells. If they are scored, badly scuffed or appear to have been seized, new shells must be installed. Always renew the bearing shells as a set. If they are badly damaged, check the corresponding balancer shaft journals. Evidence of extreme heat, such as discolouration, indicates that lubrication failure has occurred. Be sure to thoroughly check the oil pump and pressure relief valve as well as all oil holes and passages before reassembling the engine.

7 Give the balancer shaft journals a close visual examination, paying particular attention where damaged bearings have been discovered. If the journals are scored or pitted in any way, a new balancer shaft will be required.

8 No specifications are available for the

26.2 Lift the balancer shaft (arrowed) out without disturbing the crankshaft

Main journal code	Housing code		
	A – (37.000 to 37.006 mm)	B – (37.006 to 37.012 mm)	C – (37.012 to 37.018 mm)
1 – (34.007 to 34.013 mm)	Yellow	Green	Brown
2 – (34.001 to 34.007 mm)	Green	Brown	Black
3 – (33.995 to 34.001 mm)	Brown	Black	Blue

2•46 Engine, clutch and transmission

balancer shaft journals or the bearing shells. Even if both appear to be in perfect condition, if the engine has covered a high mileage and has been stripped for overhaul, install new bearing shells. Always renew all of the shells at the same time.

Installation

9 Clean the backs of the bearing shells and the bearing seats in both crankcase halves. If new shells are being fitted, ensure that all traces of protective grease are cleaned off using paraffin (kerosene). Wipe the shells and crankcase halves dry with a lint-free cloth. Make sure all the oil passages and holes are clear, and blow them through with compressed air if it is available.
10 Press the bearing shells into their seats, aligning the tab on each shell with the groove in the crankcase **(see illustration 25.12)**. Make sure the bearings are seated correctly and take care not to touch any shell's bearing surface with your fingers. Lubricate each shell with a 50/50 mixture of molybdenum disulphide grease and clean engine oil **(see illustration)**.
11 Lower the balancer shaft into position in the upper crankcase, making sure all bearings remain in place **(see illustration)**.
12 Reassemble the crankcase halves (see Section 20).

27 Camchain and guide/tensioner blades – removal, inspection and installation

Note: *To remove the camchain and guides the engine must be removed from the frame.*

Removal

1 Remove the engine from the frame (see Section 5), remove the cylinder head (see Section 10) and separate the crankcase halves (see Section 20).
2 Remove the crankshaft (see Section 25) and lift the camchain off the shaft.
3 Lift the front guide blade (orange) and rear tensioner blade (black) out of the upper crankcase half. Note how they fit with the bowed face bearing against the camchain and their lower ends resting in slots in the upper crankcase half **(see illustrations)**.

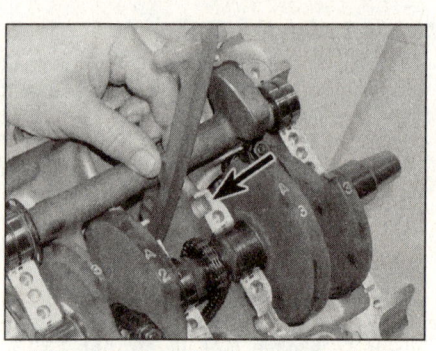

Inspection

4 Check the camchain for signs of wear such as highly polished surfaces on the sideplates. Lay the chain on the work surface and compress and stretch sections of the chain to check for wear between the links. Flex the chain and check for binding between links. If the chain is worn in any respect, it must be renewed.
5 Check the camchain for signs of damage, and if any is found, inspect the sprockets on the crankshaft and the camshafts **(see illustration)**. Renew the camchain together with the damaged components.

6 No specifications are available for service limits for the camchain. If the engine has covered a high mileage and has been stripped for overhaul it is advisable to renew the chain.
7 Inspect the camchain guide blade and tensioner blade, including the upper guide on the cylinder head, for grooves, cracking and other damage. Check the tensioner blade where it makes contact with the camchain tensioner plunger. Renew the guides/blades if any wear is found.

Installation

8 Install the front and rear camchain blades (see Step 3). Fit the camchain over the crankshaft sprocket and lower the crankshaft into position in the upper crankcase.
9 Ensure the camchain is laying correctly over the crankshaft sprocket. Pull the camchain through the engine block and secure it with a length of wire to prevent it

26.10 Lubricate the bearing shells . . .

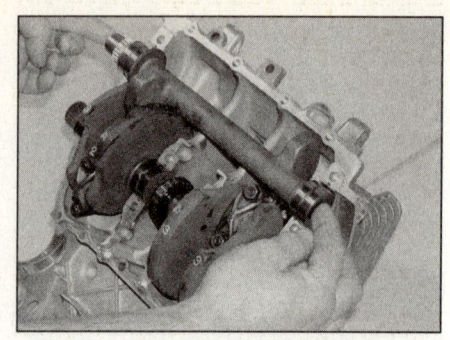

26.11 . . . and avoid touching the journals on reassembly

27.3a Note location (arrowed) of front guide blade . . .

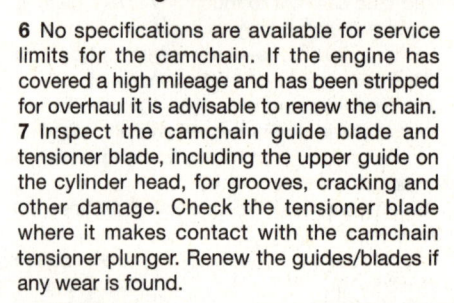

27.3b . . . and rear tensioner blade

27.5 Inspect crankshaft camchain sprocket for wear and damage

falling back into the crankcase during engine reassembly.

28 Transmission shafts and bearings – removal and installation

Note: *To remove the transmission shafts the engine must be removed from the frame.*

Removal

1 Remove the engine from the frame (see Section 5) and separate the crankcase halves (see Section 20). **Note:** *Before separating the crankcase halves make sure the transmission is in neutral.*
2 Unscrew the two bolts securing the input shaft bearing retainer plate to the right-hand side of the crankcase and remove the plate **(see illustration)**.

28.2 Unscrew bolts (arrowed) to remove retainer plate

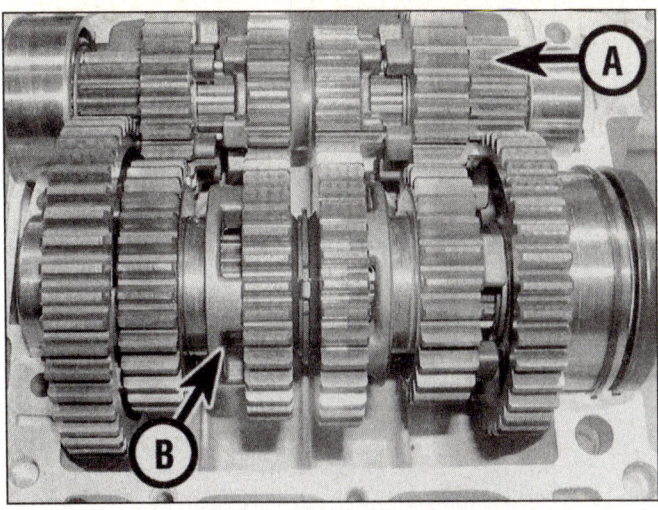

Engine, clutch and transmission 2•47

28.3a Input shaft (A) and output shaft (B)

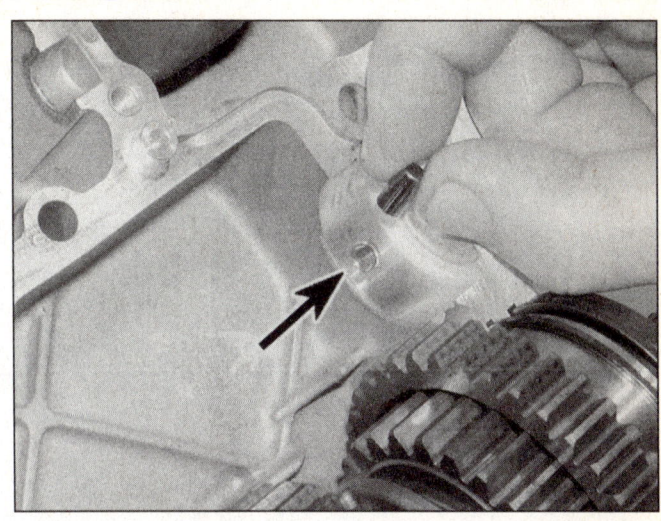

28.3b Location of input shaft needle bearing dowel (arrowed)

3 Note how the input shaft and the output shaft mesh together **(see illustration)**. Lift the shafts out of the casing. If they are stuck in the casing, use a soft-faced hammer and gently tap on the ends of the shafts to free them. Remove the dowel for the needle bearing on the left-hand end of the input shaft from the crankcase if it is loose, noting how it fits **(see illustration)**. If it is not in its hole in the crankcase, remove it from the bearing on the shaft.

4 Referring to *Tools and Workshop Tips* (Section 5) in the Reference Section, check the bearings at both ends of the shafts. Renew the bearings if necessary, noting that the bearing on the left-hand end of the output shaft is not available separately and the shaft itself must be renewed. The shafts have to be disassembled for bearing removal and for inspection for wear or damage generally (see Section 29).

5 Remove and check the condition of the output shaft oil seal and renew it if it is worn or damaged **(see illustration)**. Apply some grease to the inside of the seal on installation. It is advisable to renew the seal as a matter of course.

Installation

6 Clean the bearing seats in both halves of the crankcase with solvent and wipe them dry with a lint-free cloth. Lower the output shaft into the crankcase. Ensure that the bearings locate correctly in their seatings and that the locating ring around the bearing and the lip of the oil seal on the left-hand end of the shaft fit into the slots in the casing **(see illustration)**.

7 If removed, fit the input shaft needle bearing dowel into its hole in the upper crankcase. Lower the input shaft into position, making sure it locates correctly onto the dowel **(see illustration)**.

8 Make sure both transmission shafts are correctly seated and their related pinions are correctly engaged.

9 Install the transmission input shaft bearing retainer plate onto the right-hand side of the crankcase. Apply a suitable non-permanent thread-locking compound to the threads of the bolts and tighten them securely **(see illustration)**.

Caution: If the transmission shaft bearings are not correctly seated the crankcase halves will not join correctly.

10 Lubricate the selector fork grooves with a 50/50 mixture of molybdenum disulphide grease and clean engine oil. Lubricate the gears, shafts and bushings with clean engine oil.

11 Ensure the gears are in the neutral position and check the shafts are free to rotate easily and independently (ie the input shaft can turn whilst the output shaft is held stationary) before proceeding further.

12 Reassemble the crankcase halves (see Section 20).

28.5 Check the condition of the output shaft oil seal and grease on reassembly

28.6 Ensure the correct location of the bearing ring (A) and oil seal lip (B)

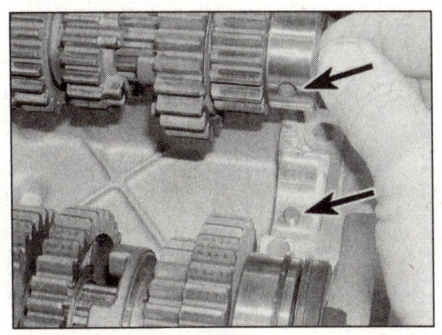

28.7 Ensure input shaft bearing locates on dowel

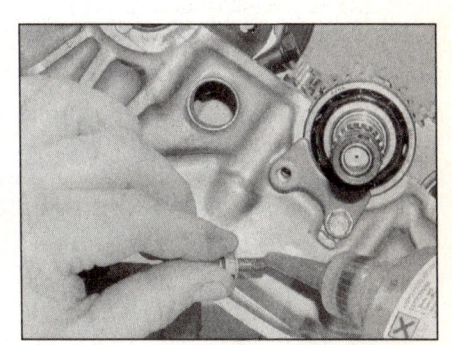

28.9 Use locking compound on bearing retainer plate bolts

2•48 Engine, clutch and transmission

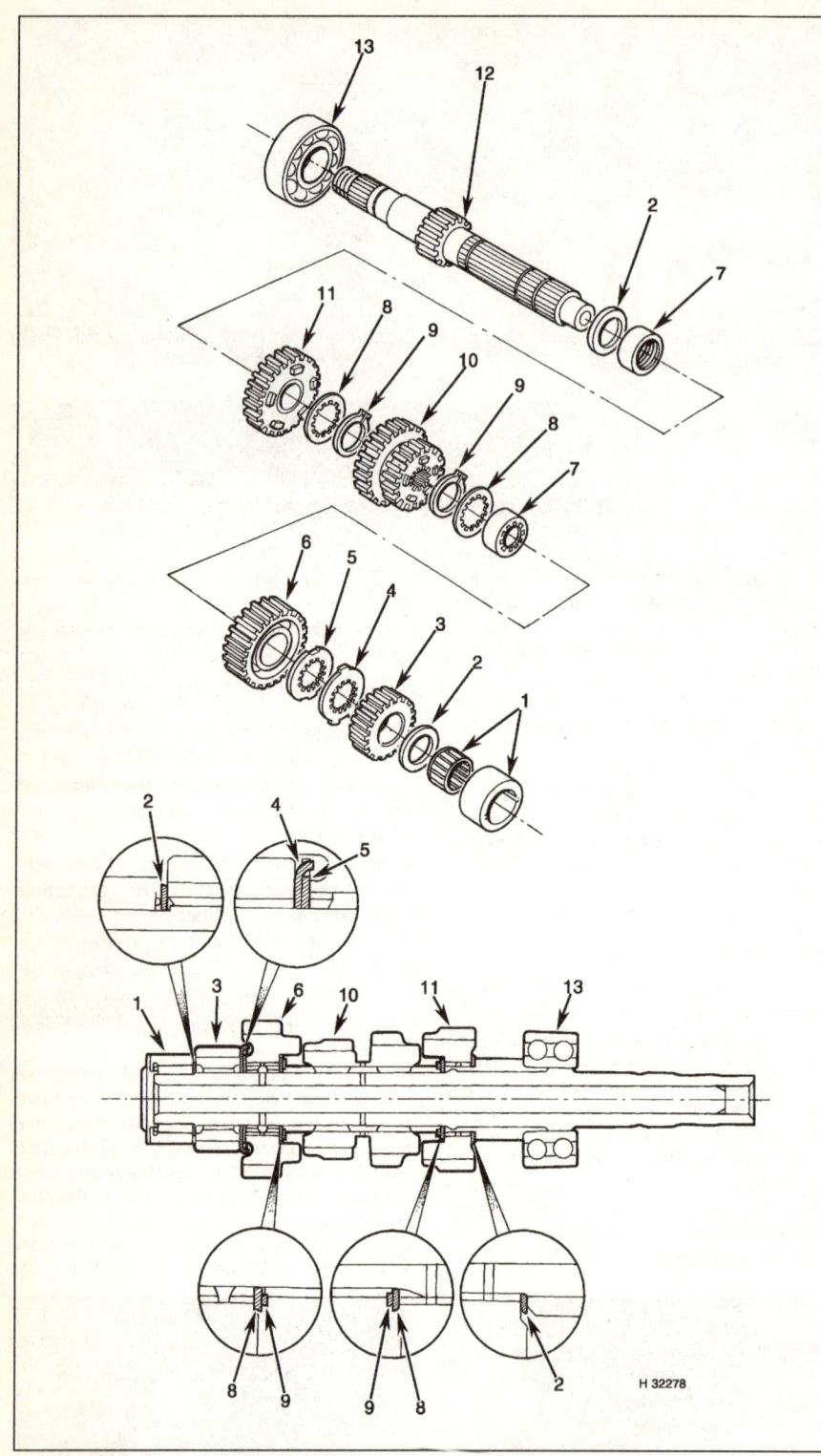

29.1a Transmission input shaft components

1. Needle roller bearing and cage
2. Thrust washer
3. 2nd gear pinion
4. Lockwasher
5. Slotted splined washer
6. 6th gear pinion
7. Bush
8. Splined washer
9. Circlip
10. 3rd/4th gear pinion
11. 5th gear pinion
12. Input shaft with integral 1st gear pinion
13. Ball bearing

29 Transmission shafts – disassembly, inspection and reassembly

1 Remove the transmission shafts from the casing (see Section 28). Always disassemble the transmission shafts separately to avoid mixing up the components **(see illustrations)**.

> **HAYNES HiNT** When disassembling the transmission shafts, place the parts on a long rod or thread a wire through them to keep them in order and facing in the proper direction.

Input shaft

Disassembly

2 Slide the needle roller bearing cage and bearing off the left-hand end of the shaft **(see illustration 29.23)**.

3 Slide the thrust washer and the 2nd gear pinion off the shaft **(see illustrations 29.22b and a)**.

4 Note how the tabs on the lock washer fit into the slotted splined washer and remove the lockwasher **(see illustration 29.21c)**.

5 Turn the slotted splined washer to align it with the splines on the shaft and slide it off the shaft **(see illustration 29.21b)**.

6 Slide the 6th gear pinion and its bush, and the splined washer off the shaft **(see illustrations 29.20c, b and a)**.

7 Remove the circlip securing the combined 3rd/4th gear pinion. Do not expand the ends of the circlip any further than is necessary to slide it down the length of the shaft, then slide the pinion off the shaft **(see illustrations 29.19b and a)**.

8 Remove the circlip securing the 5th gear pinion, then slide the splined washer, the pinion and its bush, and the thrust washer off the shaft **(see illustrations 29.18f, e, d, c, b and a)**. The 1st gear pinion is integral with the shaft.

9 If necessary, press the caged ball bearing off the shaft using a suitable tool.

Inspection

10 Wash all of the components in clean solvent and dry them off.

11 Check the gear teeth for cracking, chipping, pitting and other obvious wear or damage. Any pinion that is damaged must be renewed. Inspect the dogs and the dog holes in the gears for cracks, chips, and excessive wear especially in the form of rounded edges. Make sure mating gears engage properly. Renew the paired gears as a set if necessary.

12 Check for signs of scoring or bluing on the pinions, bushes and shaft. This could be caused by overheating due to inadequate

Engine, clutch and transmission 2•49

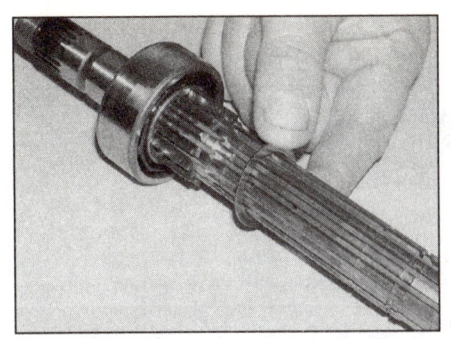

29.18a Slide on the thrust washer . . .

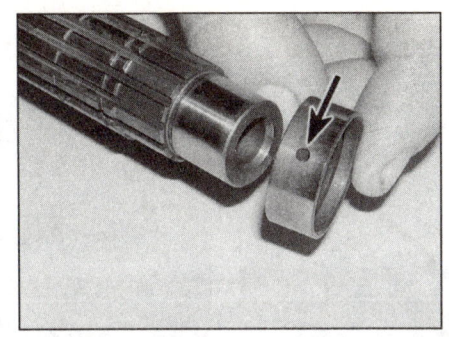

29.18b . . . 5th gear pinion bush, oil hole (arrowed) . . .

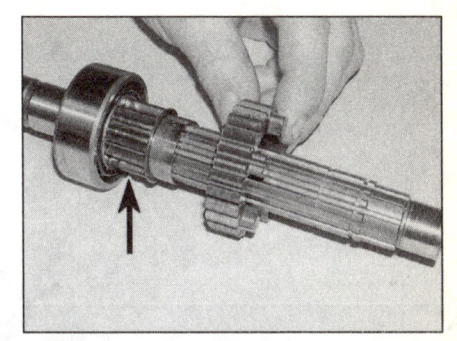

29.18c . . . 5th gear pinion, integral 1st gear (arrowed) . . .

29.18d . . . and the splined washer . . .

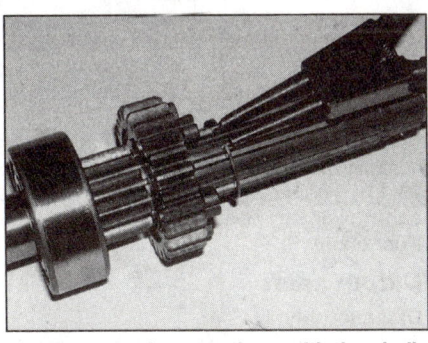

29.18e . . . and secure them with the circlip

29.18f Ensure circlip (arrowed) locates correctly in its groove

lubrication. Check that all the oil holes and passages are clear. Renew any damaged pinions or bushes.

13 Check that each pinion moves freely on the shaft or bush but without undue freeplay. Check that each bush moves freely on the shaft but without undue freeplay. If the necessary measuring equipment is available, the gear, bush and shaft dimensions can be checked and compared with the Specifications at the beginning of this chapter. The shaft is unlikely to sustain damage unless the engine has seized, placing an unusually high loading on the transmission, or the machine has covered a very high mileage. Check the surface of the shaft, especially where a pinion turns on it, and renew the shaft if it has scored or picked up, or if there are any cracks or wear. Damage of any kind can only be cured by renewal.

14 Check the washers and circlips and renew any that are bent or appear weakened or worn. Use new ones if in any doubt. It is good practice to renew the washers and circlips as a matter of course when overhauling the gearshaft.

15 Check the bearings referring to *Tools and Workshop Tips* (Section 5) in the Reference Section.

Reassembly

16 During reassembly, apply clean engine oil to the mating surfaces of the shaft, pinions and bushes, and to the bearings. When installing the circlips, do not expand the ends any further than is necessary to slide them along the shaft. Install them so that their chamfered side faces the pinion they secure (see *Correct fitting of a stamped circlip* illustration in *Tools and Workshop Tips* (Section 2) in the Reference Section). Also refer to illustration 29.1a.

If removed, press the caged ball bearing onto the right-hand end of the shaft.

18 Slide the thrust washer onto the left-hand end of the shaft, followed by the 5th gear pinion bush, aligning the oil hole in the bush with the hole in the shaft, and the 5th gear pinion, with the pinion dogs facing away from the integral 1st gear **(see illustrations)**. Slide the splined washer onto the shaft, then fit the circlip, making sure that it locates correctly in the groove on the shaft **(see illustrations)**.

19 Slide the combined 3rd/4th gear pinion onto the shaft with the larger 4th gear pinion facing the 5th gear pinion. Ensure the oil hole in the pinion aligns with the oil hole in the shaft **(see illustration)**. Fit the circlip, making sure it is locates correctly in its groove on the shaft **(see illustration)**.

20 Slide the splined washer onto the shaft, followed by the 6th gear pinion bush, aligning

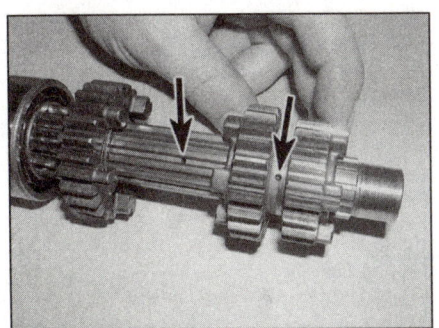

29.19a Slide on 3rd/4th gear pinion, ensuring oil holes (arrowed) align . . .

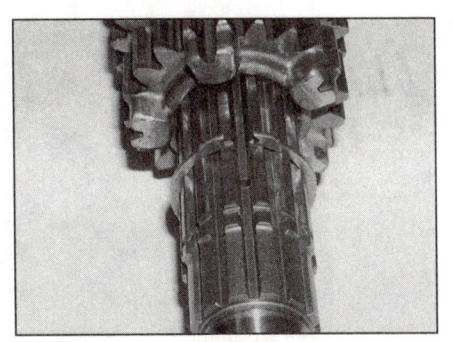

29.19b . . . and secure it with the circlip

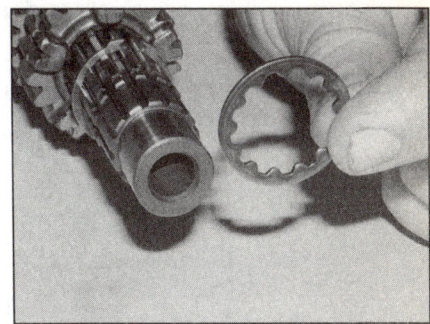

29.20a Slide on the splined washer . . .

2•50 Engine, clutch and transmission

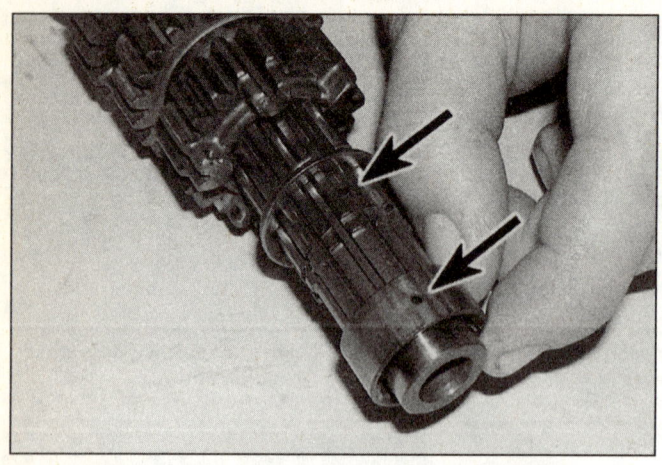

29.20b ...6th gear pinion bush, oil holes arrowed...

29.20c ...6th gear pinion...

the oil hole in the bush with the hole in the shaft, and the 6th gear pinion, making sure the dogs on the pinion face the 3rd/4th gear pinion **(see illustrations)**.

21 Slide the slotted splined washer onto the shaft and locate it in its groove, then turn it in the groove so that the splines on the washer align against the splines on the shaft and secure the washer in the groove. Slide the lock washer onto the shaft so that the tabs on the lockwasher locate in the slots on the outside edge of the splined washer **(see illustrations)**.

22 Slide the 2nd gear pinion and the thrust washer onto the shaft **(see illustrations)**.

Check that all components have been correctly installed **(see illustration 29.1a)**.

23 Fit the needle roller bearing and its cage over the end of the shaft **(see illustration)**.

Output shaft

Disassembly

24 Remove the caged ball bearing from the right-hand end of the shaft, referring to *Tools and Workshop Tips* (Section 5) in the Reference Section if required **(see illustration 29.40)**. Slide the thrust washer off the shaft, followed by the 1st gear pinion and its needle roller bearing, the thrust washer and the 5th gear

pinion **(see illustrations 29.39c, b and a, and 29.38b and a)**.

25 Remove the circlip securing the 4th gear pinion, then slide the splined washer, the pinion and its splined bush off the shaft **(see illustrations 29.37d, c, b and a)**.

26 Slide the lockwasher off the shaft. Note how the tabs on the lock washer fit into the slotted splined washer. Turn the slotted splined washer to align it with the splines on the shaft and slide it off the shaft **(see illustration 29.36b and a)**.

27 Slide the 3rd gear pinion and its splined bush, followed by the splined washer, off the shaft **(see illustration 29.35c, b and a)**.

28 Remove the circlip securing the 6th gear

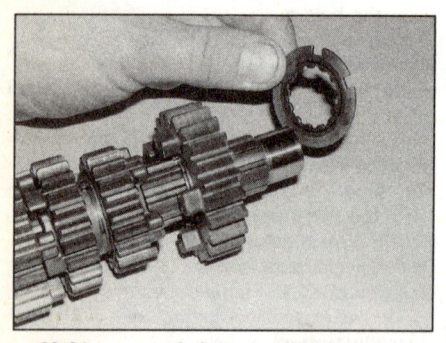

29.21a ...and slotted splined washer

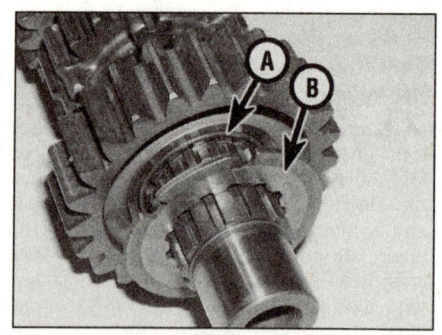

29.21b Locate slotted splined washer (B) in shaft groove (A)...

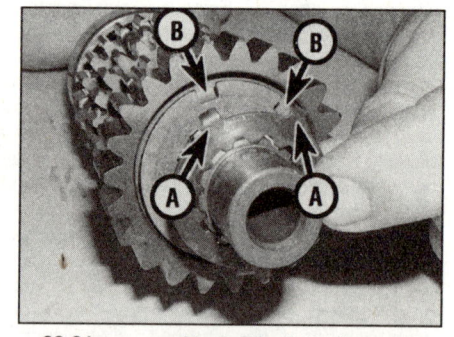

29.21c ...and install lockwasher with tabs (A) in splined washer slots (B)

29.22a Slide on the 2nd gear pinion...

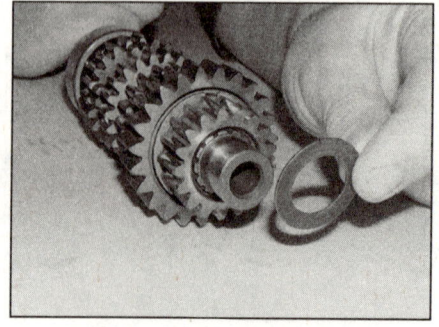

29.22b ...and the thrust washer

29.23 Fit the needle bearing

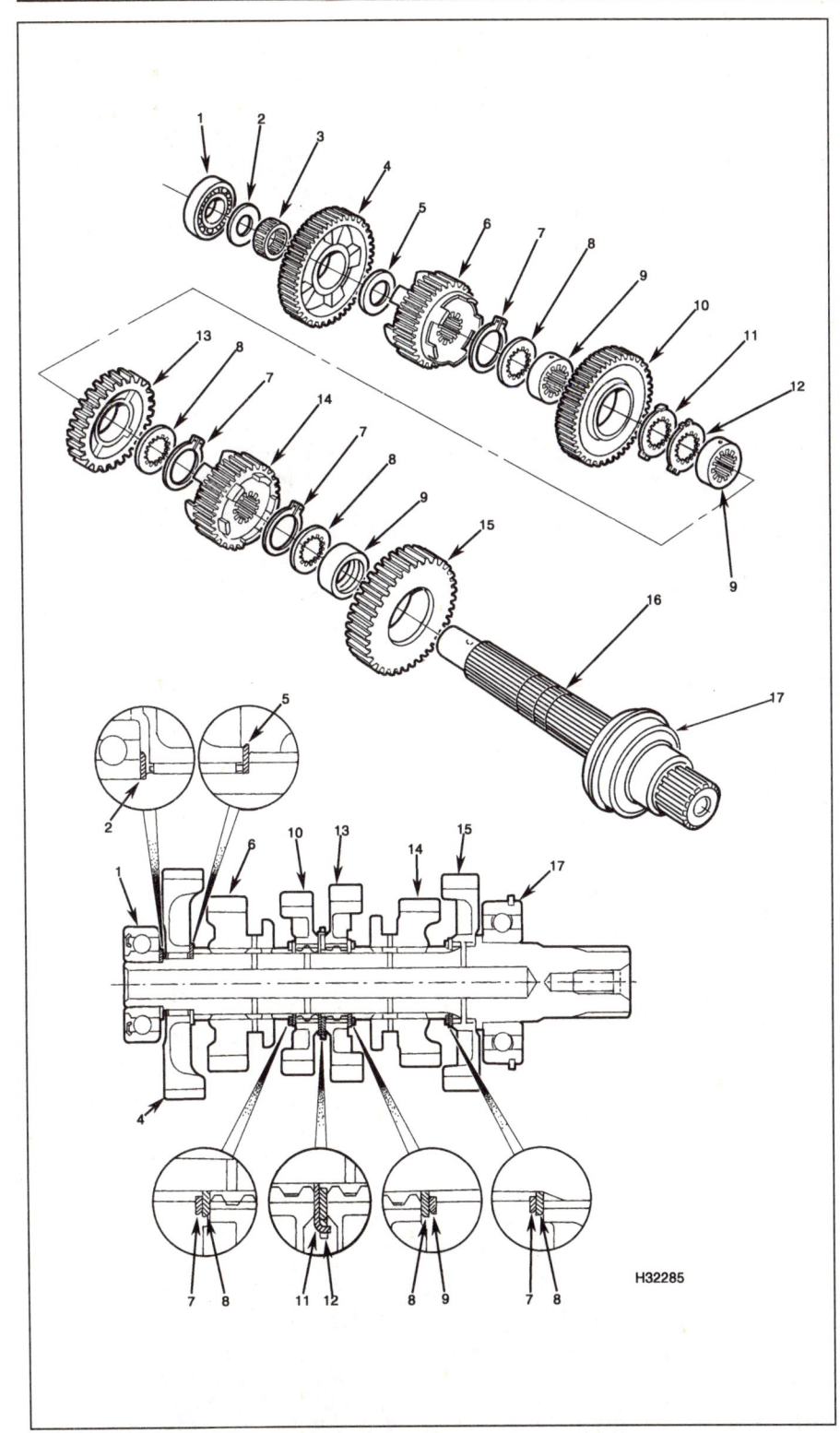

Engine, clutch and transmission 2•51

pinion, then slide the pinion off the shaft **(see illustrations 29.34b and a)**.

29 Remove the circlip securing the 2nd gear pinion, then slide the splined washer, the pinion and its bush off the shaft **(see illustrations 29.33d, c, b and a)**.

30 The remaining caged ball bearing is an integral part of the output shaft and cannot be removed.

Inspection

31 Refer to Steps 10 to 15 above.

Reassembly

32 During reassembly, apply clean engine oil to the mating surfaces of the shaft, pinions and bushes, and to the bearings. When installing the circlips, do not expand the ends any further than is necessary to slide them along the shaft. Install them so that the chamfered side faces the pinion they secure (see *Correct fitting of a stamped circlip* illustration in *Tools and Workshop Tips* in the Reference section). Also refer to illustration 29.1b.

33 Slide the 2nd gear pinion bush, the 2nd gear pinion and the splined washer onto the shaft, then fit the circlip, making sure it is locates correctly in its groove on the shaft **(see illustrations)**.

34 Slide the 6th gear pinion with its selector fork groove facing away from the 2nd gear pinion, onto the shaft. Ensure that the oil hole in the pinion aligns with the hole in the shaft, then fit the circlip, making sure it is locates correctly in its groove on the shaft **(see illustrations)**.

35 Slide the splined washer and the 3rd gear

29.33a Slide on the 2nd gear pinion bush . . .

29.33b . . . the 2nd gear pinion . . .

29.1b Transmission output shaft components

1 Ball bearing
2 Thrust washer
3 Needle roller bearing
4 1st gear pinion
5 Thrust washer
6 5th gear pinion
7 Circlip
8 Splined washer
9 Bush
10 4th gear pinion
11 Lockwasher
12 Slotted splined washer
13 3rd gear pinion
14 6th gear pinion
15 2nd gear pinion
16 Output shaft
17 Ball bearing

2•52 Engine, clutch and transmission

29.33c . . . and the splined washer . . .

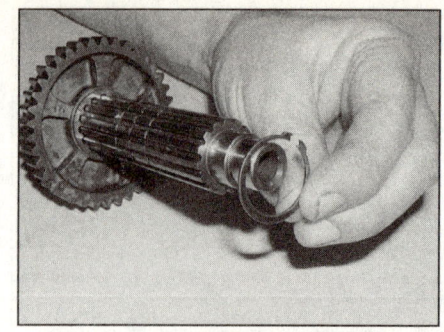

29.33d . . . and secure them with the circlip

29.34a Slide on the 6th gear pinion, aligning the oil holes (arrowed) . . .

29.34b . . . and fit the circlip (arrowed)

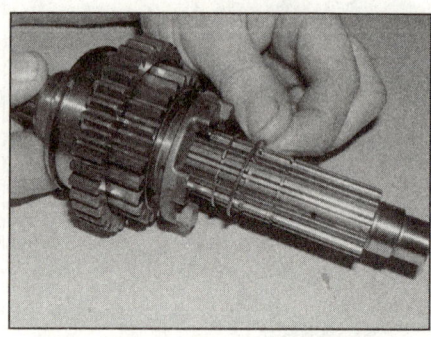

29.35a Slide on the splined washer . . .

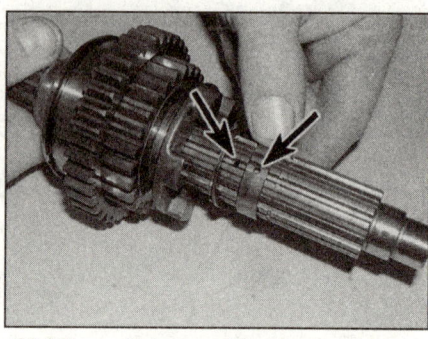

29.35b . . . 3rd gear pinion bush, aligning the oil holes (arrowed) . . .

29.35c . . . and the 3rd gear pinion

29.36a Slide on the slotted splined washer . . .

pinion bush onto the shaft, making sure the oil hole in the bush aligns with the hole in the shaft, followed by the 3rd gear pinion. The open side of the 3rd gear pinion faces the 6th gear pinion (see illustrations).

36 Slide the slotted splined washer onto the shaft and locate it in its groove (see illustration). Turn it in the groove so that the splines on the washer locate against the splines of the shaft and secure the washer in the groove. Slide the lockwasher onto the shaft, so that the tabs on the lockwasher locate into the slots on the outside edge of the splined washer (see illustration).

37 Slide the 4th gear pinion bush onto the

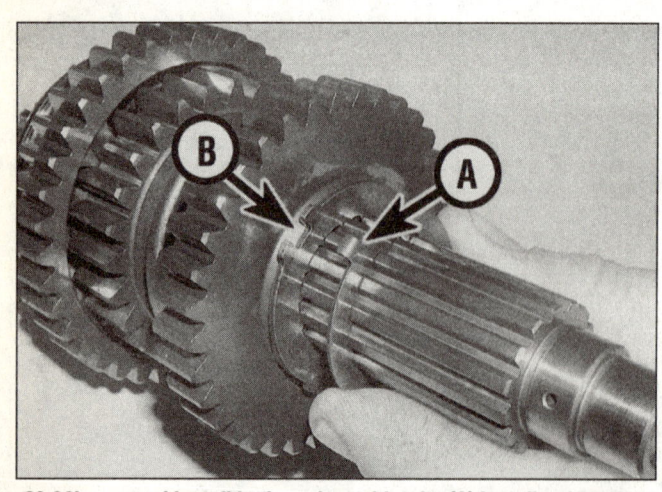

29.36b . . . and install lockwasher with tabs (A) in splined washer slots (B)

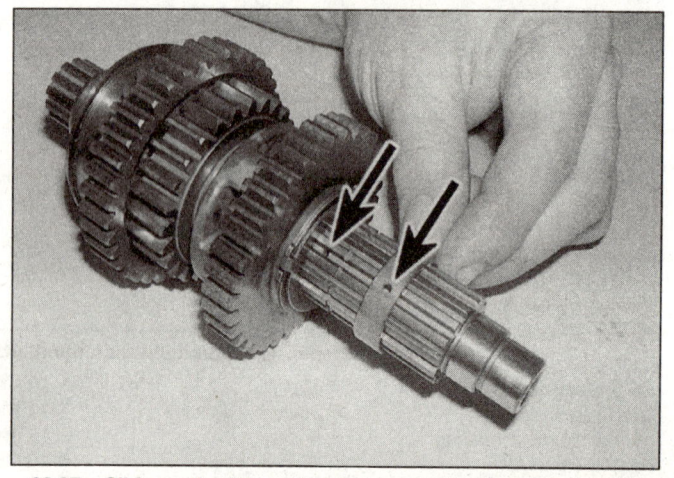

29.37a Slide on the 4th gear pinion bush, aligning the oil holes (arrowed) . . .

Engine, clutch and transmission 2•53

29.37b ... the 4th gear pinion ...

29.37c ... and the splined washer ...

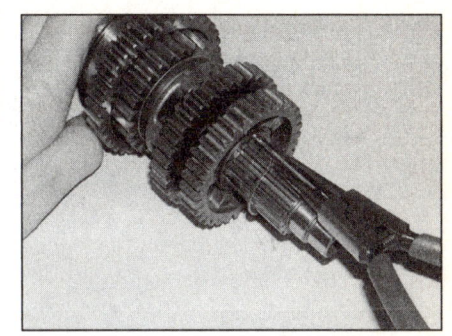

29.37d ... and secure them with the circlip

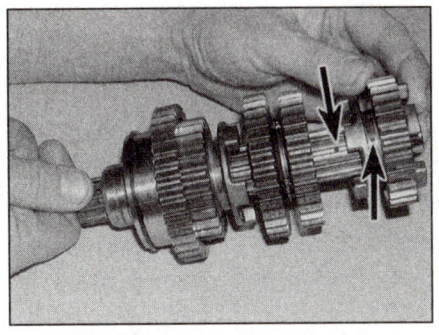

29.38a Slide on the 5th gear pinion, aligning the oil holes (arrowed) ...

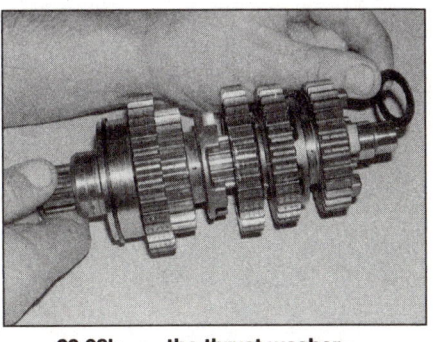

29.38b ... the thrust washer ...

29.39a ... the 1st gear pinion needle bearing ...

shaft, making sure the oil hole in the bush aligns with the hole in the shaft, followed by the 4th gear pinion (recessed side facing away from the 3rd gear pinion) and the splined washer. Fit the circlip, making sure it is locates correctly in its groove on the shaft **(see illustrations)**.

38 Slide the 5th gear pinion onto the shaft with its selector fork groove facing the 4th gear pinion. Ensure that the oil hole in the pinion aligns with the hole in the shaft. Fit the thrust washer **(see illustrations)**.

39 Slide the 1st gear pinion needle roller bearing onto the shaft, followed by the 1st gear pinion (the open side of the gear faces the 5th gear pinion) and the thrust washer **(see illustrations)**.

40 Fit the caged ball bearing onto the right-hand end of the shaft, referring to *Tools and Workshop Tips* (Section 5) in the Reference Section if required **(see illustration)**.

41 Check that all components have been correctly installed **(see illustration 29.1b)**. Install the transmission shafts in the casing (see Section 28).

30 Selector drum and forks – removal, inspection and installation

Note: To remove the selector drum and forks the engine must be removed from the frame.

Removal

1 Remove the engine from the frame (see Section 5), remove the clutch (see Section 15) the oil sump (see Section 18) and the gearchange external mechanism (see Section 19). The selector drum and forks are located in the lower crankcase half.

2 Make sure the transmission is in neutral. Note the position of the selector drum against the neutral switch **(see illustration 30.4)**.

3 Before removing the selector forks, note that each fork is lettered for identification. The right-hand fork has an R, the centre fork a C, and the left-hand fork an L **(see illustration 30.5c)**. These letters face the right-hand side of the engine. If no letters are visible, mark them using a felt pen.

4 Note how the forks locate in the grooves on the input shaft 3rd/4th gear pinion and output shaft 5th and 6th gear pinions. Note how the

29.39b ... the 1st gear pinion ...

29.39c ... and the thrust washer

29.40 Fit the caged ball bearing

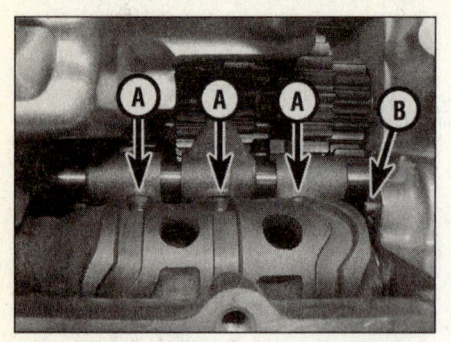

30.4 Note position of selector fork guide pins (A) and neutral switch pin (B)

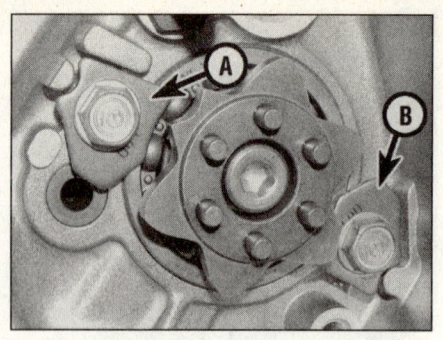

30.5a Selector fork retainer plate (A), selector drum retainer plate (B)

30.5b Withdrawing the selector fork shaft from the casing

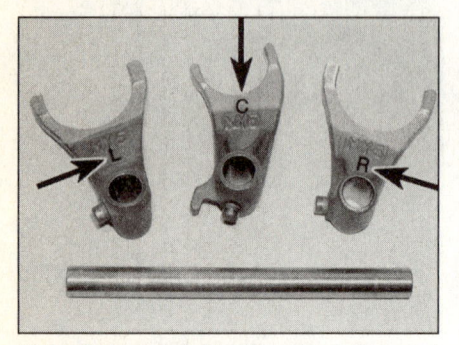

30.5c Seletor fork identification letters and fork shaft

30.6 Remove the selector drum

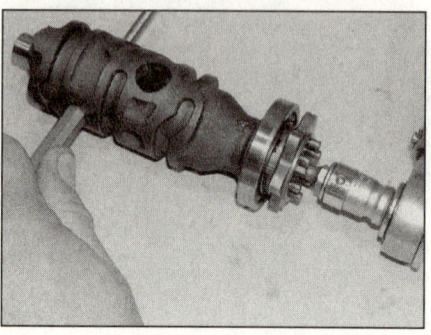

30.7a Unscrew the cam retaining bolt . . .

guide pins on the forks locate in the grooves in the selector drum **(see illustration)**.

5 Unscrew the selector fork shaft retainer bolt, and on CB models remove the retainer plate, noting how it fits **(see illustration)**. Support the selector forks and withdraw the shaft from the casing, then remove the forks **(see illustration)**. Once removed, slide the forks back onto the shaft in their correct order and way round **(see illustration)**.

6 Unscrew the selector drum retainer bolt, and on CB models remove the retainer plate, noting how it fits **(see illustration 30.5a)**. Slide the selector drum out of the crankcase **(see illustration)**.

7 If the selector cam has not already been removed from the end of the selector drum, pass a steel rod through the drum to hold it while unscrewing the cam retaining bolt **(see illustration)**. The cam locates on a pin in the end of the selector drum. Remove the pin for safekeeping **(see illustration)**.

8 If necessary, pull the caged ball bearing off the end of the selector drum **(see illustration)**.

Inspection

9 Inspect the selector forks for any signs of wear or damage, especially around the fork ends where they engage with the groove in the pinion. Check that each fork fits correctly in its pinion groove. Check closely to see if the forks are bent. If the forks are in any way damaged they must be renewed.

10 Measure the thickness of the fork ends and compare the readings to the specifications at the beginning of this Chapter **(see illustration)**. Renew the forks if they are worn beyond their specifications.

11 Check that the forks fit correctly on their shaft. They should move freely with a light fit but no appreciable freeplay. Measure the internal diameter of the fork bores and the corresponding diameter of the fork shaft. Renew the forks and/or shaft if they are worn beyond their specifications. Check that the fork shaft holes in the casing are not worn or damaged.

12 The selector fork shaft can be checked for trueness by rolling it along a flat surface. A bent rod will cause difficulty in selecting gears and make the gearchange action heavy. Renew the shaft if it is bent.

13 Inspect the selector drum grooves and selector fork guide pins for signs of wear or damage. If either component shows signs of wear or damage the selector fork(s) and drum must be renewed.

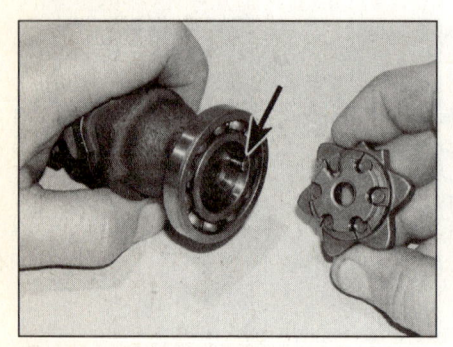

30.7b . . . and remove the cam locating pin for safekeeping

30.8 Remove the bearing if necessary

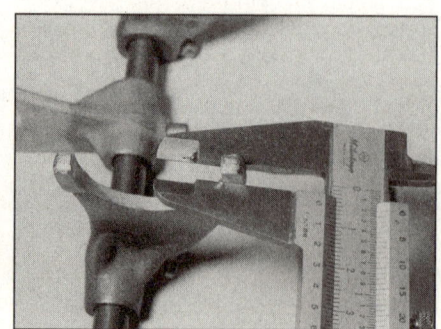

30.10 Measuring the selector fork end thickness

Engine, clutch and transmission 2•55

30.16a Assemble the selector forks on the gear pinions . . .

30.16b . . . then slide the selector drum into place

31.9 Inspect the cylinder walls for scratches and scoring

14 Check the selector drum bearing referring to *Tools and Workshop Tips* (Section 5) in the Reference Section.

Installation

15 If removed, press the selector drum bearing onto the drum. Install the pin in the end of the gear selector drum and install the cam, locating the pin in the cut-out in the back of the cam. Clean the threads of the centre bolt, then apply a suitable non-permanent thread-locking compound. Install the bolt and tighten it to the specified torque setting (see Step 7).
16 Slide the selector forks off their shaft and locate the forks in the grooves in the gear pinions (see Steps 3 and 4) **(see illustration)**. Push the forks forward and slide the selector drum into position in the crankcase **(see illustration)**. Make sure the drum end locates into its bore in the casing, and position it so that the neutral contact is against the neutral switch (see Step 2).
17 Apply a suitable non-permanent thread-locking compound to the selector drum retainer bolt. On CB models fit the retainer plate with the OUT mark facing out **(see illustration 30.5a)**. Install and tighten the bolt.
18 Lubricate the selector fork shaft with clean engine oil and slide it through the crankcase and each fork in turn, and into its bore, locating the guide pin on the end of each fork into its groove in the drum as you do **(see illustration 30.5b)**.
19 Apply a suitable non-permanent thread-locking compound to the selector fork shaft retainer bolt. On CB models fit the retainer plate with the OUT mark facing out **(see illustration 30.5a)**. Install and tighten the bolt.
20 Lubricate the selector forks and fork grooves with a 50/50 mixture of molybdenum disulphide grease and clean engine oil. Lubricate the gears, shafts and bushings with clean engine oil.
21 Ensure the gears are in the neutral position and check the shafts are free to rotate easily and independently (ie. the input shaft can turn whilst the output shaft is held stationary).
22 Install the gearchange mechanism (see Section 19) and check gearchange operation while turning the input shaft.

23 Refit the oil sump (see Section 18) and the clutch (see Section 15)

31 Crankcase halves and cylinder bores – inspection and servicing

Crankcase halves

1 After the crankcase halves have been separated, remove the crankshaft, connecting rods and pistons, transmission shafts, selector drum and forks, and, if required, the oil pressure switch, neutral switch, on CBF models the speed sensor, and the cooling system union, referring to the relevant Sections of this Chapter, to Chapter 9 for the oil pressure switch, neutral switch and speed sensor, and to Chapter 3 for the coolant union.
2 The crankcases should be cleaned thoroughly with solvent and dried with compressed air. All oil passages should be blown out with compressed air.
3 Inspect the bearing seats for signs of damage, especially if an engine or transmission bearing has overheated or seized (see Section 21). If bearing shells or a ball bearing cage are not a precise fit in their seats, ask your Honda dealer for a suitable bearing locking compound which will overcome small amounts of wear. Otherwise the crankcase halves will have to be renewed as a set.
4 All traces of old gasket sealant should be removed from the mating surfaces. Minor damage to the surfaces can be cleaned up with a fine sharpening stone or grindstone.
Caution: Be very careful not to nick or gouge the crankcase mating surfaces or oil leaks will result. Check both crankcase halves very carefully for cracks and other damage.
5 Small cracks or holes in aluminium castings may be repaired with an epoxy resin adhesive as a temporary measure. Permanent repairs can only be effected by argon-arc welding, and only a specialist in this process is in a position to advise on the economy or practical aspect of such a repair. Note that there are, however, kits available for low temperature welding. If any damage is found that can't be repaired, renew the crankcase halves as a set.
6 Damaged threads can be economically reclaimed using a thread insert, which is easily fitted after drilling and re-tapping the affected thread. There are a few types of thread insert available, of varying quality and cost.
7 Sheared studs or bolts can usually be removed with stud or screw extractors. A stud extractor should be used if part of the stud is above the surface. Otherwise use a screw extractor, which consists of a tapered, left thread screw of very hard steel. These are inserted into a hole pre-drilled centrally in the stud, and usually succeed in dislodging the most stubborn stud or screw.

> **HAYNES HiNT** *Refer to Tools and Workshop Tips for details of installing a thread insert and using screw extractors.*

8 Install all components and assemblies, referring to the relevant Sections of this Chapter and to Chapters 9 and 3, before reassembling the crankcase halves.

Cylinder bores

9 Check the cylinder walls carefully for scratches and score marks **(see illustration)**.
10 Using a precision straight-edge and a feeler gauge set to the warpage limit listed in the specifications at the beginning of this Chapter, check the block gasket mating surface for warpage. Refer to *Tools and Workshop Tips* in the Reference section for details of how to use the straight-edge. If warpage is excessive the block/crankcase must be renewed.
11 Using telescoping gauges and a micrometer (see *Tools and Workshop Tips*), check the dimensions of each cylinder to assess the amount of wear, taper and ovality. Measure near the top (but below the level of the top piston ring at TDC), centre and bottom (but above the level of the oil ring at BDC) of the bore. Measure both parallel to and across the crankshaft axis in each case and calculate the average cylinder dimension at each point

2•56 Engine, clutch and transmission

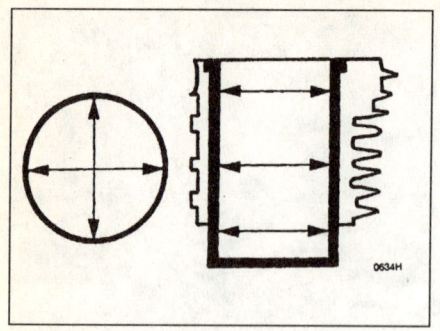

31.11 Measure the cylinder bore in the directions shown with a telescoping gauge

(see illustration). Compare the results to the specifications at the beginning of this Chapter. If the cylinders are worn, oval or tapered beyond the service limit, or badly scratched, scuffed or scored, the cylinder block must be rebored. Oversize pistons are available (see Section 23).

12 If the precision measuring tools are not available, take the upper crankcase to a Honda dealer or motorcycle engineer for assessment and advice.

13 If the cylinders are in good condition and the piston-to-bore clearance is within specifications, the cylinders should be honed (de-glazed). To perform this operation you will need the proper size flexible hone with fine stones, or a bottle-brush type hone, plenty of light oil or honing oil, some clean rags and an electric drill motor.

14 Clamp the block/crankcase securely so that the bores are horizontal rather than vertical. Mount the hone in a drill motor, compress the stones and insert the hone into the cylinder. Thoroughly lubricate the cylinder, then turn on the drill and move the hone up and down in the cylinder at a pace which produces a fine cross-hatch pattern on the cylinder wall with the lines intersecting at an angle of approximately 60°. Be sure to use plenty of lubricant and do not take off any more material than is necessary to produce the desired effect. Do not withdraw the hone from the cylinder while it is still turning. Switch off the drill and continue to move it up and down in the cylinder until it has stopped turning, then compress the stones and withdraw the hone. Wipe the oil from the cylinder and repeat the procedure on the other cylinder. Remember, do not take too much material from the cylinder wall.

15 Wash the cylinders thoroughly with warm soapy water to remove all traces of the abrasive grit produced during the honing operation. Be sure to run a brush through the oil and coolant passages and flush them with running water. After rinsing, dry the cylinders thoroughly and apply a thin coat of light, rust-preventative oil to all machined surfaces.

16 If you do not have the equipment or desire to perform the honing operation, take the block to a Honda dealer or motorcycle engineer.

32 Initial start-up after overhaul

1 Make sure the engine oil level and coolant level are correct (see *Daily (pre-ride) checks*). Turn the fuel tap to the OFF position.

2 Pull the plug caps off the spark plugs and insert a spare spark plug into each cap. Position the spare plugs so that their bodies are earthed (grounded) against the engine. Turn on the ignition switch and crank the engine over with the starter until the oil pressure warning light goes off (which indicates that oil pressure exists). Turn off the ignition. Remove the spare spark plugs and reconnect the plug caps.

3 Make sure there is fuel in the tank, then turn the fuel tap to the ON or RES position as required, and set the choke.

4 Start the engine and allow it to run at a moderately fast idle until it reaches operating temperature.

⚠ **Warning: If the oil pressure indicator light doesn't go off, or it comes on while the engine is running, stop the engine immediately and refer to Chapter 1, Section 33 for the oil pressure checking procedure.**

5 Check carefully that there are no oil or coolant leaks and make sure the transmission and controls, especially the brakes, function properly before road testing the machine. Refer to Section 33 for the recommended running-in procedure.

6 Upon completion of the road test, and after the engine has cooled down completely, recheck the valve clearances (see Chapter 1) and check the engine oil and coolant levels (see *Daily (pre-ride) checks*).

33 Recommended running-in procedure

1 Treat the machine gently for the first few miles to make sure oil has circulated throughout the engine and any new parts installed have started to seat.

2 Even greater care is necessary if new pistons/rings or new crankshaft or connecting rod bearings have been installed. This means greater use of the transmission and a restraining hand on the throttle until at least 600 miles (1000 km) have been covered. There's no point in keeping to any set road speed – the main idea is to keep from labouring the engine and to gradually increase performance up to the 1000 mile (1600 km) mark. These recommendations can be lessened when only a partial overhaul has been done, although it depends on the nature of the work carried out and which components have been renewed. Experience is the best guide, since it's easy to tell when an engine is running freely. The table below showing maximum engine speed limitations, which Honda provide for new motorcycles, can be used as a guide.

3 If a lubrication failure is suspected, stop the engine immediately and try to find the cause. If an engine is run without oil, even for a short period of time, severe damage will occur.

Up to 600 miles (1000 km)	600 rpm max	Vary throttle position/speed
600 to 1000 miles (1000 to 1600 km)	8000 rpm max	Vary throttle position/speed. Use full throttle for short bursts
Over 1000 miles (1600 km)	10,500 rpm max	Do not exceed tachometer red line

Chapter 3
Cooling system

Contents

Coolant hoses – removal and installation.................... 9
Coolant level check .... see Daily (pre-ride) checks
Coolant reservoir – removal and installation 3
Coolant temperature gauge and sensor – check and
  replacement 5
Cooling fan and cooling fan switch – check and replacement...... 4
Cooling system checks .... see Chapter 1
Cooling system draining, flushing and refilling......... see Chapter 1
General information 1
Pressure cap – check...................................... 2
Radiator – removal and installation 7
Thermostat and thermostat housing – removal,
  check and installation.................................. 6
Water pump – check, removal and installation 8

Degrees of difficulty

| Easy, suitable for novice with little experience | Fairly easy, suitable for beginner with some experience | Fairly difficult, suitable for competent DIY mechanic | Difficult, suitable for experienced DIY mechanic | Very difficult, suitable for expert DIY or professional |

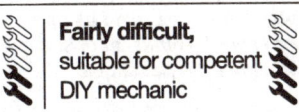

Specifications

Coolant
Mixture type .... 50% distilled water, 50% corrosion inhibited ethylene glycol anti-freeze
Capacity .... See Chapter 1

Pressure cap
Cap valve opening pressure................................... 16 to 20 psi (1.1 to 1.4 Bar)

Fan switch
Cooling fan cut-in temperature .... 98 to 102°C
Cooling fan cut-out temperature .... 93 to 97°C

Coolant temperature sensor
Resistance
  @ 50°C .... 133 to 179 ohms
  @ 80°C .... 47 to 57 ohms
  @ 120°C .... 14 to 18 ohms

Thermostat
Opening temperature.. 80 to 84°C
Valve lift .... 8 mm (min) @ 95°C

Torque settings
Cooling fan switch .... 18 Nm
Coolant filler neck-to-thermostat housing bolts............... 12 Nm
Coolant temperature sender .... 9 Nm
Radiator mounting bolt.. 12 Nm
Reservoir tank mounting bolt.................................. 12 Nm
Thermostat housing cover bolts................................ 12 Nm
Water pump drain bolt... 12 Nm
Water pump body-to-crankcase bolts............................ 9 Nm
Water pump cover bolts.. 12 Nm

3•2 Cooling system

1 General information

The cooling system uses a water/antifreeze coolant to carry away excess energy in the form of heat. The cylinders are surrounded by a water jacket from which the heated coolant is circulated by thermo-syphonic action in conjunction with a water pump, driven by the oil pump. The hot coolant passes upwards to the thermostat and through to the radiator. The coolant then flows across the radiator core, where it is cooled by the passing air, to the water pump and back to the engine where the cycle is repeated.

A thermostat is fitted in the system to prevent the coolant flowing through the radiator when the engine is cold, therefore accelerating the speed at which the engine reaches normal operating temperature. A coolant temperature sender mounted in the thermostat housing transmits to the temperature gauge on the instrument panel. A thermostatically-controlled cooling fan is fitted behind the radiator to aid cooling in extreme conditions.

The complete cooling system is partially sealed and pressurised, the pressure being controlled by a spring-loaded valve contained in the spring-loaded pressure cap. By pressurising the coolant the boiling point is raised, preventing premature boiling in adverse conditions. The overflow pipe from the system is connected to a reservoir into which excess coolant is expelled under pressure. The discharged coolant automatically returns to the radiator when the engine cools.

Coolant is routed from a union at the back of the cylinder head to the carburettors to prevent icing in extreme conditions and is returned via a hose to the water pump.

Warning: Do not remove the pressure cap from the filler neck when the engine is hot. Scalding hot coolant and steam may be blown out under pressure, which could cause serious injury. When the engine has cooled, place a thick rag, like a towel, over the pressure cap; slowly rotate the cap anti-clockwise to the first stop. This procedure allows any residual pressure to escape. When the steam has stopped escaping, press down on the cap while turning it anti-clockwise and remove it.

Warning: Do not allow antifreeze to come in contact with your skin or painted surfaces of the motorcycle. Rinse off any spills immediately with plenty of water. Antifreeze is highly toxic if ingested. Never leave antifreeze lying around in an open container or in puddles on the floor; children and pets are attracted by its sweet smell and may drink it. Check with the local authorities about disposing of used antifreeze. Many communities will have collection centres which will see that antifreeze is disposed of safely.

Caution: At all times use the specified type of antifreeze, and always mix it with distilled water in the correct proportion. The antifreeze contains corrosion inhibitors which are essential to avoid damage to the cooling system. A lack of these inhibitors could lead to a build-up of corrosion which would block the coolant passages, resulting in overheating and severe engine damage. Distilled water must be used as opposed to tap water to avoid a build-up of scale which would also block the passages.

3.1a The coolant reservoir (arrowed) is located under the air filter housing on CB models

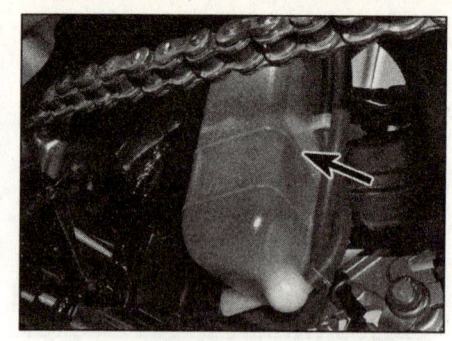

3.1b The coolant reservoir (arrowed) – CBF models

2 Pressure cap – check

1 If problems such as overheating or loss of coolant occur, check the entire system as described in Chapter 1. The filler cap opening pressure should be checked by a Honda dealer with the special tester required to do the job. If the cap is defective, renew it.

3 Coolant reservoir – removal and installation

Removal

1 The coolant reservoir is located behind the engine unit, on the right-hand side under the air filter housing on CB models, and on the left-hand side on CBF models **(see illustrations)**. On CB models remove the battery (see Chapter 9), tool box, rear mudguard and rear wheel (see Chapter 7).

2 Release the clip securing the breather hose (coming out of the top of the reservoir) and detach the hose **(see illustration)**.

3 Place a suitable container underneath the reservoir, then release the clip securing the reservoir. Detach the hose and allow the coolant to drain into the container.

4 Unscrew the reservoir mounting bolt and remove the reservoir, noting how it locates.

Installation

5 Installation is the reverse of removal. Make

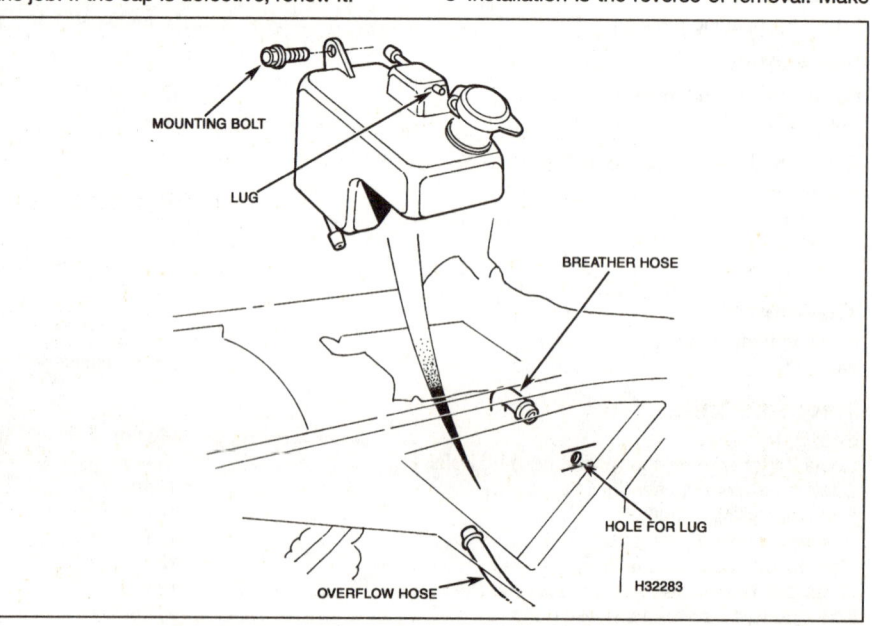

3.2 Coolant reservoir mounting details – CB models

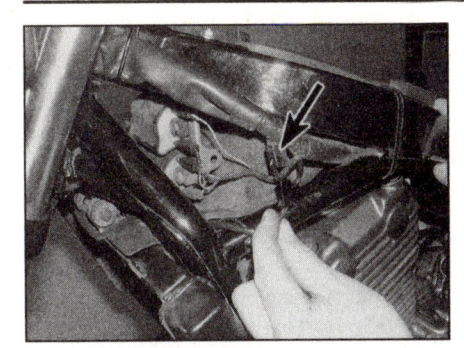

Cooling system 3•3

4.3 Disconnect the fan motor wiring connector (arrowed)

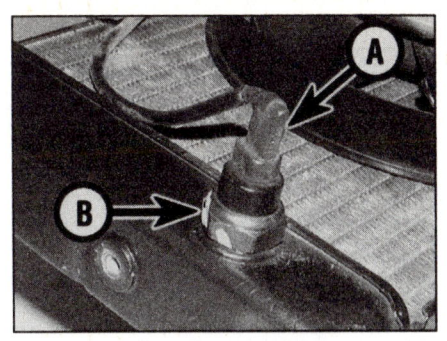

4.5a Disconnect the wiring connector (A) from the fan switch (B)

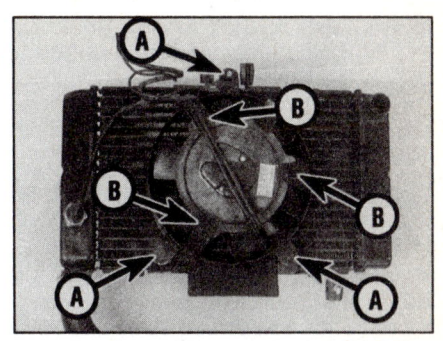

4.5b Fan assembly mounting bolts (A), fan shroud nuts (B)

sure the hoses are correctly installed and secured with their clips. On completion refill the reservoir as described in Chapter 1.

4 Cooling fan and cooling fan switch – check and replacement

Cooling fan

Check

1 If the engine is overheating and the cooling fan isn't coming on, first check the coolant level (see *Daily (pre-ride) checks*). If the level is correct, check the fan circuit fuse (see Chapter 9) and then the fan switch as described in Steps 10 to 14 below.

2 If the fan does not come on (and the fan switch is good), the fault lies in either the cooling fan motor or the relevant wiring. Test all the wiring and connections as described in Chapter 9.

3 To test the cooling fan motor, first remove the fuel tank (see Chapter 4). Trace the fan motor wiring from the motor and disconnect it at the 2-pin connector behind the steering head **(see illustration)**. Using a 12 volt battery and two jumper wires, connect the battery positive (+ve) lead to the black/blue motor wire and the battery negative (-ve) lead to earth. Once connected the fan should operate. If it does not, then the motor is faulty and should be renewed.

Replacement

⚠ *Warning: The engine must be completely cool before carrying out this procedure.*

4 Remove the radiator (see Section 7).
5 Disconnect the wiring connector from the fan switch **(see illustration)**. Unscrew the three bolts securing the fan shroud and fan assembly to the radiator, noting that one bolt also secures the earth (ground) cable, and remove the fan **(see illustration)**.
6 Unscrew the three nuts securing the fan assembly to the shroud.
7 Hold the fan with a rag to prevent damage to the blades and unscrew the fan centre nut. Remove the fan from the motor shaft, noting how it fits on the shaft.
8 Installation is the reverse of removal. Do not forget to attach the earth (ground) cable.
9 Install the radiator (see Section 7).

Cooling fan switch

Check

10 If the engine is overheating and the cooling fan isn't coming on, first check the coolant level (see *Daily (pre-ride) checks*). If the level is correct, check the fan circuit fuse (see Chapter 9). If the fuse is blown, check the fan circuit for a short to earth (see the wiring diagrams at the end of this book).
11 If the fuse is good, remove the radiator left-hand side panel (R, T, V, W, X, Y and 2 models – Chapter 8), and disconnect the wiring connector from the fan switch on the left-hand side of the radiator **(see illustration)**. Using a jumper wire if necessary, connect the wire to earth (ground) and turn the ignition ON. If the fan comes on, the switch or the switch connection is defective and must be renewed. If it does not come on, the fan motor should be tested (see Step 3).
12 If the fan stays on all the time, disconnect the wiring connector and turn the ignition ON. The fan should stop. If it does, the switch is defective and must be renewed. If it doesn't, check the wiring between the switch and the fan for a short to earth, and the fan motor itself.
13 If the fan works but is suspected of cutting in at the wrong temperature, a more comprehensive test of the switch can be made as follows. Remove the switch (see Steps 15 to 16). Fill a small heatproof container with coolant and place it on a stove. Connect the positive (+ve) probe of an ohmmeter to the terminal of the switch and the negative (-ve) probe to the switch body. Using some wire or other support, suspend the switch in the coolant so that just the sensing portion and the threads are submerged **(see illustration)**. Also place a thermometer capable of reading temperatures up to 110°C in the coolant so

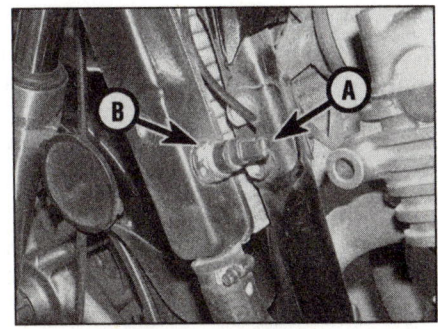

4.11 Disconnect the wiring connector (A) from the fan switch (B)

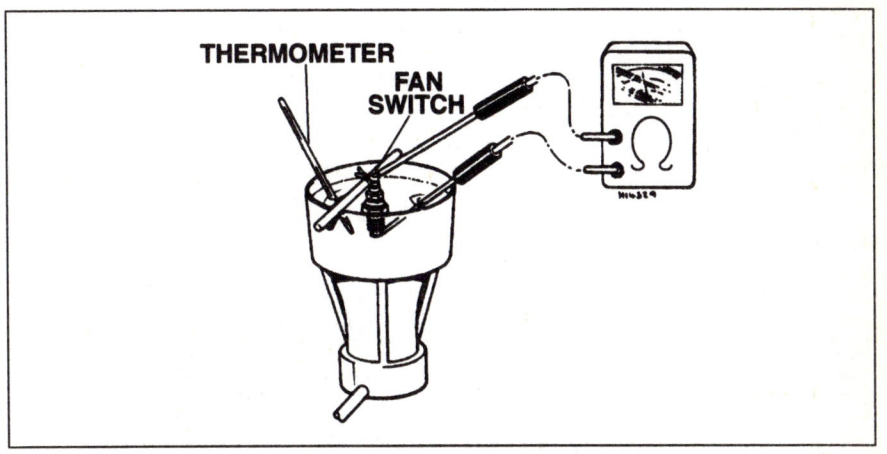

4.13 Arrangement for testing the fan switch

3•4 Cooling system

that its bulb is close to the switch. **Note:** *None of the components should be allowed to directly touch the container.*

14 Initially the ohmmeter reading should be very high indicating that the switch is open (OFF). Heat the coolant, stirring it gently.

> ⚠ **Warning:** *This must be done very carefully to avoid the risk of personal injury.*

When the temperature reaches around 98 to 102°C the meter reading should drop to around zero ohms, indicating that the switch has closed (ON). Now turn the heat off. As the temperature falls below 93 to 97°C the meter reading should show infinite (very high) resistance, indicating that the switch has opened (OFF). If the meter readings obtained are different, or they are obtained at different temperatures, then the fan switch is faulty and must be renewed.

Replacement

> ⚠ **Warning:** *The engine must be completely cool before carrying out this procedure.*

15 Drain the cooling system (see Chapter 1).
16 Disconnect the wiring connector from the fan switch on the left-hand side of the radiator **(see illustration 4.11)**. Unscrew the switch and withdraw it from the radiator. Discard the O-ring as a new one must be used.
17 Apply a suitable sealant to the switch threads, then install the switch using a new O-ring and tighten it to the torque setting specified at the beginning of this Chapter. Take care not to overtighten the switch as the radiator could be damaged.
18 Reconnect the switch wiring and refill the cooling system (see Chapter 1).

5 Coolant temperature gauge and sender – check and replacement

Coolant temperature gauge

Check

1 The circuit consists of the sender mounted in the thermostat housing under the fuel tank, and the gauge mounted in the instrument panel. If the gauge malfunctions, first check the coolant level (see *Daily (pre-ride) checks*). If the level is correct, check that the battery is fully charged and that the fuses are all good.
2 If the gauge is still not working, remove the fuel tank (see Chapter 4). Disconnect the wire from the sender and turn the ignition switch ON **(see illustration)**. The temperature gauge needle should be on the 'C' on the gauge. Now earth the sender wire on the engine. The needle should swing immediately over to the 'H' on the gauge. If the needle moves as described, the sender is defective and must be renewed.
Caution: *Do not earth the wire for any longer than is necessary to take the reading, or the gauge may be damaged.*

3 If the needle movement is still faulty, or if it does not move at all, the fault lies in the wiring or the gauge itself. Check all the relevant wiring and wiring connectors (see Chapter 9). If all appears to be well, the gauge is defective and must be renewed.

Replacement

4 See Chapter 9.

Temperature gauge sender

Check

5 Remove the fuel tank (see Chapter 4). The sender is mounted in the thermostat housing.
6 Disconnect the sender wiring connector **(see illustration 5.2)**. Using a continuity tester, check for continuity between the sender body and earth (ground). There should be continuity. If there is no continuity, check that the thermostat mounting is secure.
7 Remove the sender (see Steps 9 and 10 below). Fill a small heatproof container with coolant and place it on a stove. Connect the positive (+ve) probe of an ohmmeter to the terminal on the sender and the negative (-ve) probe to the sender body **(see illustration 4.13)**. Using some wire or other support, suspend the sender in the coolant so that just the sensing portion and the threads are submerged. Also place a thermometer capable of reading temperatures up to 120°C in the coolant so that its bulb is close to the sender. **Note:** *None of the components should be allowed to directly touch the container.*
8 Heat the coolant to approximately 40°C and keep the temperature constant for 3 minutes before continuing the test. Then increase the heat gradually, stirring the coolant gently.

> ⚠ **Warning:** *This must be done very carefully to avoid the risk of personal injury.*

As the temperature of the coolant rises, the resistance of the sender should fall. Check that the correct resistance is obtained at the temperatures specified at the beginning of this Chapter. If the meter readings obtained are widely different, or they are obtained at different temperatures, then the sender is faulty and must be renewed.

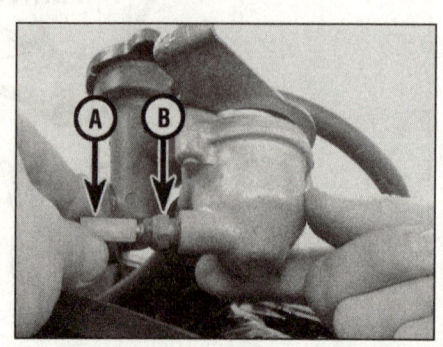

5.2 Wiring connector (A) and temperature sender (B)

Replacement

> ⚠ **Warning:** *The engine must be completely cool before carrying out this procedure.*

9 Partially drain the cooling system sufficient to remove the sender without spilling coolant over the engine and cycle parts (see Chapter 1).
10 Disconnect the sender wiring connector. Unscrew the sender and remove it from the thermostat housing.
11 Apply a suitable sealant to the sender threads, then install it into the thermostat housing and tighten it to the torque setting specified at the beginning of this Chapter. Connect the sender wiring.
12 Refill the cooling system (see Chapter 1).
13 Install the fuel tank (see Chapter 4).

6 Thermostat and thermostat housing – removal, check and installation

Removal

> ⚠ **Warning:** *The engine must be completely cool before carrying out this procedure.*

1 The thermostat is automatic in operation and should give many years service without requiring attention. In the event of a failure, the valve will probably jam open, in which case the engine will take much longer than normal to warm up. Conversely, if the valve jams shut, the coolant will be unable to circulate and the engine will overheat. Neither condition is acceptable, and the fault must be investigated promptly.
2 The thermostat is located in the thermostat housing, which is mounted to the frame under the fuel tank.
3 Partially drain the cooling system sufficient to remove the thermostat and thermostat housing without spilling coolant over the engine and cycle parts (see Chapter 1).
4 Release the clip securing the overflow hose to the radiator filler neck and disconnect the hose **(see illustration)**. Unscrew the mounting bracket nut (CB models) or bolts (CBF models)

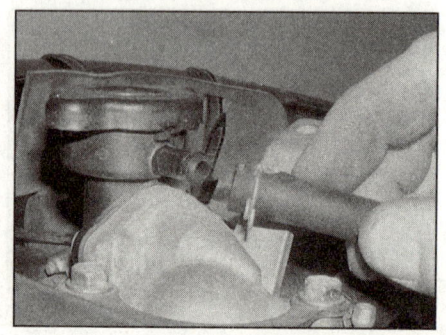

6.4a Disconnect the overflow hose from the filler neck

Cooling system 3•5

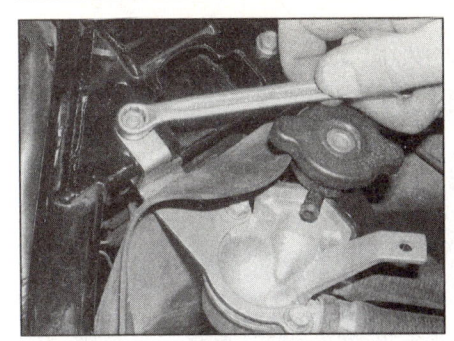

6.4b On CB models unscrew the housing bracket nut

6.4c Mounting bracket bolts (arrowed) – CBF models

6.5a Unscrew the thermostat housing cover bolts . . .

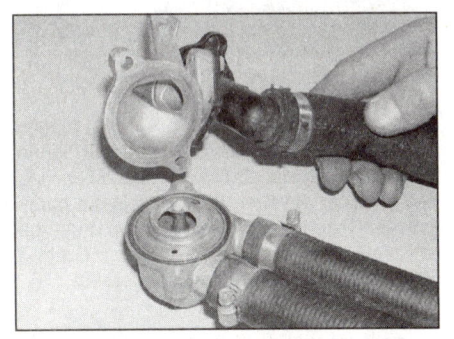

6.5b . . . lift off the cover . . .

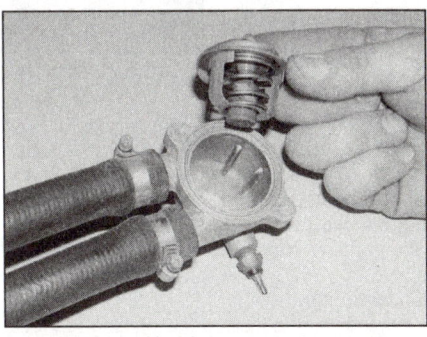

6.5c . . . and withdraw the thermostat

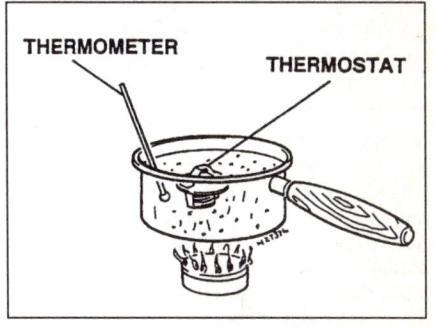

6.8 Arrangement for testing the thermostat

(see illustrations). On CB models release the bracket from the rubber shroud. If required for added freedom of movement disconnect the temperature sender wiring connector and free the coolant hose(s) from any clips.

5 Unscrew the two thermostat cover bolts and lift the cover off the housing. Withdraw the thermostat, noting how it fits (see illustrations). Discard the cover O-ring as a new one must be used.

6 To remove the thermostat housing, disconnect the temperature sender wire and slacken the clips securing the hoses to the filler neck and the thermostat housing and detach the hoses. If necessary, unscrew the two bolts that secure the filler neck to the thermostat cover and discard the O-ring as a new one must be fitted.

Check

7 Examine the thermostat visually before carrying out the test. If it remains in the open position at room temperature, it should be renewed.

8 Fill a small, heatproof container with cold water and place it on a stove. Using a piece of wire, suspend the thermostat in the water. Place a thermometer in the water so that the bulb is close to the thermostat (see illustration). Heat the water, noting the temperature when the thermostat opens, and compare the result with the specifications given at the beginning of this Chapter. Also check the amount the valve opens after it has been heated at 95°C for a few minutes and compare the measurement to the specifications. If the readings obtained differ from those given, the thermostat is faulty and must be renewed.

9 In the event of thermostat failure, as an emergency measure only, it can be removed and the machine used without it. **Note:** *Take care when starting the engine from cold as it will take much longer than usual to warm up. Ensure that a new unit is installed as soon as possible.*

Installation

10 Installation is the reverse of removal, using new O-rings. Make sure the thermostat seats correctly with the hole in the thermostat facing the left-hand side of the housing on CB models and the rear of the housing on CBF models **(see illustration)**.

11 Tighten the bolts to the torque settings specified at the beginning of this Chapter. On CB models thread the bracket through the rubber shroud **(see illustration)**. Tighten the bracket mounting nut or bolts and the hose clips. Connect the temperature sender wire.

12 Refill the cooling system (see Chapter 1).

13 Install the fuel tank (see Chapter 4).

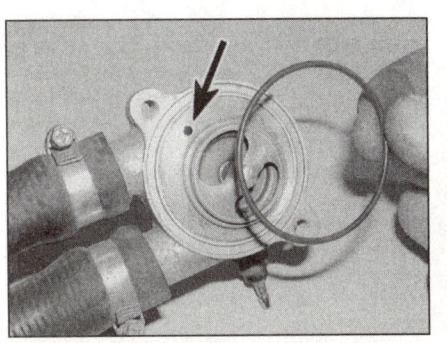

6.10 Align hole in thermostat as described and fit new O-ring

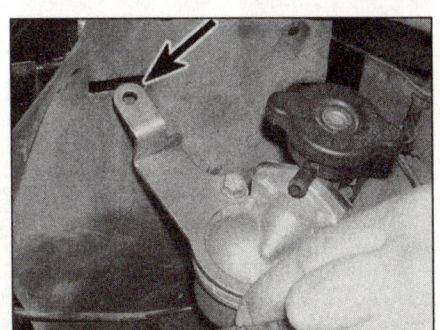

6.11 On CB models thread the bracket through the hole in the rubber shroud (arrowed)

7 Radiator – removal and installation

Removal

⚠️ **Warning: The engine must be completely cool before carrying out this procedure.**

1 On R, T, V, W, X, Y and 2 models remove the two radiator side panels; on SW, SX, SY and S2 models remove the fairing (see Chapter 8). Drain the cooling system (see Chapter 1). Unclip the stone guard from the front of the radiator.

2 Slacken the clips securing the main hoses

3•6 Cooling system

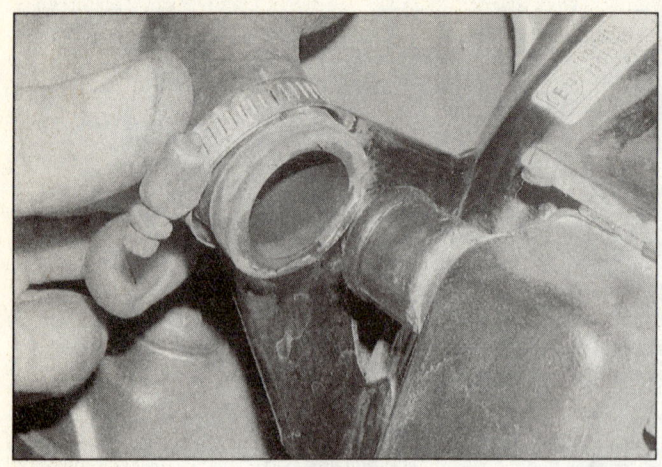

7.2 Detach the radiator hoses

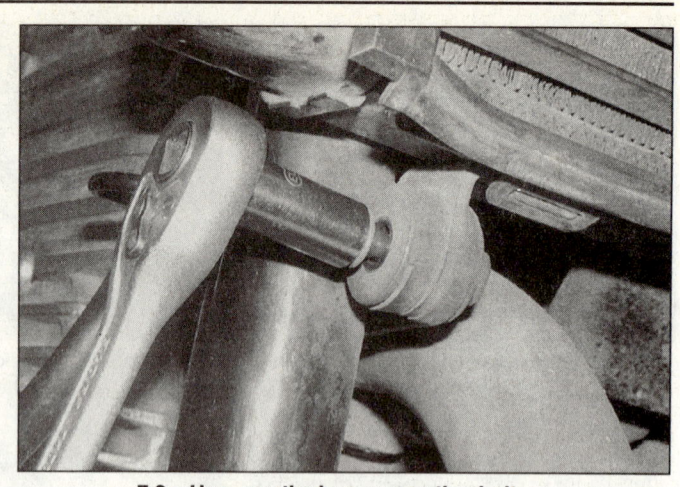

7.3a Unscrew the lower mounting bolt . . .

to the top right-hand side and bottom left-hand side of the radiator and detach the hoses **(see illustration)**. Disconnect the fan wiring connector **(see illustration 4.3)**.

3 Unscrew the radiator lower mounting bolt(s), noting the arrangement of the collar and rubber bush(es) **(see illustration)**. Unscrew the radiator upper mounting bolt, noting the arrangement of the collar and rubber bush **(see illustrations)**.

4 Pull the radiator to the left off the upper mounting lug on CB models, and to the right off the lower mounting lug on CBF models, and remove the radiator. On CB models take care not to damage the rubber shroud which is held in place by the radiator brackets **(see illustrations)**.

5 If necessary, separate the cooling fan (see Section 4) from the radiator.

6 Check the stone guard and the radiator for signs of damage and clear any dirt or debris that might obstruct air flow and inhibit cooling (see Chapter 1). If the radiator fins are badly damaged or broken the radiator must be renewed. Also check the rubber mounting grommets, and renew them if necessary.

Installation

7 Installation is the reverse of removal, noting the following.
 a) Ensure the rubber grommets and collars are correctly installed with the mounting bolts and that on CB models the rubber shroud is hooked over the upper radiator brackets.
 b) Ensure the stone guard is held securely by its four mounting clips.
 c) Make sure that the fan wiring is correctly connected.
 d) Ensure the coolant hoses are in good condition (see Chapter 1), and are securely retained by their clamps, using new ones if necessary.
 e) On completion refill the cooling system as described in Chapter 1.

7.3b . . . then the upper mounting bolt – CB models

7.3c Radiator mounting bolts (A) and lug (B) – CBF models

7.4a Radiator upper mounting lug (arrowed) – CB models

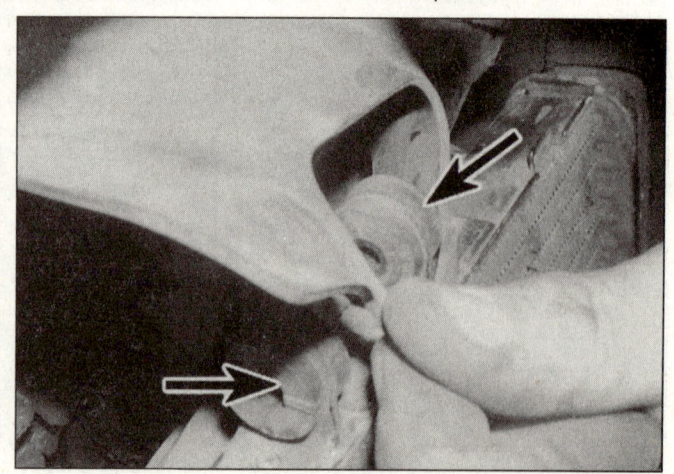

7.4b On CB models the rubber shroud locates over radiator brackets (arrowed)

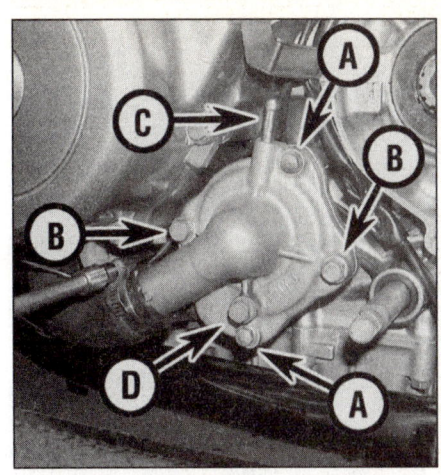

Cooling system 3•7

8.1 Pump to crankcase bolts (A), pump cover bolts (B), carburettor heater hose connection (C) and cooling system drain bolt (D)

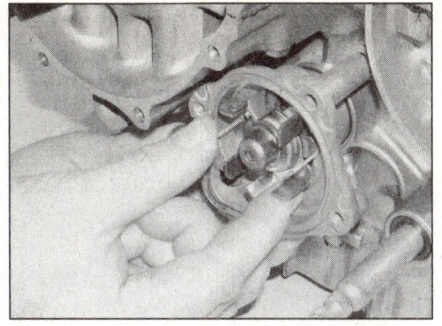

8.6 Checking for freeplay on the pump impeller

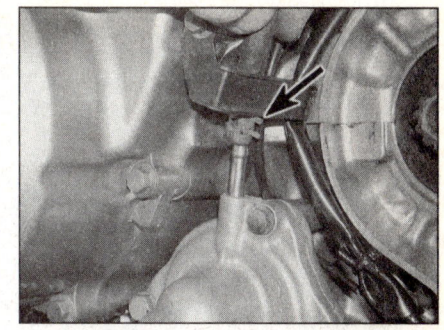

8.8a Detach the carburettor heater hose . . .

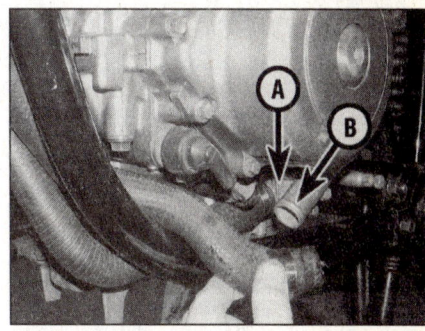

8.8b . . . and the hoses from the cylinder block (A) and radiator (B)

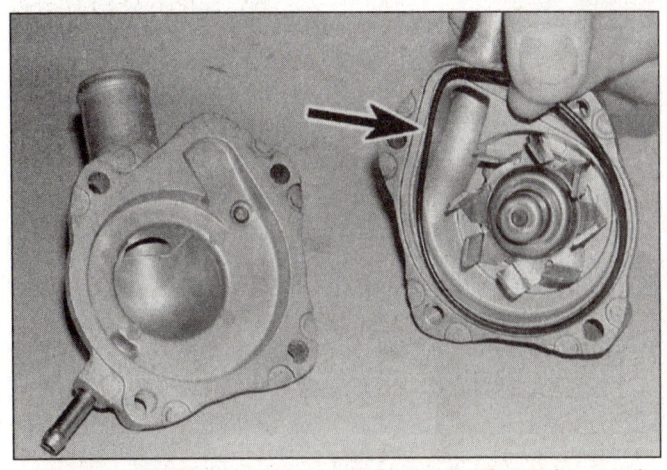

8.8c Remove the pump cover and discard the O-ring (arrowed)

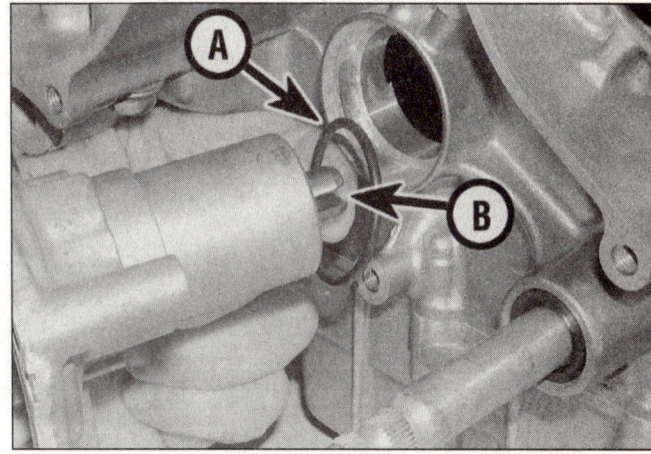

8.9 Pump body O-ring (A), slot in impeller shaft (B)

8 Water pump – check, removal and installation

Check

1 The water pump is located on the lower left-hand side of the engine (see illustration). Visually check the area around the pump for signs of leakage.

2 To prevent leakage of water from the cooling system to the lubrication system and vice versa, two seals are fitted on the pump shaft. On the underside of the pump body there is also a drainage hole. If either seal fails, this hole should allow the coolant or oil to escape and prevent the oil and coolant mixing.

3 The pump body is only available as a complete assembly. Therefore, if on inspection the drainage hole shows signs of leakage, the pump must be removed and renewed.

Removal

4 Drain the cooling system (see Chapter 1). Remove the front sprocket cover to allow easier access to the hoses. Place a suitable container below the water pump to catch any residual oil as the water pump is removed.

5 To remove the pump cover, unscrew the four bolts securing the cover to the pump (the upper and lower bolts also secure the pump to the crankcase), and remove the cover (see illustration 8.1). There is no need to detach the hoses unless you want to. Note the position of each bolt as they are different lengths. Discard the cover O-ring as a new one must be used.

6 Check for freeplay between the pump impeller and the pump body (see illustration). If there is excessive movement the pump must be renewed. Also check for corrosion or a build-up of scale in the pump body and clean or renew the pump as necessary.

7 To remove the pump body, slacken the clamp securing the coolant hose to the pump body and detach the hose. Carefully draw the pump from the crankcase, noting how it fits. Remove the O-ring from the rear of the pump body and discard it as a new one must be used.

8 To remove the whole pump as an assembly, slacken the clamps securing the coolant hoses to the pump cover and body, noting that access to the inner main hose is restricted (detach it from the engine if it proves stubborn). Detach the hoses, noting which fits where (see illustrations). Unscrew the two bolts securing the pump assembly to the crankcase (see Step 5) and carefully draw the pump out, noting how it fits. Remove the O-ring from the rear of the pump body and discard it as a new one must be used. Separate the cover from the pump if required and discard the O-ring (see illustration).

Installation

9 Apply a smear of clean engine oil to the pump body O-ring and install it onto the rear of the pump body, then carefully press the pump body into the crankcase, aligning the slot in the impeller shaft with the oil pump shaft (see illustration).

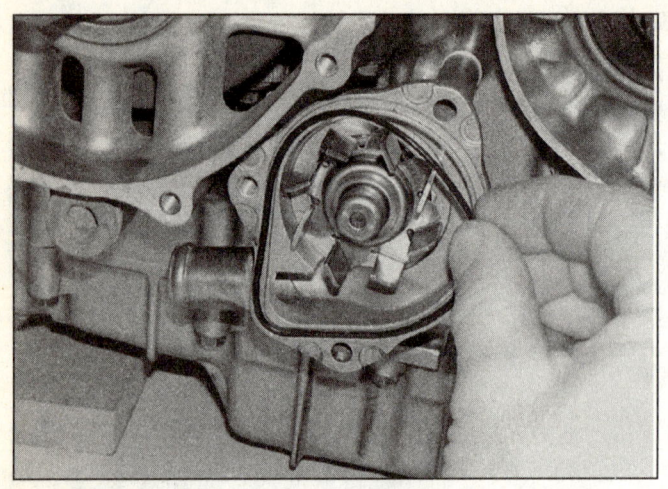

8.11 Fit a new O-ring into the groove in the pump body

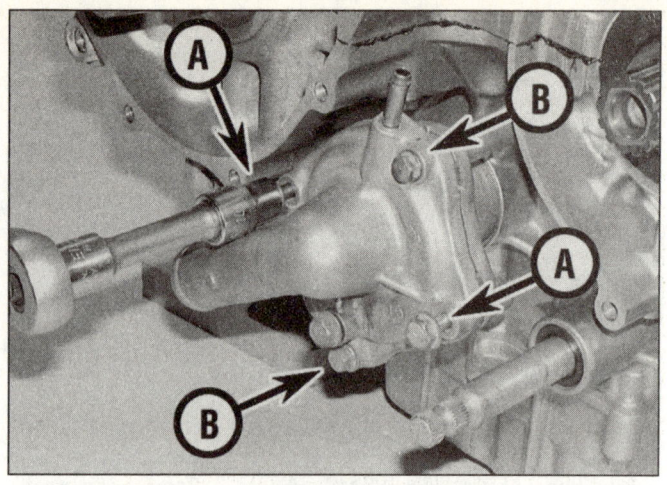

8.12 Pump body bolts (A) are shorter than pump-to-crankcase bolts (B)

10 Attach the coolant hose to the pump body and secure it with the clip.
11 Install a new O-ring into its groove in the pump body **(see illustration)**.
12 Install the pump cover onto the pump, then install the bolts and tighten them to the torque settings specified at the beginning of this Chapter. Make sure the different length bolts are in their correct locations **(see illustration)**.
13 If detached, attach the coolant hoses to the pump cover and secure them with their clips **(see illustrations 8.8a and b)**.
14 If the pump was removed as a whole, install the two bolts securing the pump to the crankcase and tighten them to the torque settings specified at the beginning of this Chapter. Attach the coolant hoses to the pump and secure them with their clips **(see illustrations 8.8a and b)**.
15 Install the front sprocket cover and refill the cooling system (see Chapter 1).

9 Coolant hoses – removal and installation

Removal

1 Before removing a hose, drain the cooling system (see Chapter 1).
2 Use a screwdriver or small socket to slacken the larger-bore hose clips, then slide them back along the hose and clear of the union spigot. The smaller-bore hoses are secured by spring clips which can be expanded by squeezing their ends together with pliers **(see illustrations)**.

Caution: The radiator unions are fragile. Do not use excessive force when attempting to remove the hoses.

3 If a hose proves stubborn, release it by rotating it on its union before working it off. If all else fails, cut the hose with a sharp knife then slit it at each union so that it can be peeled off in two pieces. Whilst this means renewing the hose, it is preferable to buying a new radiator.
4 The coolant pipe inlet union at the front of the cylinder block and the outlet union at the rear of the cylinder head can be removed by unscrewing the retaining bolts **(see illustrations)**.

Installation

5 Slide the clips onto the hose and then work the hose onto its respective union.

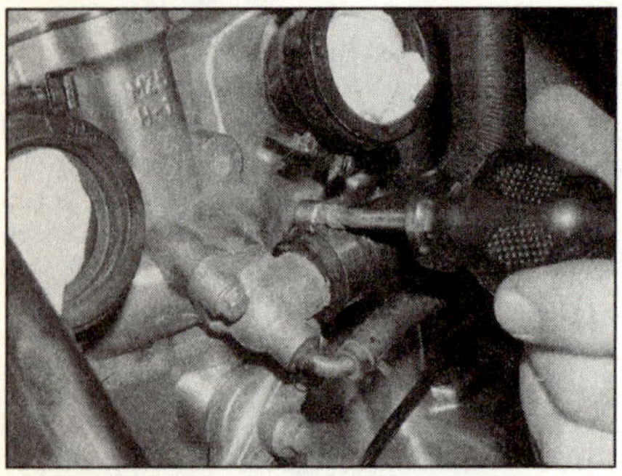

9.2a Release clips on large-bore hoses with a screwdriver . . .

9.2b . . . or a small socket

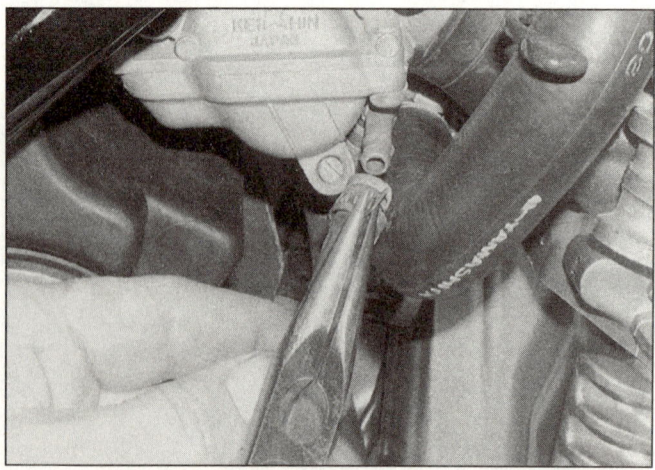

9.2c Clips on small-bore hoses can be released with pliers

Cooling system 3•9

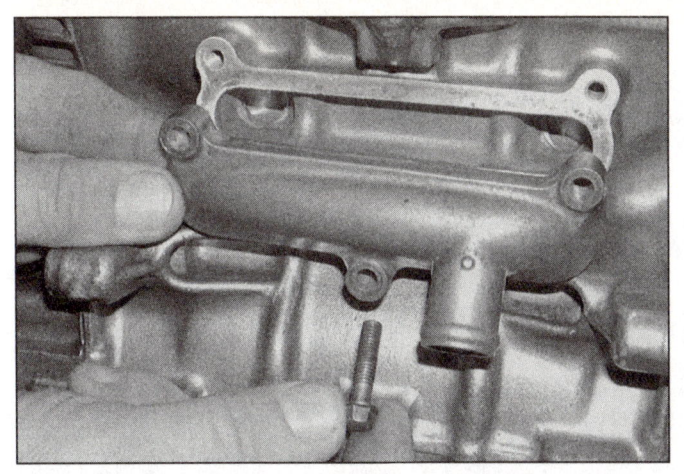

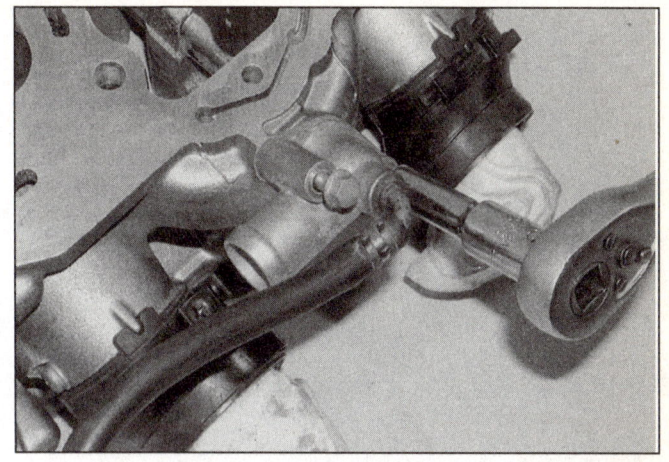

9.4a Union on the front of the block is retained by three bolts . . .

9.4b . . . and on the back of the head by two bolts

> **HAYNES HiNT** *If the hose is difficult to push on its union, it can be softened by soaking it in very hot water, or alternatively a little soapy water can be used as a lubricant.*

6 Rotate the hose on its unions to settle it in position before sliding the clips into place and tightening them securely.

7 If either the inlet union to the cylinder block or the outlet union from the cylinder head has been removed, fit a new O-ring, then install the union and tighten the mounting bolts securely **(see illustration)**.

8 Refill the cooling system (see Chapter 1).

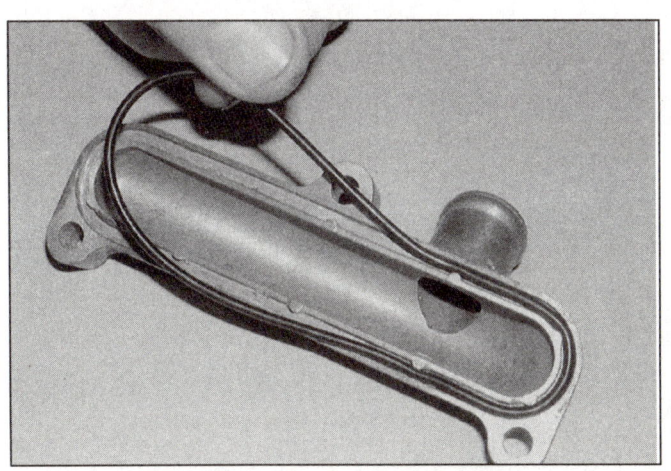

9.7a Fit new O-rings to the front union . . .

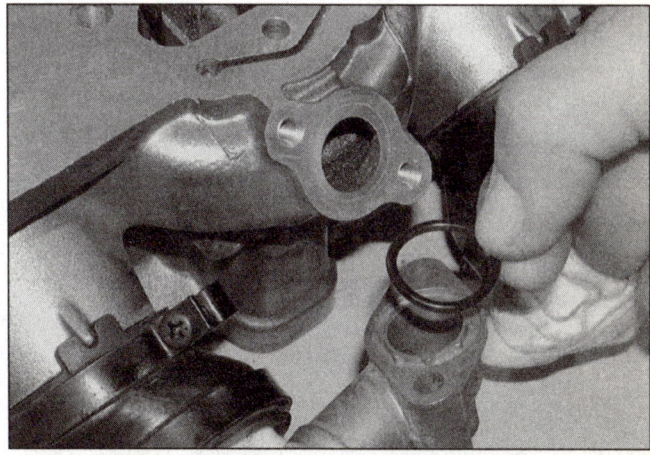

9.7b . . . and the rear union before reassembly

Chapter 4
Fuel and exhaust systems

Contents

Air filter – cleaning and renewal .... see Chapter 1
Air filter housing – removal and installation 12
Air/fuel mixture adjustment – general information 4
Carburettor overhaul – general information 5
Carburettor synchronisation .... see Chapter 1
Carburettors – removal and installation 6
Carburettors – disassembly, cleaning and inspection 7
Carburettors – reassembly, and float height check 8
Carburettors – separation and joining 9
Choke cable – removal and installation 11
Exhaust system – removal and installation 13
Fuel hoses – check and renewal .... see Chapter 1
Fuel system – check .... see Chapter 1
Fuel tank – cleaning and repair 3
Fuel tank, fuel tap and level sensor – removal and installation 2
General information and precautions 1
Idle speed – check .... see Chapter 1
PAIR system – check, removal and installation 15
Sub-air cleaner .... see Chapter 1
Throttle and choke cables – check and adjustment .... see Chapter 1
Throttle cables – removal and installation 10
Throttle position sensor – check, removal and installation
 (CBF models) 14

Degrees of difficulty

| Easy, suitable for novice with little experience | Fairly easy, suitable for beginner with some experience | Fairly difficult, suitable for competent DIY mechanic | Difficult, suitable for experienced DIY mechanic | Very difficult, suitable for expert DIY or professional |

Specifications

Fuel
Grade .... Unleaded, minimum 91 RON (Research Octane Number)
Fuel tank capacity
  CB models .... 18 litres
  CBF models .... 19 litres

Carburettors
Type .... Keihin VP 34 mm
Pilot screw initial setting (turns out) – see text
  CB models .... 2 1/2 turns out
  CBF models .... 2 turns out
Float height .... 13.7 mm
Idle speed .... 1300 rpm (± 100 rpm)
Pilot jet 38
Main jet
  CB models 122
  CBF models 128

Throttle position sensor – CBF models
Input voltage .... 4.7 to 5.3 volts
Resistance .... 4 to 6 K ohms @ 20°C

Torque settings
Carburettor joining nut (upper) .... 10 Nm
Carburettor joining nut (lower) .... 5 Nm
Exhaust downpipe nuts .... 13 Nm
Exhaust and silencer mounting bolts
  CB models .... 27 Nm
  CBF models .... 22 Nm
Fuel level sensor .... 2 Nm
Silencer clamp bolt .... 21 Nm

4•2 Fuel and exhaust systems

1 General information and precautions

General information

The fuel system consists of the fuel tank, fuel tap, filter, carburettors, fuel hoses and control cables.

The fuel filter is part of the tap and is fitted inside the fuel tank.

The carburettors used on all models are CV types with one carburettor for each cylinder. For cold starting, a choke lever or knob is mounted on the left-hand handlebar, and is connected to the carburettors by a cable. To prevent carburettor icing, coolant is routed via a small-bore hose from the union on the back of the cylinder head to the carburettor bodies and then back to the cooling system via a union on the water pump cover.

Air is drawn into the carburettors through an air filter which is housed under the carburettors.

The exhaust system is a two-into-one design.

Many of the fuel system service procedures are considered routine maintenance items and for that reason are included in Chapter 1.

Precautions

⚠ **Warning:** *Petrol (gasoline) is extremely flammable, so take extra precautions when you work on any part of the fuel system. Don't smoke or allow open flames or bare light bulbs near the work area, and don't work in a garage where a natural gas-type appliance is present. If you spill any fuel on your skin, rinse it off immediately with soap and water. When you perform any kind of work on the fuel system, wear safety glasses and have a fire extinguisher suitable for a class B type fire (flammable liquids) on hand.*

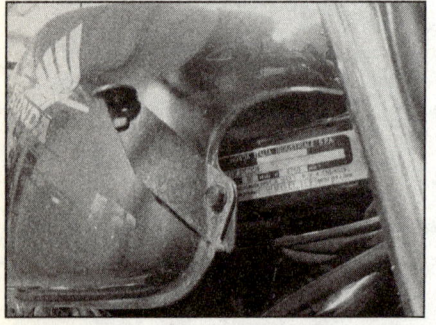

Always perform service procedures in a well-ventilated area to prevent a build-up of fumes.

Never work in a building containing a gas appliance with a pilot light, or any other form of naked flame. Ensure that there are no naked light bulbs or any sources of flame or sparks nearby.

Do not smoke (or allow anyone else to smoke) while in the vicinity of petrol (gasoline) or of components containing it. Remember the possible presence of vapour from these sources and move well clear before smoking.

Check all electrical equipment belonging to the house, garage or workshop where work is being undertaken (see the Safety first! section of this manual). Remember that certain electrical appliances such as drills, cutters etc. create sparks in the normal course of operation and must not be used near petrol (gasoline) or any component containing it. Again, remember the possible presence of fumes before using electrical equipment.

Always mop up any spilt fuel and safely dispose of the rag used.

Any stored fuel that is drained off during servicing work must be kept in sealed containers that are suitable for holding petrol (gasoline), and clearly marked as such; the containers themselves should be kept in a safe place. Note that this last point applies equally to the fuel tank if it is removed from the machine; also remember to keep its filler cap closed at all times.

Read the Safety first! section of this manual carefully before starting work.

2 Fuel tank and fuel tap – removal and installation

⚠ **Warning:** *Refer to the precautions given in Section 1 before starting work.*

Fuel tank

Removal

Note: *If it is necessary to drain the tank, connect a length of fuel hose to the tap and place the free end of the hose in a suitable container below the level of the fuel in the tank. On CB models turn the tap to the RES position. Draw vacuum in the vacuum hose; fuel should flow from the tank. Clamp the vacuum hose until sufficient fuel has been drawn from the tank, then turn the tap OFF and disconnect the fuel hose.*

1 Make sure the fuel filler cap is secure. On CB models make sure the fuel tap is in the OFF position. Remove the seat(s) and the frame side panels; on R, T, V, W, X, Y and 2 models remove the radiator side panels and on SW, SX, SY and S2 models detach the lower edge of the fairing from the fuel tank (see Chapter 8) **(see illustration)**. Place a rag under the fuel tap to catch any residual fuel as the hose is detached.

2 On CB models release the clip securing the fuel hose to the tap and detach the hose **(see illustration)**. Release the clip securing the vacuum hose to the left-hand inlet duct and detach the hose **(see illustration)**.

3 On CBF models disconnect the fuel level sensor wiring connector **(see illustration)**. Release the clip securing the fuel hose to the tap and detach the hose. Release the clip securing the vacuum hose to the tap and detach the hose.

4 Unscrew the bolt securing the rear of the tank to the frame and lift the tank to gain access to the clips securing the drain hose and breather hose under the right-hand side of the tank. Release the clips and detach the hoses **(see illustrations)**.

2.1 On SW, SX, SY and S2 models, release the screws holding the fairing to the brackets on the front of the tank

2.2a Detach the fuel hose from the tap . . .

2.2b . . . and the vacuum hose from the inlet manifold

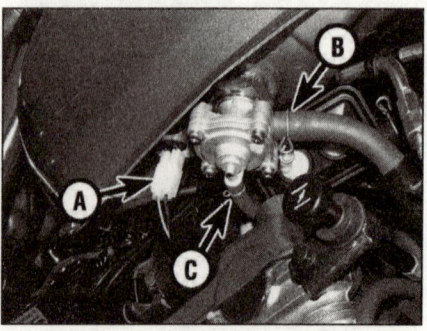

2.3 Fuel level sensor wiring connector (A), fuel hose (B), vacuum hose (C)

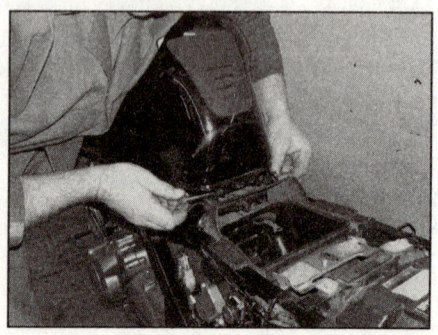

2.4a Unscrew the bolt at the back of the tank . . .

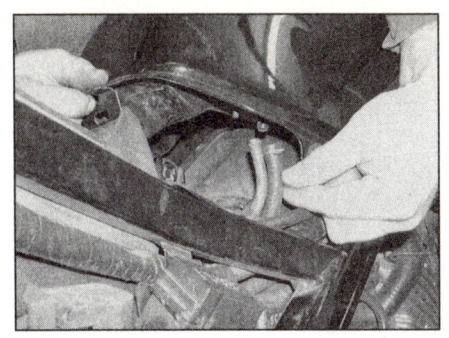

Fuel and exhaust systems 4•3

2.4b . . . and detach the drain and breather hoses

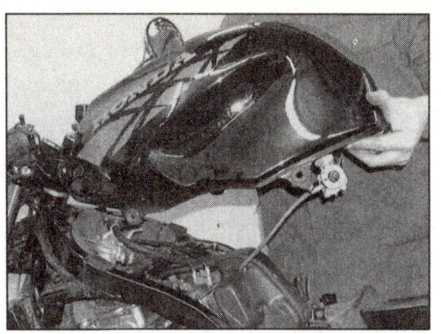

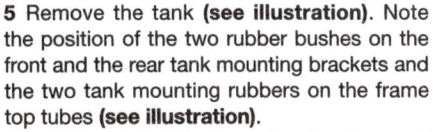

2.5a Remove the tank . . .

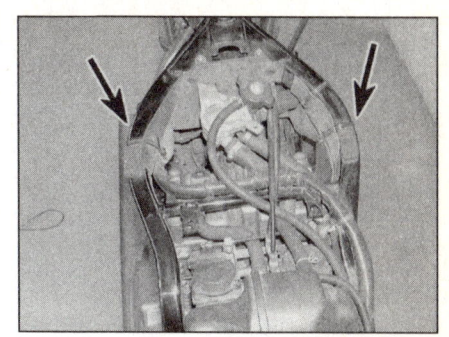

2.5b . . . noting the mounting rubbers (arrowed) on the frame tubes

5 Remove the tank **(see illustration)**. Note the position of the two rubber bushes on the front and the rear tank mounting brackets and the two tank mounting rubbers on the frame top tubes **(see illustration)**.

6 Inspect the tank mounting bushes and rubbers for signs of damage or deterioration and renew them if necessary **(see illustration)**.

Installation

7 Check that the tank mounting bushes are fitted, then carefully lower the fuel tank into position.

8 Fit the drain hose and breather hose and install the hose clips **(see illustration)**. Ensure the hoses are secure in their guides on the side of the air filter housing **(see illustration)**.

9 Install the tank rear mounting bolt and washer and tighten it securely **(see illustration)**.

10 On CB models fit the fuel hose onto its union on the tap and secure it with the clip **(see illustration)**. Fit the vacuum hose onto the left-hand inlet duct and secure it with the clip **(see illustration)**.

11 On CBF models fit the vacuum hose and fuel hose onto the tap and secure them with the clips **(see illustration 2.3)**. Connect the fuel level sensor wiring connector.

12 Turn the fuel tap ON. Start the engine and check that there is no sign of fuel leakage.

13 Install the seat and the frame side panels (see Chapter 8).

Fuel tap

14 The tap should not be removed from the tank unless the filter needs cleaning or the O-ring between the tap and tank is leaking. Remove the tank (see Steps 1 to 5) and refer to Chapter 1, Section 11 for details of these operations.

15 If fuel flow problems are experienced and this is not due to blockage of the filter, the tap diaphragm could be faulty or the vacuum pipe could be at fault; check that the vacuum pipe at it connection points are air tight before dismantling the tap.

16 To check the operation of the tap diaphragm, remove the fuel tank and move it to the work bench. Connect lengths of hose to the fuel and vacuum outlets on the tap and place the free end of the fuel hose into a suitable container below the level of fuel in the tank. Apply suction to the vacuum hose, on CB models with the tap lever in the ON and RES positions consecutively. Fuel should flow from the fuel pipe, on CB models in both tap positions if the level of fuel inside the tank is sufficient. If fuel doesn't flow the diaphragm is

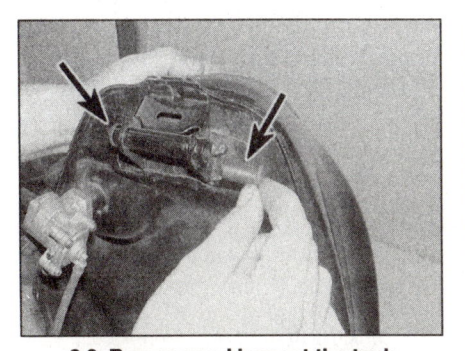

2.6 Remove and inspect the tank mounting bushes (arrowed)

2.8a Secure the drain and breather hoses to the tank with their clips (arrowed) . . .

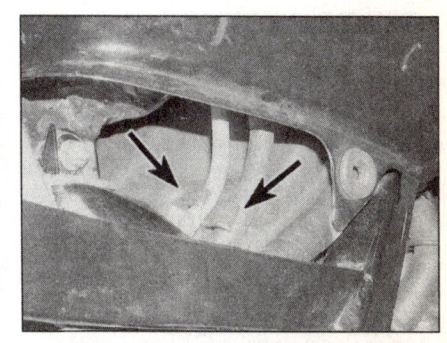

2.8b . . . and secure the hoses in the guides (arrowed) on the air filter housing

2.9 Install the tank mounting bolt and tighten it securely

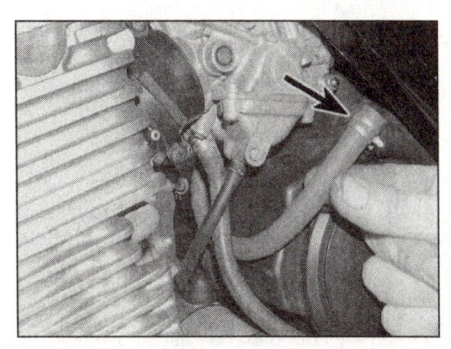

2.10a Fit the fuel hose and secure it with the clip (arrowed)

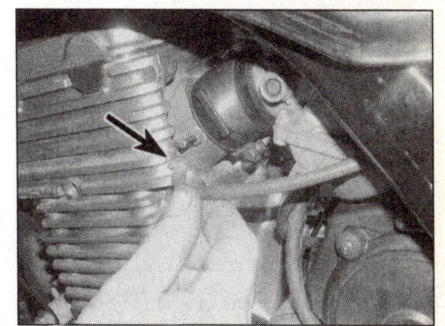

2.10b Fit the vacuum hose and secure it with the clip (arrowed)

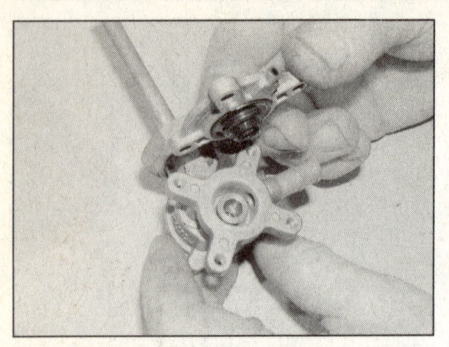

2.16 Remove the four screws to separate the tap and access the diaphragm

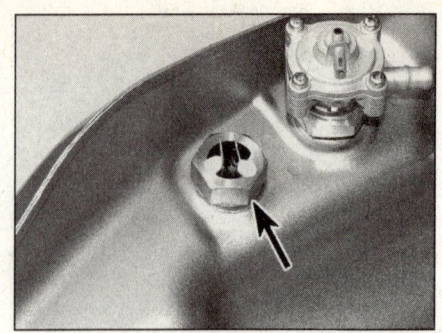

2.21 Unscrew and remove the sensor (arrowed)

4.1 Pilot screw on No. 1 carburettor (arrowed)

faulty and should be renewed. Turn the tap OFF and remove the tap from the tank as described in Chapter 1, Section 11. Remove the four screws to detach the diaphragm from the tap **(see illustration)**.

Fuel level sensor (CBF models)

Check

17 The low fuel warning light should come on when the volume of fuel in the tank drops to around 3.5 litres. If other warning lights are also faulty, check the power input to the instrument cluster (see Chapter 9).

18 To check the sensor, disconnect its wiring connectors **(see illustration 2.3)**. Using a jumper wire connect between the two wiring connectors. Turn the ignition switch ON – the low fuel warning light should come on. If it does the sensor is faulty and must be replaced with a new one. If it doesn't check the black/light green wire between the sensor and the instrument cluster for continuity, then check the green/black wire for continuity to earth. Repair or replace the wiring as required.

19 If the warning light does not go out when the tank is filled, disconnect the sensor wiring connectors. Turn the ignition switch ON – the low fuel warning light should not come on. If it does check the black/light green wire between the sensor and the instrument cluster for a short to earth. Repair or replace the wiring as required.

Removal and installation

20 Drain and remove the fuel tank.
21 Unscrew and remove the sensor **(see illustration)**. Discard the O-ring.
22 Fit a new O-ring onto the sensor, then fit the sensor into the tank and tighten it to the torque setting specified at the beginning of the Chapter. After installing and filling the tank check for leakage around the sensor.

3 Fuel tank – cleaning and repair

1 All repairs to the fuel tank should be carried out by a professional who has experience in this critical and potentially dangerous work. Even after cleaning and flushing of the fuel system, explosive fumes can remain and ignite during repair of the tank.

2 If the fuel tank is removed from the bike, it should not be placed in an area where sparks or open flames could ignite the fumes coming out of the tank. Be especially careful inside garages where a natural gas-type appliance is located, because the pilot light could cause an explosion.

4 Air/fuel mixture adjustment – general information

1 If the engine runs extremely rough at idle or continually stalls, and if a carburettor overhaul does not cure the problem (and it definitely is a carburation problem – see Section 6), the pilot screws may require adjustment. It is worth noting at this point that unless you have the experience to carry this out it is best to entrust the task to a motorcycle dealer, tuner or fuel systems specialist. The pilot screws are fairly accessible on this engine **(see illustration)**, although an adjuster tool (long thin flexible drive screwdriver with an angled end) can be obtained if access is difficult.

2 Before adjusting the pilot screws, warm the engine up to normal working temperature. Screw in the pilot screw on both carburettors until they seat lightly, then back them out to the number of turns specified (see this Chapter's Specifications). This is the base position for adjustment.

3 Start the engine and reset the idle speed to the correct level (see Chapter 1). Working on one carburettor at a time, turn the pilot screw by a small amount either side of this position to find the point at which the highest consistent idle speed is obtained. When you've reached this position, reset the idle speed to the specified amount (see Chapter 1). Repeat on the other carburettor.

4 Due to the increased emphasis on controlling exhaust emissions in certain world markets, regulations have been formulated which prevent adjustment of the air/fuel mixture. On such models the pilot screw positions are pre-set at the factory and in some cases have a limiter cap fitted to prevent tampering. Where adjustment is possible, it can only be made in conjunction with an exhaust gas analyser to ensure that the machine does not exceed the emissions regulations. **Note:** *On some carburettors a special Honda service tool (pt. no. 07KMA – MS60101) is required to engage the pilot screw heads.*

5 Carburettor overhaul – general information

1 Poor engine performance, hesitation, hard starting, stalling, flooding and backfiring are all signs that carburettor maintenance may be required.

2 Keep in mind that many so-called carburettor problems are really not carburettor problems at all, but mechanical problems within the engine or ignition system malfunctions. Try to establish for certain that the carburettors are in need of maintenance before beginning a major overhaul.

3 Check the fuel tap and filters, the fuel hoses, the inlet manifold joint clamps, the air filter, the ignition system, the spark plugs and carburettor synchronisation before assuming that a carburettor overhaul is required.

4 Most carburettor problems are caused by dirt particles, varnish and other deposits which build up in and block the fuel and air passages. Also, in time, gaskets and O-rings shrink or deteriorate and cause fuel and air leaks which lead to poor performance.

5 When overhauling the carburettors, disassemble them completely and clean the parts thoroughly with a carburettor cleaning solvent and dry them with filtered, unlubricated compressed air. Blow through the fuel and air passages with compressed air to force out any dirt that may have been loosened but not removed by the solvent. Once the cleaning process is complete, reassemble the carburettors using new gaskets and O-rings.

6 Before disassembling the carburettors, make sure you have all necessary gaskets and O-rings, some carburettor cleaner, a supply of clean rag, some means of blowing out the carburettor passages and a clean place to work. It is recommended that only one carburettor be overhauled at a time to avoid mixing up parts.

Fuel and exhaust systems 4•5

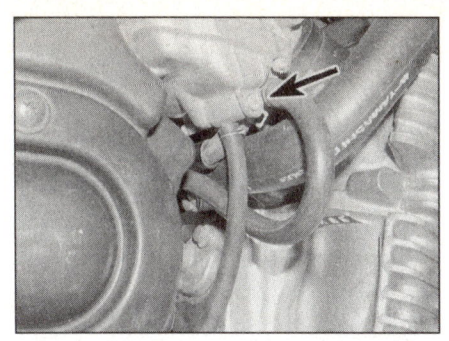

6.1 Carburettor drain screw (arrowed)

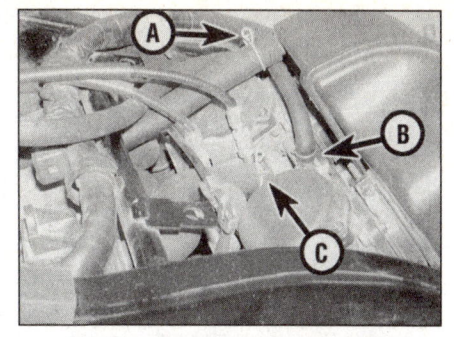

6.2 Crankcase breather hose (A), carburettor breather hose (B) and sub-air cleaner hose (C)

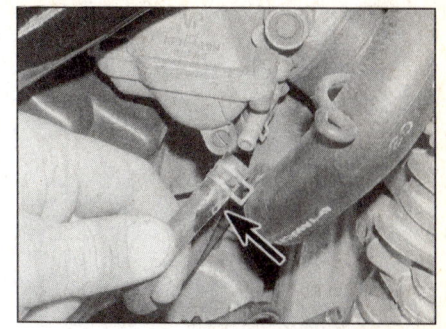

6.5 Carburettor heater hose (arrowed)

6 Carburettors – removal and installation

⚠ **Warning:** *Refer to the precautions given in Section 1 before starting work.*

Removal – CB models

1 Remove the fuel tank (see Section 2). Place a suitable container beneath the carburettor drain hoses underneath the exhaust system, unscrew the carburettor drain screws and allow fuel to empty from the system **(see illustration)**. Tighten the carburettor drain screws. Release the clips on the drain hoses and detach the hoses from the carburettors.
2 Release the clips on the crankcase breather hose between the valve cover and the air filter housing and remove the hose **(see illustration)**.
3 Release the clip on the carburettor breather hose and detach the hose **(see illustration 6.2)**.
4 Release the clip on the sub-air cleaner hose and remove the sub air cleaner assembly **(see illustration 6.2)**.

5 Release the clips from the carburettor heater hoses and detach the hoses, being prepared to catch any residual coolant **(see illustration)**. Clamp or plug the hoses from the cylinder head union and the water pump to prevent undue coolant loss.
6 Detach the choke cable from the carburettors (see Section 11, Step 2).
7 Slacken the clamps on the cylinder head inlet stubs and the air filter housing, then ease the carburettors off the inlet stubs, lifting them upwards while at the same time tilting the air filter housing back **(see illustration)**. The carburettors are a tight fit on the stubs and have to be completely free of them before they can be separated from the air filter housing. **Note:** *Keep the carburettors as upright as possible to prevent the possibility of the piston diaphragms being damaged*
8 Detach the throttle cables from the carburettors (see Section 10, Steps 3 and 4).
9 Release the clip securing the fuel supply hose to the inlet union on the carburettors and detach the hose.

Caution: *Stuff clean rag into each cylinder head inlet after removing the carburettors to prevent anything from falling in.*

10 If required, slacken the clamps securing the inlet stubs to the cylinder head and remove the stubs. The stubs are clearly marked; note which way round they fit and the location lugs on the underside of the head **(see illustrations)**.

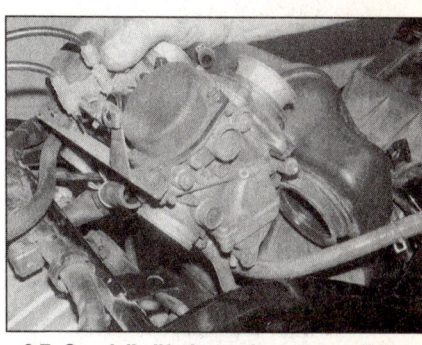

6.7 Carefully lift the carburettors off the inlet stubs

Installation – CB models

11 Installation is the reverse of removal, noting the following.
 a) Check for cracks or splits in the cylinder head inlet stubs, and renew them if necessary.
 b) If removed, make sure the inlet stubs are installed so that the slot in the underside of the stub locates over the lug on the underside of the head **(see illustration 6.10b)**.

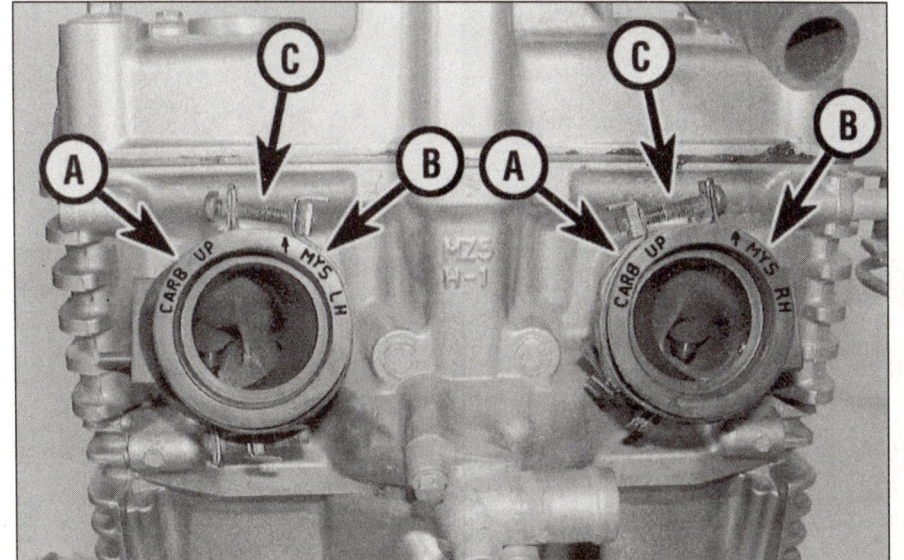

6.10a Inlet stub markings Carb UP (A), LH/RH (B) and carburettor clip screws at top of stub (C)

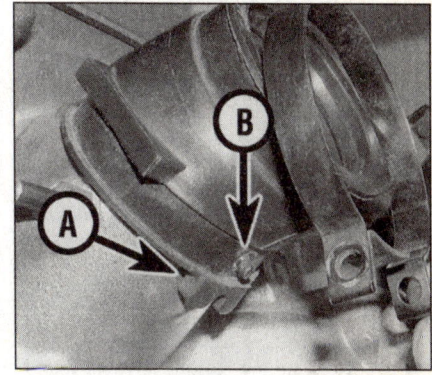

6.10b Lug to locate inlet stub on head (A) and lug for clip on stub (B)

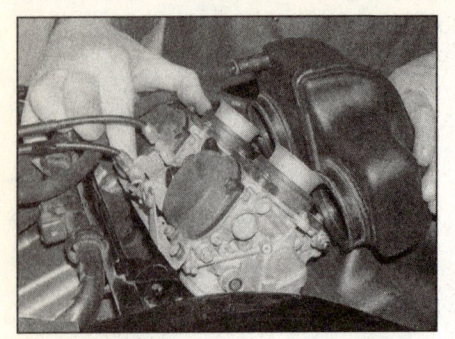

6.11a Fit carburettors into air filter housing

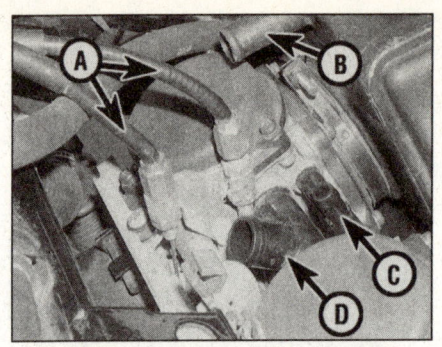

6.11b Throttle cables (A), crankcase breather hose connection (B), carburettor breather hose connection (C) and sub-air cleaner hose connection (D)

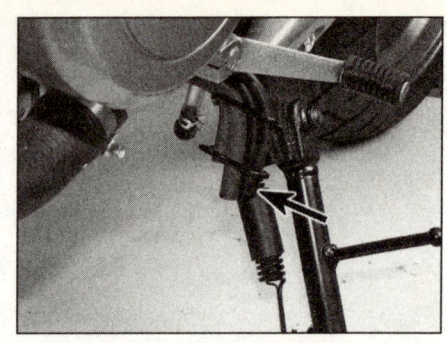

6.13a Place a container below the drain hose (arrowed) . . .

c) Fit the carburettors into the air filter housing before fitting them into the inlet stubs **(see illustration)**.
d) Make sure the carburettors are fully engaged with the inlet stubs and the clamps are securely tightened.
e) Make sure all cables and hoses are correctly routed and secured and not trapped or kinked **(see illustration)**.
f) Refer to Section 10 for installation of the throttle cables and section 11 for the choke cable. Check the operation of the cables and adjust them as necessary (see Chapter 1).
g) Check idle speed and carburettor synchronisation and adjust as necessary (see Chapter 1).

Removal – CBF models

12 Remove the fuel tank (see Section 2). Remove the front sprocket cover (see Chapter 6). Remove the air filter (see Chapter 1). Remove the battery (see Chapter 9). Drain the coolant (see Chapter 1).
13 Place a suitable container beneath the carburettor drain hose, unscrew the carburettor drain screws and allow fuel to empty from the system **(see illustrations)**. Tighten the carburettor drain screws. Unscrew the hose holder bolt and remove the holder. Release the clips on the drain hoses and detach the hoses from the carburettors.
14 Release the clips on the crankcase breather hoses between the valve cover and the air filter housing and remove the hoses **(see illustration)**.
15 Release the clip on the carburettor breather hose and detach the hose, noting its routing under the guide on the air filter housing **(see illustration 6.14)**.
16 Release the clip on the sub-air cleaner hose and remove the sub air cleaner assembly **(see illustration 6.14)**.
17 Release the clips from the carburettor heater hoses and detach the hoses from the cylinder head and the water pump.
18 Remove the choke cable (see Section 11).
19 Detach the throttle cables from the carburettors (see Section 10, Steps 3 and 4).
20 Unscrew the hose guide bolt on the right-hand side, noting the collar **(see illustration)**. Release the clamps and detach the coolant hose from the thermostat housing and the PAIR hose from the air filter housing **(see illustrations)**.
21 Disconnect the throttle position sensor wiring connector **(see illustration)**.

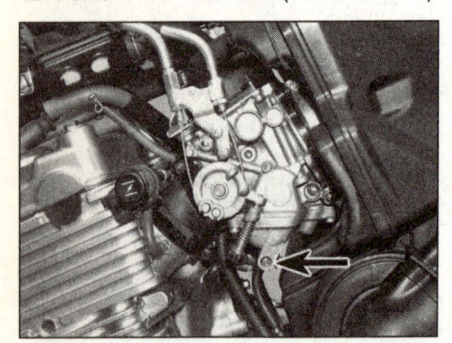

6.13b . . . then slacken the drain screw (arrowed) on each carburettor

6.14 Crankcase breather hose (A), carburettor breather hose (B), sub-air cleaner hose (C)

6.20a Unscrew the bolt (arrowed)

6.20b Release the clamps (arrowed) and detach the coolant hose from the thermostat housing . . .

6.20c . . . and the PAIR hose from the air filter housing

6.21 Disconnect the TP sensor wiring connector (arrowed)

Fuel and exhaust systems 4•7

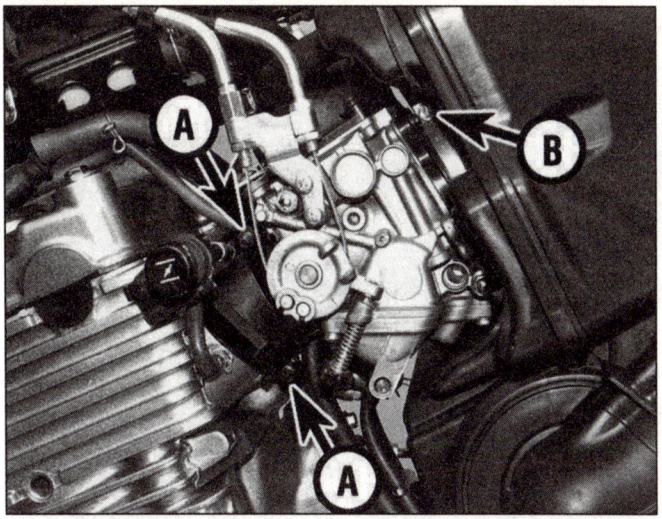

6.22a Inlet stub clamps (A), air filter housing clamp (B)

6.22b Air filter housing bolt (arrowed)

22 Slacken both clamps on each cylinder head inlet stub **(see illustration)**. Unscrew the air filter housing mounting bolt on the left-hand side. Slide the air filter housing and carburettor assembly back as far as possible, then remove the inlet stubs from between the engine and the carburettors. Now slacken the clamps on the air filter housing side and remove the carburettors. **Note:** *Keep the carburettors as upright as possible to prevent the possibility of the piston diaphragms being damaged.*
Caution: Stuff clean rag into each cylinder head inlet after removing the carburettors to prevent anything from falling in.

Installation – CBF models

23 Installation is the reverse of removal, noting the following.
a) Check for cracks or splits in the cylinder head inlet stubs, and replace them with new ones if necessary.
b) Fit the carburettors into the air filter housing before fitting them into the inlet stubs.
c) Make sure the inlet stubs are installed so that the slot in the underside of the stub locates over the lug on the underside of the head **(see illustration 6.10b)**.
d) Make sure the carburettors are fully engaged with the inlet stubs and the clamps are securely tightened.
e) Make sure all cables and hoses are correctly routed and secured and not trapped or kinked.
f) Refer to Section 10 for installation of the throttle cables and section 11 for the choke cable. Check the operation of the cables and adjust them as necessary (see Chapter 1).
g) Refill the cooling system (see Chapter 1).
h) Check the idle speed and carburettor synchronisation and adjust as necessary (see Chapter 1).

7 Carburettors – disassembly, cleaning and inspection

⚠️ **Warning: Refer to the precautions given in Section 1 before starting work.**

Disassembly

1 Remove the carburettors (see Section 6).
Note: *Do not separate the carburettors unless absolutely necessary; each carburettor can be dismantled sufficiently for all normal cleaning and adjustments while in place on the mounting brackets. Dismantle the carburettors separately to avoid interchanging parts. On CBF models do not remove the throttle position sensor unless absolutely necessary – refer to Section 14.*

2 Unscrew the heater hose union retaining screws and detach the unions, noting how they fit against the carburettor bodies **(see illustration)**.

3 Unscrew and remove the top cover retaining screws. Lift off the cover and remove the spring from inside the vacuum piston **(see illustration)**.

4 Carefully peel the diaphragm away from its sealing groove in the carburettor and withdraw the diaphragm and piston assembly. Note how the tab on the diaphragm fits in the recess in the carburettor body **(see illustration)**.
Caution: Do not use a sharp instrument to displace the diaphragm as it is easily damaged.

5 To remove the jet needle holder from the piston, thread a 4 mm screw into the top of the holder (one of the top cover retaining screws is ideal), then grasp the screw head with a pair of pliers and carefully draw the holder out of the piston. Note the spring that fits inside the holder. Discard the O-ring on

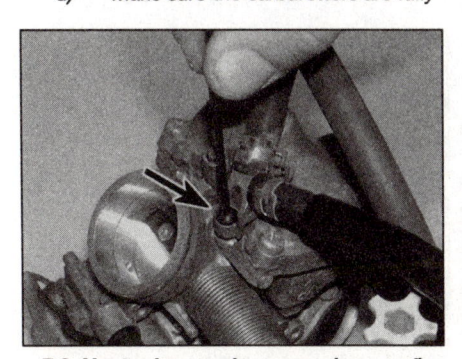

7.2 Heater hose union screw (arrowed)

7.3 Lift off the top cover and remove the spring

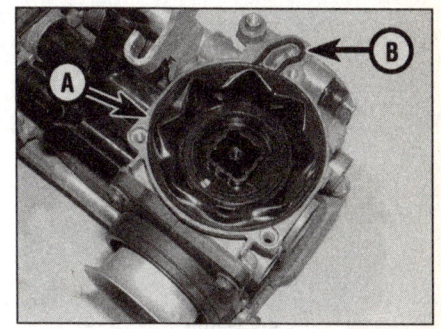

7.4 Diaphragm sealing groove (A) and tab recess (B)

4•8 Fuel and exhaust systems

7.5a Thread a 4 mm screw (arrowed) into the jet needle holder . . .

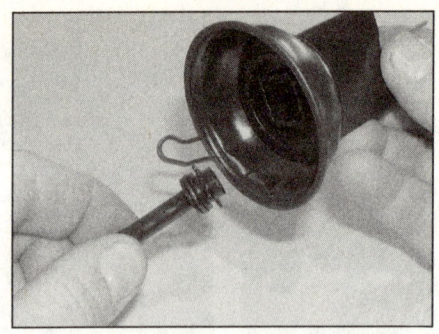

7.5b . . . and pull the holder out of the piston

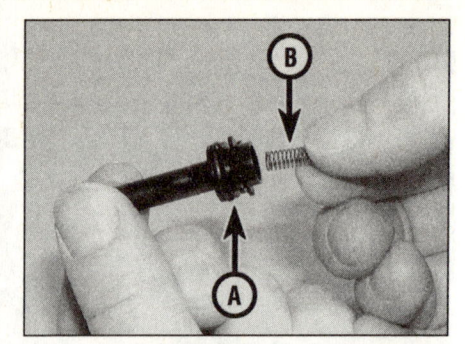

7.5c Remove the spring (B) and discard the O-ring (A)

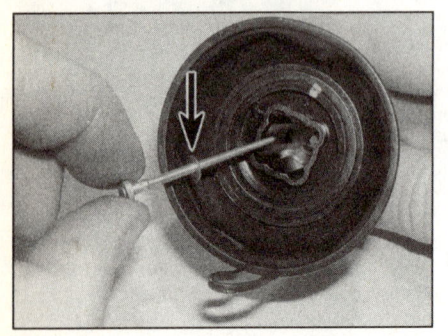

7.6 Remove the jet needle from the piston with the washer (arrowed)

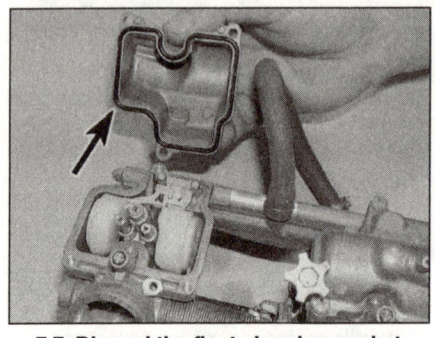

7.7 Discard the float chamber gasket (arrowed)

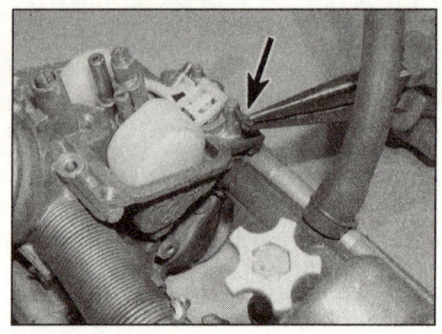

7.8a Withdraw the float pin (arrowed)

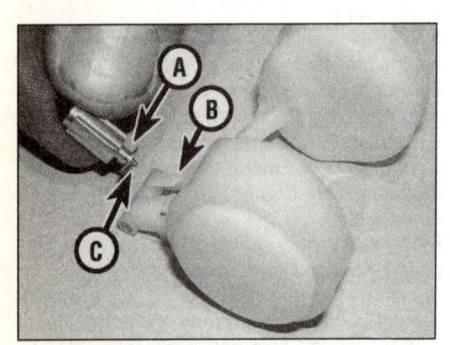

7.8b Shoulder (A) on float valve fits onto tab (B) on float. Note valve pin (C)

7.9 Main jet and jet holder (A), pilot jet (B) and pilot screw (C)

7.13a Remove choke linkage bar screws (arrowed) . . .

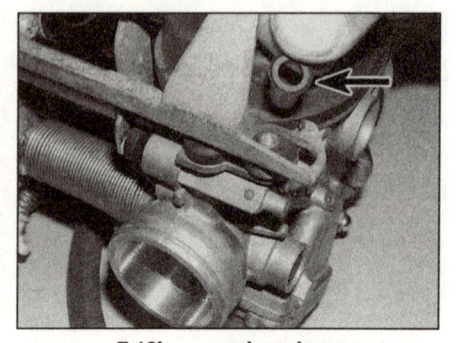

7.13b . . . and washers

the holder as a new one must be used **(see illustrations)**.
Caution: Do not push the needle holder out of the piston by pushing up on the needle.

6 Push the jet needle up from the bottom of the piston and withdraw it from the top **(see illustration)**. Note the washer that fits on the needle between the head of the needle and the piston.

7 Remove the screws securing the float chamber to the base of the carburettor and remove it. Remove the gasket and discard it as a new one must be used **(see illustration)**.

8 Using a pair of thin-nose pliers, carefully withdraw the float pin **(see illustration)**. If necessary, displace the pin using a small punch or a nail. Remove the float and unhook the float valve, noting how it fits onto the tab on the float **(see illustration)**.

9 Unscrew and remove the main jet from the base of the jet holder **(see illustration)**.

10 Unscrew and remove the jet holder **(see illustration 7.9)**.

11 Unscrew and remove the pilot jet **(see illustration 7.9)**.

12 The pilot screw can be removed if required, but note that its setting will be disturbed (see Section 4). Unscrew and remove the pilot screw along with its spring, washer and O-ring **(see illustration 7.9)**. Discard the O-ring as a new one must be used.

13 To remove a choke plunger, first remove the screws and washers securing the choke plunger linkage bar to the carburettors **(see illustrations)**. Lift off the bar and remove the collars and the

Fuel and exhaust systems 4•9

7.13c Note how return spring locates on bar

7.13d Remove linkage bar collars ...

7.13e ... and return spring

return spring, noting how they fit **(see illustrations)**.

14 Unscrew the choke plunger nut and withdraw the plunger and spring from the carburettor body, noting how they fit **(see illustration)**. Take care not to lose the spring when removing the nut **(see illustration)**.

Cleaning

Caution: Use only a petroleum based solvent for carburettor cleaning. Don't use caustic cleaners.

15 Submerge the metal components in the solvent for approximately thirty minutes (or longer, if the directions recommend it). After the carburettor has soaked long enough for the cleaner to loosen and dissolve most of the varnish and other deposits, use a nylon-bristled brush to remove the stubborn deposits. Rinse it again, then dry it with compressed air.

16 Use a jet of compressed air to blow out all of the fuel and air passages in the main and upper body, not forgetting the air jets in the carburettor inlet.

Caution: Never clean the jets or passages with a piece of wire or a drill bit, as they will be enlarged, causing the fuel and air metering rates to be upset.

Inspection

17 Check the operation of the choke plunger. If it doesn't move smoothly, inspect the plunger assembly and linkage components and renew any that are worn, damaged or bent **(see illustration)**.

18 If removed from the carburettor, check the tapered portion of the pilot screw and the spring for wear or damage **(see illustration)**. Renew them if necessary.

19 Check the carburettor body, float chamber and top cover for cracks, distorted sealing surfaces and other damage **(see illustration)**. If any defects are found, renew the faulty component, although renewal of the entire carburettor will probably be necessary (check with a Honda dealer on the availability of separate components).

7.14a Unscrew choke plunger nut ...

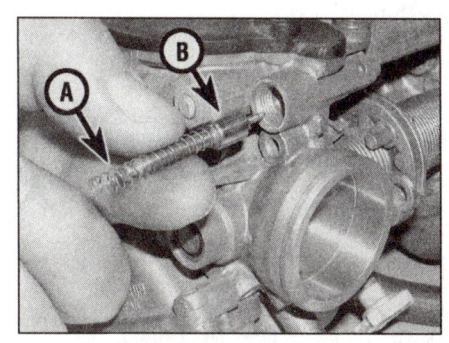

7.14b ... and withdraw spring (A) and plunger (B)

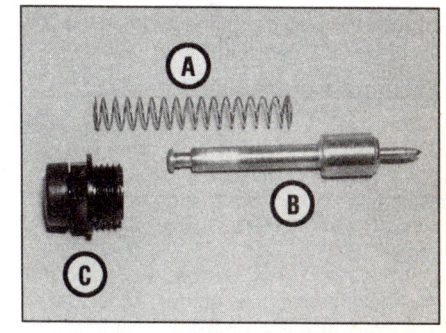

7.17 Choke plunger spring (A), choke plunger (B) and plunger nut (C)

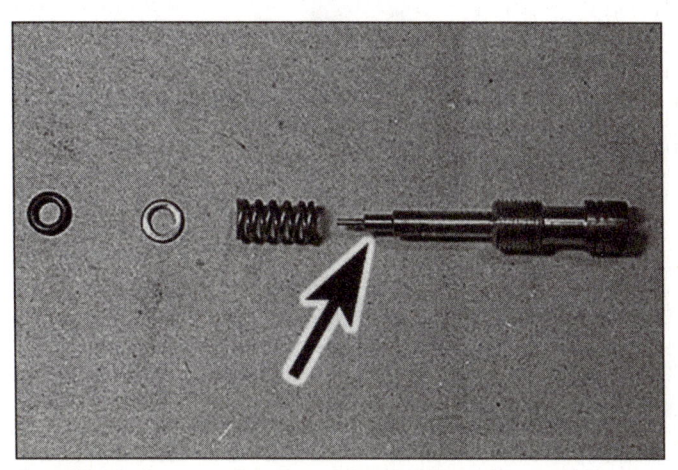

7.18 Check the tapered portion of the pilot screw (arrowed) for wear

7.19 Inspect the carburettor body for wear and damage

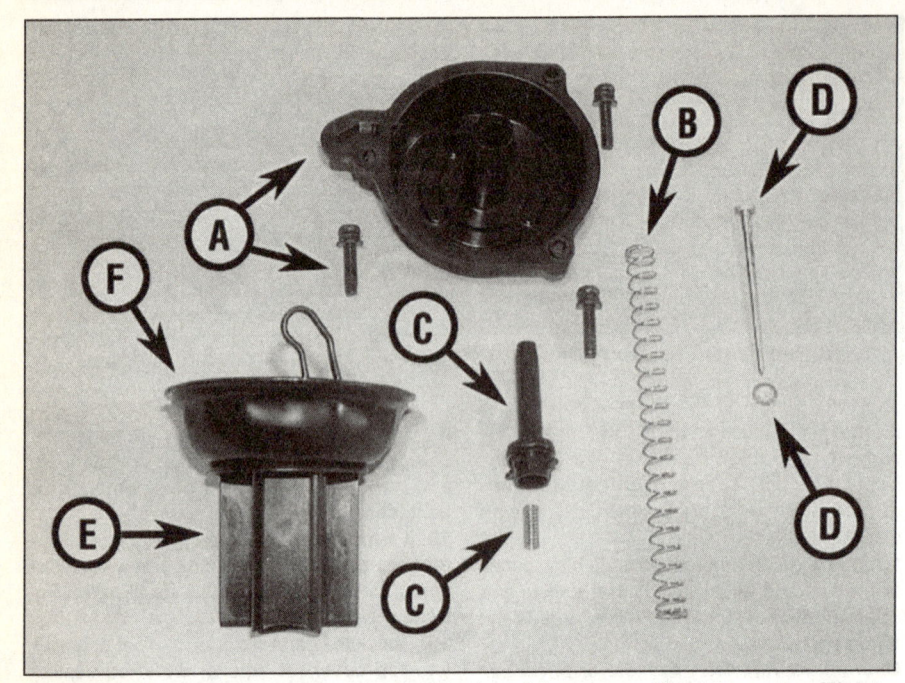

4•10 Fuel and exhaust systems

7.20 Carburettor components – Top cover and retaining screws (A), piston spring (B), jet needle holder and spring (C), jet needle and washer (D), piston (E) and diaphragm (F)

7.25 Throttle shaft return spring (A) and butterfly valve (B)

20 Check the piston diaphragm for splits, holes and general deterioration **(see illustration)**. Holding it up to a light will help to reveal problems of this nature.
21 Insert the piston in the carburettor body and check that the piston moves up-and-down smoothly. Check the surface of the piston for wear. If it's worn excessively or doesn't move smoothly in the guide, renew the components as necessary.
22 Check the jet needle for straightness by rolling it on a flat surface such as a piece of glass. Renew it if it's bent or if the tip is worn.
23 Check the tip of the float valve and the valve seat. If either has grooves or scratches in it, or is in any way worn, they must be renewed as a set. Gently push down on the rod on the top of the valve then release it – if it doesn't spring back, renew the valve **(see illustration 7.8b)**.
24 Check the float for damage. This will usually be apparent by the presence of fuel inside the float. If the float is damaged, it must be renewed.
25 Operate the throttle shaft to make sure the throttle butterfly valve opens and closes smoothly. If it doesn't, cleaning the throttle linkage may help. Otherwise, renew the carburettor **(see illustration)**.
26 Clean the threads of the synchronising screw with solvent and a brush without disturbing the screw. When dry, lightly grease the screw threads **(see illustration)**.

8 Carburettors – reassembly and float height check

Warning: *Refer to the precautions given in Section 1 before proceeding.*

Note: *When reassembling the carburettors, be sure to use new O-rings and gaskets. Do not overtighten the carburettor jets and screws as they are easily damaged.*

1 If removed, install the choke plunger assemblies into the carburettor bodies.
2 Fit the choke return spring to the left-hand carburettor. Fit the choke linkage bar collars, then locate the bar, making sure the arms fit onto the plungers, and fit the plastic washers and secure the bar with the screws **(see illustration)**.
3 Locate the cranked end of the spring in the notch on the bar **(see illustration 7.13c)**. Check the operation of the choke linkage **(see illustration)**.
4 If removed, install the pilot screw (see Section 4).
5 Install the pilot jet and the jet holder into the carburettor, then screw the main jet into the end of the jet holder **(see illustration)**.
6 Fit the float needle valve onto the float, then

7.26 Carburettor synchronising screw (arrowed)

8.2 Ensure the choke linkage bar locates correctly on the return spring and collar (A) and choke plunger (B)

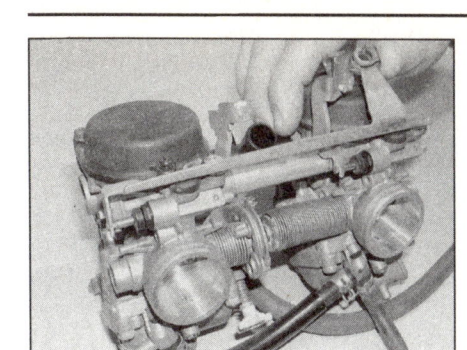

8.3 Check the operation of the choke linkage

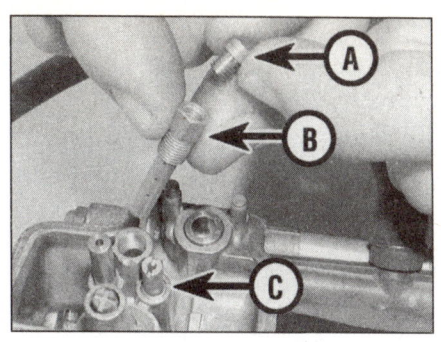

8.5 Main jet (A), jet holder (B) and pilot jet (C)

position the float assembly in the carburettor, making sure the needle valve locates in its seat, and install the pin, making sure it is secure **(see illustration)**.

7 To check the float height, hold the carburettor so the float hangs down, then tilt it back until the needle valve is just seated, but not so far that the needle's spring-loaded tip is compressed. Measure the distance between the gasket face (with the gasket removed) and the bottom of the float with an accurate ruler **(see illustration)**. The correct setting is given in the Specifications at the beginning of this Chapter. The float height is not adjustable, so if it is incorrect the float may be damaged and should be renewed.

8 With the float height checked, fit a new gasket onto the float chamber, making sure it is seated properly in its groove, and install the chamber onto the carburettor **(see illustration 7.7)**.

9 Fit the washer onto the jet needle and insert the needle into the piston **(see illustration 7.6)**.

10 Fit a new O-ring into the groove in the needle holder and ensure the spring is inside the holder. Using a 4 mm screw as on disassembly, insert the holder into the centre of the piston and push it down until the O-ring is felt to locate in its groove in the piston **(see illustration 7.5c, b and a)**. Remove the 4 mm screw.

11 Insert the piston assembly into the carburettor body and lightly push it down, ensuring the needle is correctly aligned with the jet **(see illustration)**.

12 Align the tab on the diaphragm with the recess in the carburettor body, then press the diaphragm outer edge into its groove, making sure it is correctly seated and that the tab locates in the recess around the air hole **(see illustration 7.4)**. Check the diaphragm is not creased, and that the piston moves smoothly up and down in its guide.

13 Install the spring into the needle holder **(see illustration)**. Fit the top cover to the carburettor, making sure the top end of the spring locates into the centre recess inside the cover, and aligning the protrusion on the cover with the tab on the diaphragm **(see illustration 7.3)**. Tighten the cover screws securely.

14 Install the carburettors (see Section 6).

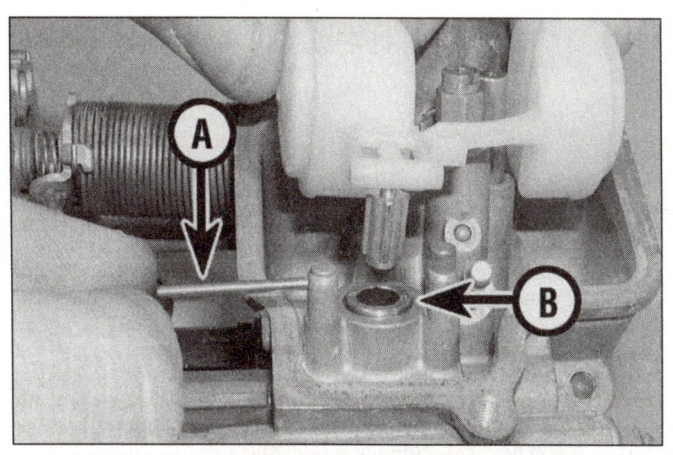

8.6 Carburettor float pin (A) and float valve seat (B)

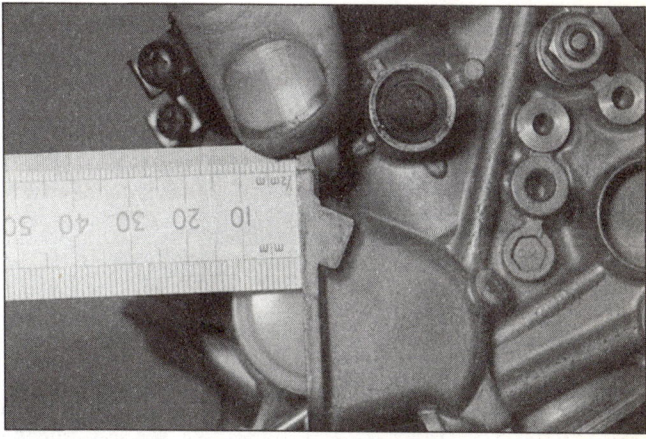

8.7 Measuring float height

8.11 Ensure that needle (A) is aligned with jet (B) when fitting piston assembly

8.13 Install the spring into the needle holder

4•12 Fuel and exhaust systems

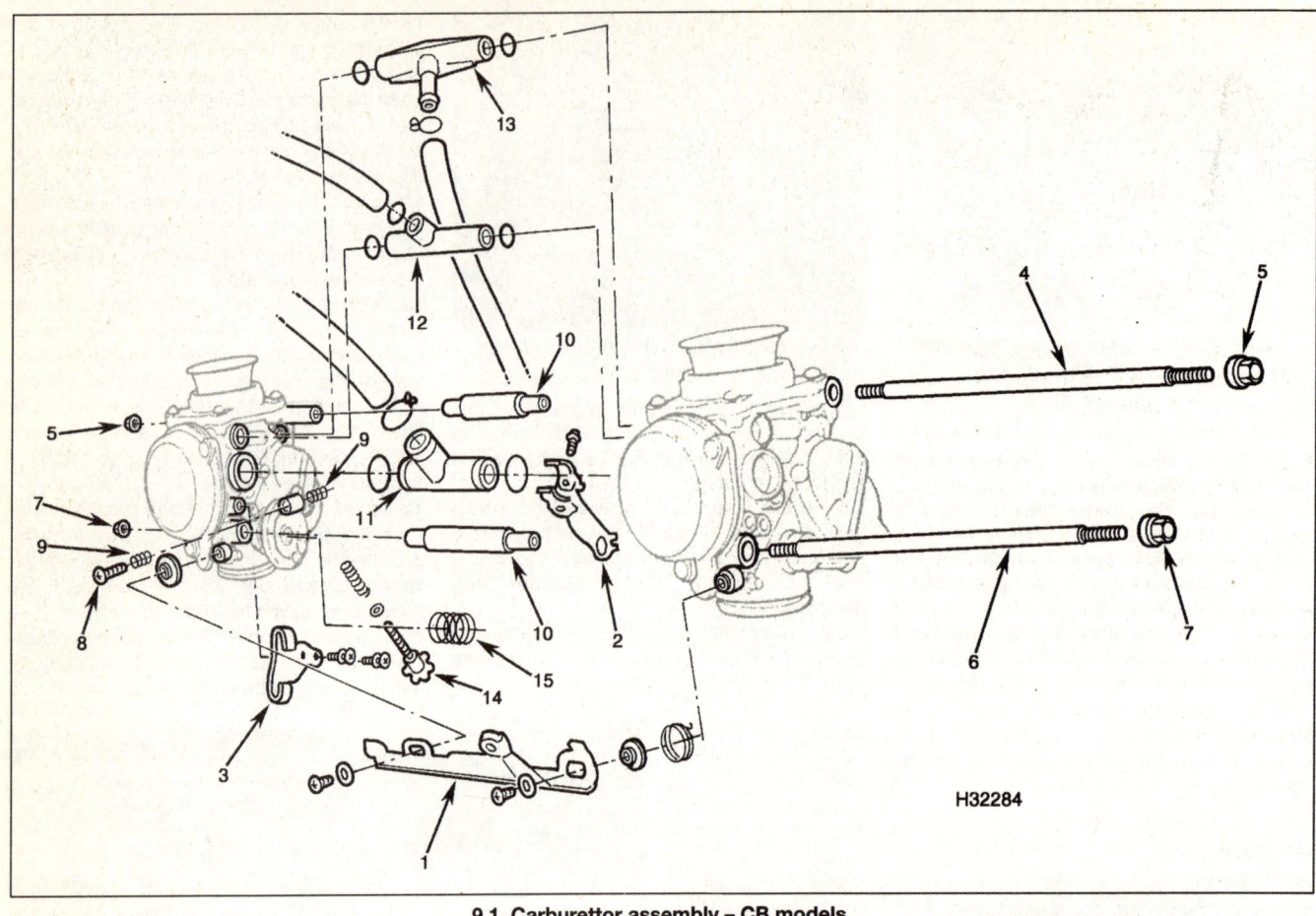

9.1 Carburettor assembly – CB models

1 Choke linkage bar
2 Choke cable clamp
3 Throttle cable bracket
4 Lower joining stud
5 5 mm nuts
6 Upper joining stud
7 6 mm nuts
8 Synchronising screw
9 Synchronising screw springs
10 Joining stud spacers
11 Sub-air cleaner joint
12 Breather joint
13 Fuel joint pipe
14 Idle speed adjuster
15 Throttle linkage spring

9 Carburettors – separation and joining

Warning: Refer to the precautions given in Section 1 before proceeding

Separation

1 The carburettors do not need to be separated for normal overhaul. If you need to separate them (to fit a new carburettor body, for example), refer to the following procedure **(see illustration)**.

2 Remove the carburettors (see Section 6). Mark the body of each carburettor with its cylinder location to ensure that it is positioned correctly on reassembly.

3 Unscrew the heater hose union retaining screws and detach the unions, noting how they fit against the carburettor bodies **(see illustration 7.2)**.

4 Remove the screws and washers securing the choke plunger linkage bar to the carburettors. Lift off the bar and remove the collars and the return spring, noting how they fit **(see illustrations 7.13a, b, c, d and e)**.

5 Alternately loosen the nuts on the left-hand end of the threaded studs which join the carburettors together. Remove the nuts and withdraw the studs **(see illustration)**.

6 Make a careful note of how the carburettor synchronising screw springs and throttle linkage springs are arranged to ensure that they are fitted correctly on reassembly. Also note the arrangement of the sub-air cleaner, carburettor breather and fuel supply joint pipes **(see illustration)**.

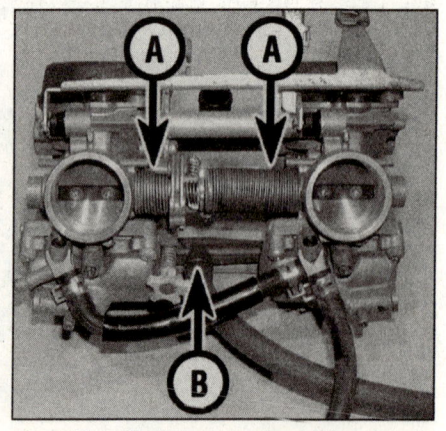

9.5 Carburettor joining stud nuts (arrowed)

9.6a Throttle linkage springs (A) and fuel supply joint pipe (B)

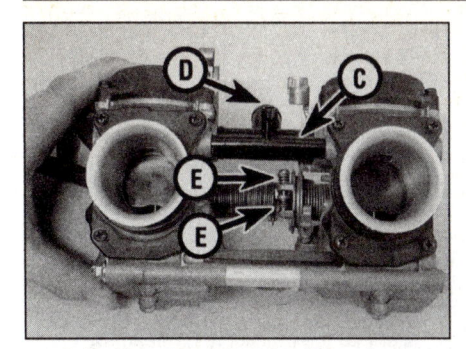

Fuel and exhaust systems 4•13

9.6b Breather joint pipe (C), sub-air cleaner joint pipe (D) and synchronising springs (E)

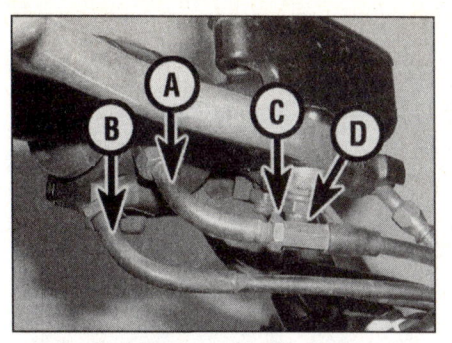

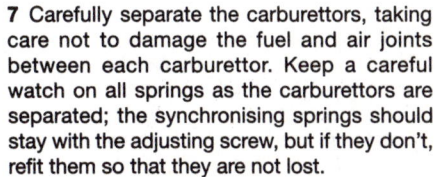

10.2a Accelerator cable (A), decelerator cable (B), lock nut (C) and adjuster (D) – CB models

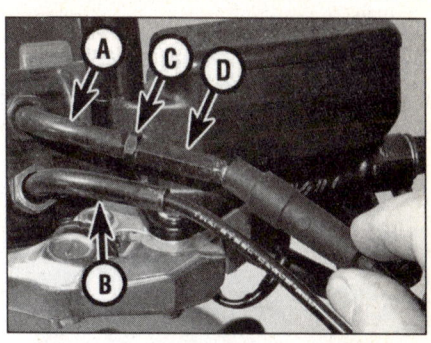

10.2b Accelerator cable (A), decelerator cable (B), locknut (C) and adjuster (D) – CBF models

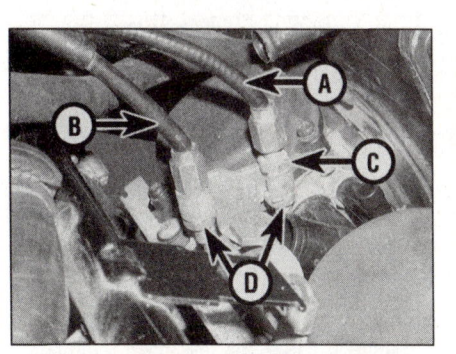

10.3a Accelerator cable (A), decelerator cable (B), adjuster locknut (C) and lower locknuts (D) – CB models

10.3b Accelerator cable (A), decelerator cable (B) – CBF models

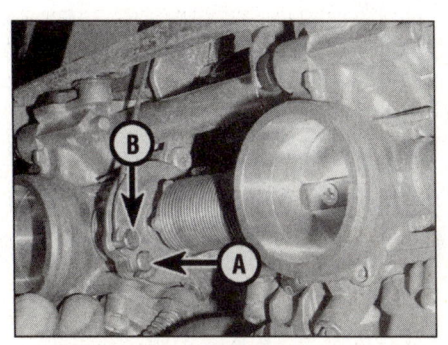

10.4 Accelerator cable nipple (A) and decelerator cable nipple (B) – CB models

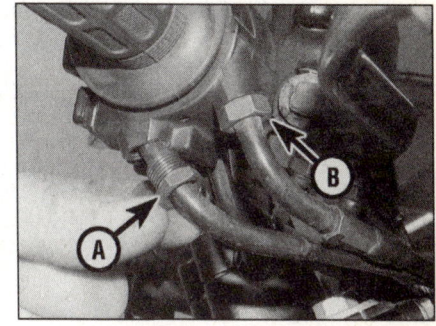

10.6 Decelerator cable elbow nut (A) and accelerator cable nut (B)

7 Carefully separate the carburettors, taking care not to damage the fuel and air joints between each carburettor. Keep a careful watch on all springs as the carburettors are separated; the synchronising springs should stay with the adjusting screw, but if they don't, refit them so that they are not lost.

8 Pull out the fuel and air vent joint pipes and discard the O-rings as new ones must be used.

9 Detach the carburettor joining stud spacers. Note how the choke cable bracket is held against the left-hand carburettor body by the upper spacer.

Joining

10 Assembly is the reverse of the disassembly procedure. Fit the fuel and air vent joint pipes and the stud spacers all to one carburettor first before joining the two together. Use new O-rings on the fuel and air vent joints, and smear them with clean engine oil.

11 Tighten the nuts on the carburettor joining studs to the torque settings specified at the beginning of this Chapter. **Note:** *The upper joining nuts (nearest the engine) are 6 mm and the lower joining nuts are 5 mm thread diameter. Take great care when tightening to avoid damaging the carburettors.*

12 Check the operation of both the choke and throttle linkages ensuring that both operate smoothly and return quickly under spring pressure before installing the carburettors.

13 Check carburettor synchronisation (see Chapter 1).

10 Throttle cables – removal and installation

Warning: Refer to the precautions given in Section 1 before proceeding.

Removal

1 On CB models remove the fuel tank (see Section 2). Whilst it is possible to detach the throttle cables with the carburettors in place, there is a limited amount of space to work in. If required, displace the carburettors to improve access (see Section 6).

2 On CB models the accelerator cable is the front cable in the twistgrip, the decelerator cable is the rear one **(see illustration)**. On CBF models the accelerator cable is the top cable in the twistgrip, the decelerator cable is the bottom one **(see illustration)**. Slacken the locknut on the accelerator cable adjuster and thread the adjuster fully in, then tighten the locknut against it. This resets the adjuster to the start of its range.

3 At the carburettor end, slacken the rear (accelerator) cable adjuster locknut (above the bracket) then unscrew the lower nut off the adjuster **(see illustrations)**.

4 Slip the adjuster out of the bracket and detach the cable nipple from the carburettor throttle cam **(see illustration)**.

5 Now unscrew the front (decelerator) cable holder nut fully and slip the holder out of the bracket and detach the cable nipple from the throttle cam **(see illustrations 10.3 and 10.4)**. Withdraw the cables from the bike noting the correct routing of each cable.

6 Unscrew the rear (decelerator) cable elbow nut from the throttle twistgrip, and slacken the front (accelerator) cable nut **(see illustration)**.

7 Remove the two handlebar switch screws

4•14 Fuel and exhaust systems

10.7a Separate the halves of the twistgrip housing . . .

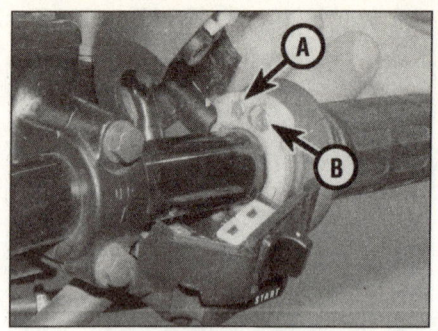

10.7b . . . and detach the accelerator cable nipple (A) and decelerator cable nipple (B)

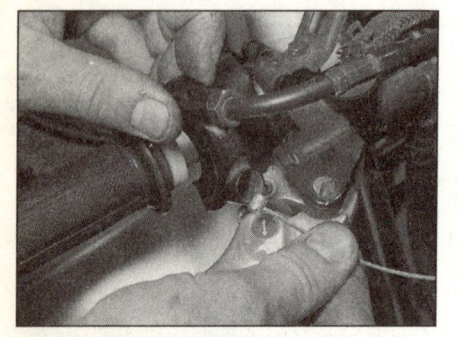

10.8a Withdraw the decelerator cable . . .

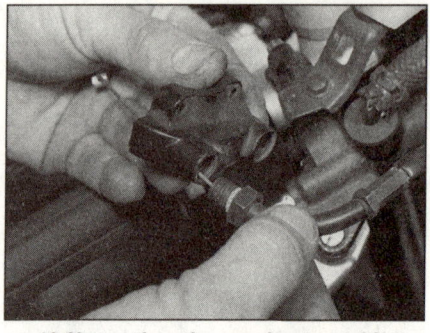

10.8b . . . then the accelerator cable

and pull off the upper half of the switch housing then detach the cable nipples from the pulley (see illustration).

8 Detach the lower half of the switch housing from the handlebar and withdraw the decelerator elbow and cable (see illustration). Thread the switch housing off the accelerator elbow and withdraw the cable (see illustration). Mark each cable to ensure it is connected correctly on installation.

Installation

9 Fit the accelerator cable elbow into the front (CB models) or top (CBF models) socket of the lower or front half of the switch housing and thread the housing onto it (see illustration 10.8b). Lubricate the cable nipple with multi-purpose grease and install it

in the throttle pulley. Fit the decelerator cable into the rear or bottom socket and tighten the nut (see illustration 10.8a). Lubricate the cable nipple with multi-purpose grease and install it in the pulley (see illustration 10.7b). Fit the lower or front half of the switch housing over the pulley and onto the handlebar.

10 Fit the upper or rear half of the switch housing onto the handlebar, making sure the pin inside the housing locates in the hole in the handlebar, and install the screws, tightening them securely. Ensure the twistgrip turns freely and adjust the alignment of the cable elbows.

11 Feed the cables through to the carburettors, making sure they are correctly routed. The cables must not interfere with any other component and should not be kinked or bent sharply.

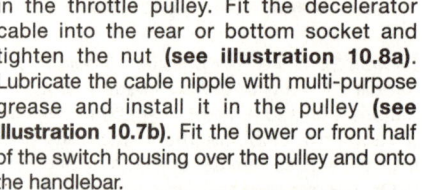

12 Lubricate the decelerator cable nipple with multi-purpose grease and fit it into the front socket on the carburettor throttle cam (see illustration 10.4). Fit the cable holder into the bracket, thread the nut onto the holder below the bracket and tighten the nut against the bracket (see illustration 10.3).

13 Lubricate the accelerator cable nipple with multi-purpose grease and fit it into the rear socket on the carburettor throttle cam (see illustration 10.4). Fit the accelerator cable adjuster into the bracket and thread the lower nut onto the adjuster. Turn the adjuster locknut (above the bracket) until the specified amount of cable freeplay is obtained (see procedure in Chapter 1). Tighten the lower nut against the bracket.

14 Operate the throttle to check that it opens and closes freely. Turn the handlebars from lock to lock to make sure the cables don't cause the steering to bind.

15 Install the carburettors (if displaced), and the fuel tank (see Sections 6 and 2).

16 Start the engine and check that the idle speed does not rise as the handlebars are turned. If it does, the throttle cables are routed incorrectly. Correct the problem before riding the motorcycle.

11 Choke cable – removal and installation

CB models

Removal

1 Remove the fuel tank (see Section 2).

2 Slacken the choke outer cable bracket screw and free the cable from the bracket on the left-hand carburettor, then detach the inner cable nipple from the choke linkage lever (see illustration). Withdraw the cable from the machine noting the correct routing.

3 Unscrew the cable elbow nut from the left handlebar switch housing (see illustration). Remove the handlebar switch housing screws and separate the two halves of the switch to

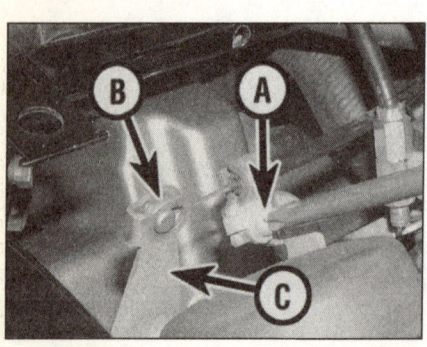

11.2 Cable bracket screw (A), inner cable nipple (B) and choke linkage lever (C)

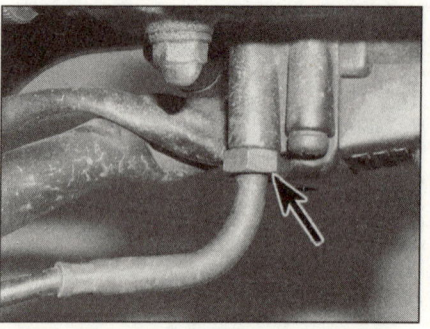

11.3a Choke cable elbow nut (arrowed)

11.3b Remove the screws and split the switch housing

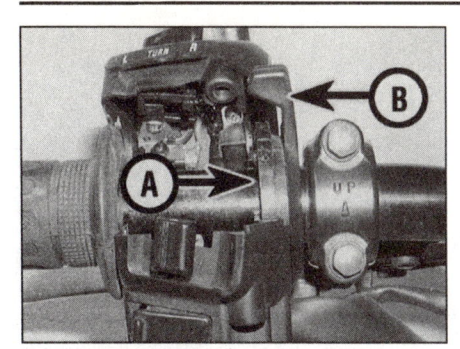

Fuel and exhaust systems 4•15

11.4 Choke cable nipple attachment point (A) and pulley/lever (B)

11.11a Unscrew the bolt (arrowed) . . .

11.11b . . . then free the cable from the holder (A) and lever (B)

gain access to the cable nipple **(see illustration)**.
4 Detach the nipple from the pulley and withdraw the cable from the housing **(see illustration)**.

Installation

5 Fit the choke cable elbow into the socket in the left handlebar switch housing and tighten the nut. Lubricate the cable nipple with multi-purpose grease and install it in the pulley.
6 Fit the two halves of the switch housing together, install the housing screws and tighten them securely. Ensure the lever turns freely.
7 Feed the cable through to the carburettors, making sure it is correctly routed. The cable must not interfere with any other component and should not be kinked or bent sharply.
8 Lubricate the cable nipple with multi-purpose grease and attach it to the choke linkage lever then fit the outer cable into its bracket, making sure there is a small amount of freeplay in the inner cable, and tighten the bracket screw **(see illustration 11.2)**.
9 Check the operation of the choke cable (see Chapter 1).
10 Install the fuel tank (see Section 2).

CBF models

11 Unscrew the choke knob bolt, noting the collar **(see illustration)**. Free the outer cable from the holder, then detach the inner cable end from the choke linkage lever and remove the cable **(see illustration)**.

12 Installation is the reverse of removal. Lubricate the cable end with multi-purpose grease. Check the operation of the choke.

12 Air filter housing – removal and installation

CB models

Removal

1 Remove the fuel tank (see Section 2), and the carburettors (see Section 6).
2 Remove the bolts securing the air filter cover to the filter housing and remove the cover and the filter element (see Chapter 1).
3 Release the hoses from the guides on the upper right-hand side of the filter housing. Release the drain hose from the bracket on the back of the exhaust system and pull the filter housing forward to release the U-shaped mounting bracket below the intake duct from the frame tube **(see illustration)**.
4 Lift the housing out of the frame between the top frame tubes. Release the clip securing the breather hose to the filter housing and detach the hose.

Installation

5 Installation is the reverse of removal. Check the condition of the breather hose(s) and drain hose and renew them if necessary.
6 Route the breather hose behind the engine

unit before installing the filter housing **(see illustration)**.
7 Make sure the mounting bracket locates correctly on the frame tube **(see illustration 12.3)**. If the bracket is damaged, it can be replaced as a separate item from the filter housing.

CBF models

8 Remove the carburettors (see Section 6).
9 Manoeuvre the air filter housing out of the frame.
10 Installation is the reverse of removal.

13 Exhaust system – removal and installation

> **Warning:** If the engine has been running the exhaust system will be very hot. Allow the system to cool before carrying out any work.

Note: The exhaust system can be removed as a complete assembly. The silencer and downpipe assembly can be separated if required, but this is best done after the complete system has been removed, rather than doing it in situ.

Removal

1 Remove the drain and breather tubes from the bracket on the back of the exhaust system below the engine unit **(see illustration)**.

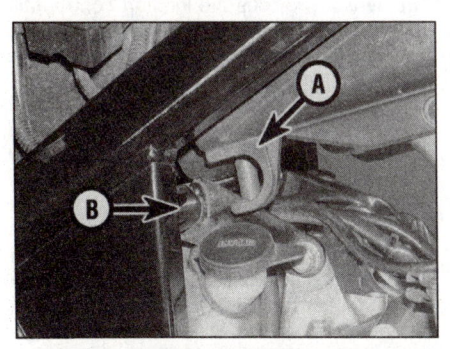

12.3 U-shaped bracket (A) locates on frame tube (B)

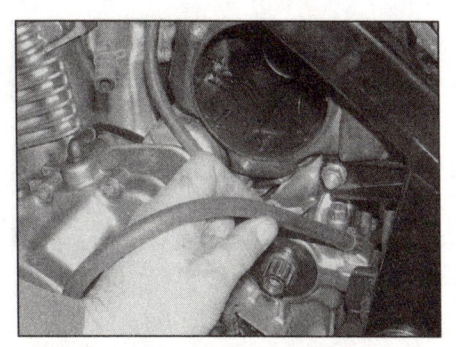

12.6 Route the breather hose behind the engine unit

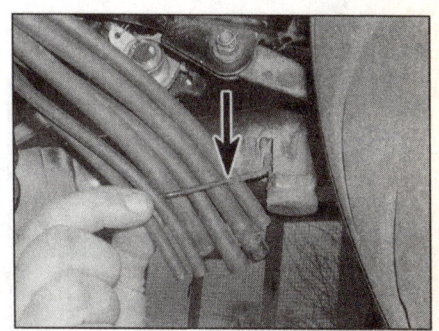

13.1 Remove tubes from bracket (arrowed)

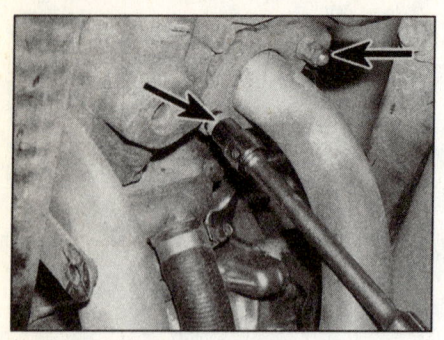

4•16 Fuel and exhaust systems

13.2 Unscrew the downpipe nuts (arrowed)

13.3a Unscrew the nut on the exhaust pipe mounting bolt . . .

13.3b . . . and remove the bolt

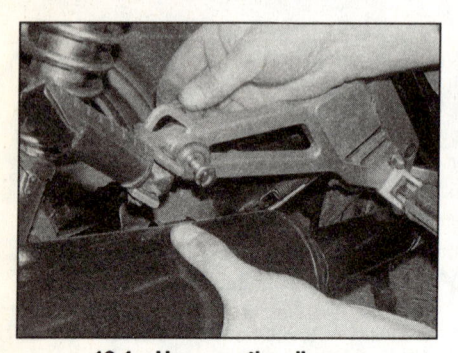

13.4a Unscrew the silencer mounting nut . . .

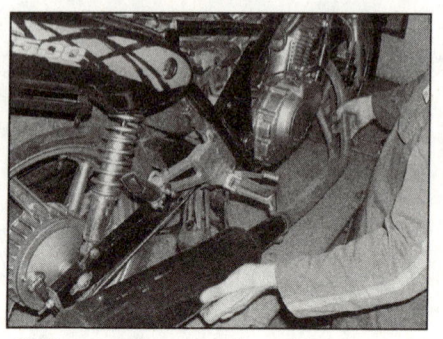

13.4b . . . and remove the exhaust system

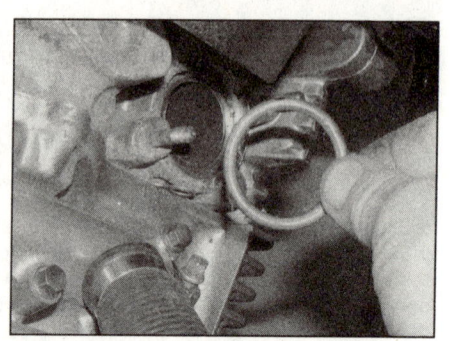

13.5 Discard the old exhaust port gaskets

2 Unscrew the nuts securing the downpipes to the cylinder head **(see illustration)**.
3 Unscrew the nut and remove the washer and bolt securing the exhaust pipe to the frame bracket **(see illustrations)**.
4 Unscrew the nut that secures the silencer to the frame and remove the washer, but leave the bolt in place **(see illustration)**. Support the system, then withdraw the bolt and manoeuvre the downpipes away from the cylinder head.
5 Remove the gaskets from the exhaust ports and discard them as new ones must be used **(see illustration)**.
6 If required, remove the clamp to separate the silencer from the exhaust pipe. Discard the joint gasket as a new one must be used.

Installation

7 Installation is the reverse of removal, noting the following:
a) Clean the jointing surfaces of the exhaust ports and the downpipes **(see illustration)**. Smear the port gaskets with grease to hold them in place while fitting the exhaust system.
b) Use new gaskets in each exhaust port and between the exhaust pipe and the silencer, if separated **(see illustration)**.
c) Clean the cylinder head studs and lubricate them with a suitable copper-based grease before reassembly.
d) Check the condition of the rubber bushings on the exhaust pipe and silencer brackets and renew them if they are damaged or deteriorated.
e) Leave all fasteners finger-tight until the entire system has been installed, making alignment easier. Tighten the mounting nuts and bolts to the torque settings specified at the beginning of this Chapter. Tighten the downpipe nuts first.
f) Run the engine and check that there are no exhaust gas leaks.

14 Throttle position sensor – check, removal and installation (CBF models)

1 The throttle position sensor (TPS) is mounted on the right-hand side of the carburettors and is keyed to the throttle shaft. The sensor provides the ignition control unit with information on throttle position and rate of opening or closing.

Check

2 Release the clamp and detach the PAIR system hose from the air filter housing **(see illustration 6.20c)**. Disconnect the sensor's wiring connector **(see illustration 6.21)**. Start the engine and set the idle speed to 2000 rpm (see Chapter 1). Reconnect the sensor wiring connector – the engine speed should increase
3 Disconnect the wiring connector. Using an ohmmeter set to the K ohm scale, measure the resistance between the yellow/black and blue/green wire terminals on the sensor. If the

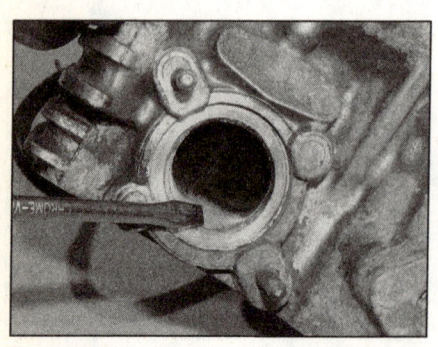

13.7a Clean carbon from the exhaust port jointing surface . . .

13.7b . . . and renew all gaskets where the system has been separated

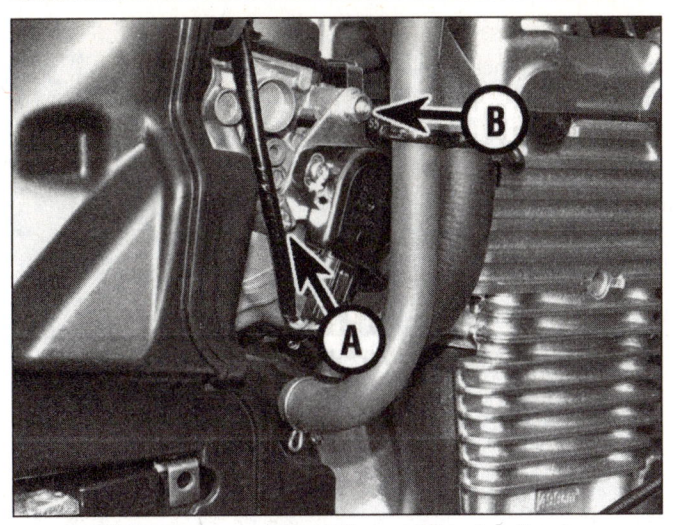

14.8 Undo the screw (A) and the nut (B) to free the sensor mounting plate

15.5 Check the air hose (A) and vacuum hoses (B)

reading is not as specified at the beginning of the Chapter, the sensor is faulty.

4 Now connect the meter between the red/yellow and blue/green terminals and open and close the throttle – the resistance should increase as the throttle is opened, and decrease again as it is closed. If it doesn't, the sensor is faulty.

5 To check the input voltage to the sensor, connect the positive (+) lead of a voltmeter to the yellow/black wire terminal on the sensor wiring connector, and the negative (–) lead to the blue/green terminal. Turn the ignition switch ON and check that a voltage of 4.7 to 5.3 volts is present. If it isn't, there is a fault in the wiring between the ignition control unit and sensor connector or in the control unit itself.

6 Disconnect the ICU connector (see Chapter 5), then use a multimeter set to resistance or a continuity tester, and check for continuity between the terminals on the loom side of the sensor wiring connector and the corresponding terminals on the ignition control unit connector. There should be continuity between each terminal. If not, this is probably due to a damaged or broken wire between the connectors; pinched or broken wires can usually be repaired. Also check the connectors for loose or corroded terminals, and check the sensor itself for cracks and other damage.

Replacement

7 The throttle sensor is mounted on the right-hand end of the carburettor assembly. Release the clamp and detach the PAIR system hose from the air filter housing **(see illustration 6.20c)**. Disconnect the sensor's wiring connector **(see illustration 6.21)**.
8 Undo the screw and the nut and remove the sensor with its mounting plate, noting how it keys onto the end of the throttle shaft **(see illustration)**. Do not loosen the actual sensor mounting screws.

9 Install the sensor, locating the flat on the throttle shaft in the sensor so that it is keyed to it, then fit and tighten the nut and the screw.
10 Reconnect the sensor wiring connector. Fit the PAIR hose onto the air filter housing and secure it with the clamp.

15 Pulse secondary air injection (PAIR) system – CBF models only

General information

Note: *The PAIR system is fitted to CB500-Y, SY, 2 and S-2 models in certain Europe markets.*

1 To reduce the amount of unburnt hydrocarbons released in the exhaust gases, a pulse secondary air injection (PAIR) system is fitted. The system consists of the control valve assembly, which is mounted on the front of the engine unit, a vacuum hose from each inlet duct, the air supply hose from the air filter housing, and the air feed pipes from the control valve to each exhaust duct in the cylinder head.
2 Under normal running conditions the control valve is open. With the valve open, whenever there is a negative pulse in the exhaust system, filtered air is drawn from the air filter housing through the control valve and into the exhaust ports in the cylinder head. This fresh air promotes the burning of any excess fuel present in the exhaust gases, so reducing the amount of harmful hydrocarbons released into the atmosphere via the exhaust gases.
3 When manifold depression reaches a certain level, usually if the throttle is closed when engine speed is high, the vacuum created in the hoses closes the control valve and the air supply is cut off. This prevents exhaust popping on over-run.

4 The control valve assembly is fitted with a pair of one-way reed valves to prevent the exhaust gases passing through the control valve and into the air filter housing.

Check

5 First make sure the air supply hose and vacuum hoses are in good condition and are secure on each end **(see illustration)**.
6 Remove the air filter (see Chapter 1). Make sure that the air supply port in the filter housing is clear and clean.
7 Disconnect the air supply hose from the right-hand side of the air filter housing and the vacuum hose from the front of the three-way connector **(see illustration 6.20c)**. Plug the connector to prevent air being drawn into the inlet ducts. Start the engine and make sure that air is being drawn into the air supply hose by placing your thumb over its end and feeling for suction. Now gradually apply a vacuum to the vacuum hose – when it reaches 390 mmHg the vacuum should close the valve and air should no longer be drawn into the supply hose. Check that the vacuum level is maintained and does not bleed away. If the valve does not close, or is permanently closed, or if the vacuum does not hold, and all the hoses are in good condition, replace the control valve with a new one.
8 To inspect the reed valves first remove the control valve, then clean off all dirt and debris from the outside. Undo the reed valve cover screws and remove the cover and reed valve from each end of the control valve. Check the valve for damage and any build-up of carbon. Make sure there is no gap between the reed and its rubber seat – if there is and it is caused by carbon deposits clean them off and then check again for any gap. Make sure there is a sufficient amount of spring in the reed. The reed valves are not available individually so if there is any damage a new control valve is required.

Removal

9 Slacken the retaining clips and disconnect the vacuum hose and air hose from the control valve **(see illustration)**.
10 Unscrew the four bolts and release the air feed pipes from the front of the cylinder head. Remove the gaskets – new ones must be used.
11 Undo the two bolts securing the control valve bracket to the front of the crankcase and remove the valve complete with air feed hoses and pipes.
12 If necessary, unbolt the control valve from the mounting bracket.
13 Inspect the pipes and hoses for signs of cracks and splits and replace damaged components.

Installation

14 Ensure the air feed pipe and cylinder head mating surfaces are clean and dry. If removed fit the control valve onto the mounting bracket.
15 Fit a new gasket onto each of the pipe heads and keep them in place by inserting the bolts.
16 Position the control valve assembly and tighten the feed pipe bolts onto the cylinder head. Fit the control valve bolts and tighten them.
17 Connect the vacuum hose and air supply hose to the control valve and secure them with the retaining clips.

15.9 PAIR system control valve (arrowed)

Chapter 5
Ignition system

Contents

General information 1	Pulse generator coil assembly – check, removal
Ignition control unit – check, removal and installation 5	and installation 4
Ignition (main) switch – check, removal and installation . see Chapter 9	Sidestand switch – check and replacement........... see Chapter 9
Ignition HT coils – check, removal and installation 3	Spark plugs – gap check and renewal .... see Chapter 1
Ignition system – check 2	Throttle position sensor, check, removal and
Ignition timing – general information and check 6	installation.................................... see Chapter 4
Neutral switch – check and replacement........... see Chapter 9	

Degrees of difficulty

Easy, suitable for novice with little experience	Fairly easy, suitable for beginner with some experience	Fairly difficult, suitable for competent DIY mechanic	Difficult, suitable for experienced DIY mechanic	Very difficult, suitable for expert DIY or professional

Specifications

General information
Cylinder numbering (from left-hand to right-hand side of the bike).... 1-2
Spark plugs .... see Chapter 1

Ignition timing
At idle .... 14° BTDC
Full advance... 33° BTDC

Pulse generator coil
Resistance .... 450 to 550 ohms @ 20°C

Ignition HT coils
Primary winding resistance .... 2.6 to 3.2 ohms @ 20°C
Secondary winding resistance
  with plug caps and leads .... 17.2 to 22.8 K-ohms @ 20°C
  without plug caps and leads .... 13.5 to 16.5 K-ohms @ 20°C

5•2 Ignition system

1 General information

All models are fitted with a fully transistorised electronic ignition system, which due to its lack of mechanical parts is totally maintenance free. The system comprises a rotor, pulse generator coil, ignition control unit and ignition HT coils (refer to the wiring diagrams at the end of Chapter 9 for details).

The ignition triggers, which are on the rotor mounted on the right-hand end of the crankshaft, magnetically operate the pulse generator coil as the crankshaft rotates. The pulse generator coil sends a signal to the ignition control unit which then supplies the ignition HT coils with the power necessary to produce a spark at the plugs.

The system uses two coils which are mounted under the fuel tank. On CB models the lower coil supplies no. 1 cylinder spark plug and the upper one supplies no. 2 cylinder spark plug. On CBF models the left-hand coil supplies No. 1 cylinder and the right-hand No. 2.

The ignition control unit incorporates an electronic advance system controlled by signals generated by the ignition triggers and the pulse generator coil.

The ignition system incorporates a safety interlock circuit which will cut the ignition if the sidestand is put down whilst the engine is running and in gear, or if a gear is selected whilst the engine is running and the sidestand is down. It also prevents the engine from being started if the sidestand is down and the engine is in gear unless the clutch lever is pulled in.

Because of their nature, the individual ignition system components can be checked but not repaired. If ignition system troubles occur, and the faulty component can be isolated, the only cure for the problem is to renew the part. Keep in mind that most electrical parts, once purchased, cannot be returned. To avoid unnecessary expense, make very sure the faulty component has been positively identified before buying a new part.

Note that there is no provision for adjusting the ignition timing on these models.

2 Ignition system – check

Warning: The energy levels in electronic systems can be very high. On no account should the ignition be switched on whilst the plugs or plug caps are being held. Shocks from the HT circuit can be most unpleasant. Secondly, it is vital that the engine is not turned over or run with any of the plug caps removed, and that the plugs are soundly earthed (grounded) when the system is checked for sparking. The ignition system components can be seriously damaged if the HT circuit becomes isolated.

1 As no means of adjustment is available, any failure of the system can be traced to failure of a system component or a simple wiring fault. Of the two possibilities, the latter is by far the most likely. In the event of failure, check the system in a logical fashion, as described below.

2 Disconnect the HT leads from the spark plugs. Connect each lead to a spare spark plug and lay each plug on the engine with the threads earthed (grounded) **(see illustration)**. If necessary, hold each spark plug with an insulated tool.

Warning: Do not remove any of the spark plugs from the engine to perform this check – atomised fuel being pumped out of the open spark plug hole could ignite, causing severe injury!

3 Having observed the above precautions, check that the kill switch is in the RUN position and the transmission is in neutral, then turn the ignition switch ON and turn the engine over on the starter motor. If the system is in good condition a regular, fat blue spark should be evident at each plug electrode. If the spark appears thin or yellowish, or is non-existent, further investigation will be necessary. Before proceeding further, turn the ignition OFF and remove the key as a safety measure.

4 The ignition system must be able to produce a spark which is capable of jumping a particular size gap. Honda do not provide a specification, but a healthy system should produce a spark capable of jumping at least 6 mm. A simple testing tool can be made to test the minimum gap across which the spark will jump (see **Tool Tip**) or alternatively it is possible to buy an ignition spark gap tester tool **(see illustration)**.

5 Connect the HT lead from one coil to the protruding electrode on the test tool, and clip the tool to a good earth (ground) on the engine or frame **(see illustration)**. Check that the kill switch is in the RUN position, turn the ignition switch ON and turn the engine over on the starter motor. If the system is in good condition a regular, fat blue spark should be seen to jump the gap between the nail ends. Repeat the test for the other coil. If the test results are good the entire ignition system can be considered good. If the spark appears thin or yellowish, or is non-existent, further investigation will be necessary.

6 Ignition faults can be divided into two categories, namely those where the ignition system has failed completely, and those which are due to a partial failure. The likely faults are listed below, starting with the most probable source of failure. Work through the list systematically, referring to the subsequent

TOOL TIP

A simple spark gap testing tool can be made from a block of wood, a large alligator clip and two nails, one of which is fabricated so that a spark plug cap or bare HT lead end can be connected to its end. Make sure the gap between the two nail ends is 6 mm.

2.2 Earth the spark plug and operate the starter – bright blue sparks should be visible

2.4 Using a spark gap tester – note earth wire (arrowed)

2.5 Connect the tester as shown – when the engine is cranked sparks should jump the gap between the nails

Ignition system 5•3

sections for full details of the necessary checks and tests. **Note:** *Before checking the following items ensure that the battery is fully charged and that all fuses are in good condition.*
a) Loose, corroded or damaged wiring connections, broken or shorted wiring between any of the component parts of the ignition system (see Chapter 9).
b) Faulty HT lead or spark plug cap, faulty spark plug, dirty, worn or corroded plug electrodes, or incorrect gap between electrodes.
c) Faulty ignition (main) switch or engine kill switch (see Chapter 9).
d) Faulty neutral, clutch or sidestand switch (see Chapter 9).
e) Faulty pulse generator coil or damaged rotor.
f) Faulty ignition HT coil(s).
g) Faulty ignition control unit.
h) Faulty throttle position sensor (CBF500).

7 If the above checks don't reveal the cause of the problem, have the ignition system tested by a Honda dealer equipped with a special diagnostic tester.

3 Ignition HT coils – check, removal and installation

Check

1 In order to determine conclusively that the ignition coils are defective, they should be tested by a Honda dealer equipped with a peak voltage tester.
2 However, the coils can be checked visually (for cracks and other damage) and the primary and secondary coil resistance can be measured with a multimeter. If the coils are undamaged, and if the resistance readings are as specified at the beginning of this Chapter, they are probably capable of proper operation.

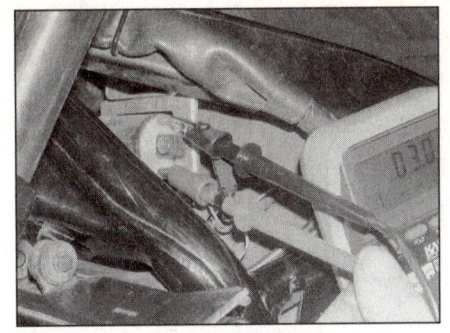

3.6 Measuring primary resistance between the primary circuit terminals on the coil

3 Disconnect the battery negative (-ve) lead (see Chapter 9). Remove the fuel tank (see Chapter 4).
4 The coils are mounted on a bracket behind the steering head on CB models and under the main frame spine on CBF models **(see illustration 3.10a or b)**.
5 Disconnect the primary circuit electrical connectors from the coil being tested and the HT lead from the spark plug. Note the locations of all wires before disconnecting them.
6 Set the meter to the ohms x 1 scale and measure the resistance between the primary circuit terminals **(see illustration)**. This will give a resistance reading for the primary windings of the coil and should be consistent with the value given in the Specifications at the beginning of this Chapter.
7 To check the condition of the secondary windings, set the meter to the K-ohm scale. Connect one meter probe to the spark plug cap and the other probe to the negative (-ve) black/white wire terminal on the coil **(see illustration)**. If the reading obtained is not within the range shown in the Specifications, unscrew the cap from the end of the HT lead and repeat the test. If the reading is now as specified, renew the spark plug cap. If the reading is still outside the specified range, it is likely that the coil is defective.

3.7 Measuring secondary resistance between the plug cap and the negative (-ve) coil terminal

8 Should any of the above checks not produce the expected result, have your findings confirmed by a Honda dealer (see Step 1). If the coil is confirmed to be faulty, it must be renewed; the coil is a sealed unit and cannot be repaired.

Removal

9 Disconnect the battery negative (-ve) lead (see Chapter 9). Remove the fuel tank (see Chapter 4).
10 The coils are mounted on a bracket behind the steering head on CB models and under the main frame spine on CBF models **(see illustrations)**.
11 Disconnect the primary circuit electrical connectors from the coil and disconnect the HT lead from the spark plug. Mark the locations of all wires before disconnecting them.
12 Unscrew the coil mounting bolts, on CBF models counter-holding and retrieving the nuts as you do, and remove the coil with the two spacers between it and the bracket **(see illustration 3.10a or b)**. Note the routing of the HT leads.

Installation

13 Installation is the reverse of removal. Don't

3.10a Coil location and mounting bolts (arrowed) – CB models

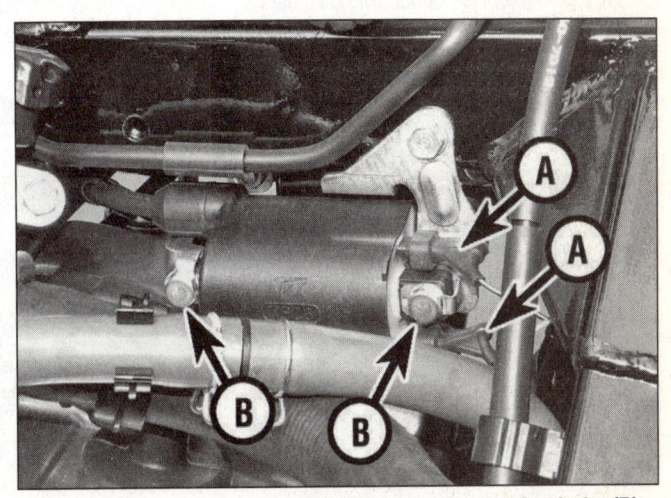

3.10b Primary wiring connectors (A) and coil mounting bolts (B) – CBF models

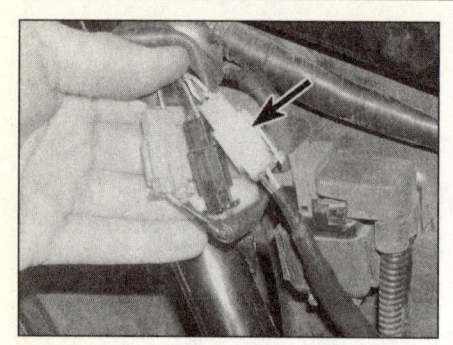

4.2a Pulse generator coil wiring connector – CB models

4.2b Pulse generator coil wiring connector – CBF models

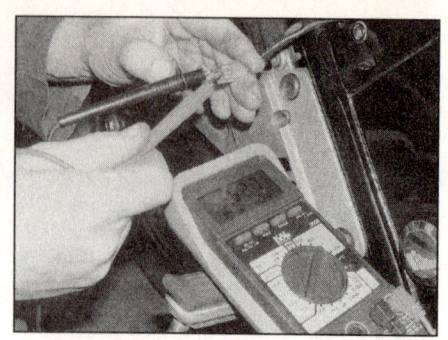

4.2c Measuring resistance between terminals of the wiring connector

forget to replace the spacers between the coil and the mounting bracket. Make sure the wiring connectors and HT leads are securely connected.

4 Pulse generator coil assembly – check, removal and installation

Check

1 Disconnect the battery negative (-ve) lead. Remove the right-hand frame side panel on CB models, and the left-hand panel on CBF models (see Chapter 8).
2 Trace the pulse generator coil wiring back from the right-hand engine cover and disconnect it at the white 4-pin connector on CB models or blue 2-pin connector on CBF models inside the rubber boot on the right-hand side of the frame on CB models and on the left on CBF models **(see illustrations)**. Using a multimeter set to the ohms x 100 scale, measure the resistance between the white/yellow and yellow terminals on the pulse generator coil side of the connector **(see illustration)**.
3 Compare the reading obtained with that given in the Specifications at the beginning of this Chapter. The pulse generator coil must be renewed if the reading obtained differs greatly from that given, particularly if the meter indicates a short circuit (no measurable resistance) or an open circuit (infinite, or very high resistance).

4 If the pulse generator coil is thought to be faulty, first check that this is not due to a damaged or broken wire from the coil to the connector; pinched or broken wires can usually be repaired.
5 If the reading is satisfactory, reconnect the 4-pin connector and check the wiring between the connector and the ignition control unit.
6 Refer to Section 5 for access and disconnect the ignition control unit multi-pin connector **(see illustration)**. Identify the pulse generator terminals (white/yellow and yellow). Measure the resistance between the terminals of the connector as before. If the reading is not within the specifications inspect the wires for damage or breakage.

Removal

7 Disconnect the battery negative (-ve) lead. Remove the right-hand frame side panel on CB models, and the left-hand panel on CBF models (see Chapter 8).
8 Trace the pulse generator coil wiring back from the right-hand engine cover and disconnect it at the white 4-pin connector on CB models or blue 2-pin connector on CBF models inside the rubber boot on the right-hand side of the frame on CB models and on the left on CBF models **(see illustration 4.2a or b)**. Feed the wiring through to the cover, noting its routing.
9 Drain the engine oil (see Chapter 1). Remove the engine right-hand cover as described in Chapter 2, Section 15, Step 3; the clutch cable can remain attached. Unscrew the bolts securing the pulse generator coil **(see illustration)**. While the cover is off, examine the triggers on the timing rotor for signs of damage and renew it if necessary (see Chapter 2, Section 16).
10 On CB models, because the pulse generator coil shares its wiring sub-harness with the oil pressure and neutral switches, these must also be disconnected. Unscrew the bolts securing the front sprocket cover and remove it and the chain guide. Unscrew the bolt securing the wiring guide and remove the guide.
11 Disconnect the neutral switch and the oil pressure switch wiring connectors (see Chapter 2, Section 5). Feed the wiring to the right-hand side of the bike, noting its routing, and remove the loom and pulse generator coil complete.

Installation

12 Installation is the reverse of removal, noting the following:
a) Apply a suitable non-permanent locking compound to the bolts securing the pulse generator coil and tighten them securely.
b) Apply a suitable sealant to the wiring grommet and fit it into its recess **(see illustration 15.37 in Chapter 2)**.
c) When fitting the engine cover, use a new gasket and check that the two dowels and clutch pushrod are in place. Refill the engine with oil (see Chapter 1 and Daily (pre-ride) checks).
d) On CB models ensure the neutral switch and oil pressure switch wiring is clipped behind the wiring guide on the left-hand side of the engine.
e) Ensure the wiring is not routed where it will rub against the engine casing, air filter housing or frame.

5 Ignition control unit – check, removal and installation

Check

1 If the tests shown in the preceding Sections have failed to isolate the cause of an ignition fault, it is possible that the ignition control unit itself is faulty. No details are available for

4.6 Separate the multi-pin connector from the control unit

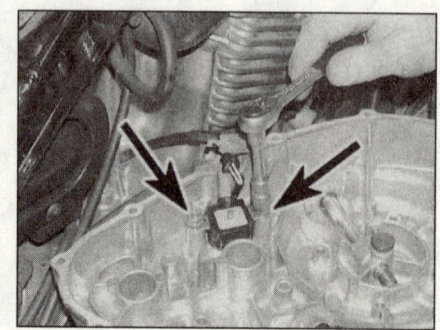

4.9 Pulse generator coil is retained by two bolts

testing the unit on home workshop equipment. Take the motorcycle to a Honda dealer for testing of the peak voltage.

Removal

2 On CB models remove the seat and the seat cowling (see Chapter 8) – the control unit is secured to the right-hand side of the frame, alongside the rear mudguard **(see illustration)**. On CBF models remove the fuel tank (see Chapter 4) – the control unit is mounted behind the air filter housing **(see illustration)**.
3 Disconnect the battery negative (-ve) lead (see Chapter 9).
4 Disconnect the multi-pin wiring connector from the ignition control unit.
5 Remove the ignition control unit from its rubber sleeve, or lift the sleeve and unit together off the sleeve's mounting lugs, and remove the unit.

Installation

6 Installation is the reverse of removal. Make sure the wiring connector is correctly and securely connected and that the rubber sleeve is securely fixed to the frame.

6 Ignition timing – general information and check

General information

1 Since no provision exists for adjusting the ignition timing and since no component is subject to mechanical wear, there is no need for regular checks; only if investigating a fault such as a loss of power or a misfire, should the ignition timing be checked.
2 The ignition timing is checked dynamically (engine running) using a stroboscopic lamp. The inexpensive neon lamps should be adequate in theory, but in practice may produce a pulse of such low intensity that the timing mark remains indistinct. If possible, one of the more precise xenon tube lamps should

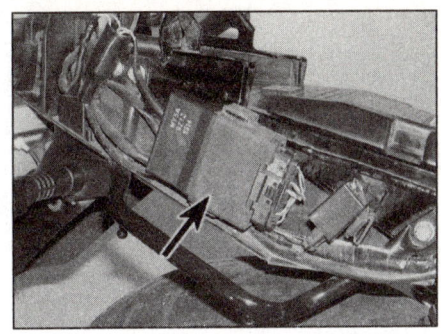

5.2a Ignition control unit (arrowed) – CB models

be used, powered by an external source of the appropriate voltage. **Note:** *Do not use the motorcycle's own battery as an incorrect reading may result from stray impulses within the machine's electrical system.*

Check

3 Warm the engine up to normal operating temperature then stop it.
4 Unscrew the timing inspection plug from the left-side engine cover **(see illustration)**. Discard the cover O-ring as a new one must be used.
5 The timing mark on the alternator rotor which indicates the firing point at idle speed for cylinder no. 1 is a line with the letter F next to it. The static timing mark with which this should align is the notch in the timing inspection hole **(see illustration 6.4)**.

> **HAYNES HINT**
> *The timing marks can be highlighted with white paint to make them more visible under the stroboscope light.*

6 Connect the timing light to the no. 1 cylinder HT lead as described in the manufacturer's instructions.
7 Start the engine and aim the light at the static timing mark.
8 With the machine idling, the line next to the

5.2b Ignition control unit (arrowed) – CBF models

letter F should align with the static timing mark.
9 Slowly increase the engine speed whilst observing the timing mark. The timing mark should move anti-clockwise, increasing in relation to the engine speed until it reaches full advance (no identification mark).
10 As already stated, there is no means of adjusting the ignition timing on these machines. If the ignition timing is incorrect, or suspected of being incorrect, one of the ignition system components is at fault, and the system must be tested as described in the preceding Sections of this Chapter.
11 When the check is complete, install the timing inspection plug using a new O-ring and smear it and the cover threads with engine oil.

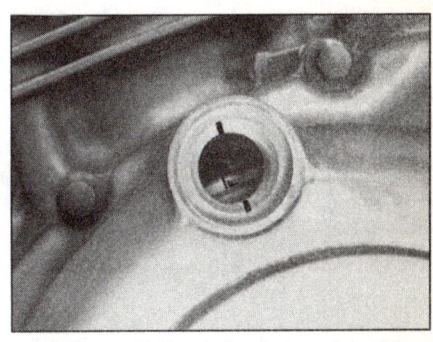

6.4 Unscrew the timing inspection plug

Notes

Chapter 6
Frame, suspension and final drive

Contents

Drive chain – removal, cleaning and installation 14
Drive chain and sprockets – check, adjustment
  and lubrication .... see Chapter 1
Footrests, brake pedal and gearchange lever –
  removal and installation 3
Forks – disassembly, inspection and reassembly 7
Forks – oil change .... see Chapter 1
Forks – removal and installation............................. 6
Frame – inspection and repair 2
General information 1
Handlebars and levers – removal and installation 5
Handlebar switches – check .... see Chapter 9
Handlebar switches – removal and installation ...... see Chapter 9
Rear shock absorber – removal, inspection and installation 10
Stands – check and lubrication .... see Chapter 1
Stands – removal and installation 4
Sidestand switch – check and replacement.......... see Chapter 9
Sprockets – check, renewal and installation 15
Sprocket coupling/rubber damper – check and renewal......... 16
Steering head bearings – freeplay check and
  adjustment................................ see Chapter 1
Steering head bearings – inspection and renewal 9
Steering head bearings – lubrication .... see Chapter 1
Steering stem – removal and installation 8
Suspension – adjustments.................................. 11
Suspension – check........................... see Chapter 1
Swingarm – inspection and bearing renewal 13
Swingarm and drive chain slider – removal and installation 12

Degrees of difficulty

| Easy, suitable for novice with little experience | Fairly easy, suitable for beginner with some experience | Fairly difficult, suitable for competent DIY mechanic | Difficult, suitable for experienced DIY mechanic | Very difficult, suitable for expert DIY or professional |

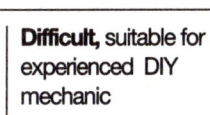

Specifications

Note: *Where applicable, models are identified by their production code letter – refer to 'Identification numbers' at the front of this manual for details.*

Front forks

Fork oil type .... 10W fork oil
Fork oil capacity
  CB models
    R models.. 320 cc
    T models.. 309 cc
    V models.. 313 cc
    W, SW, X, SX, Y, SY, 2 and S2 models .... 320 cc
  CBF models .... 452.2 to 457.5 cc
Fork oil level*
  CB models
    R models.. 150 mm
    T models.. 165 mm
    V models.. 160 mm
    W, SW, X, SX, Y, SY, 2 and S2 models .... 150 mm
  CBF models .... 148 mm
Fork spring free length (min)
  CB models
    R models
      Standard.................................... 308.3 mm
      Service limit................................. 305.2 mm
    T models
      Standard.................................... 509.9 mm
      Service limit................................. not available
    W, SW, X, SX, Y, SY, 2 and S2 models
      Standard.................................... 409.1 mm
      Service limit................................. not available
  CBF models
    Standard....................................... 336.6 mm
    Service limit .... 330 mm
Fork tube runout limit............................... 0.2 mm

*Oil level is measured from the top of the tube with the fork spring removed and the leg fully compressed.

Rear suspension
Shock absorber spring free length
   R, T and V models
      Standard .... 234.5 mm
      Service limit .... 232.2 mm
   W, SW, X, SX, Y, SY, 2 and S2 models
      Standard .... 243.7 mm
      Service limit .... 241.3 mm

Final drive
Chain type
   CB models .... RK 525SMOZ5 or DID 525V8 (108 links)
   CBF models .... DID 525V8 (116 links)
Joining link pin projection from side plate (unstaked)
   RK type chain .... 1.2 to 1.4 mm
   DID type chain .... 1.15 to 1.55 mm
Joining link staked ends diameter, both types .... 5.50 to 5.80 mm

Torque settings
Rider's footrest bracket bolts
   CB models .... 45 Nm
   CBF models .... 26 Nm
Brake pedal bracket-to-footrest bracket cap nut ...... 27 Nm
Brake pedal bracket-to-footrest bracket bolt .... 12 Nm
Side stand pivot nut .... 40 Nm
Side stand pivot bolt
   CB models .... 10 Nm
   CBF models .... 15 Nm
Side stand switch bolt .... 10 Nm
Top yoke fork clamp bolts .... 23 Nm
Lower yoke fork clamp bolts .... 40 Nm
Handlebar clamp bolts .... 27 Nm
Front brake master cylinder clamp bolts .... 12 Nm
Clutch lever pivot bolt .... 1 Nm
Clutch lever pivot nut .... 6 Nm
Fork top bolt .... 23 Nm
Fork drain plug .... 7.5 Nm
Damper rod Allen bolt .... 20 Nm
Steering stem nut .... 105 Nm
Steering head bearing adjuster nut .... 25 Nm
Shock absorbers – CB models
   Upper mounting bolts .... 27 Nm
   Lower mounting bolts .... 38 Nm
Shock absorber – CBF models
   Mounting bolt nuts .... 42 Nm
Swingarm pivot bolt nut – CB models .... 90 Nm
Swingarm pivot bolt nut – CBF models .... 118 Nm
Pivot bracket bolt nuts – CBF models .... 69 Nm
Front sprocket bolt .... 55 Nm
Rear sprocket nuts .... 100 Nm

1 General information

CB models use a full cradle twin spar steel frame. The left-hand frame downtube is detachable to ease engine removal. CBF models have a box-section steel spine frame that uses the engine as a stressed member.

Front suspension is by a pair of oil-damped telescopic forks.

At the rear, a steel, box-section swingarm acts on twin shock absorbers on CB models and on a single shock absorber on CBF models. The shock absorber(s) is/are adjustable for spring pre-load.

The drive to the rear wheel is by chain.

2 Frame – Inspection and repair

1 The frame should not require attention unless accident damage has occurred. In most cases, frame renewal is the only satisfactory remedy for such damage. A few frame specialists have the jigs and other equipment necessary for straightening the frame to the required standard of accuracy, but even then there is no simple way of assessing to what extent the frame may have been over stressed.

2 On a high mileage bike, the frame should be examined closely for signs of cracking or splitting at the welded joints. Loose engine mount bolts can cause ovaling or fracturing of the mounting tabs. Minor damage can often be repaired by specialist welding, depending on the extent and nature of the damage.

3 Remember that a frame which is out of alignment will cause handling problems. If misalignment is suspected as the result of an accident, it will be necessary to strip the machine completely so the frame can be thoroughly checked.

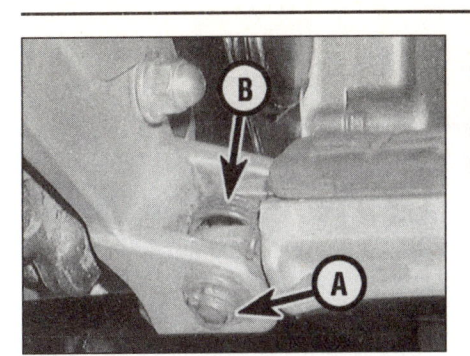

Frame, suspension and final drive 6•3

3.1 Remove the split pin and washer (A) to withdraw pivot pin. Note fitting of spring (B)

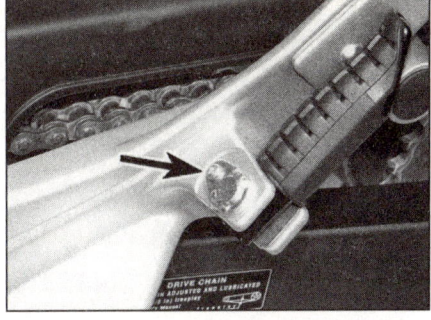

3.3 Undo the screw (arrowed) then withdraw the pivot pin

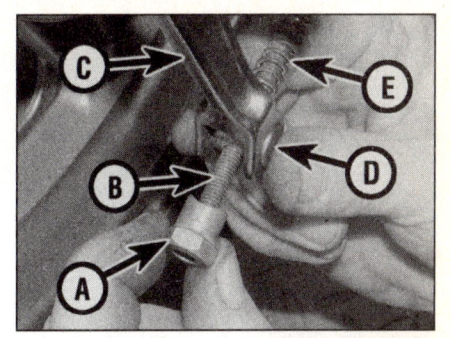

3.8 Adjuster (A), brake rod (B), brake arm (C), trunnion (D) and spring (E)

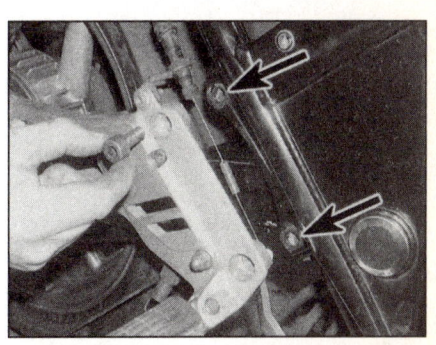

3 Footrests, brake pedal and gearchange lever – removal and installation

Rider's footrests

Removal

1 Remove the split pin and washer from the footrest pivot pin, then withdraw the pivot pin and remove the footrest, noting the fitting of the return spring (see illustration). The footrest rubbers are secured by two bolts on the underside of the footrest and can be renewed separately.

Installation

2 Installation is the reverse of removal. Apply a small amount of copper-based grease to the pivot pin. Use a new split pin and bend its ends securely.

Passenger footrests

Removal

3 Remove the split pin and washer from the bottom of the footrest pivot pin. Undo the screw, then withdraw the pivot pin and remove the footrest, noting the fitting of washer(s) on the end of the footrest (see illustration). The footrest rubbers are a push fit on the footrest and can be renewed separately.

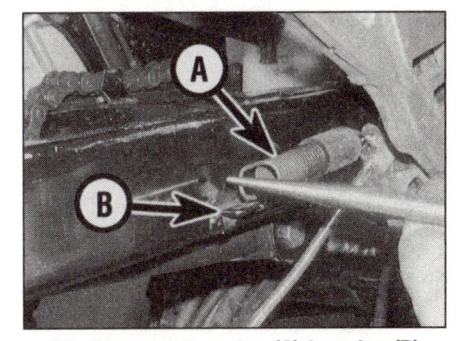

3.9 Disconnect spring (A) from lug (B)

Installation

4 Installation is the reverse of removal. Apply a small amount of copper-based grease to the pivot pin. Use a new split pin and bend its ends securely.

Brake pedal – R and T models

Removal

5 Remove the right-hand frame side panel (see Chapter 8).
6 Trace the wiring from the brake light switch and disconnect it at the black 2-pin connector.
7 Remove the silencer mounting nut and bolt (see Chapter 4, Section 13).
8 Unscrew and remove the rear brake rod adjuster nut and disconnect the rod from the brake arm (see illustration). Note the position

3.10 Footrest bracket attachment points on frame (arrowed)

of the spring. Remove the trunnion from the brake arm, thread it onto the brake rod and screw the adjuster nut onto the rod for safekeeping.
9 Disconnect the brake pedal return spring from the lug on the swingarm (see illustration).
10 Unscrew the footrest bracket bolts and remove the bracket (see illustration).
11 Disconnect the brake light switch spring and the pedal return spring from the lugs on the pedal (see illustration). Remove the split pin from the clevis pin securing the brake pedal to the brake rod. Remove the clevis pin and separate the rod from the pedal.
12 The pedal pivots on the footrest bracket. Remove the circlip and washer from the pedal pivot and remove the pedal (see illustrations).

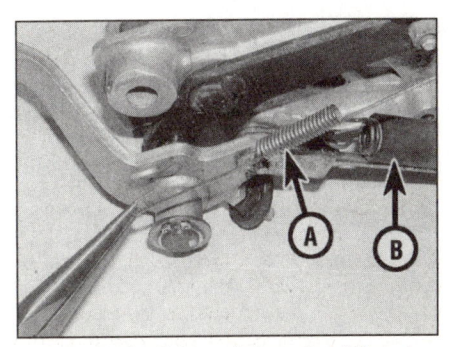

3.11 Brake light switch spring (A) and pedal return spring (B)

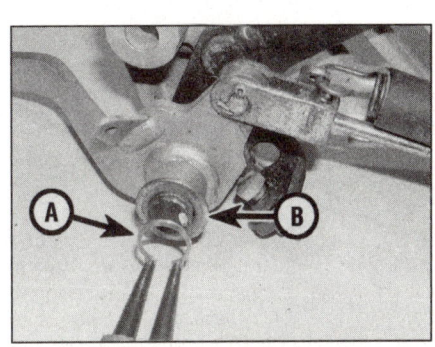

3.12a Remove circlip (A) and washer (B) . . .

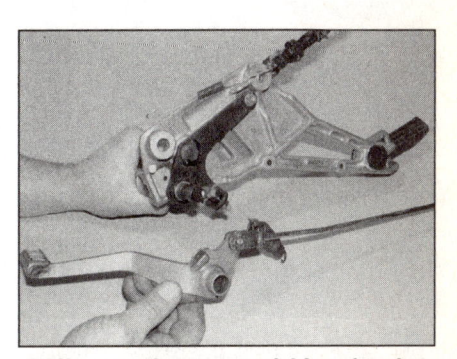

3.12b . . . and remove pedal from bracket

6•4 Frame, suspension and final drive

3.14 Lubricate pedal pivot before reassembly

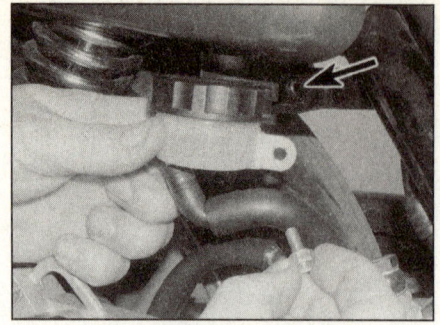

3.18 Remove the brake fluid reservoir from the frame bracket (arrowed)

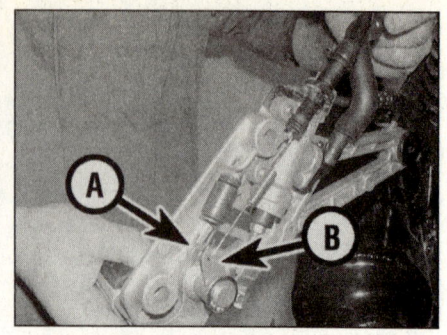

3.20 Pedal return spring (A) and brake light switch spring (B)

13 If necessary, unscrew the cap nut and nut/bolt securing the brake pedal bracket to the footrest bracket and separate them.

Installation
14 Installation is the reverse of removal, noting the following:
a) Apply molybdenum disulphide or copper-based grease to the brake pedal pivot **(see illustration)**.
b) Ensure the circlip is a firm fit in its groove on the brake pedal pivot. Don't forget the washer.
c) Fit a new split pin in the clevis pin and bend the split pin ends securely.
d) Tighten all nuts and bolts to the torque settings specified at the beginning of this Chapter.
e) Reset the brake pedal freeplay and check the operation of the brake light switch. If the pedal stop bolt was disturbed, reset it (Chapter 1).

Brake pedal – all other models

Removal
15 Remove the right-hand frame side panel (see Chapter 8).
16 Trace the wiring from the brake light switch and disconnect it at the black 2-pin connector.
17 Remove the silencer mounting nut and bolt (see Chapter 4, Section 13).
18 Ensure the rear brake fluid reservoir cap is secure and unscrew the bolt securing the reservoir to the frame **(see illustration)**. Wrap a rag around the reservoir to prevent accidental spillage and rest the reservoir in an upright position against the bike.
19 Unscrew the footrest bracket bolts and lift the bracket away from the bike, taking care not to strain the brake hose **(see illustration 3.10)**.
20 Disconnect the pedal return spring and the brake light switch spring from the lugs on the pedal **(see illustration)**.
21 Remove the split pin from the clevis pin securing the brake pedal to the brake master cylinder pushrod. Remove the clevis pin and separate the rod from the pedal **(see illustration)**.
22 The pedal pivots on the footrest bracket. Remove the circlip and washer from the pedal pivot and remove the pedal **(see illustration 3.21)**.

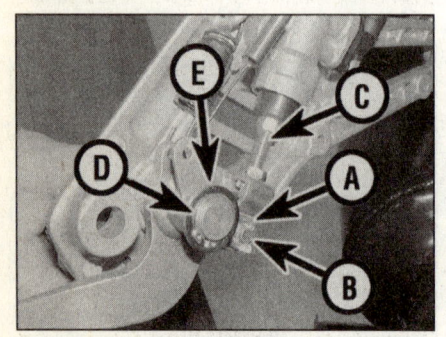

Installation
23 Installation is the reverse of removal, noting the following:
a) Apply molybdenum disulphide or copper-based grease to the brake pedal pivot **(see illustration 3.14)**.
b) Ensure the circlip is a firm fit in its groove on the brake pedal pivot. Don't forget the washer.
c) Use a new split pin on the clevis pin securing the brake pedal to the master cylinder pushrod and bend the split pin ends securely.
d) Tighten all bolts to the torque settings specified at the beginning of this Chapter.
e) Check the operation of the rear brake and the brake light switch (see Chapter 1).

Gearchange lever

Removal
24 Note the alignment punch mark on the gearchange shaft **(see illustration)**. If no mark is visible, make your own so that the lever can be correctly aligned with the shaft on installation. Unscrew the gearchange lever pinch bolt and remove the lever from the shaft.

Installation
25 Installation is the reverse of removal. Align the lever slot with the punchmark in the end of the shaft.
26 Ensure that the pinch bolt is tightened securely.

4 Stands – removal and installation

Sidestand
1 The sidestand is attached to a bracket on the left-hand side of the frame. Two springs, one inside the other, ensure that the stand is held in the retracted or extended position.
2 Support the bike on its centre stand and unhook the stand springs from the frame.
3 Unscrew the sidestand switch retaining bolt and remove the sidestand switch, noting how it fits **(see illustration)**.
4 Unscrew the nut securing the stand on the pivot bolt then unscrew the pivot bolt from the inside of the bracket and remove the stand.

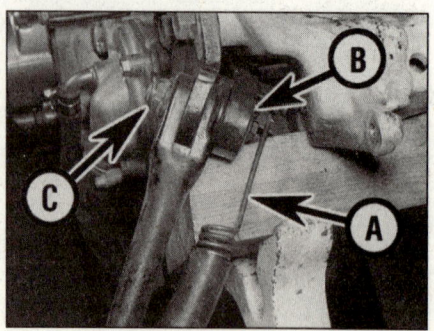

3.21 Split pin (A), clevis pin (B), master cylinder pushrod (C). Circlip (D) and washer (E) retain pedal

3.24 Gear lever alignment punchmark (arrowed)

4.3 Stand spring (A), sidestand switch (B) and pivot bolt nut (C)

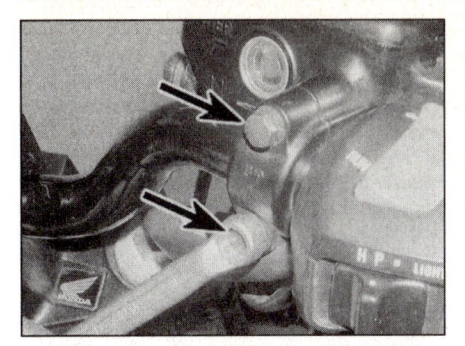

Frame, suspension and final drive 6•5

5.2 Master cylinder clamp bolts (arrowed)

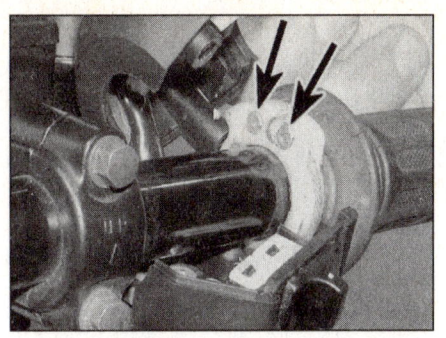

5.3 Separate the two halves of the switch and detach the throttle cables from the pulley (arrowed)

5.4 Unscrew the clutch lever clamp bolts (arrowed) and separate the two halves of the switch

5 On installation apply grease to the pivot bolt shank and tighten the nut securely.

6 Locate the tab on the inside of the sidestand switch in the hole in the stand and align the switch body with the stand spring post. Reconnect the sidestand springs and check that they hold the stand up securely when not in use – an accident is almost certain to occur if the stand extends while the machine is in motion.

7 Check the operation of the sidestand switch (see Chapter 1, Section 17).

Centre stand

8 Support the bike on its sidestand. Unhook the two centre stand return springs, on CBF models noting how the spring plate fits.

9 On CB models remove the split pin from the right-hand end of the stand pivot tube. On CBF models unscrew the pivot retaining bolt. Withdraw the pivot tube and remove the stand.

10 Clean all old grease and road dirt from the pivot tube and mounting lugs. Apply fresh grease to the pivot points and slide the tube in from the left side on CB models and from the right on CBF models. On CB models use a new split pin to secure the tube and bend its ends securely. On CBF models tighten the pivot retaining bolt.

11 Reconnect the springs, on CB models making sure that the spring sleeve is in place, and on CBF models fitting the spring plate with them. Check that the stand is held securely retracted by its springs.

5 Handlebars and levers – removal and installation

Handlebars

Removal

1 Lift the cover, where fitted, on the rear view mirror mounting lock nuts, slacken the locknuts and unscrew the mirrors from the handlebar brackets.

2 Ensure the brake fluid reservoir cap is secure. Wrap rag around the reservoir to prevent accidental spillage. Unscrew the master cylinder clamp bolts and remove the clamp **(see illustration)**. Position the master cylinder assembly clear of the handlebar, making sure no strain is placed on the hydraulic hose and the brake light switch wiring. Keep the master cylinder reservoir upright to prevent possible fluid leakage.

3 Unscrew the two screws from the combined throttle twistgrip/handlebar switch and free the two halves of the switch unit from the handlebar **(see illustration)**. Note the punchmark reference on the handlebar for reassembly. Detach the throttle cables from the throttle pulley – if necessary, slacken the throttle cable adjuster (see Chapter 1, Section 14). Position the switch unit away from the handlebar.

4 Unscrew the two clutch lever clamp bolts and remove the clamp **(see illustration)**.

Position the lever clear of the handlebar, making sure the cable and the clutch switch wiring is not unduly bent or strained.

5 Undo the left-hand switch housing screws and free the two halves of the switch unit from the handlebar **(see illustration 5.4)**. Note the punchmark reference on the handlebar for reassembly. On CB models detach the choke cable from the lever.

6 Unscrew the handlebar end-weight retaining screws, then remove the end-weights **(see illustration)**. Slide the throttle grip off the right-hand end of the bar and the hand grip off the left-hand end. If the left-hand grip is stuck in place, slit the grip open using a sharp blade and replace it with a new one on assembly. On CB models slide the choke lever pulley off the handlebar.

7 The left and right-hand internal handlebar weight assemblies are held in place by spring retainers. Thread the end-weight screw into the internal weight assembly, depress the spring retainer through the hole in the handlebar with a small screwdriver and withdraw the assembly.

8 Remove the plugs from the tops of the handlebar clamp bolts, unscrew the bolts and remove the clamps and the handlebars. Note the punch mark reference on the clamps and the handlebars for reassembly **(see illustrations)**.

9 On SW, SX, SY and S2 models, the handlebar clamps are rubber mounted in the top yoke. The clamps are secured by self-locking nuts on the underside of the yoke

5.6 Handlebar end-weight screw (arrowed)

5.8a Handlebar clamp punch marks (arrowed)

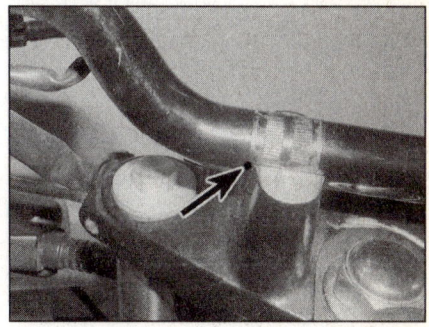

5.8b Handlebar punch mark (arrowed)

6•6 Frame, suspension and final drive

5.9 Handlebar clamp self-locking nut on SW, SX, SY and S2 models

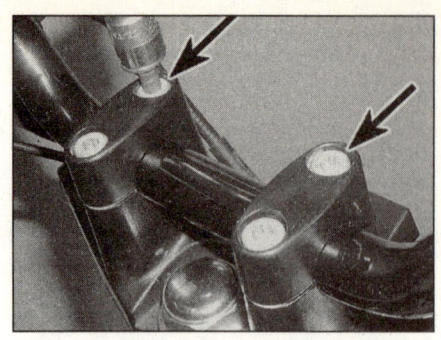

5.10a Tighten the front bolts (arrowed) first

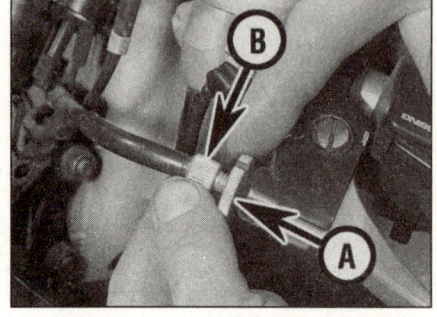

5.10b Ensure lever clamps with UP facing up

(see illustration); if the mountings are removed, new nuts should be used on reassembly.

Installation

10 Installation is the reverse of removal, noting the following.
a) Align the punch mark on the handlebars with the split in the clamp (see illustration 5.8b). Fit the handlebar clamps with the punch marks to the front (see illustration 5.8a). Tighten the front handlebar clamp bolts to the specified torque setting before the rear ones (see illustration).
b) On CB models do not forget to install the choke lever pulley on the left-hand end of the bar first.

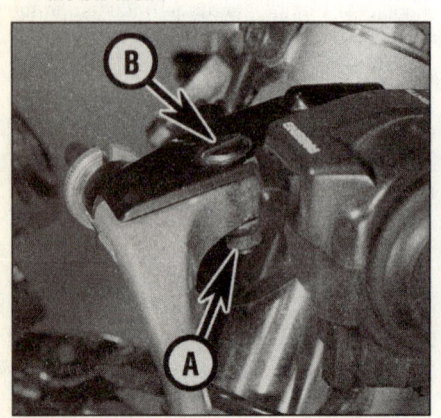

5.11a Slacken the lockring (A) and thread in the adjuster (B) . . .

c) Make sure the front brake and clutch lever assembly clamps are installed with the UP mark facing up and so that the clamp joint at the top aligns with the punchmark on the handlebar (see illustration).
d) Make sure the pin in the lower half of each switch housing locates in the hole in the handlebar.
e) If removed, apply a suitable non-permanent locking compound to the handlebar end-weight retaining screws. If new grips are being fitted, secure them using a suitable adhesive to the handlebar (left-hand grip) or to the throttle twistgrip (right-hand grip).
f) Tighten all bolts to the torque settings specified at the beginning of this Chapter.

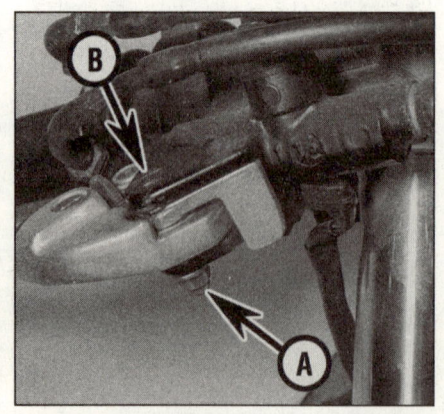

Handlebar levers

Removal – clutch lever

11 Slacken the clutch cable adjuster lockring and thread the adjuster fully into the bracket to provide maximum freeplay in the cable (see illustration). Unscrew the lever pivot bolt locknut, then withdraw the pivot bolt and remove the lever, detaching the cable nipple as you do so (see illustration).

Removal – brake lever

12 Unscrew the lever pivot bolt locknut, then withdraw the pivot bolt and remove the lever (see illustration).

Installation

13 Installation is the reverse of removal. Apply grease to the pivot bolt shafts and the contact areas between the lever and its bracket, and to the clutch cable nipple. Adjust the clutch cable freeplay (see Chapter 1).

6 Forks – removal and installation

Removal

Caution: Although not strictly necessary, before removing the forks it is recommended that the fairing (CB500-SW, SX, SY and S2 models) and radiator side panels (R, T, V, W, X, Y and 2 models) are removed (see Chapter 8). This will prevent accidental damage to the paintwork.

1 Remove the front wheel (see Chapter 7) and the front mudguard (see Chapter 8).
2 Remove each fork leg individually.
3 Unscrew the bolts securing the front brake caliper bracket to the right-hand fork slider and remove the caliper. Secure the caliper to the bike with a cable tie to ensure no strain is placed on the hydraulic hose. Discard the caliper bracket bolts as new ones must be fitted on reassembly (see Chapter 7).
4 Slacken but do not remove the fork clamp bolt in the top yoke (see illustration). If the fork legs are going to be disassembled or the fork oil is going to be changed, slacken the fork top bolt while the leg is still clamped in the bottom yoke.

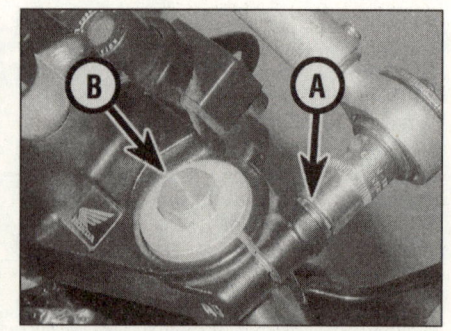

5.11b . . . then unscrew the nut (A) and the pivot bolt (B)

5.12 Unscrew the nut (A) and the pivot bolt (B)

6.4 Slacken the top yoke clamp bolt (A). Fork top bolt (B)

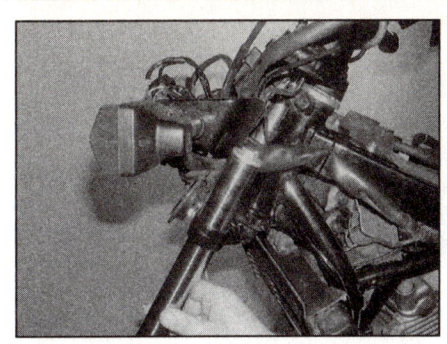

Frame, suspension and final drive 6•7

6.5 Slide the fork leg out of the bottom yoke

5 Slacken but do not remove the fork clamp bolt in the bottom yoke and remove the fork leg by twisting it and pulling it downwards **(see illustration)**.

HAYNES HINT: *If the fork legs are seized in the yokes, spray the area with penetrating oil and allow time for it to soak in before trying again.*

Installation

6 Remove all traces of corrosion from the fork tubes and the yokes. Install each fork leg individually. Slide the leg up through the bottom yoke into the top yoke until the top edge of the fork tube is level with the top edge of the yoke. Tighten the fork clamp bolt in the bottom yoke to the torque setting specified at the beginning of this Chapter.
7 If the fork legs have been disassembled or the fork oil changed, tighten the fork top bolt to the specified torque setting, and then tighten the fork clamp bolt in the top yoke to the specified torque setting.
8 Install the front brake caliper on the right-hand fork slider with new bolts and tighten the bolts to the specified torque setting.
9 Install the front mudguard (see Chapter 8) and the front wheel (see Chapter 7).
10 Check the operation of the front forks and brakes before taking the machine out on the road.

7 Forks – disassembly, inspection and reassembly

Disassembly

1 Always dismantle the fork legs separately to avoid interchanging parts. Store all components in separate, clearly marked containers **(see illustration)**.
2 Before dismantling the fork leg, it is advised that the damper rod bolt be slackened at this stage. Invert the fork leg and compress the fork tube in the slider so that the spring exerts maximum pressure on the damper rod head, then slacken the damper rod bolt in the base of the fork slider **(see illustration 7.7)**.

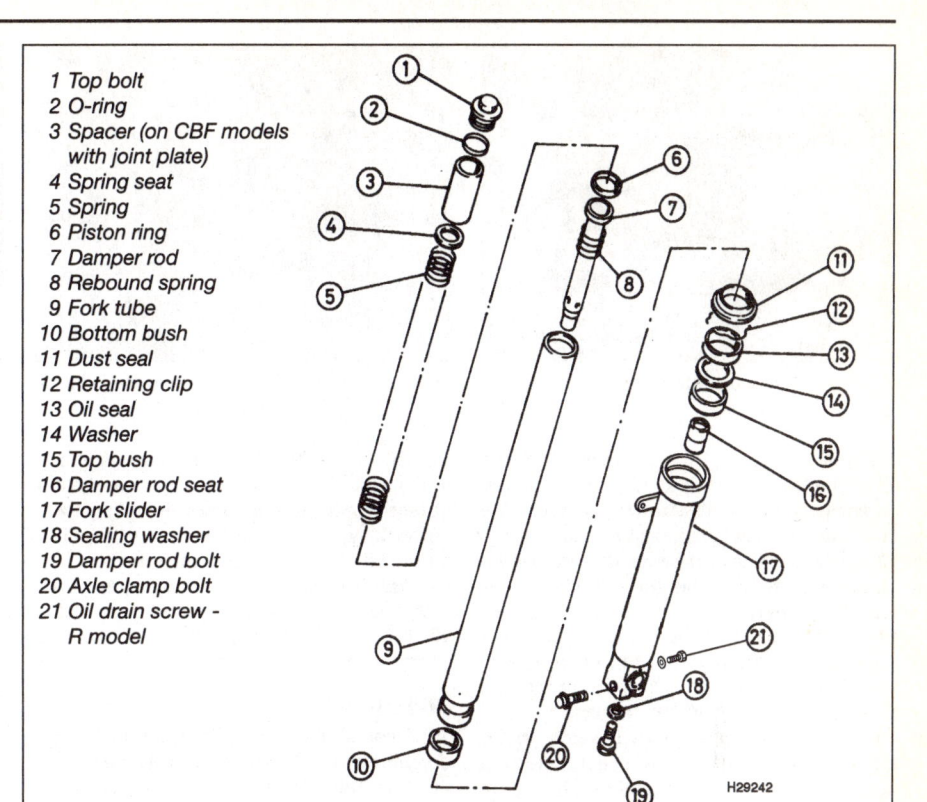

1 Top bolt
2 O-ring
3 Spacer (on CBF models with joint plate)
4 Spring seat
5 Spring
6 Piston ring
7 Damper rod
8 Rebound spring
9 Fork tube
10 Bottom bush
11 Dust seal
12 Retaining clip
13 Oil seal
14 Washer
15 Top bush
16 Damper rod seat
17 Fork slider
18 Sealing washer
19 Damper rod bolt
20 Axle clamp bolt
21 Oil drain screw - R model

7.1 Front fork components

3 If the fork top bolt was not slackened with the fork on the bike, carefully clamp the fork tube in a vice equipped with soft jaws, taking care not to overtighten or score its surface, and slacken the top bolt.
4 Unscrew the fork top bolt from the top of the fork tube and discard the O-ring as a new one must be fitted on reassembly.

⚠ **Warning: The fork spring is pressing on the fork top bolt with considerable pressure. Unscrew the bolt very carefully, keeping a downward pressure on it and release it slowly as it is likely to spring clear. It is advisable to wear some form of eye and face protection when carrying out this operation.**

5 Slide the fork tube down into the slider and withdraw the spacer, on CBF models noting the joint plate fitted in the top, the spring seat and the spring from the tube **(see illustrations**

7.26a, b and c). Note which way up the spring is fitted.
6 Invert the fork leg over a suitable container and pump the fork vigorously to expel as much fork oil as possible.
7 Remove the previously slackened damper rod bolt and its copper sealing washer from the bottom of the slider **(see illustration)**. Discard the sealing washer as a new one must be used on reassembly. If the damper rod bolt was not slackened before dismantling the fork, it may be necessary to re-install the spring, spring seat, spacer and top bolt to prevent the damper rod from turning. Alternatively, a long metal bar passed down through the fork tube and pressed hard into the damper rod head quite often suffices.
8 Withdraw the damper rod from inside the fork tube. Remove the rebound spring from the damper rod **(see illustration)**.

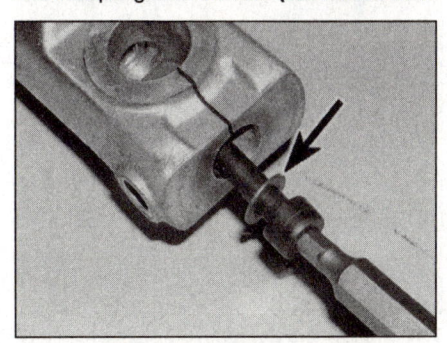

7.7 Remove the damper rod bolt and sealing washer (arrowed)

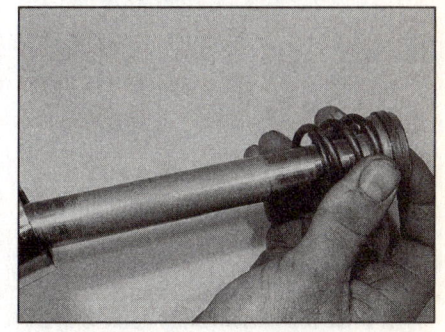

7.8 Withdraw the damper rod and rebound spring from the tube

7.9 Prise out the dust seal . . .

7.10 . . . then carefully remove the oil seal retaining clip

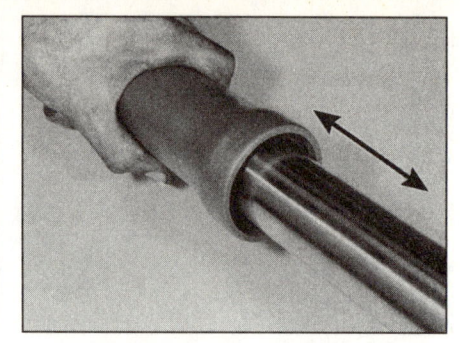

7.11 Pull the fork tube and slider apart to drive the top bush out

9 Carefully prise out the dust seal from the top of the slider to gain access to the oil seal retaining clip **(see illustration)**. Discard the dust seal as a new one must be used.

10 Carefully remove the retaining clip, taking care not to scratch the surface of the tube **(see illustration)**.

11 To separate the tube from the slider it is necessary to displace the top bush and oil seal. The bottom bush will not pass through the top bush, and this can be used to good effect. Push the tube gently inwards until it stops against the damper rod seat. Take care not to do this forcibly or the seat may be damaged. Then pull the tube sharply outwards until the bottom bush strikes the top bush. Repeat this operation until the top bush and seal are tapped out of the slider **(see illustration)**.

12 With the tube removed, slide off the oil seal and its washer, noting which way up they fit **(see illustration)**. Discard the oil seal as a new one must be used. The top bush can then be slid off the upper end of the tube.

Caution: Do not remove the bottom bush from the tube unless it is to be renewed.

13 Tip the damper rod seat out of the slider, noting which way up it fits.

Inspection

14 Clean all parts in solvent and blow them dry with compressed air, if available. Check the fork tube for score marks, scratches, flaking of the chrome finish and excessive or abnormal wear. Look for dents in the tube and renew the tube in both forks if any are found. Check the fork seal seat for nicks, gouges and scratches. If damage is evident, leaks will occur. Also check the oil seal washer for damage or distortion and renew it if necessary.

15 Check the fork tube for runout using V-blocks and a dial gauge, or have it done by a Honda dealer **(see illustration)**. If the amount of runout exceeds the service limit specified, the tube should be renewed.

> **Warning: If the tube is bent or exceeds the runout limit, it should not be straightened; renew it.**

16 Check the spring for cracks and other damage. Measure the spring free length and compare it to the specifications at the beginning of this Chapter. If it is defective or has sagged below the service limit, renew the springs in both forks. Never renew only one spring. Also check the rebound spring.

17 Examine the working surfaces of the two bushes; if worn or scuffed they must be renewed. To remove the bottom bush from the fork tube, prise it apart at the slit using a flat-bladed screwdriver and slide it off **(see illustration)**. Make sure the new one seats properly.

18 Check the damper rod and its piston ring for damage and wear, and renew them if necessary **(see illustration)**. Do not remove the ring from the piston unless it requires renewal

Reassembly

19 If removed, install the new piston ring into the groove in the damper rod, then slide the rebound spring onto the rod **(see illustration)**.

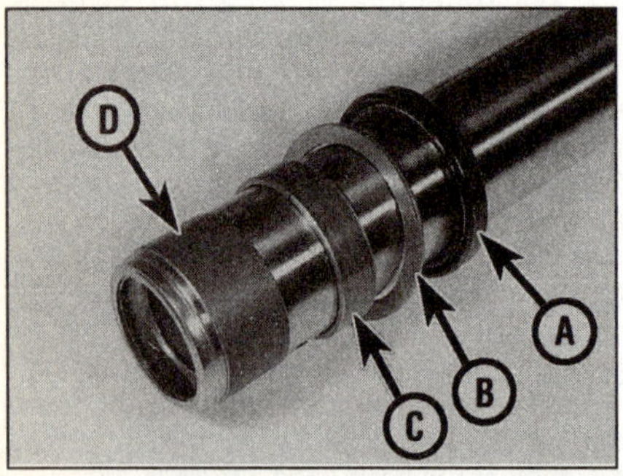

7.12 Oil seal (A), washer (B), top bush (C) and bottom bush (D)

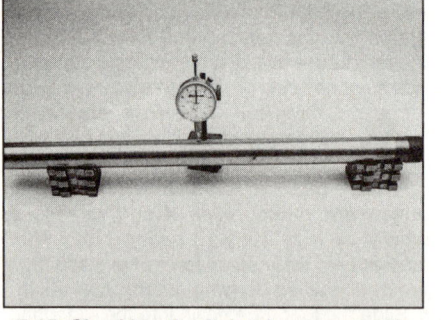

7.15 Checking the fork tube runout with V-blocks and a dial gauge

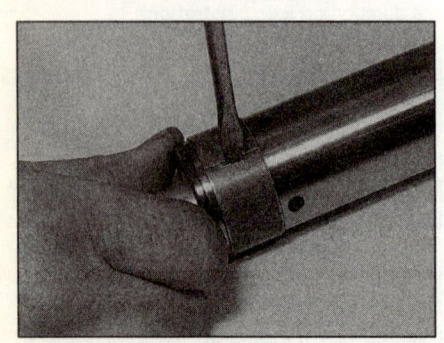

7.17 Prise the bottom bush apart with a flat-bladed screwdriver and slide it off the end of the tube

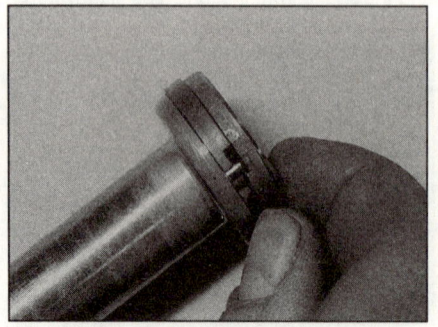

7.18 Renew the damper rod piston ring if it is worn or damaged

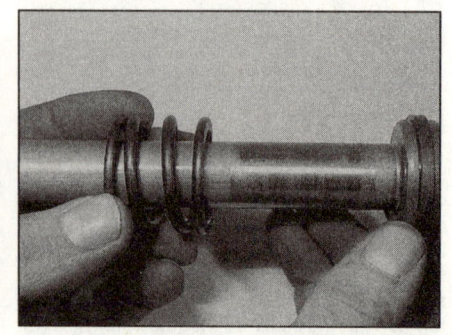

7.19a Slide the rebound spring onto the damper rod

Frame, suspension and final drive 6•9

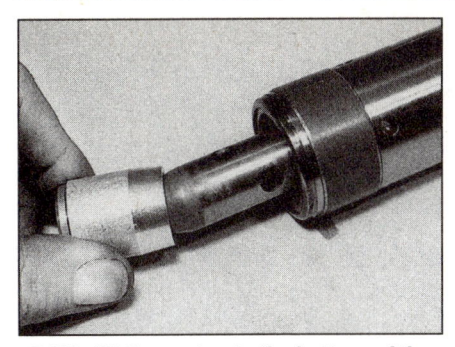

7.19b Fit the seat onto the bottom of the rod

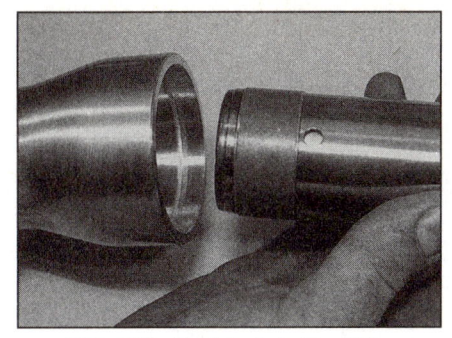

7.20 Slide the tube into the slider

7.21a Install the top bush . . .

Insert the damper rod into the fork tube and slide it into place so that it projects fully from the bottom of the tube, then install the seat on the bottom of the damper rod **(see illustration)**.

20 Oil the fork tube and bottom bush with the specified fork oil and insert the assembly into the slider **(see illustration)**. Fit a new copper sealing washer to the damper rod bolt and apply a few drops of a suitable non-permanent thread-locking compound, then install the bolt into the bottom of the slider. Tighten the bolt to the specified torque setting. If the damper rod rotates inside the tube, hold the rod with spring pressure or a metal bar as on disassembly (see Step 7).

21 Push the fork tube fully into the slider, then oil the top bush and slide it down over the tube **(see illustration)**. Press the bush squarely into its recess in the slider as far as possible, then install the oil seal washer **(see illustration)**. Either use the Honda service tool or a suitable piece of tubing to tap the bush fully into place; the tubing must be slightly larger in diameter than the fork tube and slightly smaller in diameter than the bush recess in the slider. Take care not to scratch the fork tube during this operation; if the fork tube is pushed fully into the slider any accidental scratching is confined to the area above the oil seal.

22 When the bush is seated squarely in its recess in the slider (remove the washer to check, wipe the recess clean, then reinstall the washer) install the new oil seal. Lubricate the inside of the seal with fork oil, slide it over the tube with its markings facing upwards and press it into place as described in Step 21 until the retaining clip groove is visible above the seal **(see illustration)**.

23 Once the seal is correctly seated, fit the retaining clip, making sure it is correctly located in its groove **(see illustration)**.

24 Lubricate the inside of the new dust seal then slide it down the fork tube and press it into position **(see illustration)**.

25 Slowly pour in the correct quantity of the specified grade of fork oil and carefully pump the fork at least ten times to distribute it evenly **(see illustration)**; the oil level should also be measured and adjustment made by adding or subtracting oil. Fully compress the fork tube into the slider and measure the fork oil level from the top of the tube **(see illustration)**. Add or subtract fork oil until it is at the level specified at the beginning of this Chapter.

26 Pull the fork tube out of the slider to its full extension and install the spring with its closer-wound coils at the bottom, followed by

7.21b . . . followed by the oil seal washer

7.22 Install the oil seal with markings facing up

7.23 Install the oil seal retaining clip . . .

7.24 . . . and then the dust seal

7.25a Pour the oil into the top of the tube . . .

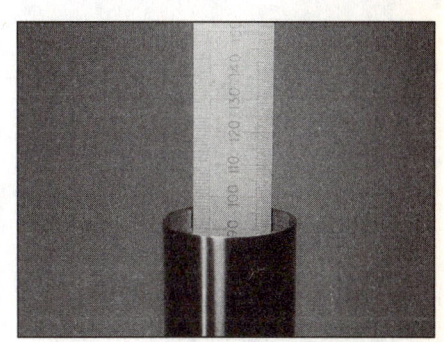

7.25b . . . then measure the oil level with the fork leg fully compressed

6•10 Frame, suspension and final drive

7.26a Install the spring . . .

7.26b . . . followed by the spring seat . . .

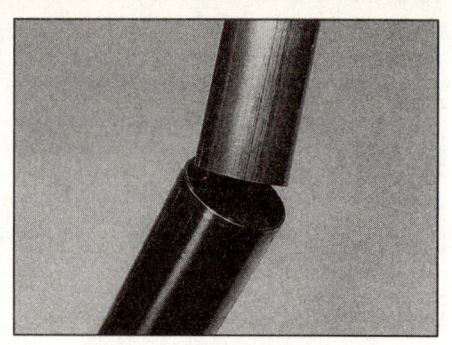

7.26c . . . and then the spacer

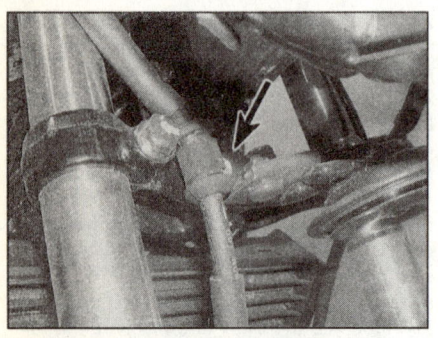

8.2 Bolt (arrowed) secures the horn, bracket and hose clamp on SW, SX and SY models

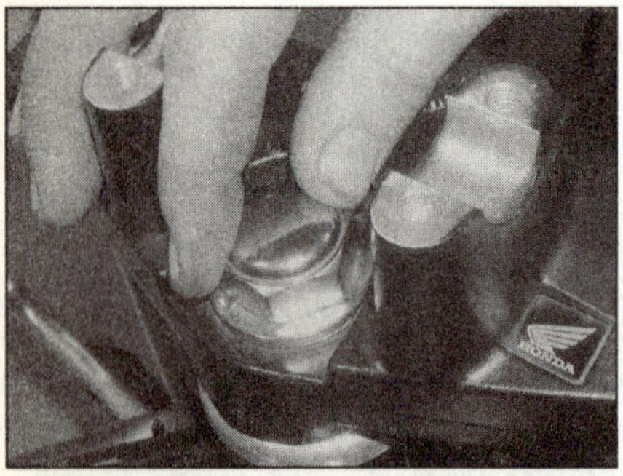

8.3a Remove the steering stem nut . . .

8.3b . . . then, on CB500 models, the washer

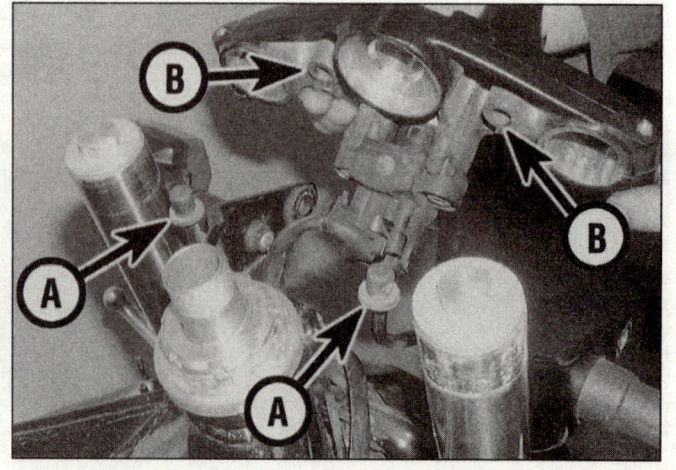

8.4 Bushes (A) locate in sockets (B) on underside of top yoke

the spring seat and the spacer (see illustrations). On CBF models fit the joint plate into the top of the spacer.

27 Fit a new O-ring to the fork top bolt and thread the bolt into the top of the fork tube.

> **Warning:** It will be necessary to compress the spring by pressing it down with the top bolt in order to engage the threads of the top bolt with the fork tube. This is a potentially dangerous operation and should be performed with care, using an assistant if necessary. Wipe off any excess oil before starting to prevent the possibility of slipping.

Keep the fork leg fully extended whilst pressing on the spring. Screw the top bolt carefully into the fork tube making sure it is not cross-threaded. **Note:** *The top bolt can be tightened to the specified torque setting when the fork has been installed in the bike and is securely held in the bottom yoke.*

28 Install the fork leg (see Section 6).

8 Steering stem – removal and installation

Removal

Caution: Although not strictly necessary, before removing the forks it is recommended that the fairing (SW, SX and SY models) and radiator side panels (R, T, V, W, X and Y models) are removed (see Chapter 8). This will prevent accidental damage to the paintwork.

1 Remove the front forks (see Section 6). Displace the handlebars (see Section 5). On R, T, V, W, X, Y and 2 models and CBF models remove the instrument cluster (see Chapter 9).

2 Unscrew the bolt securing the front brake hose clamp to the bottom yoke and remove the clamp. On SW, SX and SY models, disconnect the wiring connectors to the horn and at the same time and remove the horn, bracket and hose clamp (see illustration).

3 Except on CBF models where this has already been done during removal of the instruments, unscrew and remove the steering stem nut using a socket or spanner, and remove the washer (see illustrations). Lift the top yoke off the steering stem and place it aside, making sure no strain is placed on the ignition switch wiring.

4 On R, T, V, W, X, Y and 2 models and CBF models, the headlight and turn signal assembly locates in sockets in the top and bottom yoke. Lift the assembly clear of the bottom yoke, making sure the mounting bushes are secure on the four ends of the assembly stays (see illustration).

5 Prise the lockwasher tabs out of the notches in the locknut (see Chapter 1,

Frame, suspension and final drive 6•11

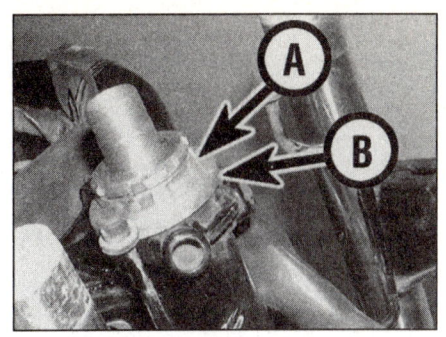

8.6 Adjuster nut (A) and bearing cover (B)

Section 19). Unscrew the locknut using either a C-spanner or a suitable drift located in one of the notches then remove the lockwasher, bending up the remaining tabs to release it from the adjuster nut if necessary. Inspect the tabs for cracks or signs of fatigue. If there are any, discard the lockwasher and use a new one; it is advisable to renew it as a matter of course.

6 Supporting the bottom yoke, unscrew the adjuster nut using either a C-spanner, a peg-spanner or a drift located in one of the notches, then remove the adjuster nut and the bearing cover from the steering stem (see illustration).

7 Gently lower the bottom yoke and steering stem out of the frame.

8 Remove the upper bearing inner race and the caged bearing from the top of the steering head (see illustration). Remove all traces of old grease from the upper and lower bearings and races and check them for wear or damage as described in Section 9. **Note:** *Do not attempt to remove the races from the frame or the lower bearing from the steering stem unless they are to be renewed.*

Installation

9 Smear a liberal quantity of grease on the bearing races in the frame. Work the grease well into both the upper and lower bearing cages.

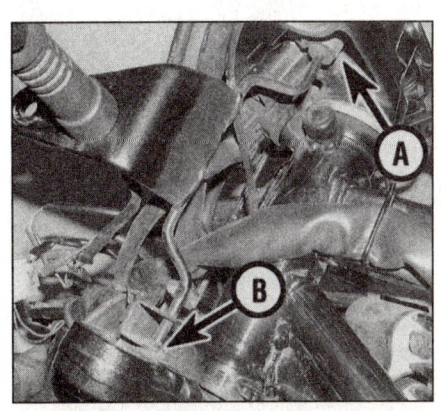

8.13 Ensure assembly locates correctly at points (A) and (B)

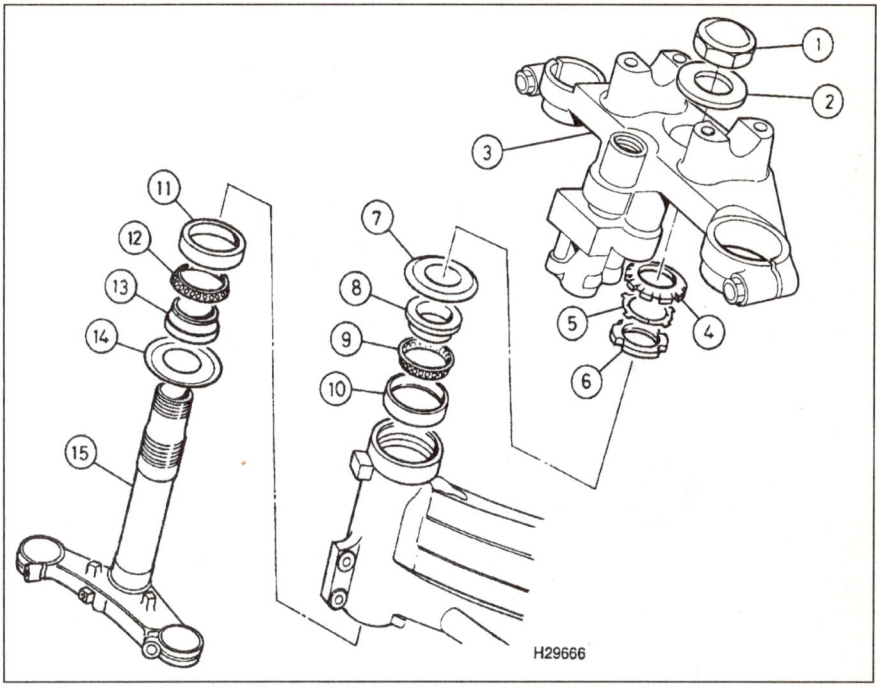

8.8 Steering stem components

1 Steering stem nut
2 Steering stem washer (CB models)
3 Top yoke
4 Locknut
5 Lockwasher
6 Adjuster nut
7 Bearing cover
8 Upper bearing inner race
9 Upper bearing
10 Upper bearing outer race
11 Lower bearing outer race
12 Lower bearing
13 Lower bearing inner race
14 Dust seal
15 Bottom yoke and steering stem

10 Carefully lift the steering stem/bottom yoke up through the frame. Install the upper bearing and its inner race in the top of the steering head, then install the bearing cover and thread the adjuster nut onto the steering stem **(see illustration 8.8)**. Tighten the nut lightly to settle the bearings, then slacken if off so that it is finger-tight. Using either the C-spanner or drift, tighten the nut a little at a time until all freeplay in the bearings is removed, yet the steering is able to move freely from lock to lock. If the Honda adapter tool (pt. no. 07946-4300101) is available you can apply the torque setting specified at the beginning of this Chapter. Now turn the steering from lock to lock five times to settle the bearings, then recheck the adjustment or the torque setting. The object is to set the adjuster nut so that the bearings are under a very light loading, just enough to remove any freeplay (see Chapter 1, Section 19).

Caution: Take great care not to apply excessive pressure because this will cause premature failure of the bearings. If the torque setting is applied and the bearings are too loose or tight, set them up according to feel.

11 Install the lockwasher on the adjuster nut and thread the locknut onto the steering stem and tighten it finger-tight. Tighten the locknut (to a maximum of 90°) to align the tabs on the lockwasher with the slots in the locknut. Hold the adjuster nut to prevent it from moving. Bend up the lockwasher tabs to secure the locknut.

12 On R, T, V, W, X, Y and 2 models and CBF models, install the headlight and turn signal assembly in the sockets in the bottom yoke, making sure the mounting bushes are in place.

13 On CBF models position the instrument cluster on its bracket. Install the top yoke, steering stem nut and washer (where fitted). On R, T, V, W, X, Y and 2 models and CBF models, ensure the headlight assembly stays locate correctly in the top yoke **(see illustration)**. Leave the steering stem nut finger-tight to allow for alignment of the fork legs when they are fitted. Install the front brake hose clamp (and horn on SW, SX, SY and S2 models) to the bottom yoke.

14 Install the front fork legs to align the top and bottom yokes and tighten the steering stem nut to the torque setting specified at the beginning of this Chapter. Complete the installation of the front forks (see Section 6). Install the handlebars (see Section 5). On R, T, V, W, X, Y and 2 models and CBF models install the instrument cluster (see Chapter 9).

15 Carry out a check and, if necessary, readjust the steering head bearing freeplay (see Chapter 1).

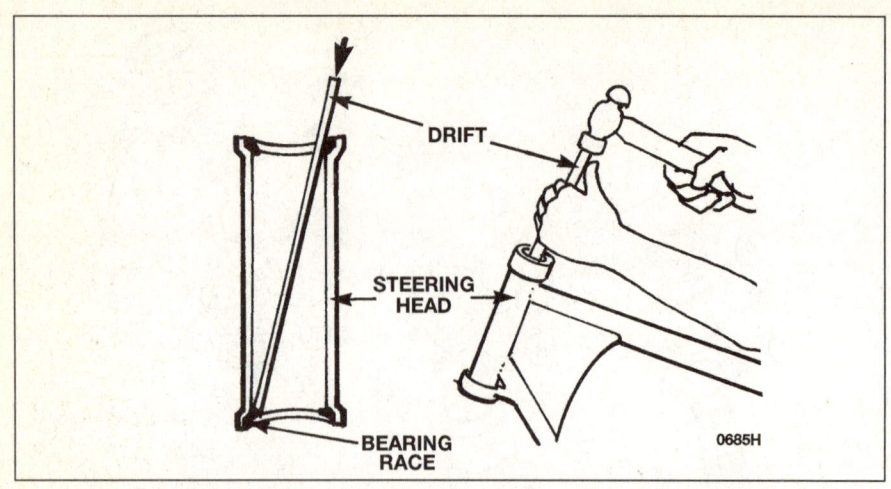

6•12 Frame, suspension and final drive

9.4 Drive the bearing outer races out with a brass drift as shown

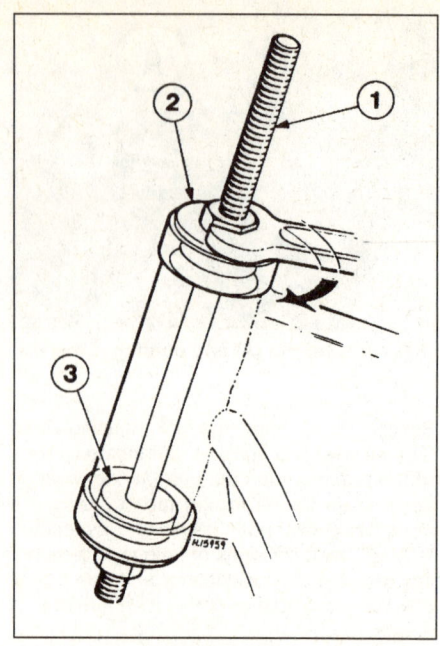

9.6 Using a drawbolt to fit the outer races into the headstock

1. *Long bolt or threaded bar*
2. *Thick washer*
3. *Guide for lower race*

9 Steering head bearings – inspection and renewal

Inspection

1 Remove the steering stem (see Section 8).
2 Remove all traces of old grease from the bearings and races and check them for wear or damage.
3 The outer races should be polished and free from indentations and pitting. Inspect the bearings for signs of wear, damage or discoloration, and examine the bearing ball retainer cage for signs of cracks or splits. Spin the bearings by hand. They should spin freely and smoothly. If there are any signs of wear on any of the above components both upper and lower bearing assemblies must be renewed as a set. Only remove the outer races from the frame headstock if they need to be renewed – do not re-use them once they have been removed.

Renewal

4 The outer races are an interference fit in the frame headstock and can be tapped out with a suitable drift **(see illustration)**. Tap firmly and evenly around each race to ensure that it is driven out squarely. It may prove advantageous to curve the end of the drift slightly to improve access.
5 Alternatively, the races can be removed using a slide-hammer type bearing extractor; these can often be hired from tool shops.
6 The new outer races can be pressed into the headstock using a drawbolt arrangement **(see illustration)**, or by using a large diameter tubular drift. Ensure that the drawbolt washer or drift (as applicable) bears only on the outer edge of the race and does not contact the working surface and that the race fits all the way into its seat. Alternatively, have the races installed by a Honda dealer.

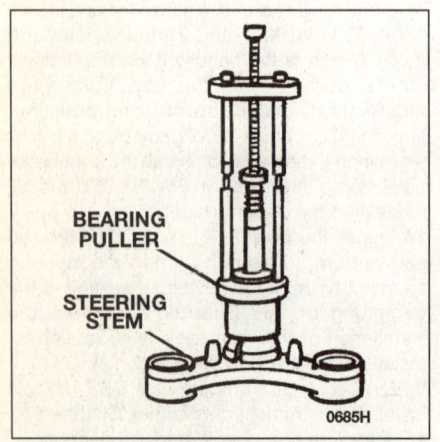

HAYNES HiNT *Installation of new bearing outer races is made much easier if the races are left overnight in the freezer. This causes them to contract slightly making them a looser fit.*

7 To remove the lower bearing inner race from the steering stem, use two screwdrivers placed on opposite sides of the race to work it free. If the bearing is firmly in place it will be necessary to use a bearing puller **(see illustration)**. Take the steering stem to a Honda dealer if necessary. Check the condition of the dust seal that fits under the inner race and renew it if it is worn or damaged.
8 Fit the new lower bearing inner race onto the steering stem. A length of tubing with an internal diameter slightly larger than the steering stem will be needed to tap the new bearing into position **(see illustration)**. Ensure that the drift bears only on the inner edge of the bearing race and does not contact its working surface.
9 Install the steering stem (see Section 8).

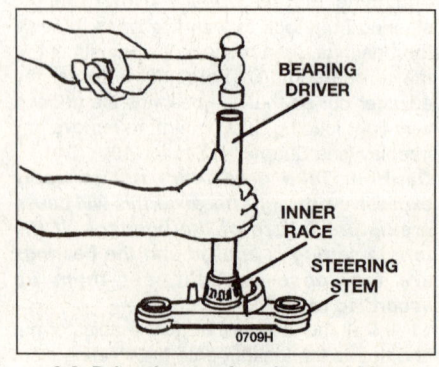

10 Rear shock absorbers – removal, inspection and installation

CB models

Removal

1 Place the machine on its centre stand. Position a support under the rear wheel so that it does not drop when the shock absorbers are removed, but also making sure that the weight of the machine is off the rear suspension so that the shocks are not compressed.
2 Remove the seat cowling (see Chapter 8) and the exhaust silencer (see Chapter 4).
3 Working on one shock at a time, unscrew the shock absorber lower mounting bolt, pull the shock towards the rear of the bike so that it is free of the lower mounting bracket, and

9.7 Using a bearing puller to remove the bearing from the steering stem

9.8 Drive the new bearing on with a suitable driver or length of pipe

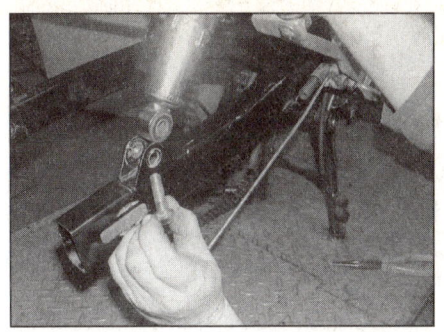

Frame, suspension and final drive 6•13

10.3a Remove the lower mounting bolt . . .

10.3b . . . then pull the shock off the upper bracket

then pull it off the upper mounting bracket (see illustrations).

Inspection

4 Inspect the damper rod for signs of wear, corrosion or pitting. Damage to the surface of the rod will wear the oil seal and lead to oil loss and lack of suspension damping.

5 Dismantling of the shock absorbers requires the use of a spring compressor. It is advised that this task is entrusted to a Honda dealer or suspension specialist. Note that apart from the mounting bolts, bushes and spacers, individual components are not available for the shock absorbers.

⚠ **Warning:** *The spring is under considerable pressure. It is advisable to wear some form of eye and face protection when carrying out this operation. Note that the shocks must be set to their minimum preload setting before dismantling.*

6 Check the action of the damper rod. Compress the rod into the damper body, then pull it out of the body; movement should be slow and progressive. If the rod binds in the body it is likely to be bent. If there is no damping to the rod movement, the damping oil has leaked out of the shock. In either case the shock in unserviceable. Individual component parts are not available and worn or damaged shock absorbers should always be renewed as a pair.

7 Check the spring for cracks and other damage. If the shock has been dismantled, measure the spring free length and compare it to the specifications at the beginning of this Chapter. If it is defective or has sagged below the service limit, the shocks should be renewed as a pair – the springs are not available separately.

8 If the shock absorber mounting bushes are a loose fit or show signs of deterioration, renew them. The bushes are a press fit in the ends of the shocks. Use a suitable lubricant when fitting new bushes and always renew bushes as a set. If necessary, withdraw the inner sleeves from the lower mounting bushes to aid removal of the bushes and renew the sleeves if worn.

Installation

9 Installation is the reverse of removal, noting the following.
a) *Apply molybdenum disulphide or copper-based grease to the spring pre-load adjusters.*
b) *Ensure the spring pre-load adjusters are adjusted equally (see Section 11).*
c) *Tighten the mounting bolts to the torque settings specified at the beginning of this Chapter.*

CBF models

Removal

10 Place the machine on its centre stand. Position a support under the rear wheel so that it does not drop when the shock absorber is removed, but also making sure that the weight of the machine is off the rear suspension so that the shock is not compressed. Remove the left-hand frame side panel (see Chapter 8).

11 On ABS models release the brake pipe holder clasp (see illustration). Depress the tab on the holder using a small screwdriver and slide it off its post on the frame (see illustration).

12 Unscrew the nut and withdraw the bolt securing the bottom of the shock absorber to the swingarm.

13 Unscrew the nut and withdraw the bolt securing the top of the shock absorber to the frame. Remove the support from under the rear wheel and drop the wheel, then manoeuvre the shock absorber out to the left-hand side (see illustration).

Inspection

14 Inspect the shock absorber for obvious physical damage and oil leakage, and the coil spring for looseness, cracks or signs of fatigue.

15 Inspect the bush in the top of the shock absorber for wear or damage (see illustration).

16 Withdraw the spacer from the bottom mounting. Check the condition of the grease seals and the needle bearing (see illustration).

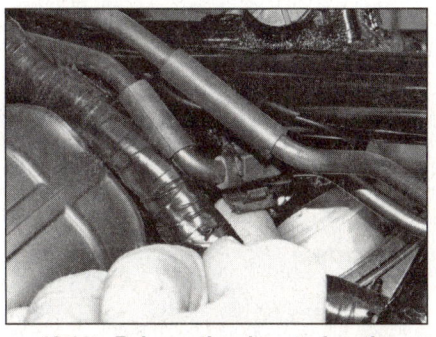

10.11a Release the clasp to free the pipe . . .

10.11b . . . then depress the tab to free the holder

10.13 Manoeuvre the shock out to the left-hand side

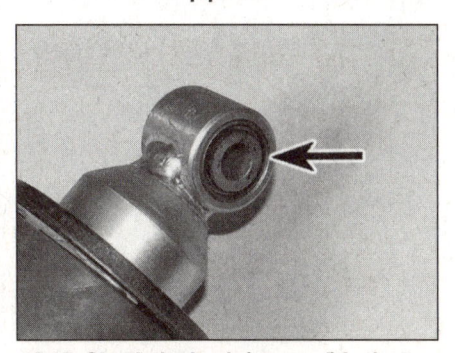

10.15 Check the bush (arrowed) in the top mount

10.16 Withdraw the spacer and check the seals and bearing (arrowed)

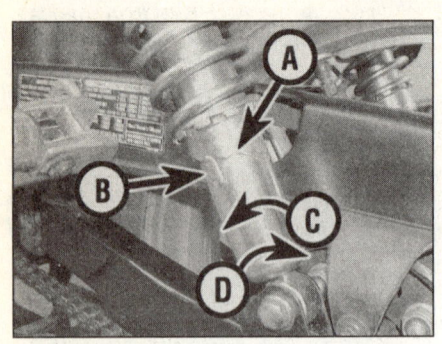

6•14 Frame, suspension and final drive

11.2 Align notches (A) with adjustment stopper (B). Turn spring seat anti-clockwise (C) to increase pre-load, clockwise (D) to reduce pre-load

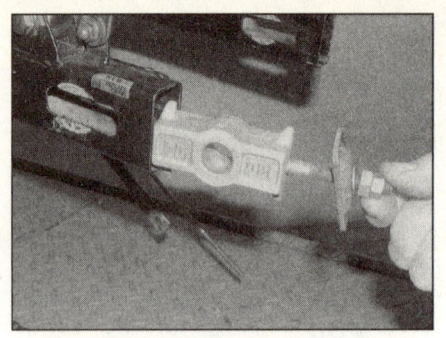

12.2 Chain tensioner assembly

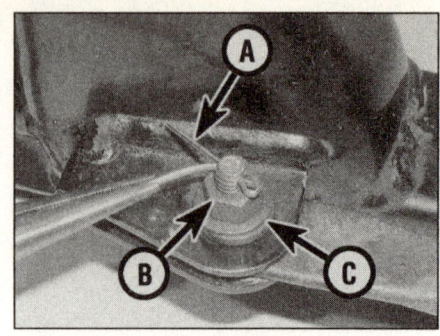

12.3 Torque arm split pin (A), nut (B) and washers (C)

If necessary lever out the grease seals, noting which way round they fit. Discard the seals as they must be replaced with new ones. Press the old bearing out and the new bearing in, referring to *Tools and Workshop Tips* in the Reference Section – make sure the bearing set depth is equal on each side (5.5 mm). Grease the bearing.

17 Smear the new seals with grease and press them squarely into place with the marked side facing out. Smear the spacer with grease and slide it through.

18 With the exception of the bottom pivot components, parts are not available for the shock absorber. If it is worn or damaged, it must be replaced with a new one.

Installation

19 Installation is the reverse of removal. Tighten the mounting bolt nuts to the torque setting specified at the beginning of the Chapter.

11 Suspension – adjustments

Front forks

1 The front forks are not adjustable. If the suspension action is poor, check the forks thoroughly (see Chapter 1). After a high mileage it may be necessary to change the fork oil (see Chapter 1) or renew the fork springs (see Section 7).

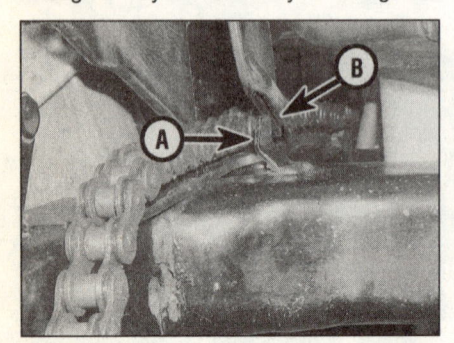

Rear shock absorber(s)

2 The rear shock absorber(s) is/are adjustable for spring pre-load. Adjustment is made using a suitable C-spanner (one is provided in the toolkit) to turn the spring seat on the bottom of the shock absorber **(see illustration)**. On CB models there are five positions – position 1 is the softest setting, position 5 is the hardest, and the standard setting is position 2. On CBF models there are seven positions – position 1 is the softest setting, position 7 is the hardest, and the standard setting is position 3. Align the setting required with the adjustment stopper.

3 To increase the pre-load, turn the spring seat anti-clockwise; to decrease the pre-load, turn the spring seat clockwise **(see illustration 11.2)**. **Note:** *Always ensure both shock absorber spring pre-load adjusters are adjusted equally.*

12 Swingarm and drive chain slider – removal and installation

CB models

Removal

1 Place the bike on its centre stand. Remove the silencer (see Chapter 4) and the rear wheel (see Chapter 7).

2 Remove the chain tensioner assembly from each end of the swingarm **(see illustration)**.

3 On R and T models, detach the brake pedal return spring from the lug on the swingarm **(see illustration 3.9)**. Although not essential for swingarm removal, remove the split pin and unscrew the nut and bolt that secure the brake torque arm to the swingarm, noting the position of the plain and spring washers **(see illustration)**. Discard the split pin.

4 On V, W, SW, X, SX, Y, SY, 2 and S2 models, unscrew the bolt securing the rear brake hose clamp to the swingarm and secure the brake caliper bracket to the frame with a cable tie to avoid placing a strain on the hose (remove the right-hand side frame side panel if necessary).

5 Unscrew the nuts securing the chainguard to the swingarm and remove the shouldered washers and the chainguard. Note the position of the washers and how the slot in the front edge of the chainguard locates over a bracket on the swingarm **(see illustration)**.

6 Place a suitable container underneath the coolant reservoir and detach the radiator overflow hose (see Chapter 3, Section 3).

7 Unscrew the shock absorber lower mounting bolts and pull the shocks free of the swingarm mounting brackets **(see illustration 10.3a)**.

8 Remove the caps from the swingarm pivot points **(see illustration)**. Unscrew the nut on the right-hand end of the swingarm pivot bolt and withdraw the bolt while supporting the swingarm **(see illustration)**.

12.5 Note how bracket (A) fits in slot (B)

12.8a Remove caps and remove the nut from the right side . . .

12.8b . . . then withdraw the pivot bolt from the left

Frame, suspension and final drive 6•15

12.9 Lower swingarm to clear coolant reservoir (arrowed)

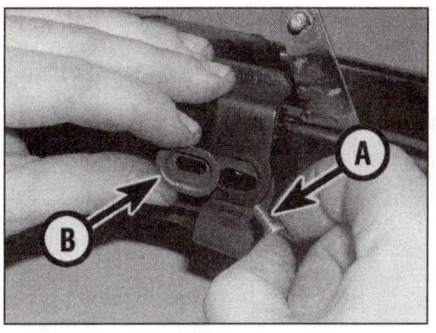

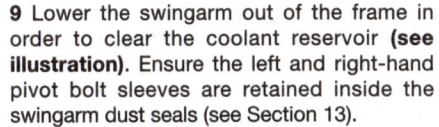

12.10a Drive chain slider retaining screw (A) and shouldered washer (B)

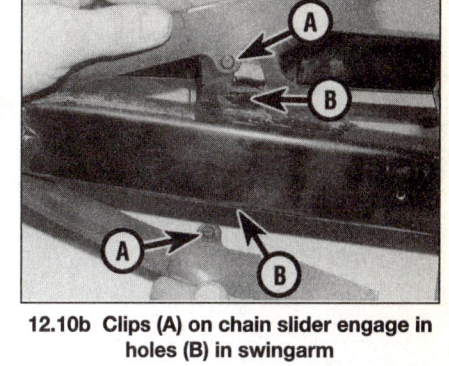

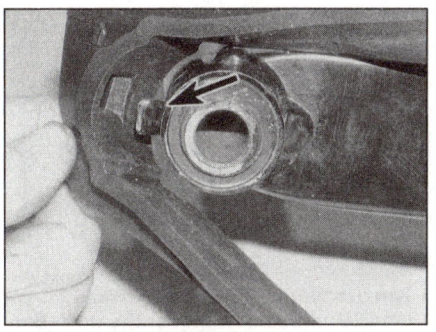

12.10b Clips (A) on chain slider engage in holes (B) in swingarm

9 Lower the swingarm out of the frame in order to clear the coolant reservoir **(see illustration)**. Ensure the left and right-hand pivot bolt sleeves are retained inside the swingarm dust seals (see Section 13).

10 If necessary, unscrew the screw securing the drive chain slider to the swingarm, remove the shouldered washer and unclip the slider from the swingarm **(see illustrations)**.

Installation

11 If removed, install the drive chain slider **(see illustrations 13.10a, b and c)** and lubricate the bearings, dust seals, sleeves and the pivot bolt with grease.

12 Offer up the swingarm, ensuring it is correctly positioned between the top and bottom runs of the drive chain and that none of the breather hoses are pinched behind the engine unit. Align the swingarm with the pivot bolt lugs and install the pivot bolt through the frame and swingarm from the left-hand side **(see illustration)**. Install the pivot bolt nut and tighten to the torque setting specified at the beginning of this Chapter. Check that the swingarm pivots freely without binding and refit the pivot caps.

13 Engage the lower ends of the shock absorbers in the swingarm mounting brackets, install the mounting bolts and tighten to the specified torque setting.

14 Connect the radiator overflow hose to the coolant reservoir and refill the reservoir (see Chapter 3).

15 Position the chainguard over the chain and align the guard with its mounting brackets, noting the alignment tab and slot on the front mounting. Install the shouldered washes and tighten the nuts securely.

16 Install the chain tensioners.

17 On R and T models, if removed, install the brake torque arm between the two halves of the bracket on the swingarm. Fit the mounting bolt from the right-hand side and then the spring washer, plain washer and nut from the left-hand side of the bracket. Tighten the nut securely and fit a new split pin. Connect the brake pedal return spring to the lug on the swingarm and install the rear wheel.

18 On V, W, SW, X, SX, Y, SY, 2 and S2 models, engage the lug on the brake caliper

12.10c Forward edge of slider engages on lug (arrowed) on swingarm

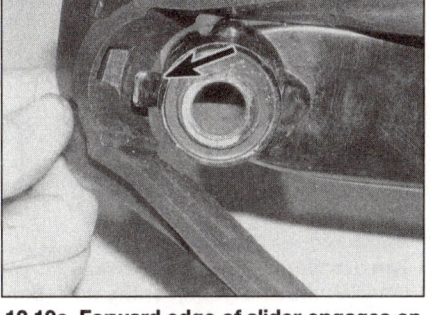

bracket with the socket on the swingarm and install the rear wheel. Install the bolt securing the rear brake hose clamp to the swingarm and tighten the bolt securely.

19 Install the silencer (see Chapter 4).

20 Check and adjust the drive chain slack (see Chapter 1). Check the operation of the rear suspension before taking the machine on the road.

CBF models

Removal

21 Place the bike on its centre stand. Remove the rear wheel (see Chapter 7).

22 Remove the chain tensioner assembly from each end of the swingarm **(see illustration 12.2)**.

23 Undo the rear brake hose guide screws

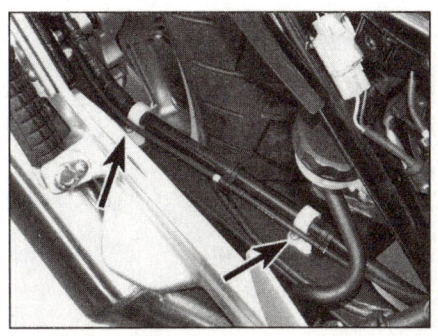

12.23 Undo the screws (arrowed) to free the hose

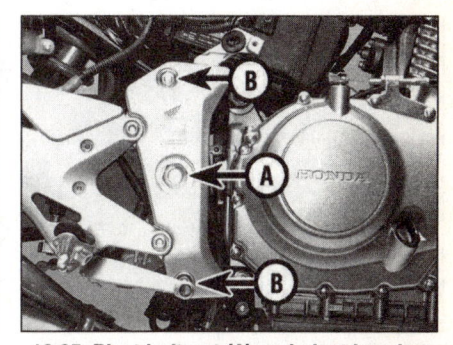

12.12 Align the swingarm with the pivot bolt lugs (arrowed)

(see illustration) and secure the brake caliper bracket to the frame with a cable-tie to avoid placing a strain on the hose (remove the right-hand frame side panel if necessary).

24 Unscrew the nut and withdraw the bolt securing the bottom of the shock absorber to the swingarm, or remove the shock absorber completely if required (see Section 10).

25 Unscrew the nut on the right-hand end of the swingarm pivot bolt and remove the washer **(see illustration)**. Slacken the nuts on the right-hand end of the pivot bracket bolts.

26 Support the swingarm, then withdraw the pivot bolt and manoeuvre the swingarm out.

27 If required, unscrew the chain guard bolts and remove the guard, noting how it locates in

12.25 Pivot bolt nut (A) and pivot bracket bolt nuts (B)

6•16 Frame, suspension and final drive

12.27a Undo the screw (arrowed) on the underside . . .

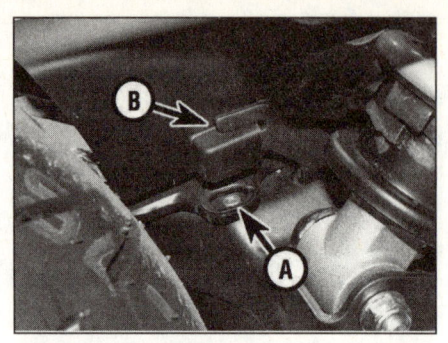

12.27b . . . and the screw (A) on the inner side. Note how the chain guard locates (B)

the bracket on the inside **(see illustration)**. Undo the screws securing the drive chain slider to the swingarm, remove the shouldered washers and unclip the slider from the swingarm **(see illustration)**.

Installation

28 Installation is the reverse of removal. If removed, install the drive chain slider and chain guard. Remove the sleeve from the left-hand pivot, and the spacer from each side of the right-hand pivot, noting which fits where. Lubricate the bearings, dust seals, sleeve, spacers and the pivot bolt with grease. Refit the spacers and sleeve. Offer up the swingarm, ensuring it is correctly positioned between the top and bottom runs of the drive chain. Tighten the pivot bolt nut, and the pivot bracket bolt nuts to the torque setting specified at the beginning of the Chapter.

Check that the swingarm pivots freely without binding before fitting the shock absorber.

13 Swingarm – inspection and bearing renewal

Inspection

1 Remove the swingarm (see Section 12).
2 On CB models remove the left and right-hand pivot bolt sleeves from the swingarm dust seals; note that the left-hand sleeve is longer than the right-hand sleeve **(see illustration)**. Lever out the seals from each side, taking care not to damage them.
3 On CBF models remove the sleeve from the left-hand pivot, and the spacer from each side of the right-hand pivot, noting which fits where. Lever out the seals from each side, taking care not to damage them.
4 Thoroughly clean all components with solvent, removing all traces of dirt, corrosion and grease. Inspect all components closely, looking for obvious signs of wear such as heavy scoring, and cracks or distortion due to accident damage. Check that the two arms of the swingarm are aligned with each other. Any damaged or worn component must be renewed.
5 If you doubt the straightness or alignment of the swingarm, lay the swingarm on the work surface and support it so that the pivot bolt tube is level (check this with a spirit level). Install the chain adjusters and the axle and check the level of the axle. If the axle is not level, the swingarm is out of true and must be renewed, although seek the advice of a frame specialist to confirm your findings.
6 Check the swingarm pivot bolt for straightness by rolling it on a flat surface such as a piece of plate glass (first wipe off all old grease and remove any corrosion using fine emery cloth). If the equipment is available, place the bolt in V-blocks and measure the runout using a dial gauge. If the bolt is bent or the runout excessive, renew it.
7 Clean the bearings with solvent and dry them with compressed air. Inspect the bearings (see *Tools and Workshop Tips (Section 5)* in the Reference section).

Bearing renewal

8 Remove the pivot bolt sleeve(s)/spacers and the dust seals (see Step 2 or 3) according to model. Remove the circlip from the right-hand side **(see illustration)**.
9 Refer to *Tools and Workshop Tips (Section 5)* in the Reference section before removing bearings. Drive out the caged ball bearing(s) in the right-hand end of the pivot tube from the opposite or inner end, according to model **(see illustration)**. On CB models remove the bearing spacer from inside the tube, noting which way round it fits. Drive out the needle roller bearing in the left-hand side from the opposite or inner end, according to model.
Note: *The needle bearing should be renewed as a matter of course if it is removed.*

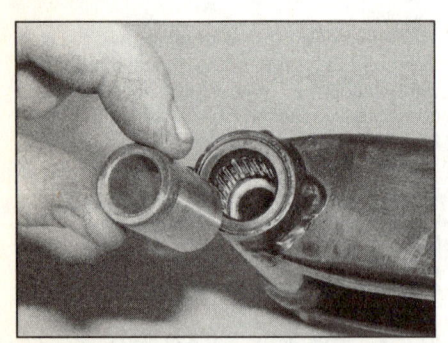

13.2a Left-hand sleeve is longer . . .

13.2b . . . than the right-hand sleeve

13.2c Remove the dust seals carefully

13.8 The right-hand caged ball bearing is retained by a circlip

13.9 Drive the caged ball bearing out from the left-hand side

Frame, suspension and final drive 6•17

13.11 Grease both bearings before reassembly

10 Clean the caged ball bearing(s) and inspect it/them for wear or damage. If a bearing does not run smoothly and freely or if there is excessive freeplay, it must be replaced with a new one.
11 On CB models install the bearing spacer from the right-hand side. Install the caged ball bearing(s) into the right-hand end of the pivot tube with the stamped side facing out, then fit the circlip and grease the bearing (see illustration).
12 When installing the new needle bearing, lubricate it with molybdenum disulphide grease.
13 Check the condition of the dust seals and renew them if they are damaged or deteriorated.

14 Drive chain – removal, cleaning and installation

Removal

Note: *The original equipment drive chain fitted to these models has a staked-type joining link which can be disassembled using either Honda service tool, Pt. No. 07HMH-MR10102, or one of several commercially-available drive chain cutting/staking tools. Such chains can be recognised by the joining link side plate's identification marks (and usually its different colour), as well as by the staked ends of the link's two pins which look as if they have been deeply centre-punched, instead of peened over as with all the other pins.*

> **Warning:** NEVER install a drive chain which uses a clip-type split link. Use ONLY the correct service tools to secure the staked-type of joining link – if you do not have access to such tools, have the chain renewed by a Honda dealer to be sure of having it securely installed.

1 The chain can be removed by splitting it at the joining link as described in Steps 2 and 3 below or by removing the rear wheel and swingarm and detaching it from the front sprocket after the sprocket has been removed from the output shaft.
2 Slacken the drive chain (see Chapter 1). Locate the joining link in a suitable position to work on by rotating the back wheel.
3 Split the chain at the joining link using the chain breaker, carefully following the manufacturer's operating instructions (see also Section 8 in *Tools and Workshop Tips* in the Reference Section). Remove the chain from the bike, noting its routing around the swingarm.

Cleaning

4 Soak the chain in kerosene (paraffin) for approximately five minutes.
Caution: *Don't use gasoline (petrol), solvent or other cleaning fluids which might damage its internal sealing properties. Don't use high-pressure water. Remove the chain, wipe it off, then blow dry it with compressed air immediately. The entire process shouldn't take longer than ten minutes – if it does, the O-rings in the chain rollers could be damaged.*

Installation

5 Unscrew the bolts securing the sprocket cover and remove it and the chain guide (see illustration).
6 Install the chain through the chainguard and around the front and rear sprockets, leaving the two ends in a convenient position to work on.
7 Refer to Section 8 in *Tools and Workshop Tips* in the Reference Section. Install the new joining link from the inside with the four O-rings correctly located between the link plates. Install the new side plate with its identification marks facing out. Measure the amount that the joining link pins project from the side plate and check they are within the measurements specified at the beginning of this Chapter for RK and DID chains. Stake the new link using the drive chain cutting/staking tool, carefully following the instructions of both the chain manufacturer and the tool manufacturer. DO NOT re-use old joining link components.
8 After staking, check the joining link and staking for any signs of cracking (see illustration). If there is any evidence of cracking, the joining link, O-rings and side plate must be renewed. Measure the diameter of the staked ends in two directions and check that they are evenly staked and within the measurements specified at the beginning of this Chapter.
9 Install the front sprocket cover and chain guide, then adjust and lubricate the chain following the procedures described in Chapter 1.

15 Sprockets – check, removal and installation

Check

1 Unscrew the bolts securing the front sprocket cover and remove it and the chain guide (see illustration 14.5).
2 Check the wear pattern on both sprockets (see Chapter 1, Section 1). If the teeth of either sprocket are worn excessively, renew the

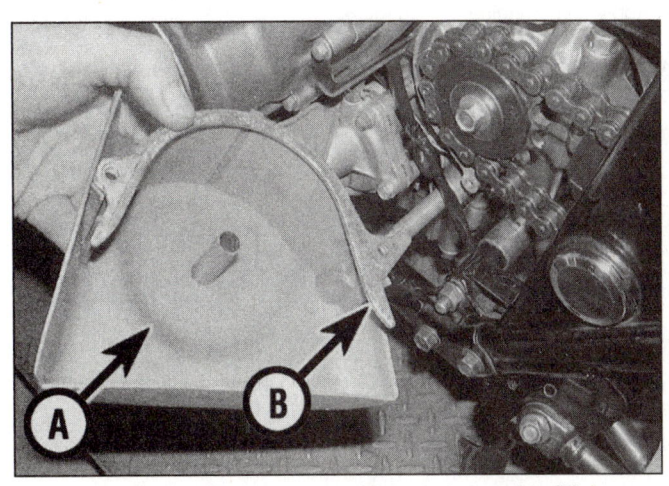

14.5 Front sprocket cover (A) and chain guide (B)

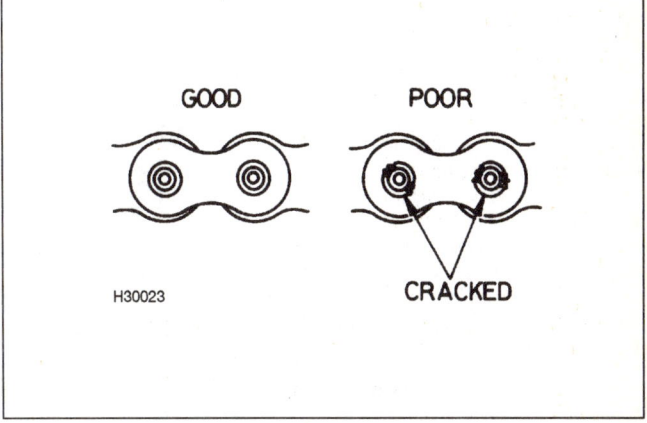

14.8 Check that the ends of the joining link are properly staked

6•18 Frame, suspension and final drive

15.5 Unscrew the sprocket bolt

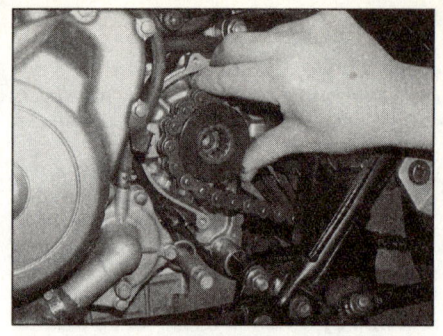

15.7 Slide the sprocket onto the shaft with the chain in place

15.11a Pull the sprocket coupling from the hub ...

15.11b ... then unscrew the sprocket nuts

15.12 Inspect the rubber damper ring

chain and both sprockets as a set. Whenever the sprockets are inspected, the drive chain should be inspected also (see Chapter 1). If you are renewing the chain, renew the sprockets as well.

3 Adjust and lubricate the chain following the procedures described in Chapter 1.

Caution: Use only the recommended lubricant.

Front sprocket

4 Unscrew the bolts securing the sprocket cover and remove it and the chain guide **(see illustration 14.5)**.

5 Have an assistant apply the rear brake, then unscrew the sprocket bolt and remove the washer **(see illustration)**.

6 Slacken the drive chain (see Chapter 1), slide the sprocket and chain off the shaft and slip the sprocket out of the chain.

7 Engage the new sprocket with the chain and slide it on the shaft with its marked side facing outwards **(see illustration)**. Install the sprocket bolt and washer and hand tighten. Adjust the chain tension (see Chapter 1).

8 Have an assistant apply the rear brake, then tighten the sprocket bolt to the torque setting specified at the beginning of this Chapter.

9 Install the front sprocket cover and chain guide, then check and lubricate the chain following the procedures described in Chapter 1.

Rear sprocket

10 Remove the rear wheel (see Chapter 7). Slacken the sprocket nuts before pulling the sprocket coupling out of the hub.

11 Lift the sprocket coupling out of the hub, unscrew the nuts securing the sprocket to the coupling and remove the sprocket, noting which way round it fits **(see illustrations)**.

12 If necessary, remove the rubber damper ring from the flange riveted to the outer face of the sprocket and renew it if it is worn or damaged **(see illustration)**.

13 Install the sprocket onto the coupling with the stamped tooth number marking and riveted flange facing out. Apply some engine oil to the threads of the nuts and tighten them hand-tight.

14 Press the coupling into the hub and tighten the nuts to the torque setting specified at the beginning of this Chapter.

15 Install the rear wheel (see Chapter 7), then adjust and lubricate the chain following the procedures described in Chapter 1.

16 Sprocket coupling/rubber damper – check and renewal

1 Remove the rear wheel (see Chapter 7).

2 Lift the sprocket coupling out of the hub **(see illustration 15.11a)**. The coupling should be a press fit between the rubber dampers with no freeplay. Check the coupling for cracks and damage, and renew it if necessary.

3 Remove the rubber dampers from the hub and check them for cracks, hardening and general deterioration and renew them if necessary **(see illustration)**.

4 Check the condition of the coupling O-ring and renew it if it is damaged or deteriorated. Smear the O-ring with grease before installing the sprocket coupling onto the hub.

5 Checking and renewal procedures for the coupling bearing are in Chapter 7.

6 Installation is the reverse of the removal procedure.

16.3 Inspect the coupling dampers for wear and deterioration

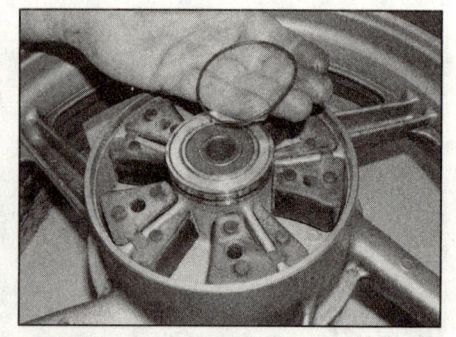

16.4 Check the condition of the coupling O-ring

Chapter 7
Brakes, wheels and tyres

Contents

ABS (Anti-lock Brake System) – operation and fault finding 19
ABS (Anti-lock Brake System) – system checks 20
Brake fluid level check . . . . see *Daily (pre-ride) checks*
Brake light switches – check and replacement . . . . see Chapter 9
Brake pad wear check . . . . see Chapter 1
Brake hoses and unions – inspection and renewal 10
Brake system – bleeding and fluid change 11
Brake system check . . . . see Chapter 1
Front brake caliper – removal, overhaul and installation 3
Front brake disc – inspection, removal and installation 4
Front brake master cylinder – removal, overhaul and installation . . . 5
Front brake pads – renewal 2
Front wheel – removal and installation 15
General information 1
Rear drum brake – removal, inspection and installation 12
Rear brake caliper – removal, overhaul and installation 7
Rear brake disc – inspection, removal and installation 8
Rear brake master cylinder – removal, overhaul
 and installation 9
Rear brake pads – renewal 6
Rear wheel – removal and installation 16
Tyres – general information and fitting 18
Tyres – pressure, tread depth and
 condition . . . . see *Daily (pre-ride) checks*
Wheels – general check . . . . see Chapter 1
Wheels – alignment check 14
Wheels – inspection and repair 13
Wheel bearings – check . . . . see Chapter 1
Wheel bearings – renewal 17

Degrees of difficulty

Easy, suitable for novice with little experience | **Fairly easy,** suitable for beginner with some experience | **Fairly difficult,** suitable for competent DIY mechanic | **Difficult,** suitable for experienced DIY mechanic | **Very difficult,** suitable for expert DIY or professional

Specifications

Front brake

Brake fluid type . . . . DOT 4
Caliper bore inside diameter
  CB500-R and T models
    Standard . . . . 27.000 to 27.050 mm
    Service limit (max) . . . . 27.060 mm
  CB500-V, W, SW, X, SX, Y, SY, 2 and S2 models
    Standard . . . . 30.000 and 32.000 mm
    Service limit (max) . . . . 30.040 and 32.040 mm
  CBF models – non ABS
    Standard . . . . 27.000 to 27.050 mm
    Service limit (max) . . . . 27.060 mm
  CBF models – with ABS
    Standard . . . . 25.400 to 25.450 mm
    Service limit (max) . . . . 25.460 mm
Caliper piston outside diameter
  CB500-R and T models
    Standard . . . . 26.935 to 26.968 mm
    Service limit (min) . . . . 26.910 mm
  CB500-V, W, SW, X, SX, Y, SY, 2 and S2 models
    Standard . . . . 29.970 and 31.970 mm
    Service limit (min) . . . . 29.910 and 31.910 mm

Front brake (continued)

Master cylinder bore inside diameter
  CB500-R and T models
    Standard... 11.000 to 11.043 mm
    Service limit (max)................................. 11.055 mm
  CB500-V, W, SW, X, SX, Y, SY, 2 and S2 models
    Standard... 13.000 mm
    Service limit (max)................................. not available
  CBF models – non ABS
    Standard... 11.000 to 11.043 mm
    Service limit (max)................................. 11.055 mm
  CBF models – with ABS
    Standard... 12.700 to 12.743 mm
    Service limit (max)................................. 12.755 mm
Master cylinder piston outside diameter
  CB500-R and T models
    Standard... 10.957 to 10.984 mm
    Service limit (min)................................. 10.945 mm
  CB500-V, W, SW, X, SX, Y, SY, 2 and S2 models
    Standard... 12.985 mm
    Service limit (min)................................. not available
Disc thickness
  CB models
    Standard... 4.80 to 5.20 mm
    Service limit (min)................................. 4.00 mm
  CBF models
    Standard... 5.80 to 6.20 mm
    Service limit (min)................................. 5.00 mm
Disc maximum runout
  CB models... 0.10 mm
  CBF models.. 0.25 mm

Rear brake – CB500-R and T models

Drum inside diameter
  Standard... 160.00 mm
  Service limit (max).................................. 161.00 mm
Shoe lining thickness
  Standard... 5.00 mm
  Service limit (min).................................. 2.10 mm

Rear brake – CB500-V, W, SW, X, SX, Y, SY, 2 and S2 models

Brake fluid type....................................... DOT 4
Caliper bore inside diameter
  Standard... 34.000 mm
  Service limit (max).................................. 34.040 mm
Caliper piston outside diameter
  Standard... 33.970 mm
  Service limit (min).................................. 33.910 mm
Master cylinder bore inside diameter
  Standard... 12.000 mm
  Service limit (max).................................. 12.043 mm
Master cylinder piston outside diameter
  Standard... 11.985 mm
  Service limit (min).................................. 11.955 mm
Disc thickness
  Standard... 5.0 mm
  Service limit (min).................................. 4.0 mm
Disc maximum runout................................... 0.10 mm

Rear brake – CBF models

Brake fluid type....................................... DOT 4
Caliper bore inside diameter
  Standard... 38.180 to 38.230 mm
  Service limit (max).................................. 38.240 mm
Master cylinder bore inside diameter
  Standard... 14.000 to 14.043 mm
  Service limit (max).................................. 14.055 mm
Disc thickness
  Standard... 4.8 to 5.2 mm
  Service limit (min).................................. 4.0 mm
Disc maximum runout................................... 0.25 mm

ABS
Wheel sensor air gap .... 0.2 to 1.2 mm

Wheels
Maximum wheel runout (front and rear)
    Axial (side-to-side) .... 0.3 mm
    Radial (out-of-round) .... 0.3 mm
Maximum axle runout (front and rear) .... 0.2 mm

Tyres

Tyre sizes	Front	Rear
CB models	110/80-17 57H	130/80-17 65H
CBF models	120/70-17 58W	160/60-17 69W

Torque settings
Front brake – CB500-R and T models
    Caliper mounting bolts .... 32 Nm
    Caliper pad retaining pin plug .... 2.5 Nm
    Caliper pad retaining pin .... 18 Nm
    Caliper slider pin (in caliper) .... 23 Nm
    Caliper slider pin (in caliper bracket) .... 13 Nm
    Caliper bleed valve .... 5.5 Nm
    Master cylinder clamp bolts .... 12 Nm
    Brake hose banjo bolts .... 35 Nm
    Disc bolts .... 43 Nm
Front brake – CB500-V, W, SW, X, SX, Y, SY, 2 and S2 models
    Caliper mounting bolts .... 32 Nm
    Master cylinder clamp bolts .... 12 Nm
    Brake hose banjo bolts .... 35 Nm
Front brake – CBF models
    Caliper mounting bolts .... 30 Nm
    Caliper pad retaining pin plug .... 3 Nm
    Caliper pad retaining pin .... 18 Nm
    Caliper slider pin (in caliper)
        Non-ABS models .... 27 Nm
        ABS models .... 23 Nm
    Caliper slider pin (in caliper bracket) .... 13 Nm
    Caliper bleed valve .... 6.0 Nm
    Master cylinder clamp bolts .... 12 Nm
    Brake hose banjo bolts .... 34 Nm
    Disc bolts .... 42 Nm
    Pulser ring bolts – ABS models .... 8 Nm
Rear brake – CB500-V, W, SW, X, SX, Y, SY, 2 and S2 models
    Master cylinder mounting bolts .... 12 Nm
    Brake hose banjo bolts .... 35 Nm
    Disc bolts .... 43 Nm
Rear brake – CBF models
    Caliper mounting bolt .... 23 Nm
    Caliper pad retaining pin .... 18 Nm
    Caliper bleed valve .... 6 Nm
    Master cylinder mounting bolts .... 12 Nm
    Brake hose banjo bolts .... 35 Nm
    Disc bolts .... 42 Nm
    Pulser ring bolts – ABS models .... 8 Nm
Front axle nut – CB500 models .... 60 Nm
Front axle bolt – CBF500 models .... 59 Nm
Front axle clamp bolt(s)
    CB models .... 27 Nm
    CBF models .... 22 Nm
Rear axle nut
    CB500 models .... 90 Nm
    CBF500 models .... 93 Nm
Rider's footrest bracket bolts .... 45 Nm
Silencer mounting bolt .... 27 Nm
ABS wheel sensor mounting bolts .... 10 Nm
ABS control unit mounting bolts
    Bottom bolts .... 12 Nm
    Left-hand side bolt .... 10 Nm
ABS control unit pipe gland nuts .... 17 Nm

7•4 Brakes, wheels and tyres

2.1a Remove the plug (arrowed) . . .

2.1b . . . then unscrew the pin . . .

2.1c . . . and withdraw the pads from the caliper – R and T models

1 General information

All CB models and CBF models without ABS are equipped with a twin piston, sliding caliper, hydraulic disc front brake. CBF models with ABS have a triple piston sliding front caliper. CB500-R and T models have a drum rear brake and all other models have a single piston, sliding caliper, hydraulic disc rear brake. The front brake caliper on CB500-R and T models and the front and rear calipers on all CBF models are made by Nissin; the front and rear brake calipers on CB500-V, W, SW, X, SX, Y, SY, 2 and S2 models are made by Brembo.

All models are fitted with cast alloy wheels designed for tubeless tyres only.

Caution: *Disc brake components rarely require disassembly. Do not disassemble components unless absolutely necessary. If a hydraulic brake line is loosened, the entire system must be disassembled, drained, cleaned and then properly filled and bled upon reassembly. Do not use solvents on internal brake components. Solvents will cause the seals to swell and distort. Use only clean brake fluid or denatured alcohol for cleaning. Use care when working with brake fluid as it can injure your eyes and it will damage painted surfaces and plastic parts.*

2 Front brake pads – renewal

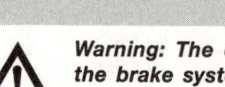

⚠ **Warning:** *The dust created by the brake system may contain asbestos, which is harmful to your health. Never blow it out with compressed air and don't inhale any of it. An approved filtering mask should be worn when working on the brakes.*

1 On CB500-R and T models and CBF models without ABS, unscrew the pad retaining pin plug. On those models and on CBF models with ABS unscrew and remove the pad retaining pin, noting how it fits, and slide the pads out of the caliper **(see illustrations)**.

2 On CB500-V, W, SW, X, SX, Y, SY, 2 and S2 models, remove the R-pin and tap the pad retaining pin out of the caliper from the right-hand side using a small punch, then withdraw the pin and slide the pads out of the caliper **(see illustrations)**.

3 Inspect the surface of each pad for contamination and check that the friction material has not worn down level with or beyond the wear limit groove in the pad edge **(see illustration)**. If either pad is worn down to, or beyond, the wear limit, is fouled with oil or grease, or heavily scored or damaged by dirt and debris, both pads must be renewed as a set. **Note:** *It is not possible to degrease the friction material; if the pads are contaminated in any way they must be renewed.*

4 Check that each pad has worn evenly at each end, and that each has the same amount of wear as the other. If uneven wear is noticed, one of the pistons is probably sticking in the caliper, in which case the caliper must be overhauled (see Section 3). **Note:** *Do not operate the brake lever while the pads are out of the caliper.*

5 If the pads are in good condition clean them carefully, using a fine wire brush which is completely free of oil and grease, to remove all traces of road dirt and corrosion. Using a pointed instrument, clean out the grooves in the friction material and dig out any embedded particles of foreign matter. Any areas of glazing may be removed using emery cloth.

6 Check the condition of the brake disc (see Section 4).

7 Remove all traces of corrosion from the pad pin. Inspect the pin for signs of damage and renew it if necessary. On CBF models with ABS make sure the stopper ring on the end of the pin is in good condition and replace it with a new one if necessary.

8 If new pads are being installed, push the

2.2a Remove the R-pin . . .

2.2b . . . and tap the pin through . . .

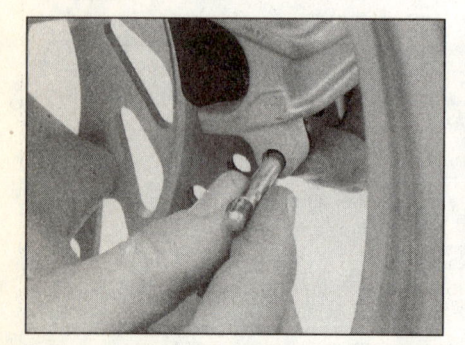

2.2c . . . to the left-hand side – V, W, SW, X, SX, Y, SY, 2 and S2 models

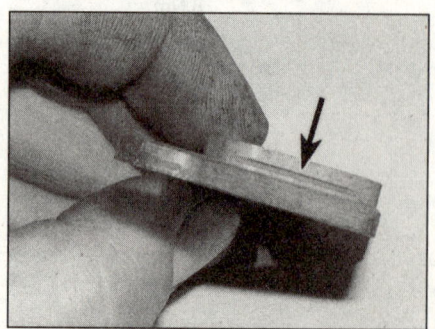

2.3 If the pads have worn to or beyond the wear limit (arrowed) they must be renewed

pistons as far back into the caliper as possible. A good way of doing this is to insert one of the old pads between the disc and the piston, then push the caliper against the pad and disc using hand pressure. Due to the increased friction material thickness of new pads, it may be necessary to remove the master cylinder reservoir cover and diaphragm and remove some fluid.

9 Smear the backs of the pads and the shank of the pad pin lightly with copper-based grease, making sure that none gets on the front or sides of the pads. On CBF models with ABS smear the pad pin stopper ring with silicone grease.

10 Installation of the pads is the reverse of removal. Make sure the pad spring is correctly positioned in the caliper mouth and the pad plate is in place on the caliper bracket **(see illustration 3.2a and b)**. Insert the pads into the caliper so that the friction material faces the disc, making sure they locate correctly against the pad spring and pad plate, then push up on the pads to align the holes in the pads with the holes in the caliper. Slide the pad retaining pin into place (see Steps 1 and 2) making sure the pin passes through the hole in each pad.

11 On CB500-R and T models and CBF models, tighten the pad retaining pin to the torque setting specified at the beginning of this Chapter. On non-models without ABS install the pad pin plug and tighten it to the specified torque.

12 On CB500-V, W, SW, X, SX, Y, SY, 2 and S2 models, insert the R-pin through the hole in the pad retaining pin.

13 Check the level of fluid in the master cylinder reservoir and top-up if necessary (see *Daily (pre-ride) checks*).

14 Operate the brake lever several times to bring the pads into contact with the disc. Check the operation of the brake before riding the motorcycle.

3 Front brake caliper – removal, overhaul and installation

Warning: If the caliper indicates the need for an overhaul (usually due to leaking fluid or sticky operation), all old brake fluid should be flushed from the system. Also, the dust created by the brake system may contain asbestos, which is harmful to your health. Never blow it out with compressed air and don't inhale any of it. An approved filtering mask should be worn when working on the brakes. Do not, under any circumstances, use petroleum-based solvents to clean brake parts. Use clean brake fluid, brake cleaner or denatured alcohol only.

Removal

1 If the caliper is just being displaced, the brake pads can be left in place and it is not necessary to disconnect the brake hose.

2 If the caliper is being overhauled, read through the entire procedure first and make sure that you have obtained all the new parts required, including some new DOT 4 brake fluid **(see illustrations)**.

3 Remove the brake pads (see Section 2). Note the alignment of the brake hose on the caliper, then unscrew the hose banjo bolt and separate the hose from the caliper (see

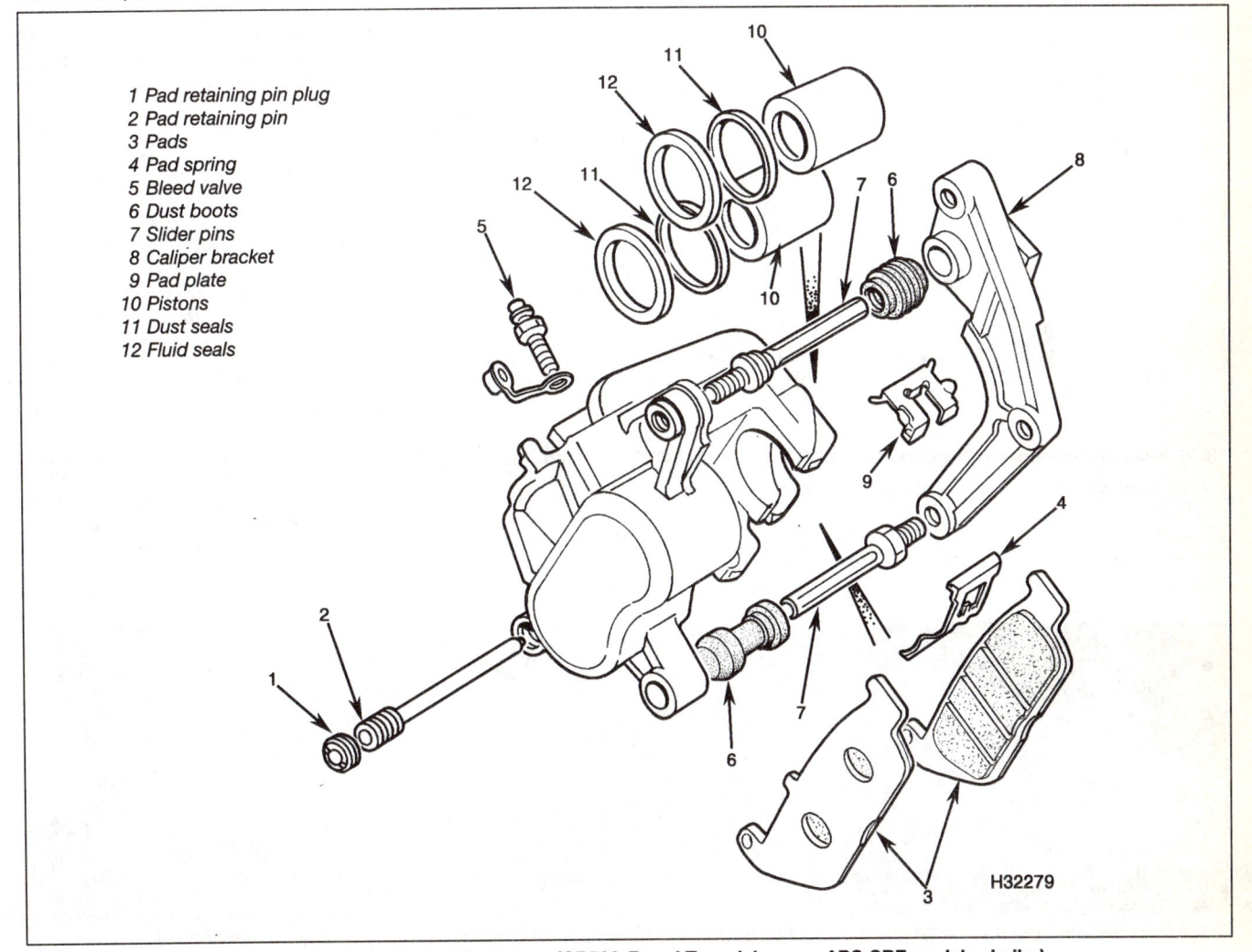

1 Pad retaining pin plug
2 Pad retaining pin
3 Pads
4 Pad spring
5 Bleed valve
6 Dust boots
7 Slider pins
8 Caliper bracket
9 Pad plate
10 Pistons
11 Dust seals
12 Fluid seals

3.2a Front brake caliper – Nissin type (CB500-R and T models – non ABS CBF models similar)

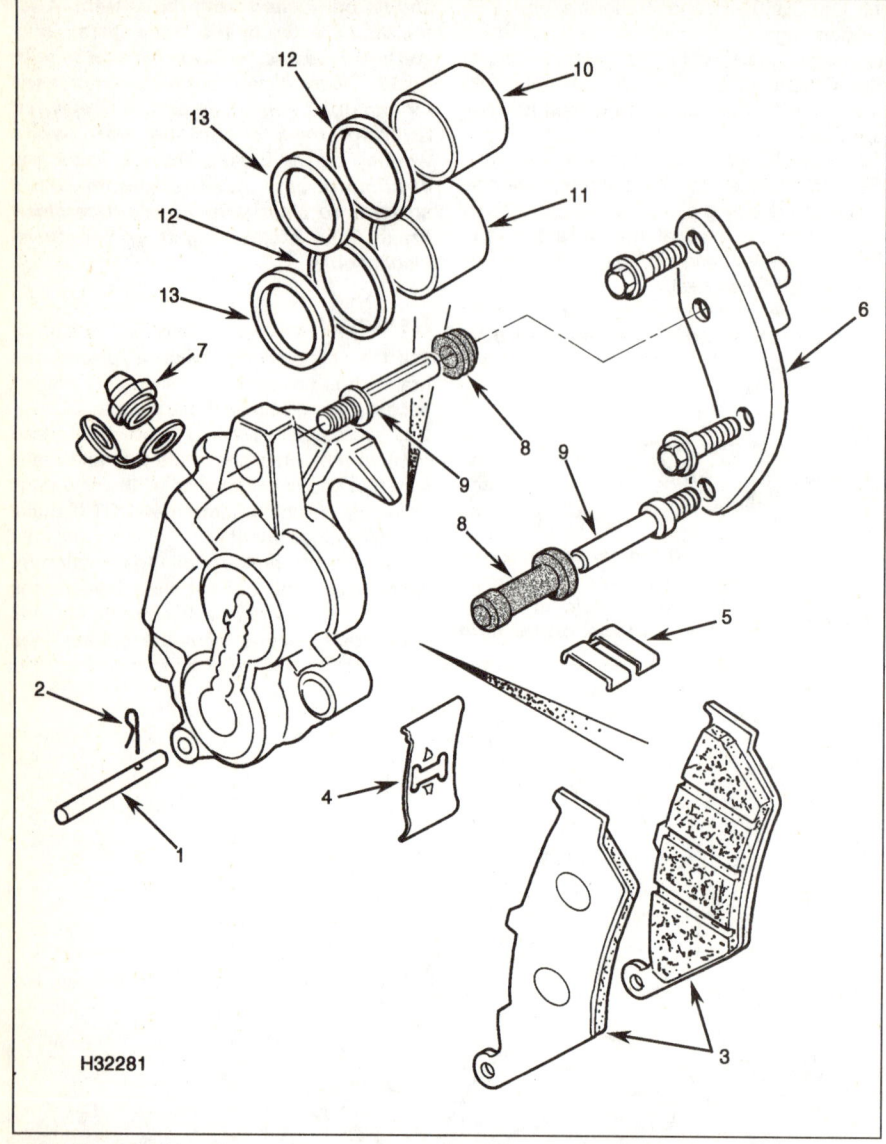

7•6 Brakes, wheels and tyres

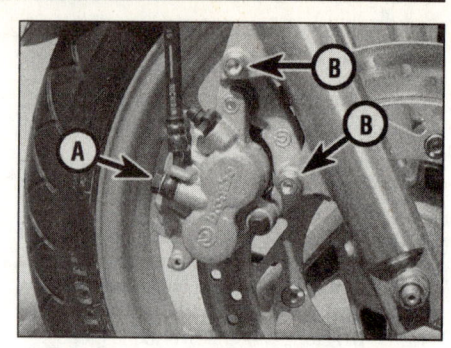

3.3 Hose banjo bolt (A), caliper mounting bolts (B)

illustration). **Note:** If you are planning to overhaul the caliper and don't have a source of compressed air to blow out the pistons, the hydraulic system can be used to force the pistons out of the body once the pads have been removed. Disconnect the hose once the pistons have been sufficiently displaced.

4 Clamp the hose or wrap a plastic bag tightly around it to prevent fluid spills and stop dirt entering the system. Discard the sealing washers as new ones must be used on installation.

5 Unscrew the caliper mounting bolts, and pull the caliper off the disc **(see illustration)**. Discard the bolts as Honda specify new ones must be fitted on reassembly – this is because the bolts come pre-treated with a locking agent. If necessary you could clean up the threads of the old bolts and apply some new locking agent to the threads.

6 If the caliper is just being displaced, unscrew the bolt securing the brake hose to the fork slider and secure the caliper to the motorcycle with a cable tie to ensure no strain is placed on the hose. **Note:** *Do not operate the brake lever while the caliper is off the disc.*

Overhaul

7 Pull the caliper off the caliper bracket, noting how it fits.

8 Remove the slider pin boots, the pad spring inside the caliper mouth and the pad plate (if loose) on the caliper bracket, noting how they fit **(see illustrations)**. Clean the exterior of the caliper with denatured alcohol or brake system cleaner. Make sure all old grease is removed from the slider pins. If the pad plate or the pad

3.2b Front brake caliper – Brembo type (CB500-V, W, SW, X, SX, Y, SY, 2 and S2 models)

1 Pad retaining pin
2 R-clip
3 Brake pads
4 Pad spring
5 Pad plate
6 Caliper bracket
7 Bleed valve
8 Dust boots
9 Slider pins
10 32 mm piston
11 30 mm piston
12 Dust seals
13 Fluid seals

3.5 Remove the caliper from the disc

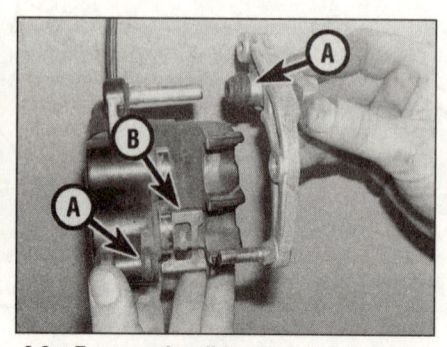

3.8a Remove the slider pin boots (A), pad spring (B) . . .

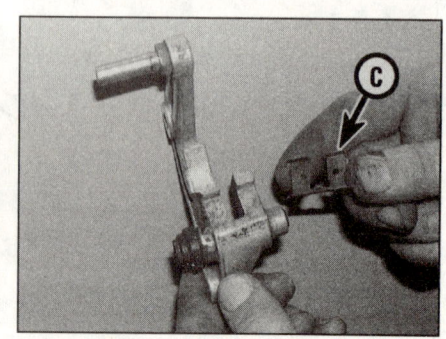

3.8b . . . and pad plate (C)

Brakes, wheels and tyres

3.10 Remove the dust seal carefully to avoid damage to the bore and seal groove

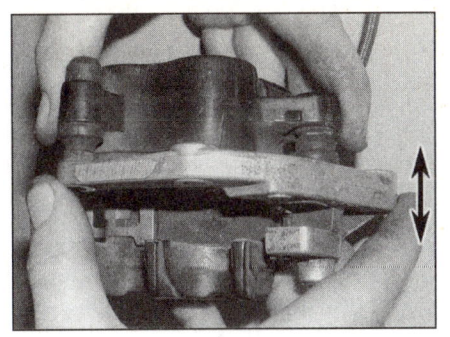

3.17 Caliper should move freely on the bracket

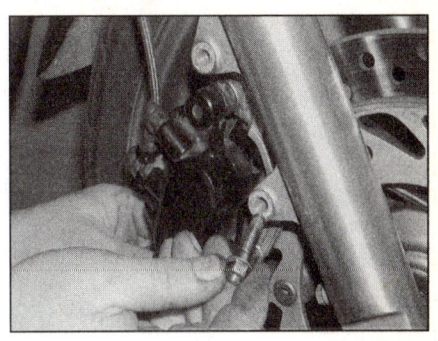

3.18 Use new caliper mounting bolts on reassembly

spring are badly corroded or worn they should be renewed.

9 If the pistons have not been displaced already and cannot be removed by hand (see Step 3), force them out using compressed air. If compressed air is used, place a wad of rag between the pistons and the caliper to act as a cushion, then use compressed air directed into the fluid inlet to force the pistons out of the body. Use only low pressure to ease the pistons out and make sure both pistons are displaced at the same time. If the air pressure is too high and the pistons are forced out, the caliper and/or pistons may be damaged. Mark each piston head and caliper bore with a felt marker to ensure that the pistons can be easily matched to their original bores on reassembly – this is particularly important on the Nissin caliper fitted to CB500-R and T and all CBF models because the bores are the same diameter.

⚠️ **Warning: Never place your fingers in front of the pistons in an attempt to catch or protect them when applying compressed air, as serious injury could result.**

10 Using a wooden or plastic tool, remove the dust seals from the caliper bores **(see illustration)**. Discard them as new ones must be used on installation. If a metal tool is being used, take great care not to damage the caliper bores. Remove and discard the piston seals in the same way.

11 Clean the pistons and bores with denatured alcohol, clean brake fluid or brake system cleaner. If compressed air is available, use it to dry the parts thoroughly (make sure it's filtered and unlubricated).

Caution: Do not, under any circumstances, use a petroleum-based solvent to clean brake parts.

12 Inspect the caliper bores and pistons for signs of corrosion, nicks, burrs and loss of plating. If surface defects are present, the caliper assembly must be renewed. If the necessary measuring equipment is available, compare the dimensions of the pistons and bores to those given in the Specifications section of this Chapter, renewing any component that is worn beyond the service limit – note that the pistons/bores on the Brembo caliper fitted to CB500-V models onwards differs by 2 mm. If the caliper is in bad shape, the master cylinder should also be checked.

13 Clean all traces of corrosion off the slider pins and their bores in the caliper and bracket. Renew the rubber dust boots if they are damaged or deteriorated. On CB500-R and T and all CBF models, if the pins are loose, remove them and clean the threads. Apply a suitable non-permanent thread locking compound and tighten them to the specified torque. Do not attempt to remove the pins on CB500-V, W, SW, X, SX, Y, SY, 2 and S2 models.

14 Lubricate the new piston seals with clean brake fluid and install them in their grooves in the caliper bores. **Note:** *On CB500-V, W, SW, X, SX, Y, SY, 2 and S2 models, the upper piston and caliper bore diameters are 2 mm larger than the lower ones (see Specifications). Match the correct seals with the bores.*

15 Lubricate the new dust seals with clean brake fluid and install them in their grooves in the caliper bores.

16 Lubricate the pistons with clean brake fluid and install them closed-end first into the caliper bores. Using your thumbs, push the pistons all the way in, making sure they enter the bore squarely.

17 Fit the pad spring and pad plate, making sure they are correctly positioned as noted on removal **(see illustration 3.8a and b)**. Apply a smear of copper-based grease to the slider pins, fit the slider pin boots and then slide the caliper onto the bracket and check that it is able to move freely **(see illustration)**.

Installation

18 If the caliper was merely displaced from the fork slider, slide it onto the brake disc, making sure the pads sit squarely each side of the disc, install new mounting bolts and tighten them to the torque setting specified at the beginning of this Chapter **(see illustration)**. Note that the caliper mounting bolt threads are coated with a dry locking compound when new and thus Honda advise that the bolts are renewed whenever they have been removed.

19 If the caliper has been overhauled, slide it over the brake disc, install new mounting bolts and tighten them to the torque setting specified at the beginning of this Chapter, then ease the caliper apart enough to fit the brake pads (see Section 2).

20 Connect the brake hose to the caliper, using new sealing washers on each side of the banjo fitting. Align the hose as noted on removal. Tighten the banjo bolt to the torque setting specified at the beginning of this Chapter. Top-up the master cylinder reservoir with new DOT 4 brake fluid, then fill and bleed the hydraulic system (see Section 11).

21 Operate the brake lever several times to bring the pads into contact with the disc. Check that there are no fluid leaks and thoroughly test the operation of the brake before riding the motorcycle.

4 Front brake disc – inspection, removal and installation

Inspection

1 Visually inspect the surface of the disc for score marks and other damage. Light scratches are normal after use and won't affect brake operation, but deep grooves and heavy score marks will reduce braking efficiency and accelerate pad wear. If the disc is badly grooved it must be machined or renewed.

2 To check disc runout, position the motorcycle its centre stand and support it so that the front wheel is raised off the ground. Mount a dial gauge on a fork leg, with the gauge plunger touching the surface of the disc about 10 mm from the outer edge **(see illustration)**. Rotate the wheel and watch the gauge needle, comparing the reading with the limit listed in the Specifications at the beginning of this Chapter. If the runout is

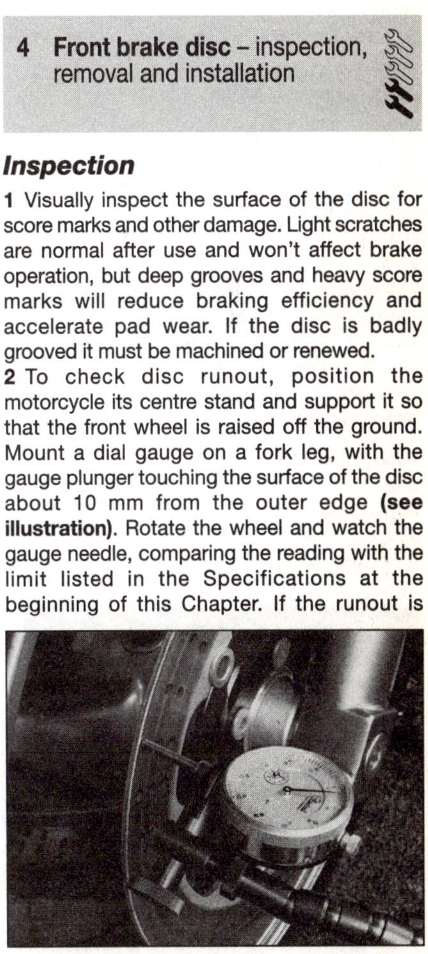

4.2 Checking the brake disc runout

7•8 Brakes, wheels and tyres

4.3a The minimum disc thickness is marked on the disc

4.3b Measuring the disc thickness

4.5 Unscrew the disc retaining bolts in a criss-cross pattern

greater than the service limit, check the wheel bearings for play (see Chapter 1). If the bearings are worn, renew them (see Section 17) and repeat this check. If the disc runout is still excessive, the disc will have to be renewed, although machining by an engineer may be possible.

3 The disc must not be machined or allowed to wear down to a thickness less than the service limit listed in this Chapter's Specifications and as marked on the disc itself **(see illustration)**. The thickness of the disc can be checked with a micrometer **(see illustration)**. If the thickness of the disc is less than the service limit, it must be renewed.

Removal

4 Remove the front wheel (see Section 15).
Caution: Do not lay the wheel down and allow it to rest on the disc – the disc could become warped.
5 Mark the relationship of the disc to the hub, so it can be installed in the same position. Unscrew the disc retaining bolts, loosening them a little at a time in a criss-cross pattern to avoid distorting the disc, then remove the disc from the wheel **(see illustration)**. Discard the bolts as Honda specify new ones must be fitted on reassembly.

Installation

6 Place the disc on the hub, making sure the marked side is on the outside. Align the previously applied reference marks if reinstalling the original disc.

7 Install the new bolts, and tighten them in a criss-cross pattern evenly and progressively to the torque setting specified at the beginning of this Chapter. Note that the new bolts are supplied with a dry coating of locking compound on their threads.
8 Clean all grease off the brake disc using acetone or brake system cleaner. If a new brake disc has been installed, remove any protective coating from its working surfaces.
9 Install the wheel (see Section 15).
10 Operate the brake lever several times to bring the pads into contact with the disc. Check the operation of the brake carefully before riding the motorcycle.

5 Front brake master cylinder – removal, overhaul and installation

1 If the brake lever does not feel firm when the brake is applied, and the hydraulic hose is in good condition and bleeding the brake does not help (see Section 11), or if the master cylinder is leaking fluid, then master cylinder overhaul is recommended.
2 Before disassembling the master cylinder, read through the entire procedure and make sure that you have obtained all the new parts required including some new DOT 4 brake fluid, some clean rags and internal circlip pliers. **Note:** *If the master cylinder is just being displaced and not completely removed from the motorcycle, do not remove the front brake lever or the brake light switch or disconnect the brake hose..*
Caution: Disassembly, overhaul and reassembly of the brake master cylinder must be done in a spotlessly clean work area to avoid contamination and possible failure of the brake hydraulic system components. To prevent damage to the paint from spilled brake fluid, always cover the fuel tank when working on the master cylinder.

Removal

3 Lift the cover, where fitted, on the rear view mirror mounting locknut, slacken the locknut and unscrew the mirror.
4 Disconnect the wiring connectors from the brake light switch **(see illustration)**. If necessary, remove the screw securing the brake light switch to the bottom of the master cylinder and remove the switch.
5 Remove the front brake lever (see Chapter 6, Section 5).
6 Unscrew the brake hose banjo bolt and separate the hose from the master cylinder, noting its alignment **(see illustration)**. Discard the sealing washers as they must be renewed. Wrap a plastic bag tightly around the end of the hose to stop dirt entering the system and secure the hose in an upright position to prevent fluid spills.
7 Unscrew the master cylinder clamp bolts, remove the clamp, then lift the master cylinder away from the handlebar **(see illustration)**.
8 Unscrew the reservoir cover retaining screws and lift off the cover, the diaphragm

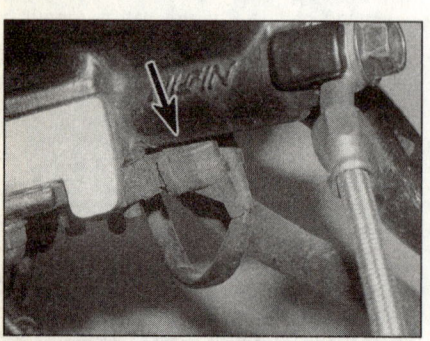

5.4 Brake light switch wiring connectors (arrowed)

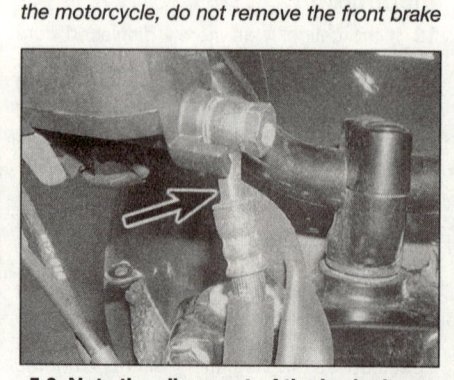

5.6 Note the alignment of the brake hose (arrowed) before removal

5.7 Unscrew the clamp bolts (arrowed) and remove the master cylinder

Brakes, wheels and tyres 7•9

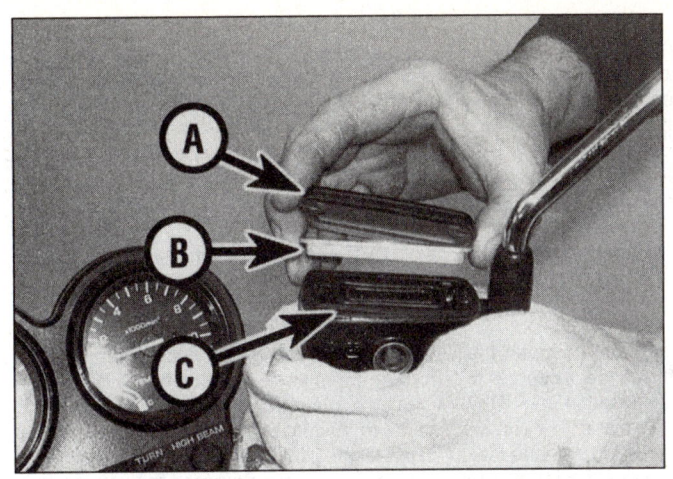

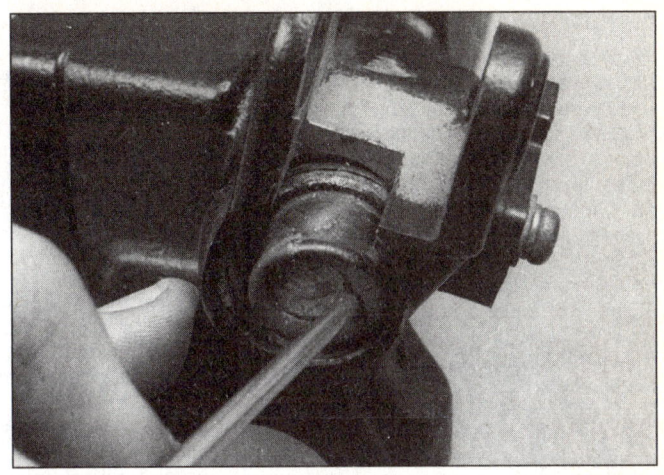

5.8 Reservoir cover (A), diaphragm plate (B) and diaphragm (C)

5.9 Remove the dust boot from the end of the master cylinder piston

plate and the diaphragm **(see illustration)**. Drain the brake fluid from the reservoir into a suitable container. Wipe any remaining fluid out of the reservoir with a clean rag. Inspect the reservoir cover rubber diaphragm and renew it if it is damaged or deteriorated.

Overhaul

9 Carefully remove the dust boot from the end of the piston, on CB500-2 and S2 models bringing the pushrod and its spring with it **(see illustration)**.

10 Using circlip pliers, remove the circlip and withdraw the piston assembly and spring, noting how they fit **(see illustration)**. If they are difficult to remove, apply low pressure compressed air to the fluid outlet inside the reservoir. Lay the parts out in the proper order to prevent confusion during reassembly.

11 Clean all parts with clean brake fluid or denatured alcohol. If compressed air is available, use it to dry the parts thoroughly (make sure it's filtered and unlubricated).

Caution: Do not, under any circumstances, *use a petroleum-based solvent to clean brake parts.*

12 Check the master cylinder bore for corrosion, scratches, nicks and score marks. If the necessary measuring equipment is available, compare the dimensions of the piston and bore to those given in the Specifications at the beginning of this Chapter. If damage or wear is evident, the master cylinder must be renewed. If the master cylinder is in poor condition, then the caliper should be checked as well. Ensure that the fluid inlet and outlet ports in the master cylinder are clear.

13 The dust boot, circlip, washer, piston assembly and spring are all included in the master cylinder overhaul kit. Use all of the new parts, regardless of the apparent condition of the old ones. If the seal and cup are not already on the piston, fit them according to the layout of the old piston assembly.

14 Fit the narrow end of the spring onto the post on the inner end of the piston. Lubricate the new piston, seal and cup with clean brake fluid.

15 Slide the piston and spring assembly into the master cylinder, making sure the lips on the cup do not turn inside out when they are slipped into the bore **(see illustration 5.10)**. Depress the piston and install the new washer and circlip, making sure the circlip locates in its groove.

16 On CB500-2 and S2 models fit the pushrod into the rubber boot so the inner rim on the boot locates in the groove in the outer end of the pushrod. Fit the spring narrow end first onto the inner end of the pushrod, then fit the assembly into the master cylinder, locating the pushrod end into the indent in the outer end of the piston, and pressing the rim of the boot against the circlip in the master cylinder. On all other models, install the new dust boot, pushing the wide rim in against the circlip in the master cylinder and locating the inner rim in the groove in the end of the piston **(see illustration 5.9)**.

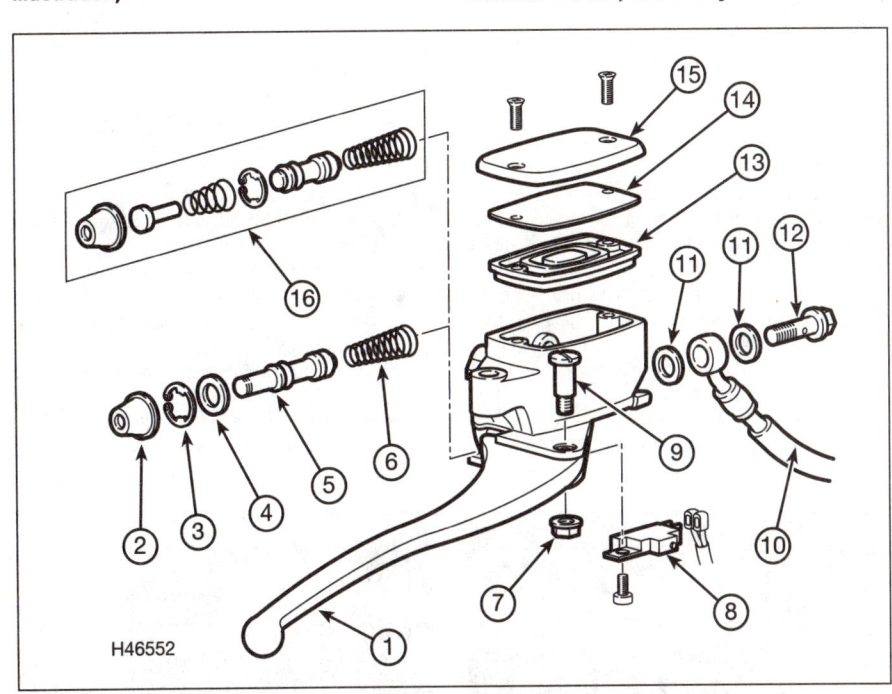

5.10 Front brake master cylinder components

1 Brake lever
2 Dust boot
3 Circlip
4 Washer
5 Piston/seals
6 Spring
7 Nut
8 Brake light switch
9 Lever pivot bolt
10 Brake hose
11 Sealing washers
12 Union bolt
13 Diaphragm
14 Diaphragm plate
15 Reservoir cover
16 Piston assembly on CB500-2 and S-2 models

7•10 Brakes, wheels and tyres

Installation

17 Attach the master cylinder to the handlebar and fit the clamp with its UP mark facing up and the clamp joint at the top aligned with the punchmark on the handlebar. Tighten the bolts to the torque setting specified at the beginning of this Chapter.

18 Connect the brake hose to the master cylinder, using new sealing washers on each side of the union, and aligning the hose as noted on removal **(see illustration 5.6)**. Tighten the banjo bolt to the torque setting specified at the beginning of this Chapter.

19 Install the brake lever (see Chapter 6, Section 5).

20 Install the brake light switch (if removed) and connect the brake light wiring **(see illustration 5.4)**.

21 Fill the fluid reservoir with new DOT 4 brake fluid, then fill and bleed the hydraulic system (see Section 11).

22 Fit the diaphragm, making sure it is correctly seated, the diaphragm plate and the cover onto the master cylinder reservoir and install the cover screws.

23 Install the rear view mirror.

24 Check the operation of the brake before riding the motorcycle.

6 Rear brake pads – renewal

⚠ **Warning: The dust created by the brake system may contain asbestos, which is harmful to your health. Never blow it out with compressed air and don't inhale any of it. An approved filtering mask should be worn when working on the brakes.**

1 On CB models remove the R-pin and tap the pad retaining pin out of the caliper from the right-hand side using a small punch **(see illustration)**. Slide the pads out of the caliper **(see illustration 2.1c)**.

2 On CBF models slacken the pad retaining pin **(see illustration 6.11c)**. Unscrew the rear mounting bolt, then pivot the back of the caliper up and withdraw the pads, noting how they locate **(see illustrations 6.11b and a)**.

3 Inspect the surface of each pad for contamination and check that the friction material has not worn level with or beyond the wear limit groove in the pad **(see illustration 2.3b)**. If either pad is worn down to, or beyond, the wear limit, fouled with oil or grease, or heavily scored or damaged by dirt and debris, both pads must be renewed as a set. Note that it is not possible to degrease the friction material; if the pads are contaminated in any way they must be renewed. **Note:** *Do not operate the brake pedal while the pads are out of the caliper.*

4 If the pads are in good condition, clean them carefully using a fine wire brush which is completely free of oil and grease to remove all traces of road dirt and corrosion. Using a pointed instrument, clean out the grooves in the friction material and dig out any embedded particles of foreign matter. Any areas of glazing may be removed using emery cloth.

5 Check the condition of the brake disc (see Section 8).

6 Remove all traces of corrosion from the pad pin. Inspect the pin for signs of damage and renew it if necessary. On CBF models make sure the stopper ring on the end of the pin is in good condition and replace it with a new one if necessary.

7 If new pads are being installed, push the pistons as far back into the caliper as possible. A good way of doing this is to insert one of the old pads between the disc and the piston, then push the caliper against the pad and disc using hand pressure. Due to the increased friction material thickness of new pads, it may be necessary to remove the master cylinder reservoir cover and diaphragm and remove some fluid.

8 Smear the backs of the pads and the shank of the pad pin lightly with copper-based grease, making sure that none gets on the front or sides of the pads. On CBF models smear the pad pin stopper ring with silicone grease.

9 Installation of the pads is the reverse of removal. Make sure the pad spring is correctly positioned in the caliper and the pad plate is clipped to the caliper bracket **(see illustration 7.2)**.

10 On CB models insert the pads into the caliper so that the friction material faces the disc, making sure they locate correctly against the pad spring and engage the pad plate, then push up on the pads to align the holes in the pads with the holes in the caliper. Slide the pad retaining pin into place making sure the pin passes through the hole in each pad. Insert the R-pin through the hole in the pad retaining pin.

11 On CBF models insert the pads into the caliper so that the friction material faces the disc, making sure they locate correctly against the pad spring and engage the pad plate **(see illustration)**. Pivot the caliper down then install the mounting bolt and tighten it to the torque setting specified at the beginning of the Chapter **(see illustration)**. Push up on the pads to align the holes with those in the caliper, then slide the pad retaining pin into place, making sure it passes through the hole in each pad, and tighten it to the specified torque **(see illustration)**.

12 Check the level of fluid in the master cylinder reservoir and top-up if necessary (see *Daily (pre-ride) checks*).

13 Operate the brake pedal several times to bring the pads into contact with the disc. Check the operation of the brake before riding the motorcycle.

7 Rear brake caliper – removal, overhaul and installation

⚠ **Warning: If the caliper indicates the need for an overhaul (usually due to leaking fluid or sticky**

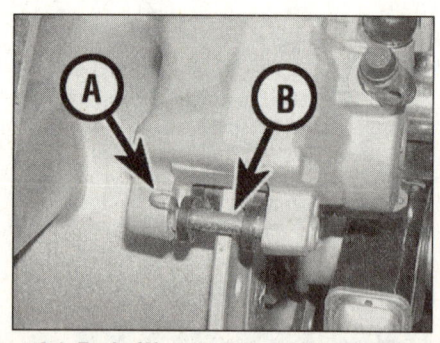

6.1 R-pin (A) and pad retaining pin (B)

6.11a Locate the pads between caliper and bracket . . .

6.11b . . . then fit the mounting bolt

6.11c Push the pads up and slide the retaining pin through

Brakes, wheels and tyres 7•11

operation), all old brake fluid should be flushed from the system. Also, the dust created by the brake system may contain asbestos, which is harmful to your health. Never blow it out with compressed air and don't inhale any of it. An approved filtering mask should be worn when working on the brakes. Do not, under any circumstances, use petroleum-based solvents to clean brake parts. Use clean brake fluid, brake cleaner or denatured alcohol only.

Removal

1 If the caliper is just being displaced, the brake pads can be left in place and it is not necessary to disconnect the brake hose.
2 If the caliper is being overhauled, read through the entire procedure first and make sure that you have obtained all the new parts required, including some new DOT 4 brake fluid.
3 Note the alignment of the brake hose on the caliper, then unscrew the hose banjo bolt and separate the hose from the caliper **(see illustration)**. **Note:** *If you are planning to overhaul the caliper and don't have a source of compressed air to blow out the piston, the hydraulic system can be used to force the piston out of the body once the pads have been removed. Disconnect the hose once the piston has been sufficiently displaced.*
4 Clamp the hose or wrap a plastic bag tightly around it to prevent fluid spills and stop dirt entering the system. Discard the sealing washers as new ones must be used on installation.
5 On CB models remove the brake pads (see Section 6) and rear wheel (see Section 16), then slide the caliper off the bracket, noting how it fits. On CBF models, if the pads have not been removed unscrew the rear mounting bolt **(see illustration 6.11b)**. Pivot the back of the caliper up and slide it off the bracket **(see illustration)** – pivot the pads out of the caliper and clear of the disc if they have not been removed.
6 If the caliper is just being displaced, unscrew the bolt securing the rear brake hose clamp to the swingarm and secure the caliper and caliper bracket to the frame with a cable tie to avoid placing a strain on the hose (remove the right-hand side frame panel if necessary). **Note:** *Do not operate the brake pedal while the caliper is off the disc.*

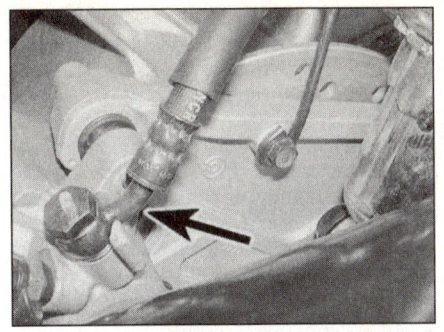

7.3 Note the alignment of the brake hose (arrowed) before removal

7.5 Pivot the caliper up and slide it off

Overhaul

7 Remove the slider pin boots, the pad spring in the caliper and the pad plate (if loose) on the caliper bracket, noting how they fit **(see illustration)**. Clean the exterior of the caliper with denatured alcohol or brake system cleaner. Make sure all old grease is removed from the slider pins. If the pad plate or the pad spring are badly corroded or worn they should be renewed, although check whether they are available separately.
8 If the piston has not been displaced already

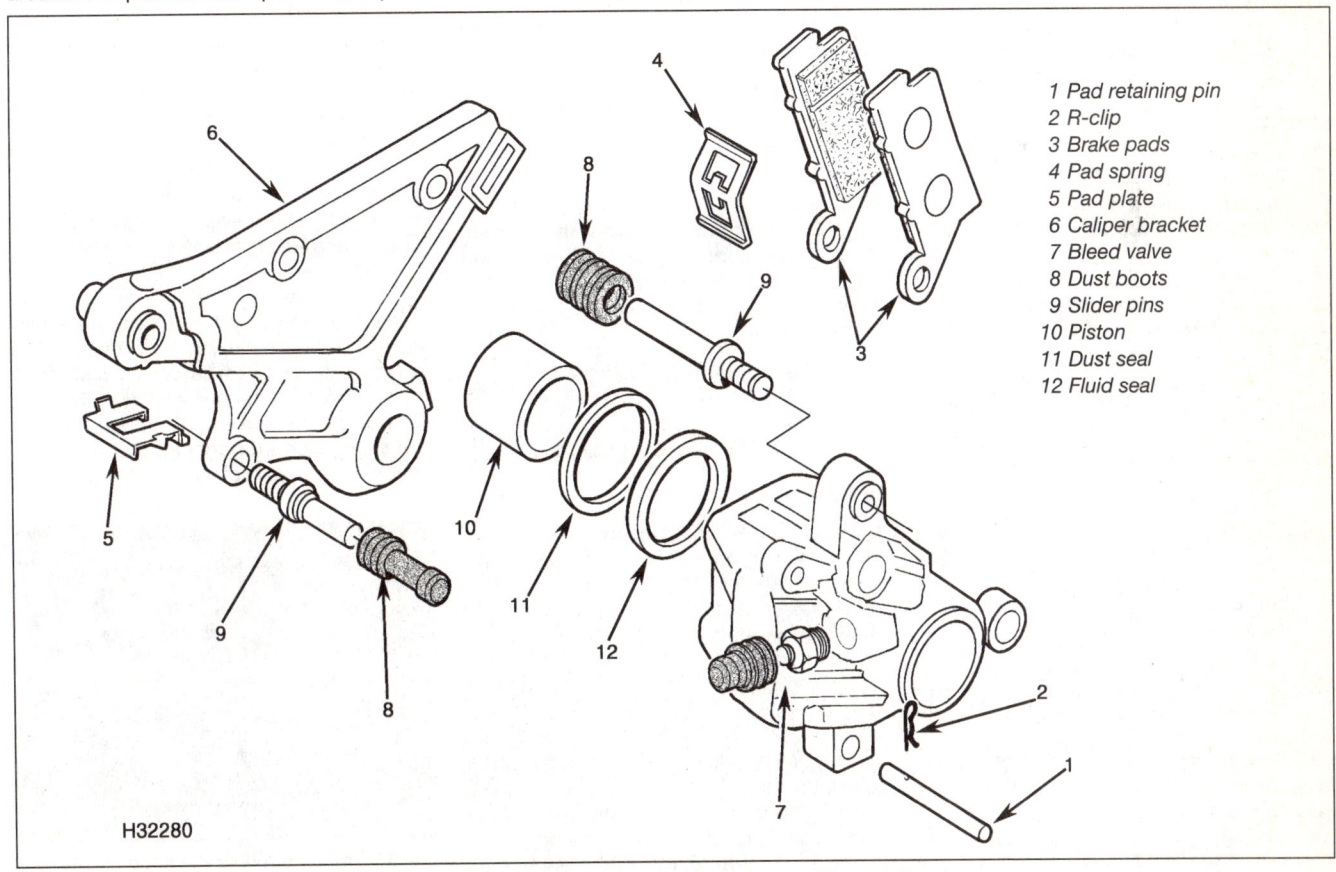

1 Pad retaining pin
2 R-clip
3 Brake pads
4 Pad spring
5 Pad plate
6 Caliper bracket
7 Bleed valve
8 Dust boots
9 Slider pins
10 Piston
11 Dust seal
12 Fluid seal

7.7 Rear brake caliper components – CB models

7•12 Brakes, wheels and tyres

7.18 Slide the caliper into the bracket – make sure the boot lip locates over the inner end of the pin

and cannot be removed by hand (see Step 3), force it out using compressed air. If compressed air is used, place a wad of rag between the piston and the caliper to act as a cushion, then use compressed air directed into the fluid inlet to force the piston out of the body. Use only low pressure to ease the piston out. If the air pressure is too high and the piston is forced out, the caliper and/or piston may be damaged.

⚠ **Warning: Never place your fingers in front of the piston in an attempt to catch or protect it when applying compressed air, as serious injury could result.**

9 Using a wooden or plastic tool, remove the dust seal from the caliper bore **(see illustration 3.10)**. Discard it as a new one must be used on installation. If a metal tool is being used, take great care not to damage the caliper bore. Remove and discard the piston seal in the same way.

10 Clean the piston and bore with denatured alcohol, clean brake fluid or brake system cleaner. If compressed air is available, use it to dry the parts thoroughly (make sure it's filtered and unlubricated).

Caution: Do not, under any circumstances, use a petroleum-based solvent to clean brake parts.

11 Inspect the caliper bore and piston for signs of corrosion, nicks and burrs and loss of plating. If surface defects are present, the caliper assembly must be renewed. If the necessary measuring equipment is available, compare the dimensions of the piston and bore to those given in the Specifications section of this Chapter, renewing any component that is worn beyond the service limit. If the caliper is in bad shape the master cylinder should also be checked.

12 Clean all traces of corrosion from the slider pins and their bores in the caliper and bracket. Renew the rubber boots if they are damaged or deteriorated.

13 Lubricate the new piston seal with clean brake fluid and install it in its groove in the caliper bore.

14 Lubricate the new dust seal with clean brake fluid and install it in its groove in the caliper bore.

15 Lubricate the piston with clean brake fluid and install it closed-end first into the caliper bore. Using your thumbs, push the piston all the way in, making sure it enters the bore squarely.

16 If removed, fit the pad spring in the caliper and the pad plate on the bracket, making sure they are correctly positioned as noted on removal. Apply a smear of copper-based grease to the slider pins, fit the slider pin boots and then slide the caliper onto the bracket and check that it is able to move freely.

Installation

17 On CB models, engage the lug on the caliper bracket with the lug on the swingarm and install the rear wheel. Fit the brake pads and check that they sit squarely each side of the disc. Install the bolt securing the rear brake hose clamp to the swingarm and tighten the bolt securely.

18 On CBF models slide the caliper back onto the bracket **(see illustration)** – if the pads are installed keep them clear of the disc until the caliper is in place, then position them correctly as you pivot the caliper down, and install the rear mounting bolt and tighten it to the torque setting specified at the beginning of the Chapter **(see illustration 6.11b)**. If the pads were removed install them (see Section 6).

19 Connect the brake hose to the caliper, using new sealing washers on each side of the banjo fitting. Align the hose as noted on removal **(see illustration 7.3)**. Tighten the banjo bolt to the torque setting specified at the beginning of this Chapter. Top-up the master cylinder reservoir with DOT 4 brake fluid and flush the hydraulic system (see Section 11).

20 Operate the brake pedal several times to bring the pads into contact with the disc. Check that there are no fluid leaks and thoroughly test the operation of the brake before riding the motorcycle.

8 Rear brake disc – removal, overhaul and installation

Inspection

1 Visually inspect the surface of the disc for score marks and other damage. Light scratches are normal after use and won't affect brake operation, but deep grooves and heavy score marks will reduce braking efficiency and accelerate pad wear. If the disc is badly grooved it must be machined or renewed.

2 To check disc runout, position the motorcycle on its centre stand and support it so that the rear wheel is raised off the ground. Mount a dial gauge on the swingarm, with the gauge plunger touching the surface of the disc about 10 mm from the outer edge **(see illustration 4.2)**. Rotate the wheel and watch the gauge needle, comparing the reading with the limit listed in the Specifications at the beginning of this Chapter. If the runout is greater than the service limit, check the wheel bearings for play (see Chapter 1). If the bearings are worn, renew them (see Section 17) and repeat this check. If the disc runout is still excessive, the disc will have to be renewed, although machining by an engineer may be possible.

3 The disc must not be machined or allowed to wear down to a thickness less than the service limit listed in this Chapter's Specifications and as marked on the disc itself **(see illustration)**. The thickness of the disc can be checked with a micrometer **(see illustration 4.3b)**. If the thickness of the disc is less than the service limit, it must be renewed.

Removal

4 Remove the rear wheel (see Section 16).
Caution: Do not lay the wheel down and allow it to rest on the disc – the disc could become warped.

5 Mark the relationship of the disc to the hub, so it can be installed in the same position. Unscrew the disc retaining bolts, loosening them a little at a time in a criss-cross pattern to avoid distorting the disc, then remove the disc from the wheel. Clean the threads of the bolts.

Installation

6 Place the disc on the hub, making sure the marked side is on the outside. Align the previously applied reference marks if reinstalling the original disc.

7 Apply a suitable non-permanent thread locking compound to the bolt threads then install the bolts and tighten them in a criss-cross pattern evenly and progressively to the torque setting specified at the beginning of this Chapter.

8 Clean all grease off the brake disc using acetone or brake system cleaner. If a new brake disc has been installed, remove any protective coating from its working surfaces.

9 Install the rear wheel (see Section 16).

10 Operate the brake pedal several times to bring the pads into contact with the disc. Check the operation of the brake before riding the motorcycle.

8.3 The minimum disc thickness is marked on the disc

Brakes, wheels and tyres 7•13

9 Rear brake master cylinder – removal, overhaul and installation

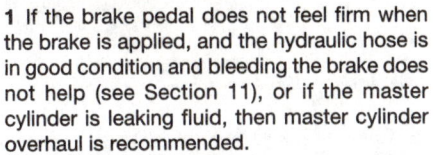

1 If the brake pedal does not feel firm when the brake is applied, and the hydraulic hose is in good condition and bleeding the brake does not help (see Section 11), or if the master cylinder is leaking fluid, then master cylinder overhaul is recommended.

2 Before disassembling the master cylinder, read through the entire procedure and make sure that you have obtained all the new parts required including some new DOT 4 brake fluid, some clean rags and internal circlip pliers. **Note:** *If the master cylinder is just being displaced and not completely removed from the motorcycle, do not drain the brake fluid or disconnect the brake hose.*

Caution: *Disassembly, overhaul and reassembly of the brake master cylinder must be done in a spotlessly clean work area to avoid contamination and possible failure of the brake hydraulic system components. To prevent damage to the paint from spilled brake fluid, always cover the surrounding components when working on the master cylinder.*

Removal

3 On CB models remove the right-hand frame side panel (see Chapter 8). On CBF models remove the left-hand frame side panel (see Chapter 8). Trace the wiring from the brake light switch and disconnect it at the black (CB) or white (CBF) 2-pin connector **(see illustration)**.

4 Unscrew the nut and bolt securing the silencer to the footrest bracket and remove the nut, bolt and washer. Place a wood block or similar under the silencer end to prevent strain on its other mounting points.

5 Unscrew the bolt securing the rear brake fluid reservoir to the frame. Unscrew the reservoir top, remove the diaphragm plate and diaphragm and drain the fluid into a suitable container. Release the clip securing the reservoir to the hose and remove the reservoir. Cover the open end of the hose to prevent dirt entering the system. Wipe any remaining fluid out of the reservoir with a clean rag. Inspect the reservoir cover diaphragm and renew it if it is damaged or deteriorated. Refit the diaphragm, diaphragm plate and top for safe keeping.

6 Unscrew the brake hose banjo bolt on the top of the master cylinder and separate the hose from the master cylinder, noting its alignment **(see illustration)**. Discard the sealing washers as they must be renewed. Wrap a plastic bag tightly around the end of the hose to stop dirt entering the system and secure the hose in an upright position to prevent fluid spills.

7 Unscrew the two footrest bracket bolts and lift the bracket away from the motorcycle **(see illustration)**.

8 Disconnect the brake light switch spring

9.3 Rear brake light switch wiring connector is one of many in the boot – CBF models

from the lug on the pedal **(see illustration 9.7)**.

9 Remove the split pin from the clevis pin securing the brake pedal to the brake master cylinder pushrod. Remove the clevis pin to separate the rod from the pedal. Discard the split pin as a new one must be used **(see illustration)**.

10 Unscrew the two bolts securing the master cylinder to the footrest bracket and remove the master cylinder and brake light switch.

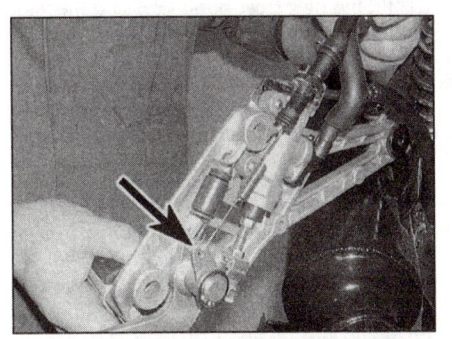

9.7 Lift the bracket away from the bike. Note brake light switch spring (arrowed)

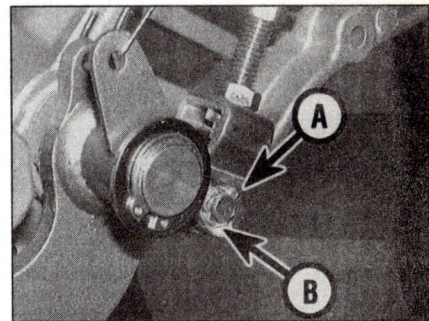

9.6 Note the alignment of the brake hose (arrowed) before removal

11 Release the clip securing the fluid reservoir hose to the master cylinder union and remove the hose. Inspect the hose for cracks or splits and renew it if necessary. If the hose clips are corroded or have weakened, use new ones **(see illustration)**.

12 Remove the screw securing the hose union to the master cylinder and detach the union and O-ring **(see illustration 9.11)**. Discard the O-ring as a new one must be used.

9.9 Remove split pin (A) and clevis pin (B) to separate brake pedal from master cylinder pushrod

1 Reservoir hose
2 Reservoir hose clamp
3 Reservoir hose union
4 O-ring
5 Clevis pin
6 Clevis
7 Clevis base nut
8 Locknut
9 Rubber boot
10 Circlip
11 Pushrod
12 Piston/seal assembly
13 Spring
14 Master cylinder
15 Split pin

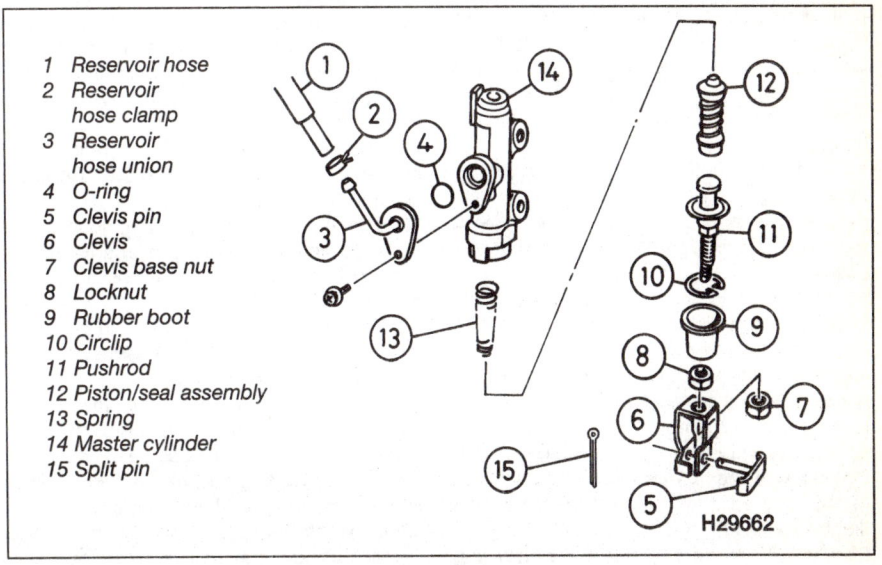

9.11 Rear master cylinder components

7•14 Brakes, wheels and tyres

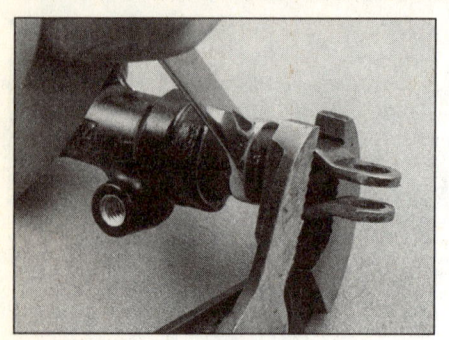

9.13 Hold the clevis and slacken the locknut

9.14 Remove the dust boot from the pushrod

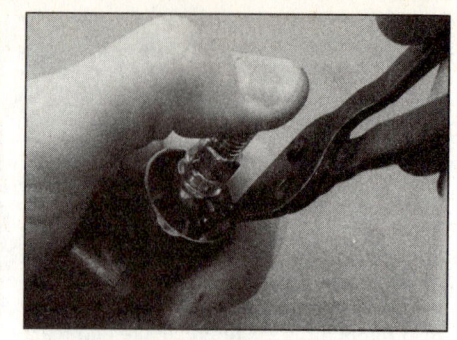

9.15 Depress the pushrod and remove the circlip

Overhaul

13 Mark the position of the clevis locknut on the pushrod (ie its distance from the end of the pushrod), then slacken the locknut and thread the clevis and the locknut off the pushrod **(see illustration)**.
14 Remove the dust boot from the base of the master cylinder to access the pushrod retaining circlip **(see illustration)**.
15 Depress the pushrod against the master cylinder spring tension, remove the circlip using circlip pliers and remove the pushrod **(see illustration)**. Withdraw the piston assembly and spring. If they are difficult to remove, apply low pressure compressed air to the fluid outlet. Lay the parts out in the proper order to prevent confusion during reassembly.
16 Clean all of the parts with clean brake fluid or denatured alcohol. If compressed air is available, use it to dry the parts thoroughly (make sure it's filtered and unlubricated).
Caution: Do not, under any circumstances, use a petroleum-based solvent to clean brake parts.
17 Check the master cylinder bore for corrosion, scratches, nicks and score marks. If the necessary measuring equipment is available, compare the dimensions of the piston and bore to those given in the Specifications section of this Chapter. If damage is evident, the master cylinder must be renewed. If the master cylinder is in poor condition, then the caliper should be checked as well.
18 The dust boot, circlip, piston assembly and spring are all included in the master cylinder overhaul kit. Use all of the new parts, regardless of the apparent condition of the old ones.
19 On CB models if the seal and cup are not already on the piston, fit them according to the layout of the old piston assembly and lubricate them with clean brake fluid. Fit the narrow end of the spring onto the post on the inner end of the piston. Slide the piston and spring assembly into the master cylinder, making sure the lips on the cup do not turn inside out as they enter the bore **(see illustration 9.11)**.
20 On CBF models lubricate the new cup with clean brake fluid and fit it over the narrow end of the spring. If the seal is not already on the piston, fit it into its groove so the wider rim faces the inner end, and lubricate it with clean brake fluid. Slide the spring wide end first into the master cylinder, then fit the piston against the cup and push it in, making sure the lips on the cup and seal do not turn inside out as they enter the bore.
21 Install and depress the pushrod, then fit the new circlip, making sure it is properly seated in the groove **(see illustration 9.15)**.
22 Fit the new dust boot, pressing the wide rim into the master cylinder and making sure the inner rim is seated properly in the groove in the pushrod **(see illustration 9.14)**.
23 Fit a new O-ring to the master cylinder hose union, then install the union and secure it with its screw.

Installation

24 Thread the clevis locknut and the clevis onto the master cylinder pushrod, position the clevis as noted on removal and tighten the nut securely **(see illustration 9.11)**. Note that changing the position of the clevis on the pushrod will alter the brake pedal height in relation to the footrest.
25 Install the master cylinder and rear brake light switch onto the footrest bracket, ensuring the clevis aligns with the brake pedal, and tighten the mounting bolts to the torque setting specified at the beginning of this Chapter. Connect the brake light switch spring to the lug on the brake pedal.
26 Install the clevis pin and secure it using a new split pin **(see illustration 9.9)**. Bend the split pin ends securely.
27 Connect the fluid reservoir hose to the master cylinder union and secure it with the clip. Cover the open end of the hose to prevent dirt entering the system.
28 Position the footrest bracket on the frame, install the mounting bolts and tighten the bolts to the specified torque setting.
29 Connect the fluid reservoir to the fluid reservoir hose, align the reservoir with the bracket on the frame and secure the hose with the clip. Temporarily secure the fluid reservoir to the frame with its retaining bolt.
30 Connect the brake hose to the master cylinder, using a new sealing washer on each side of the banjo union. Ensure that the hose is positioned so that it butts against the lug and tighten the banjo bolt to the specified torque setting.
31 Unscrew the fluid reservoir mounting bolt and pull the reservoir clear of the seat cowling to gain access to the filler cover. Remove the cover, diaphragm plate and diaphragm. Fill the fluid reservoir with new DOT 4 brake fluid, then fill and bleed the system (see Section 11).
32 Fit the diaphragm, making sure it is correctly seated, the diaphragm plate and the cover onto the reservoir and tighten the cover securely. Fix the reservoir to the frame and install the side panel.
33 Check the operation of the brake before riding the motorcycle.

10 Brake hoses and unions – inspection and renewal

Inspection

1 Brake hose and pipe condition should be checked regularly and the hose(s) renewed at the specified interval (see Chapter 1).
2 Twist and flex the hose(s) while looking for cracks, bulges and seeping fluid. Check extra carefully around the areas where the hose(s) connect to the banjo fittings, as these are common areas for hose failure **(see illustration)**.
3 Inspect the banjo fittings; if they are rusted,

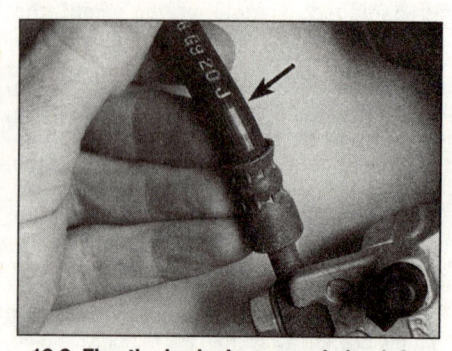

10.2 Flex the brake hoses and check for cracks, bulges and leaks

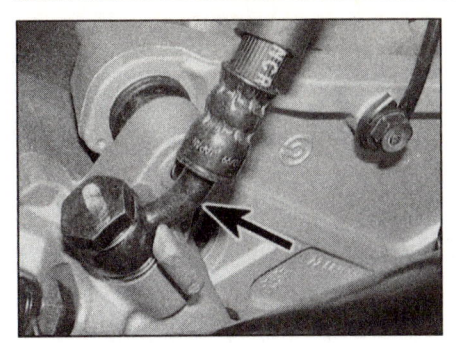

Brakes, wheels and tyres 7•15

10.3 Inspect the banjo fittings (arrowed)

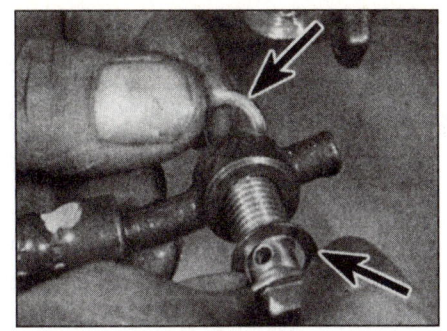

10.4 Renew washers (arrowed) on leaking banjo unions

10.5a Check the front hose-to-pipe joint . . .

10.5b . . . the rear hose-to-pipe joint . . .

10.5c . . . and the ABS control unit

cracked or damaged, renew them **(see illustration)**.
4 Inspect the banjo union connections for leaking fluid. If they leak when tightened to the specified torque setting, unscrew the banjo bolt and fit new washers **(see illustration)**.
5 On models with ABS remove the fuel tank (see Chapter 4) and check the brake pipes, the pipe joints and the ABS control unit for signs of fluid leakage and for any dents or cracks in the pipes **(see illustrations)**.

Renewal

6 The brake hose(s) have banjo union fittings on each end. Cover the surrounding area with plenty of clean rag and unscrew the banjo bolt, noting its alignment. Free the hose from any clamps or guides and remove it. Discard the sealing washers on the hose banjo unions. **Note:** *Do not operate the brake lever or pedal while a brake hose is disconnected.*

7 Position the new hose, making sure it isn't twisted or otherwise strained, and either abut the hose union against the lug on the component casting, or fit it into the slot between two lugs, where present. Otherwise align the hose or pipe as noted on removal. Install the hose banjo bolts using new sealing washers on both sides of the unions.
8 Tighten the banjo bolts to the torque settings specified at the beginning of this Chapter. Make sure the hoses are correctly aligned and routed clear of all moving components and reinstall the clamps. On models with a rear disc brake, ensure the rear brake hose is correctly routed through the guide on the caliper bracket.
9 On models with ABS the hoses join to pipes that connect the system components to the ABS hydraulic unit. The joints between the hoses and pipes, and where the pipes connect to the hydraulic unit are held by gland nuts.

There are no sealing washers. Unscrew the nuts to separates the hoses from the pipes and to detach the pipes from the hydraulic unit. When refitting them tighten the nuts to the specified torque setting if the correct tools are available.
10 Drain the old brake fluid from the hydraulic system and refill with new DOT 4 brake fluid (see Section 11). Check the operation of the brakes before riding the motorcycle.

11 Brake system – bleeding and fluid change

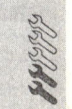

Air bleeding

1 Bleeding the brakes is simply the process of removing air from the brake fluid reservoir, the hose and the brake caliper. Bleeding is necessary whenever a brake system hydraulic connection is loosened, or after a component or hose is renewed, or after the master cylinder or caliper is overhauled. Leaks in the system may also allow air to enter, but leaking brake fluid will reveal their presence and warn you of the need for repair.
2 To bleed the brakes, you will need some new DOT 4 brake fluid, a length of clear vinyl or plastic hose that fits tightly over the caliper bleed valve, a small container partially filled with clean brake fluid, some rags and a spanner to fit the brake caliper bleed valve.
3 Cover the fuel tank and other painted components to prevent damage in the event that brake fluid is spilled.
4 If bleeding the rear brake, unscrew the bolt securing the fluid reservoir to the frame to gain access to the reservoir top.
5 Remove the reservoir top, diaphragm plate and diaphragm and slowly pump the brake lever (front brake) or pedal (rear brake) a few times, until no air bubbles can be seen floating up from the holes in the bottom of the reservoir. Doing this bleeds the air from the master cylinder end of the line. Loosely refit the reservoir cover.
6 Pull the dust cap off the caliper bleed valve **(see illustration)**. Attach one end of the clear vinyl or plastic hose to the bleed valve and submerge the other end in the brake fluid in the container **(see illustration)**.

11.6a Brake caliper bleed valve

11.6b Attach one end of the hose to the bleed valve, submerge the other in the fluid container

7•16 Brakes, wheels and tyres

11.7a Check the fluid level in the reservoir

11.7b Do not allow the fluid to fall below the lower level mark (arrowed) in the front . . .

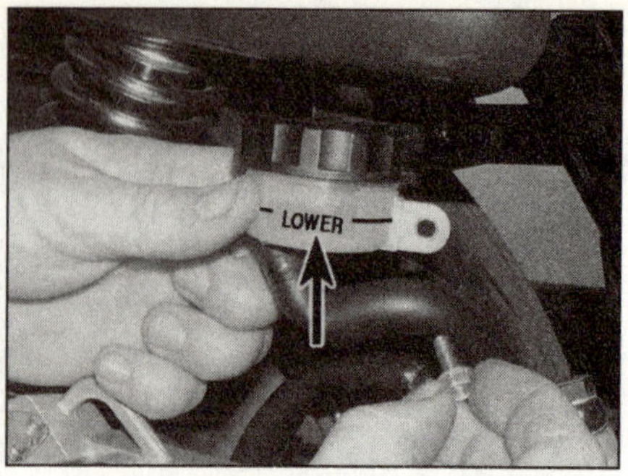

11.7c . . . or rear brake fluid reservoir

7 Remove the reservoir cover and check the fluid level. Do not allow the fluid level to drop below the lower mark during the bleeding process **(see illustrations)**.

8 Carefully pump the brake lever or pedal three or four times and hold it in (front) or down (rear) while opening the bleed valve. When the valve is opened, brake fluid will flow out of the caliper into the clear hose and the lever will move toward the handlebar or the pedal will move down. If there is air in the system there should be air bubbles in the brake fluid coming out of the caliper.

9 Retighten the bleed valve, then release the brake lever or pedal gradually. Top-up the reservoir and repeat the process until no air bubbles are visible in the brake fluid leaving the caliper and the lever or pedal is firm when applied. On completion, disconnect the hose, then tighten the bleed valve to the torque setting specified at the beginning of this Chapter and install the dust cap.

10 Top-up the reservoir, install the diaphragm, diaphragm plate and cover, wipe up any spilled brake fluid and check the entire system for leaks. Check the operation of the brakes before riding the motorcycle.

Fluid change

11 Changing the brake fluid is a similar process to bleeding the brakes and requires the same materials plus a suitable tool for siphoning the fluid out of the hydraulic reservoir. Also ensure that the container is

> **HAYNES HINT** If it's not possible to produce a firm feel to the lever or pedal the fluid may be aerated. Let the brake fluid in the system stabilise for a few hours and then repeat the procedure when the tiny bubbles in the system have settled out.

large enough to take all the old fluid when it is flushed out of the system.

12 Follow Steps 3, 4 and 6 above, then remove the reservoir top, diaphragm plate and diaphragm and siphon the old fluid out of the reservoir. Fill the reservoir with new brake fluid, then follow Step 8.

13 Retighten the bleed valve, then release the brake lever or pedal gradually. Keep the reservoir topped-up with new fluid to above the LOWER level at all times or air may enter the system and greatly increase the length of the task. Repeat the process until new fluid can be seen emerging from the bleed valve.

14 Disconnect the hose, tighten the bleed valve to the specified torque and install the dust cap.

15 Top-up the reservoir, install the diaphragm, diaphragm plate and cover, wipe up any

> **HAYNES HINT** Old brake fluid is invariably much darker in colour than new fluid, making it easy to see when all old fluid has been expelled from the system.

spilled brake fluid and check the entire system for leaks.

16 Check the operation of the brakes before riding the motorcycle.

12 Rear drum brake (R and T models) – removal, inspection and installation

Removal

1 Inspect the rear brake wear indicator (see Chapter 1, Section 3).

2 Remove the rear wheel (see Section 16) and lift out the brake backplate.

3 Unscrew the brake arm pinch bolt and remove the arm from the brake camshaft, noting the alignment punchmark **(see illustration)**.

4 Remove the brake wear indicator, noting how the wide spline aligns with the wide groove on the shaft, and the felt seal.

5 Mark the brake shoes to aid reassembly; if they are not going to be renewed they must be installed in their original positions. Note the position of the brake springs; the black spring fits next to the brake cam, the blue spring fits next to the shoe pivot posts. Note also that the open ends of the springs face away from the backplate. Mark the end of the brake cam to aid reassembly **(see illustration)**.

6 Remove the split pins from the shoe pivot posts, remove the plate, and pull the shoes off the backplate together with the brake cam **(see illustrations)**.

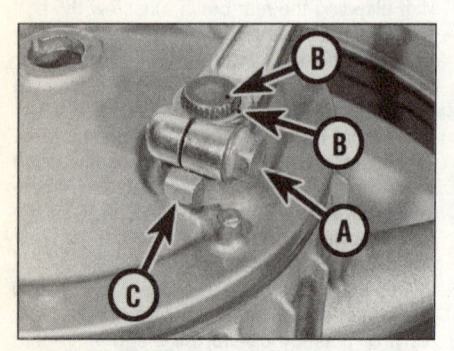

12.3 Pinch bolt (A), alignment marks (B) and wear indicator (C)

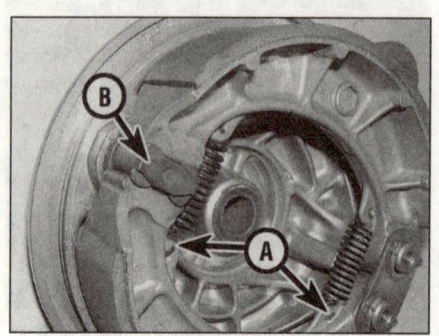

12.5 Spring ends (A) and brake cam (B)

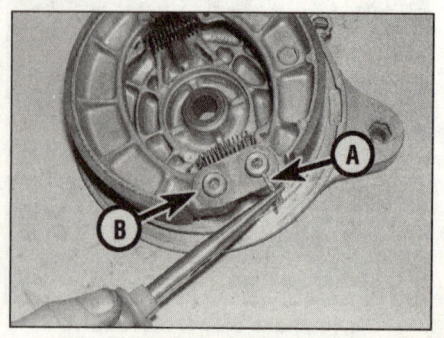

12.6a Remove the split pins (A) and plate (B) . . .

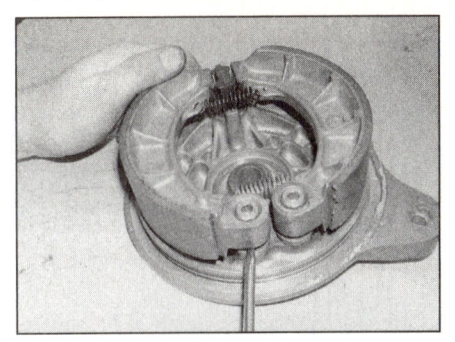

Brakes, wheels and tyres 7•17

12.6b ... then ease the shoes and cam off the backplate

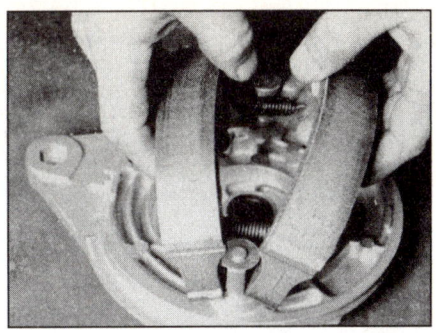

12.7 Fold the shoes as shown to release the spring tension

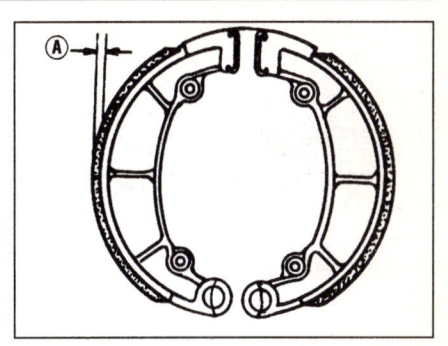

12.9 Measure the shoe lining thickness (A)

7 Fold the shoes towards each other to release the spring tension and separate the shoes (see illustration).

Inspection

8 Check the linings for wear, damage and signs of contamination from road dirt or water. If the linings are visibly defective, renew them.
9 Measure the thickness of the lining material (just the lining material, not the metal backing) and renew the shoes if the linings are worn below the minimum specified at the beginning of this Chapter at any point (see illustration).
10 Check the ends of the shoes where they contact the brake cam and the pivot posts. Renew the shoes if there's visible wear.
11 Check the lugs on the shoes where the springs locate for wear. Also make sure that the springs are not stretched and that the hooked ends are not deformed.
12 Clean all old grease from the pivot posts and the brake cam and check them for wear. Also clean any old grease from the cam pivot hole in the backplate and check that the cam is a good fit in the hole.
13 Check the inside of the brake drum for wear or damage. Measure the inside diameter of the drum at several points with a vernier caliper. If the measurements are uneven (indicating that the drum is out-of-round) or if there are scratches deep enough to snag a fingernail, have the drum turned (skimmed) to correct the surface. If the wear or damage cannot be corrected within the service limit specified at the beginning of this Chapter, the wheel must be renewed.

Installation

Caution: Do not apply too much grease otherwise there is a risk of it contaminating the brake drum and shoe linings.

14 Apply a smear of copper-based grease to the shaft of the brake cam and to the cam faces, and to the lugs on the shoes where the springs locate. Hook the springs into the shoes (see Step 5), position the shoes in a V on the brake cam (refer to Step 5 for the correct position of the cam), then fold the shoes down into position. Make sure the ends of the shoes fit correctly in the cam.
15 Apply a smear of copper-based grease to the pivot posts; locate the shaft of the brake cam in the hole in the backplate and lower the shoe assembly into place, spreading the shoe ends to fit them on the pivot posts. Install the plate on the pivot posts and fit new split pins, bending the ends securely (see illustration).
16 Lubricate the brake shaft felt seal sparingly with clean engine oil and install the brake wear indicator on the shaft splines with the pointer facing the brakeplate. Align the punch mark on the brake arm with the mark on the end of the shaft and install the brake arm. Install the pinch bolt and tighten it to the torque setting specified at the beginning of this Chapter.
17 Check the operation of the brake arm before installing the backplate in the brake drum and installing the rear wheel (see Section 16). Adjust the rear brake freeplay as described in Chapter 1.

13 Wheels – inspection and repair

1 In order to carry out a proper inspection of the wheels, it is necessary to support the motorcycle upright so that the wheel being inspected is raised off the ground. Position the motorcycle on its centre stand. Clean the wheels thoroughly to remove mud and dirt that may interfere with the inspection procedure or mask defects. Make a general check of the wheels (see Chapter 1) and tyres (see Daily (pre-ride) checks).
2 Attach a dial gauge to the fork slider or the swingarm and position its stem against the side of the rim. Spin the wheel slowly and check the axial (side-to-side) runout of the rim. In order to accurately check radial (out of round) runout with the dial gauge, the wheel would have to be removed from the machine, and the tyre from the wheel. With the axle clamped in a vice and the dial gauge positioned on the top of the rim, the wheel can be rotated to check the runout (see illustration).
3 An easier, though slightly less accurate, method is to attach a stiff wire pointer to the fork slider or the swingarm and position the end a fraction of an inch from the wheel (where the wheel and tyre join). If the wheel is true,

12.15 Fit new split pins on reassembly

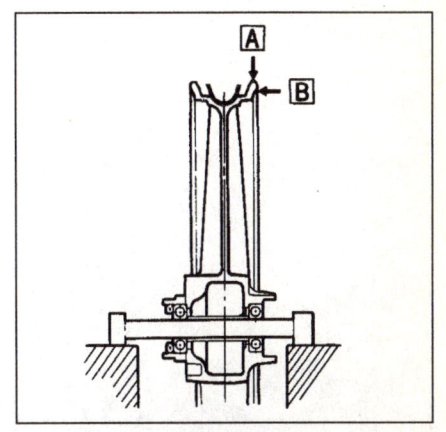

the distance from the pointer to the rim will be constant as the wheel is rotated. Note: If wheel runout is excessive, check the wheel bearings very carefully before renewing the wheel.
4 The wheels should also be visually inspected for cracks, flat spots on the rim and other damage. Look very closely for dents in the area where the tyre bead contacts the rim. Dents in this area may prevent complete sealing of the tyre against the rim, which leads to deflation of the tyre over a period of time. If damage is evident, or if runout in either direction is excessive, the wheel will have to be renewed. Never attempt to repair a damaged cast alloy wheel.

13.2 Check the wheel for radial (out-of-round) runout (A) and axial (side-to-side) runout (B)

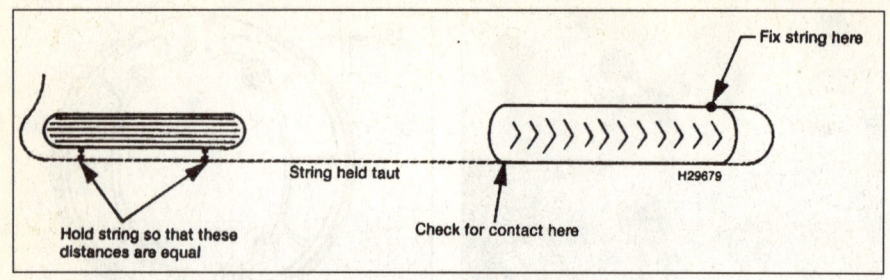

7•18 Brakes, wheels and tyres

14.5 Wheel alignment check using string

14 Wheels – alignment check

1 Misalignment of the wheels due to a bent frame or forks, can cause strange and possibly serious handling problems. If the frame or forks are at fault, repair by a specialist or renewal are the only options.
2 To check wheel alignment you will need an assistant, a length of string or a perfectly straight piece of wood and a ruler. A plumb bob or other suitable weight will also be required.
3 In order to make a proper check of the wheels, support the motorcycle on its centre stand. Measure the width of both tyres at their widest points. Subtract the smaller measurement from the larger measurement, then divide the difference by two. The result is the amount of offset that should exist between the front and rear tyres on both sides.
4 If a string is used, have your assistant hold one end of it about halfway between the floor and the rear axle, touching the rear sidewall of the tyre.
5 Run the other end of the string forward and pull it tight so that it is roughly parallel to the floor **(see illustration)**. Slowly bring the string into contact with the front sidewall of the rear tyre, then turn the front wheel until it is parallel with the string. Measure the distance from the front tyre sidewall to the string.
6 Repeat the procedure on the other side of the motorcycle. The distance from the front tyre sidewall to the string should be equal on both sides.
7 If necessary, a perfectly straight length of wood or metal bar may be substituted for the string **(see illustration)**. The procedure is the same.
8 If the distance between the string and tyre is greater on one side, or if the rear wheel appears to be out of alignment, refer to Chapter 1, Section 1 and check that the chain adjuster markings coincide on each side of the swingarm.
9 If the front-to-back alignment is correct, the wheels still may be out of alignment vertically.
10 Using the plumb bob, or other suitable weight and a length of string, check the rear wheel to make sure it is vertical. To do this, hold the string against the tyre upper sidewall and allow the weight to settle just off the floor. When the string touches both the upper and lower tyre sidewalls and is perfectly straight, the wheel is vertical. If it is not, place thin spacers under one leg of the centre stand.
11 Once the rear wheel is vertical, check the front wheel in the same manner. If both wheels are not perfectly vertical, the frame and/or major suspension components are bent.

15 Front wheel – removal and installation

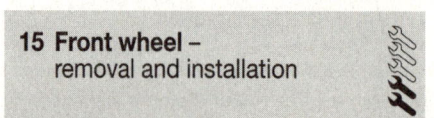

CB models

Removal

1 Put the motorcycle on its centre stand and support it under the crankcase so that the front wheel is off the ground.
2 Remove the screw that retains the speedometer cable in the speedometer drive unit on the left-hand side of the front wheel hub, withdraw the cable and secure it clear of the front wheel. Note how the rib on the drive unit locates behind the lug on the left-hand fork slider **(see illustrations)**.
3 Unscrew the axle nut from the right-hand end of the axle **(see illustration)**.

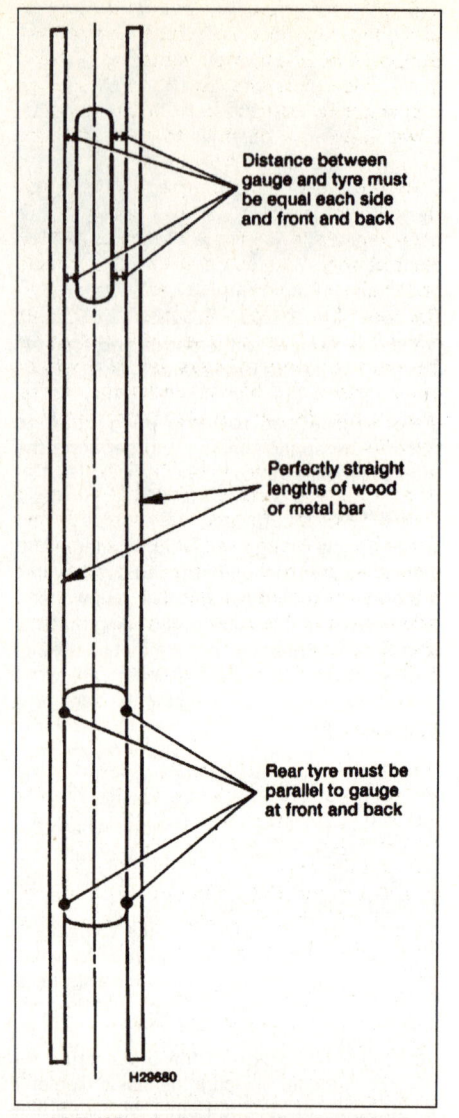

14.7 Wheel alignment check using a straight edge

15.2a Remove the screw (arrowed) and pull the cable out of the drive unit

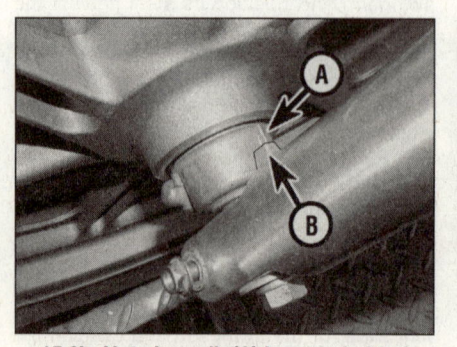

15.2b Note how rib (A) locates behind lug (B)

15.3 Unscrew the axle nut

Brakes, wheels and tyres 7•19

4 Slacken the axle clamp bolt on the bottom of the left-hand fork slider (see illustration). Support the wheel, then withdraw the axle from the left-hand side (see illustration). Use a spanner to turn the axle to aid removal. Lower the wheel carefully to clear the front mudguard withdraw it from the forks.

5 Remove the wheel spacer from the right-hand side of the hub and the speedometer drive unit the left-hand side (see illustrations). Note how the drive unit locates on the tabs of the driveplate in the hub.

Caution: Don't lay the wheel down and allow it to rest on the disc – the disc could become warped. Always lay the wheel down with wood blocks under the rim to protect the disc and prevent dirt entering the hub.

6 Check the axle for straightness by rolling it on a flat surface such as a piece of plate glass (first wipe off all old grease and remove any corrosion using wire wool). If the equipment is available, place the axle in V-blocks and measure the runout using a dial gauge. If the axle is bent or the runout exceeds the limit specified, renew it.

7 Refer to Section 17 if wheel bearing renewal is required.

Installation

8 Apply a smear of grease to the inside of the wheel spacer and the speedometer drive unit, and also to the inside edge of the grease seals. Fit the spacer into the right-hand side of the hub and the drive unit into the left-hand side (see illustration 15.5a and b). Ensure that the tabs of the driveplate locate correctly in the drive unit. Apply a thin coat of grease to the axle.

9 Manoeuvre the wheel between the fork legs, making sure the brake disc locates squarely between the pads in the caliper.

10 Lift the wheel into place, ensuring that the speedometer drive unit is correctly positioned against the lug on the fork slider (see Step 2) and that the wheel spacer remains in place. Slide the axle in from the left-hand side (see illustration 15.4b).

11 Install the axle nut and tighten it to the torque setting specified at the beginning of this Chapter, counter-holding the axle head on the other side of the machine.

12 Tighten the axle clamp bolt on the bottom of the left-hand fork slider to the specified torque setting.

13 Install the speedometer drive cable making sure it passes through the cable guide on the front mudguard; ensure the forked end of the cable engages on the spade drive from the gearbox and tighten the retaining screw securely (see illustration).

14 Apply the front brake a few times to bring the pads back into contact with the disc. Move the motorcycle off its stand, apply the front brake and pump the front forks a few times to settle all components in position.

15 Check the operation of the front brake before riding the motorcycle.

15.4a Slacken the axle clamp bolt (arrowed) . . .

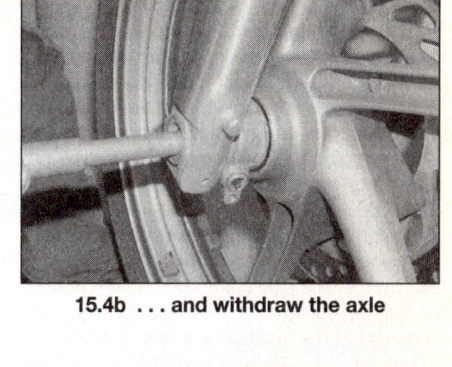

15.4b . . . and withdraw the axle

15.5a Remove the wheel spacer . . .

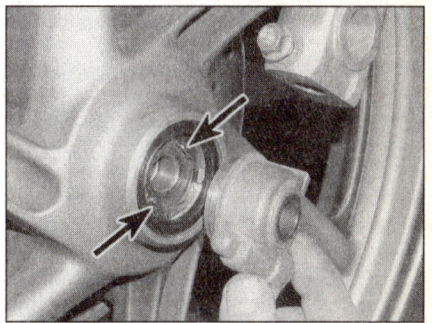

15.5b . . . and the speedometer drive unit. Note driveplate tabs (arrowed)

CBF models

Removal

16 Put the motorcycle on its centre stand and support it under the crankcase so that the front wheel is off the ground.

17 Slacken the axle clamp bolt on the bottom of the right-hand fork (see illustration).

18 Unscrew the axle bolt from the right-hand end of the axle.

19 Slacken the axle clamp bolt on the bottom of the left-hand fork (see illustration 15.4a). Support the wheel, then withdraw the axle from the left-hand side, using a screwdriver or rod through the holes as a handle if required (see illustration 15.4b). Lower the wheel carefully to clear the front mudguard withdraw it from the forks. On ABS models take care not to knock the wheel sensor.

20 Remove the wheel spacer from each side of the hub, noting which fits where (see illustration 15.5a).

Caution: Don't lay the wheel down and allow it to rest on the disc – the disc could become warped. Always lay the wheel down with wood blocks under the rim to protect the disc and prevent dirt entering the hub.

21 Check the axle is straight by rolling it on a flat surface such as a piece of plate glass (first wipe off all old grease and remove any corrosion using wire wool). If the equipment is available, place the axle in V-blocks and measure the runout using a dial gauge. If the axle is bent or the runout exceeds the limit specified, renew it.

22 Refer to Section 17 if wheel bearing renewal is required.

Installation

23 Apply a smear of grease to the inside of the wheel spacers and to the inside edge of

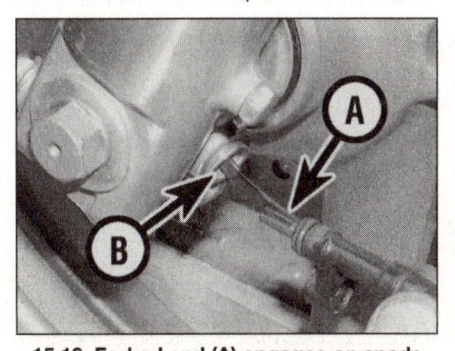

15.13 Forked end (A) engages on spade drive (B)

15.17 Axle clamp bolt (A), axle bolt (B)

7•20 Brakes, wheels and tyres

the grease seals. Fit the long spacer into the right-hand side of the hub and the short spacer into the left-hand side **(see illustration 15.5a)**. Apply a thin coat of grease to the axle.
24 Lift the wheel into place, making sure the brake disc locates squarely between the pads in the caliper and that the wheel spacers remain in place. On ABS models take care not to damage the wheel speed sensor.
25 Slide the axle in from the left-hand side and align the groove in the axle head with the outer face of the fork bottom.
26 Tighten the axle clamp bolt on the bottom of the left-hand fork to the torque setting specified at the beginning of this Chapter.
27 Install the axle bolt and tighten it to the specified torque setting **(see illustration 15.17)**.
28 Tighten the axle clamp bolt on the bottom of the right-hand fork to the specified torque setting. Now slacken the left-hand axle clamp bolt.
29 Apply the front brake a few times to bring the pads back into contact with the disc. Move the motorcycle off its stand, apply the front brake and pump the front forks a few times to settle all components in position. Tighten the left-hand axle clamp bolt to the specified torque setting.
30 Check the operation of the front brake before riding the motorcycle.

16 Rear wheel – removal and installation

CB500-R and T models

Removal

1 Position the motorcycle on its centre stand so that the wheel is off the ground
2 Unscrew and remove the rear brake rod adjuster nut and disconnect the rod from the brake arm. Note the position of the spring. Remove the brake rod pin from the brake arm, thread it onto the brake rod and screw the adjuster nut onto the rod for safekeeping.
3 Remove the split pin from the bolt securing the brake torque arm to the brake backplate, remove the nut, washer, rubber bush and bolt,

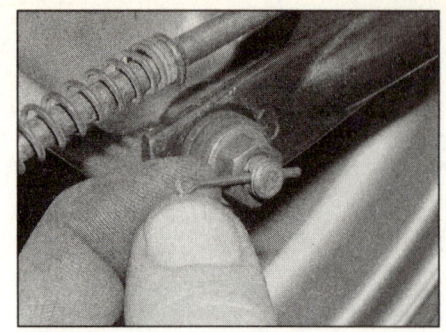

16.3a Remove the split pin and nut . . .

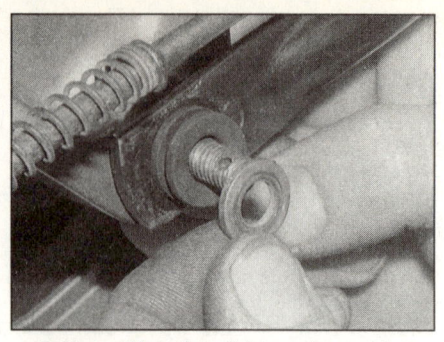

noting the order for reassembly **(see illustrations)**.
4 Unscrew the axle nut and remove the nut and the special cranked washer **(see illustration)**.
5 Support the wheel then withdraw the axle and lower the wheel to the ground **(see illustration)**.
6 Disengage the chain from the sprocket and remove the wheel from the swingarm **(see illustration)**.
7 Remove the spacers located on each side of the wheel for safekeeping, noting how they fit **(see illustration)**.
8 If necessary, withdraw the chain tensioner assemblies from the ends of the swingarm (see Chapter 6).
9 Check the axle for straightness by rolling it on a flat surface such as a piece of plate glass (if the axle is corroded, first remove the corrosion with wire wool). If the equipment is available, place the axle in V-blocks and check the runout using a dial gauge. If the axle is

16.3b . . . then the plain washer, rubber bush and bolt

bent or the runout exceeds the limit specified at the beginning of this Chapter, renew it.
10 Refer to Section 17 if wheel bearing renewal is required.

Installation

11 Apply a thin coat of grease to the axle, and also to the inside of the grease seal in the sprocket carrier and to both wheel spacers. If removed, slide the chain tensioner assemblies into the ends of the swingarm.
12 Manoeuvre the wheel between the ends of the swingarm and engage the chain on the sprocket.
13 Fit the headed spacer into the grease seal in the left-hand side of the hub, then lift the wheel into position and slide the axle into place making sure that the plain spacer is located between the brake backplate and the swingarm **(see illustration)**. Make sure that the axle has passed through both chain adjusters.

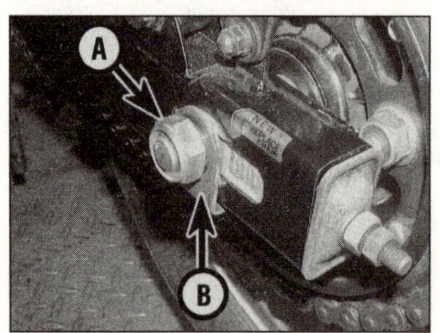

16.4 Remove the nut (A) and cranked washer (B)

16.5 Support the wheel and withdraw the axle

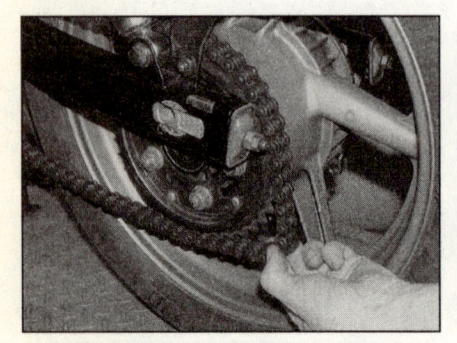

16.6 Slip the chain off the sprocket

16.7 Remove the wheel spacers, noting how they fit

16.13 Locate the spacer against the backplate on R and T models

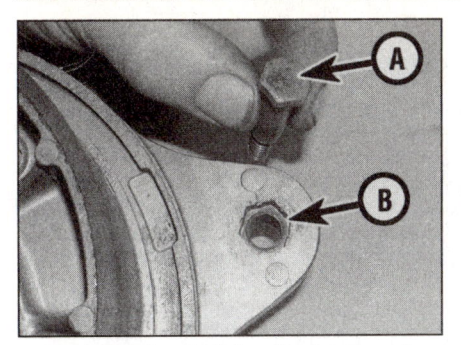

Brakes, wheels and tyres 7•21

16.15 Bolt head (A) locates in recess in backplate (B) – R and T models

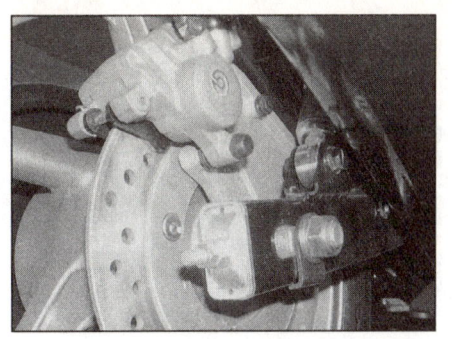

16.29 Ensure the disc is square in the caliper before installing the axle

14 Fit the cranked washer and hand tighten the axle nut. Check the chain tension (see Chapter 1) and then tighten the nut to the specified torque setting, counter-holding the axle head on the other side of the machine.
15 Align the brake torque arm with the brake backplate. Insert the bolt through the backplate first, ensuring that the bolt head locates in the recess in the backplate (see illustration), and install the bush, washer and nut and tighten the nut securely. Fit a new split pin.
16 Install the brake rod pin in the brake arm, install the rod through the pin with the spring forward of the brake arm, and screw the adjuster nut onto the rod. Adjust the rear brake (see Chapter 1).
17 Check the operation of the rear brake before riding the motorcycle.

CB500-V, W, SW, X, SX, Y, SY, 2 and S2 models and CBF models

Removal

18 Position the motorcycle on its centre stand so that the wheel is off the ground
19 Unscrew the axle nut (with the plain washer on CBF models) and the special cranked washer.
20 Support the wheel then withdraw the axle and lower the wheel to the ground.
21 Disengage the chain from the sprocket and remove the wheel from the swingarm.
22 Remove the spacers located on each side of the wheel for safekeeping, noting how they fit.
23 Note how the lug on the caliper bracket locates in the socket on the swingarm. Displace the caliper bracket from the swingarm and secure it to the frame with a cable tie, making sure no strain is placed on the hose. **Note:** *Do not operate the brake pedal while the caliper is off the disc.*
Caution: Don't lay the wheel down and allow it to rest on the disc or the sprocket – they could become warped. Always lay the wheel down with wood blocks under the rim to protect the disc and sprocket and prevent dirt entering the hub.
24 If necessary, withdraw the chain tensioner assemblies from the ends of the swingarm (see Chapter 6).
25 Check the axle for straightness by rolling it on a flat surface such as a piece of plate glass (if the axle is corroded, first remove the corrosion with wire wool). If the equipment is available, place the axle in V-blocks and check the runout using a dial gauge. If the axle is bent or the runout exceeds the limit specified at the beginning of this Chapter, renew it.
26 Refer to Section 17 if wheel bearing renewal is required.

Installation

27 Apply a thin coat of grease to the axle, and also to the inside of the grease seals and the wheel spacers. If removed, slide the chain tensioner assemblies into the ends of the swingarm.
28 Manoeuvre the wheel between the ends of the swingarm and engage the chain on the sprocket.
29 Engage the lug on the brake caliper bracket with the socket on the swingarm. Fit the headed spacers into the grease seals and lift the wheel into position, ensuring the brake disc locates squarely between the pads in the caliper (see illustration). Make sure the spacers and caliper bracket remain in place and install the axle. Check that the axle has passed through the chain adjusters, spacers and caliper bracket.
30 Fit the cranked washer with the angled section at the bottom. On CBF models fit the plain washer. Fit the axle nut and hand tighten it. Check the chain tension (see Chapter 1), and then tighten the nut to the specified torque setting, counter-holding the axle head on the other side of the machine.
31 Operate the brake pedal several times to bring the pads into contact with the disc.
32 Check the operation of the rear brake before riding the motorcycle.

17 Wheel bearings – renewal

Front wheel bearings

Note: *Always renew the wheel bearings in pairs. Never renew the bearings individually. Avoid using a high pressure cleaner on the wheel bearing area.*

1 Remove the wheel (see Section 15).
2 Set the wheel on blocks to keep the hub clear of the work surface and so as not to allow the weight of the wheel to rest on the brake disc.
3 Prise out the grease seal on each side of the hub using a flat-bladed screwdriver, taking care not to damage the rim of the hub (see illustration). Discard the seals as new ones should be fitted. Remove the speedometer driveplate from the left-hand side of the hub, noting how it fits (see illustration).
4 Using a metal rod (preferably a brass drift punch) inserted through the centre of the right-hand bearing, tap evenly around the inner race of the left-hand bearing to drive it from the hub (see illustration). The bearing spacer will also come out.

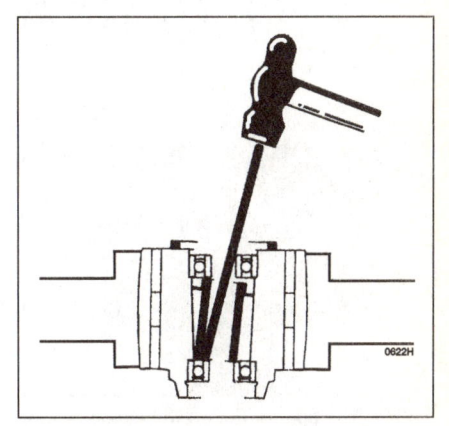

17.3a Lever out the grease seal

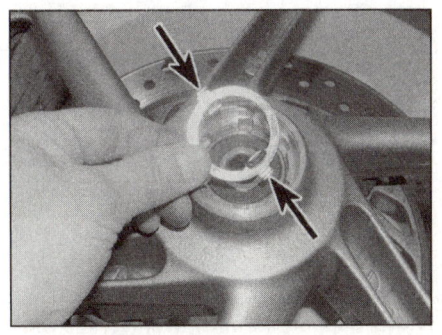

17.3b Outer tabs of driveplate (arrowed) locate in hub

17.4 Locate the rod as shown when driving out the bearing

7•22 Brakes, wheels and tyres

17.10 The seal can be driven in using a flat piece of wood

17.29 Lever out the grease seal

17.30 Drive out the spacer with a long socket

5 Lay the wheel on its other side so that the right-hand bearing faces down. Drive the bearing out of the wheel using the same technique as above.
6 If the bearings are of the unsealed type or are only sealed on one side, clean them with a high flash-point solvent (one which won't leave any residue) and blow them dry with compressed air (don't let the bearings spin as you dry them). Apply a few drops of oil to the bearing. **Note:** *If the bearing is sealed on both sides don't attempt to clean it.*
7 Hold the outer race of the bearing and rotate the inner race – if the bearing doesn't turn smoothly, has rough spots or is noisy, renew it.

HAYNES HiNT *Refer to Tools and Workshop Tips (Section 5) in the Reference section for more information about bearings.*

8 If the bearing is good and can be re-used, wash it in solvent once again and dry it, then pack it with grease.
9 Thoroughly clean the hub area, then install the right-hand bearing into its seat in the hub, with the marked or sealed side facing outwards. Using the old bearing, a bearing driver or a socket large enough to contact the outer race of the bearing, drive it in until it's completely seated.
10 Apply a smear of grease to the new seal and press it into the hub. If necessary, drive the seal into place using a seal or bearing driver, a suitable socket or a flat piece of wood **(see illustration)**.

11 Turn the wheel over and install the bearing spacer. Drive the left-hand bearing into place (see Step 9).
12 Install the speedometer driveplate, aligning the outer tabs with the recesses in the hub. Smear the new seal with grease and install the seal (see Step 10).
13 Clean the brake disc using acetone or brake system cleaner, then install the wheel (see Section 15).

Rear wheel bearings

14 Remove the rear wheel (see Section 16).
15 On R and T models, remove the brake backplate assembly.
16 Lift the sprocket coupling out of the wheel, noting how it fits.
17 Set the wheel on blocks to keep the hub clear of the work surface.
18 Prise the O-ring out of the groove on the left-hand side of the hub; discard the O-ring as a new one should be fitted.
19 On disc-braked models, prise out the grease seal on the right-hand side of the hub using a flat-bladed screwdriver, taking care not to damage the rim of the hub **(see illustration 17.3a)**. Discard the seal as a new one should be fitted.
20 On all models, using a metal rod (preferably a brass drift punch) inserted through the centre of the right-hand bearing, tap evenly around the inner race of the left-hand bearing to drive it from the hub **(see illustrations 17.4)**. The bearing spacer will also come out.
21 Lay the wheel on its other side so that the

right-hand bearing faces down. Drive the bearing out of the wheel using the same technique as above.
22 Refer to Steps 6 to 8 above and check the bearings.
23 Thoroughly clean the hub area and install the right-hand bearing, then turn the wheel over and install the bearing spacer and the left-hand bearing (see Step 9).
24 On disc-braked models, install the grease seal on the right-hand side (see Step 10).
25 Grease the new O-ring and install it in its groove on the left-hand side of the hub. Fit the sprocket coupling assembly onto the wheel.
26 Clean then install the brake drum (R and T models). Clean the brake disc (all other models) using acetone or brake system cleaner. On all models, install the wheel (see Section 16).

Sprocket coupling bearing

27 Remove the rear wheel (see Section 16). On R and T models, remove the brake backplate assembly.
28 Lift the sprocket coupling out of the wheel, noting how it fits, then prise the O-ring out of the groove on the left-hand side of the hub; discard the O-ring as a new one should be fitted.
29 Using a flat-bladed screwdriver, lever out the grease seal from the outside of the coupling **(see illustration)**.
30 Remove the spacer from the inside of the sprocket coupling bearing, noting how it fits. The spacer could be a tight fit and may have to be driven out using a suitable long socket or piece of tubing **(see illustration)**.
31 Support the coupling on blocks of wood and drive the bearing out from the inside using a bearing driver or socket.
32 Refer to Steps 6 to 8 above and check the bearing.
33 Thoroughly clean the bearing seat then install the bearing into the coupling, with the marked or sealed side facing out. Using the old bearing, a bearing driver or a socket large enough to contact the outer race of the bearing, drive it in until it is completely seated.
34 Install the spacer into the inside of the coupling bearing. If the spacer is a tight fit it will be necessary to support the inner race of the bearing when driving the spacer in **(see illustrations)**.

17.34a The spacer fits on the inside of the coupling bearing

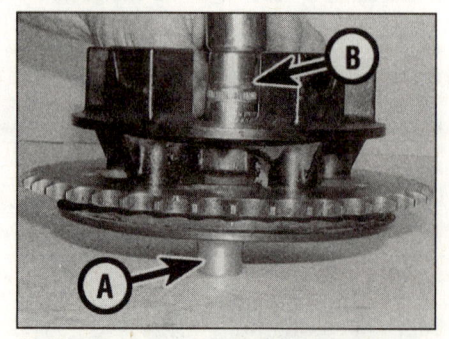

17.34b Support the inner bearing race with a suitable socket (A) while driving the spacer in (B)

Brakes, wheels and tyres 7•23

35 Install the grease seal in the outside of the coupling (see Step 10) **(see illustration)**.
36 Check the sprocket coupling/rubber damper (see Chapter 6).
37 Grease the new O-ring and install it in its groove on the left-hand side of the hub.
38 Fit the sprocket coupling assembly onto the wheel.
39 Clean then install the brake drum (R and T models). Clean the brake disc (all other models) using acetone or brake system cleaner. On all models, install the wheel (see Section 16).

18 Tyres –
general information and fitting

General information

1 The wheels fitted to all models are designed to take tubeless tyres only. Tyre sizes are given in the Specifications at the beginning of this Chapter.
2 Refer to *Daily (pre-ride) checks* at the beginning of this manual for tyre maintenance.

17.35 Install the grease seal

Fitting new tyres

3 When selecting new tyres, ensure that front and rear tyre types are compatible, the correct size and correct speed rating; if necessary seek advice from a motorcycle tyre specialist **(see illustration)**.
4 It is recommended that tyre fitting is done by a motorcycle tyre specialist rather than attempted in the home workshop. This is particularly relevant in the case of tubeless tyres because the force required to break the seal between the wheel rim and tyre bead is substantial, and is usually beyond the capabilities of an individual working with normal tyre levers without damaging the wheel. Additionally, the specialist will be able to balance the wheels after tyre fitting.
5 Note that punctured tubeless tyres can in some cases be repaired, but such repairs must be carried out by a motorcycle tyre specialist. Honda advise that following puncture repair, the motorcycle should not be ridden above 40 mph (60 km/h) for the first 24 hrs, or above 80 mph (130 km/h) thereafter.

19 ABS –
operation and fault finding

1 The ABS prevents the wheels from locking up under hard braking or on uneven road surfaces. A sensor on each wheel transmits information about the speed of rotation to the ABS control unit; if the unit senses that a wheel is about to lock, it releases brake

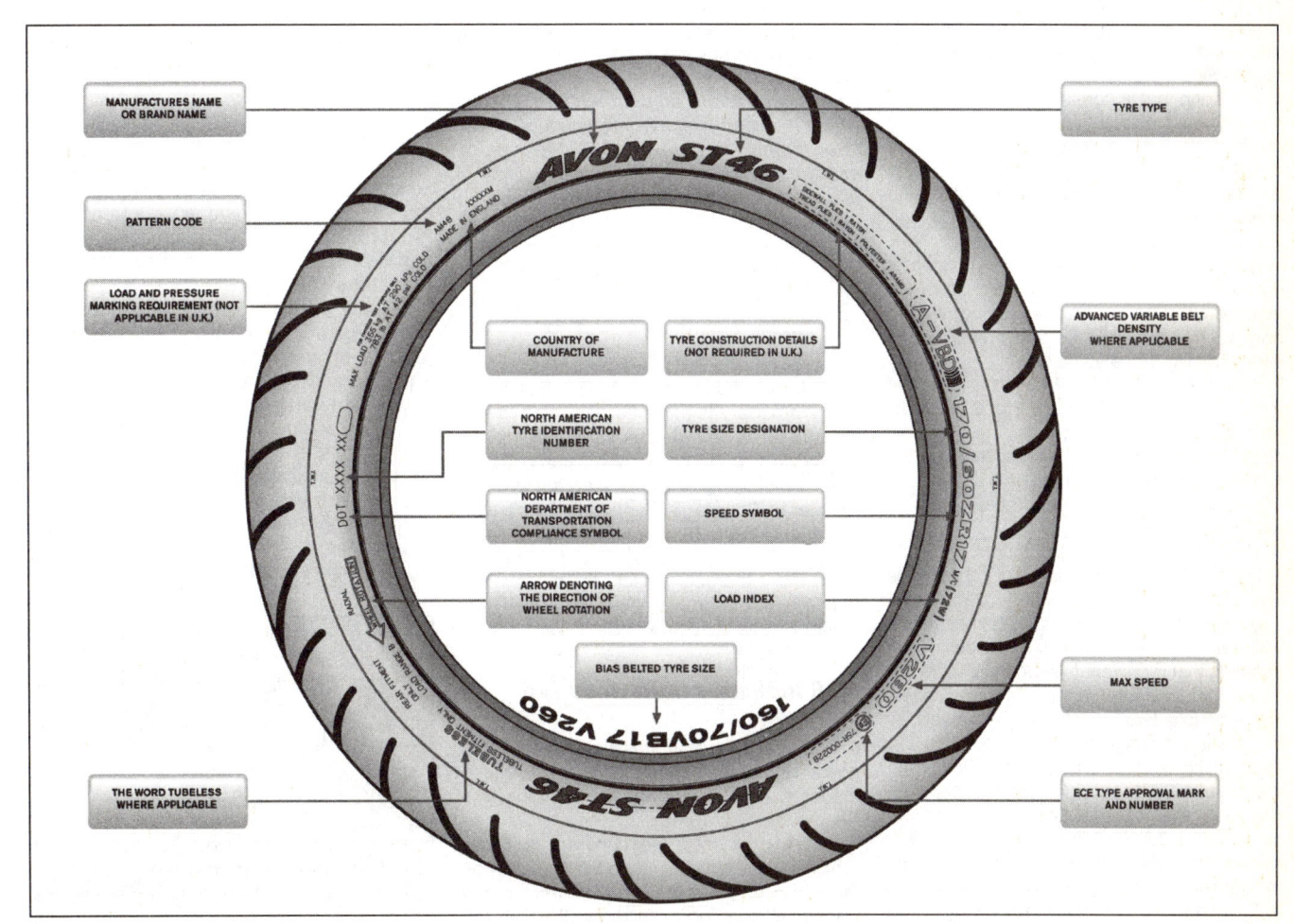

18.3 Common tyre sidewall markings

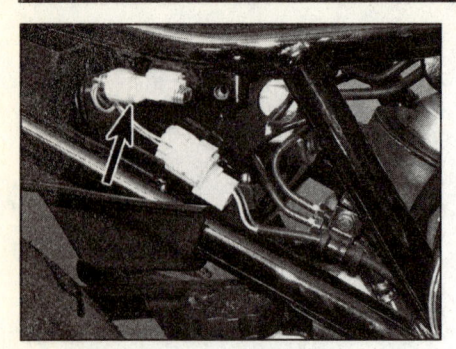

19.6 ABS service check connector (arrowed)

pressure to that wheel momentarily, preventing a skid.

2 The ABS is self-checking and is activated when the ignition (main) switch is turned on – the ABS indicator light in the instrument cluster will come on and will remain on until road speed increases above 6 mph (10 kph) at which point, if the ABS is normal, the light will go off. **Note:** *If the ABS indicator light does not come on initially there is a fault in the system – see Section 20.*

3 If the indicator light remains on, or starts flashing while the machine is being ridden, there is a fault in the system and the ABS function will be switched off – the brakes will still function, but in normal mode.

4 If a fault is indicated, details will be stored in the control unit's memory. Access the fault code(s) as follows.

5 Remove the right-hand section of the seat cowling (see Chapter 8).

6 Ensure the ignition (main) switch is OFF. Locate the white ABS service check connector and pull the connector out of its holder **(see illustration)**. Use an insulated jumper wire to connect the brown/white and green/orange wire terminals in the connector.

7 Turn the ignition (main) switch ON. The indicator light should come on for 2 seconds, then go out for 3.6 seconds, then start to flash, with the pattern of flashes denoting the fault code. If the light comes on as described but does not go out out after 3.6 seconds no fault code is stored.

8 The fault code flash patterns work as follows – a long flash or flashes of 1.3 seconds represents the first number of the 2 digit fault code. Count the number of long flashes. There will then follow a short flash or flashes of 0.3 seconds that represent the second digit of the fault code. Count the number of short flashes. For example two long flashes followed by three short flashes denotes fault code 23. Record the code and identify the fault from the table in Section 20. If there is more than one fault, the codes are given in ascending order, with a 3.6 second break between each code. The code or codes are repeated until the ignition is switched OFF.

9 Turn the ignition (main) switch OFF and remove the jumper wire when the code or codes have been recorded. **Note:** *Do not squeeze the front brake lever during this procedure.*

10 To check the ABS components see Section 20.

11 Once the fault has been corrected, reset the control unit memory as follows. Ensure the ignition (main) switch is OFF. Connect the brown/white and green/orange wire terminals in the service check connector. Hold the front brake lever on and turn the ignition (main) switch ON.

12 The ABS indicator light in the instrument cluster should come on for 2 seconds and then go off. Release the front brake lever.

13 Squeeze the brake lever when the indicator light comes on again, then release the lever when the light goes off.

14 The indicator light should now flash twice to confirm that the control unit memory has been erased, then stay on. Turn the ignition (main) switch OFF, remove the jumper wire from the service check connector and push the connector back into its holder. Install the seats (see Chapter 8). Check that the ABS is operating normally (see Step 2).

15 If the light does not flash twice, repeat the reset procedure.

Note: *The ABS indicator may diagnose a fault if tyre sizes other than those specified by Honda are fitted, if the tyre pressures are incorrect, if the machine has been run continuously over bumpy roads, if the front wheel is raised whilst riding (wheelie) or if, after riding, the machine is left on its centrestand with the engine running and the rear wheel turning.*

20 ABS – system checks

1 If a fault is indicated in the ABS, first check that the battery is fully charged, then check the ABS fuses (see Chapter 9).

2 Unless specified otherwise, carry-out all checks with the ignition (main) switch OFF.

3 Refer to Chapter 9, Section 2, for general electrical fault finding procedures and equipment.

4 If, after a thorough check, the source of a fault has not been identified, have the ABS control unit tested by a Honda dealer.

ABS indicator light does not come on

5 First check the instrument cluster fuse and power circuit, then check the wiring between the instrument cluster and the fusebox (see Chapter 9).

6 Remove the seats (see Chapter 8). Make sure the ignition is OFF, then lift the catch on the ABS control unit wiring connector and disconnect the connector **(see illustration)**. Turn the ignition ON. If the indicator light comes on, check the connector and control unit wiring terminals. If the terminals appear to be good, have the control unit tested by a Honda dealer.

7 If the light does not come on, disconnect the black instrument cluster wiring connector (see Chapter 9) and check for continuity between the red/black wire terminal on the loom side of the connector and earth (ground). There should be no continuity – if there is, check for a short circuit in the red/black wire.

8 Reconnect the black instrument cluster wiring connector and use an insulated jumper wire to connect the green wire terminal in the connector to earth (ground). Turn the ignition ON. If the indicator light comes on, check for a break in the green wire. If the light does not come on, it is likely the instrument cluster is faulty – have it checked by a Honda dealer.

ABS indicator light stays on continuously

9 Remove the seats (see Chapter 8).

10 Lift the catch on the ABS control unit wiring connector and disconnect the connector **(see illustration 20.6)**. Check for continuity between the brown/white wire terminal on the loom side of the connector and earth (ground). There should be no continuity – if there is, check for a short circuit in the brown/white wire.

11 Check for continuity between the green wire terminal in the loom side of the connector and earth (ground). There should be continuity.

12 Use an insulated jumper wire to connect the brown/white and green/orange wire terminals in the ABS service check connector **(see illustration 19.6)**. Now check for continuity between the brown/white wire terminal on the loom side of the ABS control unit wiring connector and earth (ground). There should be continuity – if there isn't, check for a break in the brown/white wire or the green/orange. Remove the jumper wire when the test is complete.

13 Use an insulated jumper wire to connect the red/black wire terminal in the instrument cluster wiring connector (with it still connected) to earth (ground) (see Chapter 9). The ABS indicator light should go off – if not, it is likely

20.6 Pull the connector catch up to release the connector

Brakes, wheels and tyres 7•25

Fault code/flashes	Faulty component – symptoms	A	B	Possible causes
Light does not come on	No voltage at instrument cluster No voltage at control unit			Damaged fuse Faulty wiring or wiring connector Damaged earth (ground) wire
Light stays on continuously	Service check connector			Faulty wiring or wiring connector Damaged earth (ground) wire
	Fail-safe relay fuse			Damaged fuse
	No voltage at instrument cluster No voltage at control unit			Faulty wiring or wiring connector Damaged earth (ground) wire
	Instrument cluster/control unit			Internal fault
11	Front wheel speed sensor	x	x	Faulty wiring or wiring connector
13	Rear wheel speed sensor	x	x	Faulty wiring or wiring connector
12	Front wheel speed sensor		x	Faulty wiring or wiring connector Dirty or damaged sensor
14	Rear wheel speed sensor		x	Faulty wiring or wiring connector Dirty or damaged sensor
21	Front wheel pulser ring		x	Damaged pulser ring
23	Rear wheel pulser ring		x	Damaged pulser ring
31, 32, 33, 34	Control unit solenoid valve	x	x	Faulty control unit
41, 42	Front brake binding Front wheel lock-up		x	Faulty brake disc or caliper Faulty speed sensor wiring or wiring connector Faulty control unit
43	Rear brake binding Rear wheel lock-up		x	Faulty brake disc or caliper Faulty speed sensor wiring or wiring connector Faulty control unit
51, 52, 53	No voltage at control unit Control unit	x	x	Damaged fuse Faulty wiring or wiring connector Internal fault
54	Fail-safe relay fuse – pre-start check No voltage at control unit Control unit	x		Damaged fuse/faulty relay Faulty wiring or wiring connector Internal fault
61	ABS main fuse Control unit	x	x	Damaged fuse Ignition voltage too low Internal fault
62	ABS main fuse Control unit		x	Damaged fuse Ignition voltage too high Internal fault
71	Tyre size		x	Incorrect tyre size/tyre pressure
81	Control unit	x	x	Internal fault

Note A: *Items checked during the pre-start inspection – during the time the ignition is switched on and the engine is started.*
Note B: *Items checked between the pre-start check and the ignition being turned off.*

the instrument cluster is faulty – have it checked by a Honda dealer.

14 If the light goes off, remove the jumper wire. Disconnect the ABS control unit wiring connector and use the jumper wire to connect the red/black wire terminal in the connector to earth (ground). The indicator light should go off – if not, check the red/black wire between the instrument cluster and ABS connector for a break.

15 If the light goes off, use the jumper wire to connect the red/black wire terminal to the green/orange wire terminal in the connector. If the light goes off it is likely the ABS control unit is faulty – have it checked by a Honda dealer. If not, check for a break in the green/orange wire.

Front wheel speed sensor and pulser ring

Note: *Before carrying out any of the checks, follow the procedure in Section 19 to reset the control unit memory, then activate the self-checking procedure. If the fault code is the result of unusual riding or conditions and the ABS is normal, the indicator light will go off. Otherwise perform the following checks.*

Check

16 Measure the air gap between the wheel sensor and the pulser ring with a feeler gauge, then compare the result with the Specification at the beginning of this Chapter **(see illustration 20.28)**. The gap is not adjustable – if it is outside the specification, check that the speed sensor and pulser ring fixings are tight, that the components are not damaged and that there is no dirt on the sensor tip or between the slots in the pulser ring. If any of the components are damaged they must be renewed.

17 Remove the seats (see Chapter 8). Lift the catch on the ABS control unit wiring connector and disconnect the connector **(see illustration 20.6)**. Check for continuity between the pink/black and green/orange wire terminals on the loom side of the connector and earth (ground). There should be no continuity – if there is continuity, go to Step 18. If there is no continuity, remove the fuel tank (see Chapter

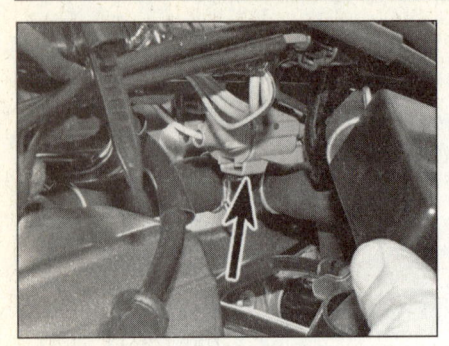

7•26 Brakes, wheels and tyres

20.17 Front wheel speed sensor wiring connector (arrowed)

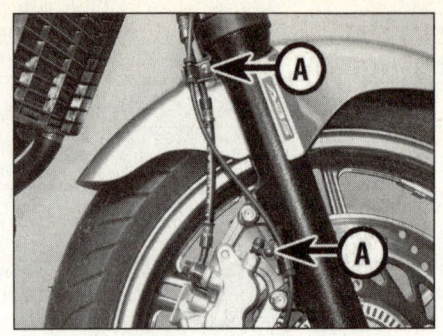

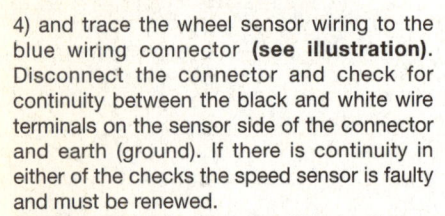

20.21a Undo the wiring guide bolts (A) . . .

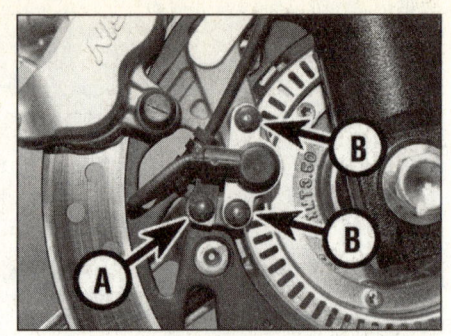

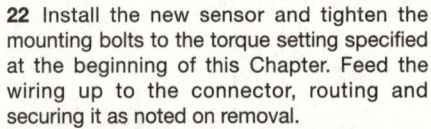

20.21b . . . and the sensor bolts (B)

4) and trace the wheel sensor wiring to the blue wiring connector **(see illustration)**. Disconnect the connector and check for continuity between the black and white wire terminals on the sensor side of the connector and earth (ground). If there is continuity in either of the checks the speed sensor is faulty and must be renewed.

18 Remove the fuel tank (see Chapter 4), then trace the speed sensor wiring to the blue wiring connector **(see illustration 20.17)**. Disconnect the connector. Check for continuity first in the pink/black wire between the control unit wiring connector and the sensor wiring connector and then in the green/orange wire and repair the break in whichever wire does not show continuity.

19 If all the checks have failed to identify the fault, replace the speed sensor with a known good one. Follow the procedure in Section 19 to reset the control unit memory, then activate the self-checking procedure. If the indicator light is no longer flashing, the original speed sensor was faulty. If the fault code reappears have the ABS control unit checked by a Honda dealer.

Removal and installation

20 Remove the fuel tank (see Chapter 4), then trace the wheel sensor wiring to the blue wiring connector **(see illustration 20.17)**. Disconnect the connector.
21 Undo the bolts securing the sensor wiring guides, and the sensor to the caliper bracket, and remove the sensor, releasing the wiring from any other clips or ties and noting its routing **(see illustrations)**.

22 Install the new sensor and tighten the mounting bolts to the torque setting specified at the beginning of this Chapter. Feed the wiring up to the connector, routing and securing it as noted on removal.
23 Check the air gap (see Step 16). Install the fuel tank (see Chapter 4).
24 To renew the pulser ring, first remove the front wheel (see Section 15).
25 Undo the Torx screws securing the ring and lift it off.
26 Ensure there is no dirt or corrosion where the ring seats on the hub – if the ring does not sit flat when it is installed the sensor air gap will be incorrect. Tighten the Torx screws securely.
27 Install the front wheel (see Section 15). Check the speed sensor air gap (see Step 16).

Rear wheel speed sensor and pulser ring

Note: *Before carrying out any of the checks, follow the procedure in Section 19 to reset the control unit memory, then activate the self-checking procedure. If the fault code is the result of unusual riding or conditions and the ABS is normal, the indicator light will go off. Otherwise perform the following checks.*

Check

28 Measure the air gap between the wheel sensor and the pulser ring with a feeler gauge, then compare the result with the Specification at the beginning of this Chapter **(see illustration)**. The gap is not adjustable – if it is outside the specification, check that the sensor and pulser ring fixings are tight, that the components are not damaged and that there is no dirt on the sensor tip or between the slots in the pulser ring. If any of the components are damaged they must be renewed.

29 Remove the seats (see Chapter 8). Lift the catch on the ABS control unit wiring connector and disconnect the connector **(see illustration 20.6)**. Check for continuity between the pink/white and green/red wire terminals on the loom side of the connector and earth (ground). There should be no continuity – if there is continuity, go to Step 30. If there is no continuity, remove the right-hand section of the seat cowling (see Chapter 8) and trace the wheel sensor wiring to the green wiring connector **(see illustration)**. Disconnect the connector and check for continuity between the black and white wire terminals on the sensor side of the connector and earth (ground). If there is continuity in either of the checks the speed sensor is faulty and must be renewed.

30 Remove the right-hand section of the seat cowling (see Chapter 8), then trace the speed sensor wiring to the green wiring connector **(see illustration 20.29)**. Disconnect the connector. Check for continuity first in the pink/white wire between the control unit wiring connector and the sensor wiring connector and then in the green/red wire and repair the break in whichever wire does not show continuity.

31 If all the checks have failed to identify the fault, replace the speed sensor with a known good one. Follow the procedure in Section 19 to reset the control unit memory, then activate the self-checking procedure. If the indicator light is no longer flashing, the original speed sensor was faulty. If the fault code reappears have the ABS control unit checked by a Honda dealer.

Removal and installation

32 Remove the right-hand section of the seat cowling (see Chapter 8), then trace the speed sensor wiring to the green wiring connector **(see illustration 20.29)**. Disconnect the connector.
33 Undo the bolts securing the speed sensor and its wiring to the caliper bracket and

20.28 Checking the wheel sensor air gap

20.29 Rear wheel speed sensor wiring connector (arrowed)

Brakes, wheels and tyres 7•27

20.33 Undo the wiring guide bolt (A) and the sensor bolts (B)

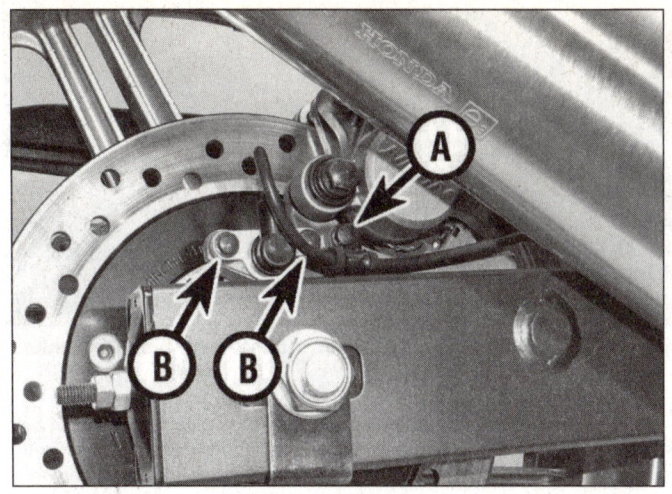

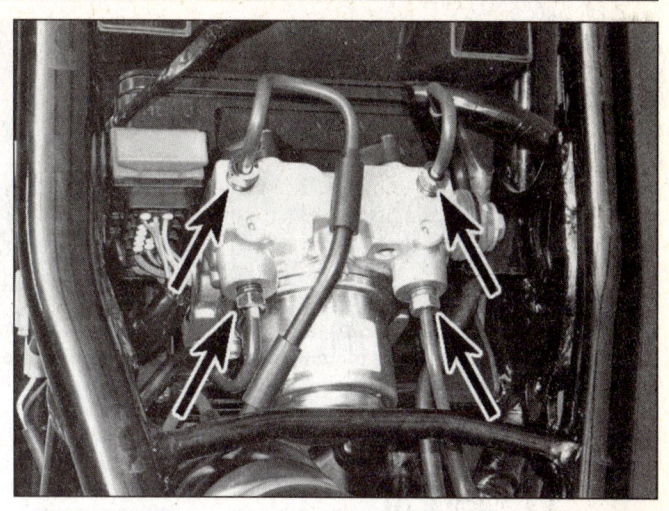

20.53 Unscrew the gland nuts (arrowed) and detach the pipes

withdraw the sensor **(see illustration)**. Release the wiring from any clips or ties, noting its routing.

34 Install the new speed sensor and tighten the mounting bolts to the torque setting specified at the beginning of this Chapter. Feed the wiring up to the green connector and secure it with the clips. Connect the green connector.

35 Check the air gap (see Step 28). Install the right-hand seat cowling (see Chapter 8).

36 To renew the pulser ring, first remove the rear wheel (see Section 16).

37 Undo the Torx screws securing the ring and lift it off.

38 Ensure there is no dirt or corrosion where the ring seats on the hub – if the ring does not sit flat when it is installed the sensor air gap will be incorrect. Tighten the Torx screws securely.

39 Install the rear wheel (see Section 16). Check the speed sensor air gap (see Step 28).

Control unit solenoid valve

40 Follow the procedure in Section 19 to reset the control unit memory, then activate the self-checking procedure. If the fault code is the result of unusual riding or conditions and the ABS is normal, the indicator light will go off. If the fault code remains it is likely the ABS control unit is faulty – have it checked by a Honda dealer.

ABS motor and fuse

41 Check the ABS motor fuse (see Chapter 9). If the fuse has blown replace it with a new one then check the wiring as follows. If the fuse hasn't blown check as follows.

42 Remove the seats (see Chapter 8). Lift the catch on the ABS control unit wiring connector and disconnect the connector **(see illustration 20.6)**. Check for battery voltage between the black/brown / red wire terminal on the loom side of the connector and earth (ground) – there should be voltage at all times i.e. with the ignition OFF. If there is no voltage, check for a break in the red wire between the connector and the fusebox connector, and if that wire is good check the wire between the fusebox and the battery. If there is voltage, follow the procedure in Section 19 to reset the control unit memory, then activate the self-checking procedure. If the fault code remains it is likely the ABS control unit is faulty – have it checked by a Honda dealer.

Fail-safe relay

43 Check the fail-safe relay fuse (see Chapter 9). If the fuse has blown replace it with a new one then check the wiring as follows. If the fuse hasn't blown check as follows.

44 Remove the seats (see Chapter 8). Lift the catch on the ABS control unit wiring connector and disconnect the connector **(see illustration 20.6)**. Check for battery voltage between the green/yellow / black wire terminal on the loom side of the connector and earth (ground) – there should be voltage at all times i.e. with the ignition OFF. If there is no voltage, check for a break in the black wire between the connector and the fusebox connector, and if that wire is good check the wire between the fusebox and the battery. If there is voltage, follow the procedure in Section 19 to reset the control unit memory, then activate the self-checking procedure. If the fault code remains it is likely the ABS control unit is faulty – have it checked by a Honda dealer.

ABS main fuse

45 Check the ABS main fuse (see Chapter 9). If the fuse has blown replace it with a new one then check the wiring as follows. If the fuse hasn't blown check as follows.

46 Remove the seats (see Chapter 8). Lift the catch on the ABS control unit wiring connector and disconnect the connector **(see illustration 20.6)**. Check the voltage between the red/brown wire terminal on the loom side of the connector and earth (ground). Honda specifies a voltage between 10 and 17 volts. If there is no voltage, check for a break in the red/brown wire. If the voltage is below the specification, check the charging system (see Chapter 9). If there is voltage, follow the procedure in Section 19 to reset the control unit memory, then activate the self-checking procedure. If the fault code remains it is likely the ABS control unit is faulty – have it checked by a Honda dealer.

Tyre size and control unit

47 Refer to the Specifications at the beginning of this Chapter for tyre sizes. Refer to *Pre-ride checks* for tyre pressures.

48 If the tyre sizes and pressures are correct, follow the procedure in Section 19 to reset the control unit memory, then activate the self-checking procedure. If the fault code remains it is likely the ABS control unit is faulty – have it checked by a Honda dealer.

ABS control unit

Note: *Before the control unit can be removed, the brake fluid must be drained from the hydraulic system. When refilling and bleeding the ABS-equipped brake system it is essential to use a vacuum-type brake bleeder kit. Alternatively, removal and installation of the control unit should be entrusted to a Honda dealer.*

49 Remove the seats (see Chapter 8).

50 Refer to the procedure in Section 11 for changing the brake fluid – siphon the fluid out of the front and rear reservoirs and pump any residual fluid out through the brake calipers, but do not refill the system at this stage.

51 Lift the catch on the ABS control unit wiring connector and disconnect the connector **(see illustration 20.6)**.

52 Cover the area around the control unit with clean rag prevent damage to paintwork in the event that brake fluid is spilled.

53 Undo the fluid pipe gland nuts and disconnect the pipes from the control unit **(see illustration)**.

7•28 Brakes, wheels and tyres

20.54 Unscrew the left-hand mounting bolt (arrowed) . . .

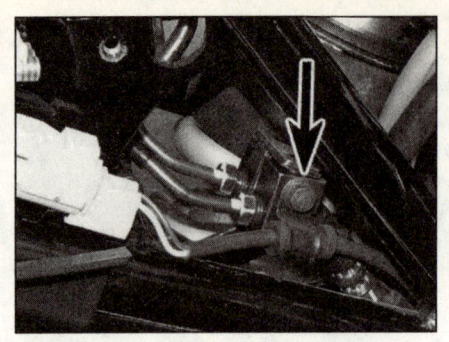

20.55 . . . the joint block bolt (arrowed) . . .

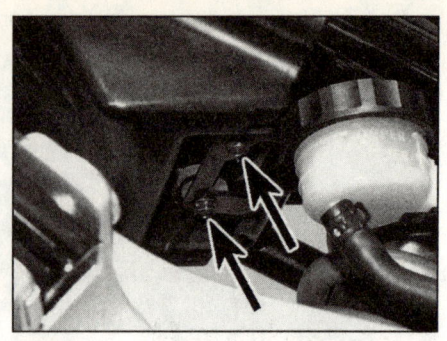

20.56 . . . and the lower mounting bolts (arrowed)

54 Undo the left-hand mounting bolt **(see illustration)**.
55 Undo the pipe-to-hose joint block bolt and release the wiring clamp **(see illustration)**.
56 Undo the lower mounting bolts and remove the brake pipe stay, then lift the control unit out **(see illustration)**.

57 Installation is the reverse of removal, noting the following:
● Tighten the mounting bolts to the torque setting specified at the beginning of this Chapter – Honda specify to use new bolts on the bottom as they come pre-treated with threadlock, but you could clean up the old bolts and apply fresh threadlock yourself if preferred.

● Smear the fluid pipe gland nut threads with new brake fluid and tighten the nuts to the torque setting specified at the beginning of this Chapter. Tighten the nuts before fitting the pipe-to-hose joint block bolt.
● Ensure the ABS control unit wiring connector is secure.
● Follow the procedure in Section 11 to refill and bleed the brake system.

Chapter 8
Bodywork

Contents

Fairing (SW, SX, SY and S2 models) – removal and installation 7
Frame side panels – removal and installation 2
Front mudguard – removal and installation 8
General information 1
Radiator side panels (R, T, V, W, X, Y and 2 models) 5
Seat(s) – removal and installation........................... 3
Seat cowling – removal and installation....................... 4
Rear view mirrors – removal and installation 6

Degrees of difficulty

| **Easy,** suitable for novice with little experience | **Fairly easy,** suitable for beginner with some experience | **Fairly difficult,** suitable for competent DIY mechanic | **Difficult,** suitable for experienced DIY mechanic | **Very difficult,** suitable for expert DIY or professional |

Specifications

Torque settings
Seat cowling side mounting bolts .... 27 Nm
Front mudguard mounting bolts
  CB models .... 10 Nm
  CBF models .... 12 Nm

1 General information

This Chapter covers the procedures necessary to remove and install the bodywork. Since many service and repair operations on these motorcycles require the removal of the bodywork, the procedures are grouped here and referred to from other Chapters.

In the case of damage to the bodywork, it is usually necessary to remove the broken component and renew it. The material that the body panels are composed of doesn't lend itself to conventional repair techniques. Note that there are however some companies that specialise in 'plastic welding' and there are a number of bodywork repair kits now available for motorcycles.

When attempting to remove any body panel, first study it closely, noting any fasteners and associated fittings, to be sure of returning everything to its correct place on installation. In some cases the aid of an assistant will be required when removing panels, to avoid the risk of damage to paintwork. Once the evident fasteners have been removed, try to withdraw the panel as described but DO NOT FORCE IT – if it will not release, check that all fasteners have been removed and try again. Where a panel engages another by means of tabs, be careful not to break the tab or its mating slot or to damage the paintwork. Remember that a few moments of patience at this stage will save you a lot of money in replacing broken fairing panels!

When installing a body panel, first study it closely, noting any fasteners and associated fittings removed with it, to be sure of returning everything to its correct place. Check that all fasteners are in good condition, including all trim nuts or clips and damping/rubber mounts; any of these must be renewed if faulty before the panel is reassembled. Check also that all mounting brackets are straight and repair or renew them if necessary before attempting to install the panel. Where assistance was required to remove a panel, make sure your assistant is on hand to help install it.

Tighten the fasteners securely, but be careful not to overtighten any of them or the panel may break (not always immediately) due to the uneven stress.

2 Frame side panels – removal and installation

CB models

1 Unscrew the retaining bolt in the lower corner of the panel. Pull the front edge of the panel away from the motorcycle to release the peg on the back of the panel from the grommet on the lower edge of the fuel tank. Draw the panel forwards to release the tab on the top

8•2 Bodywork

2.1a Unscrew the bolt (arrowed) . . .

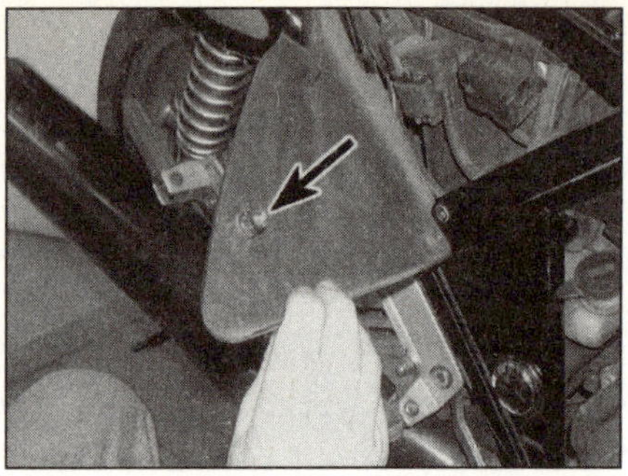

2.1b . . . and pull the peg (arrowed) out of the fixing grommet

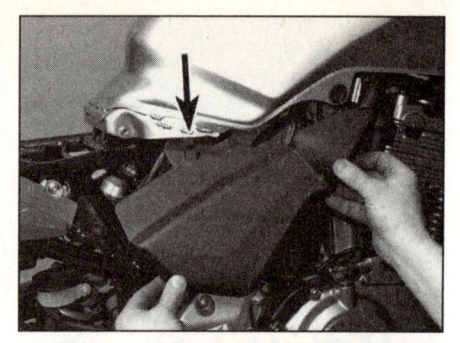

2.4 Pull the panel pegs out of the grommets – note how the tab (arrowed) locates

rear edge from the slot in the seat cowling and remove the panel **(see illustration)**.

2 Installation is the reverse of removal.

> **HAYNES HiNT** *Note that a small amount of lubricant (liquid soap or similar) applied to the mounting grommet will assist the peg to engage without the need for undue pressure.*

CBF models

3 Remove the seats (see Section 3).
4 Carefully pull the front and back of the panel away to release the pegs from the grommets **(see illustration)**. Note how the tab on the tank locates in the slot on the top edge of the panel.
5 Installation is the reverse of removal.

3 Seat(s) – removal and installation

CB models

1 To remove the seat, insert the ignition key into the seat lock and turn the key clockwise. This will release the seat lock mechanism on the front underside of the seat **(see illustration)**.
2 Lift the front of the seat and disengage the hooks at the rear of the seat under the rear cowling **(see illustration)**. Remove the seat, noting how the tabs at the back locate.
3 Installation is the reverse of removal. Make sure the hooks at the rear of the seat locate correctly into the bracket under the rear cowling. When the hooks are located, push down on the seat to engage the latch.

CBF models

4 To remove the passenger's seat, insert the ignition key into the seat lock and turn the key clockwise **(see illustration)**. Lift the rear of the seat and draw it back, noting how the tabs at the front locate under hooks on the frame.
5 To remove the rider's seat first remove the passenger's seat. Unscrew the two bolts, noting the collars **(see illustration)**. Lift the rear of the seat and draw it back, noting how the tab at the front locates **(see illustration)**.
6 Installation is the reverse of removal.

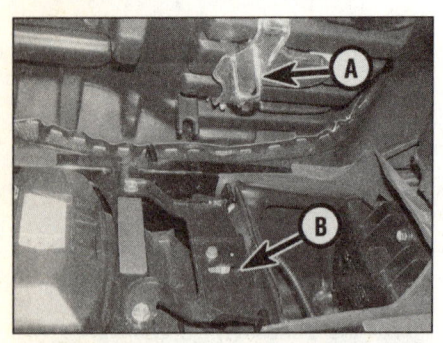

3.1 Bar (A) locates in lock mechanism (B)

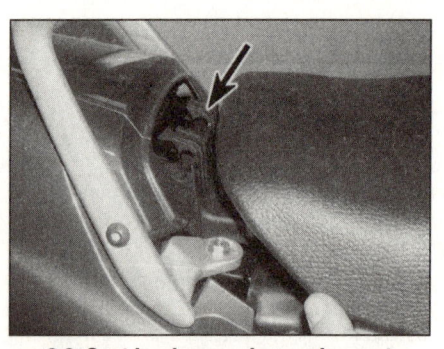

3.2 Seat hooks over bar under seat cowling at the rear

3.4 Unlock the seat and lift it

3.5a Unscrew the bolts (arrowed) . . .

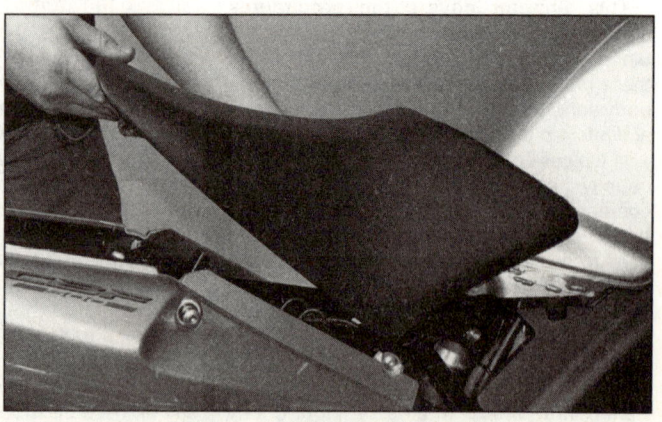

3.5b . . . and remove the seat

Bodywork 8•3

4 Seat cowling – removal and installation

CB models

Removal

1 Remove the frame side panels (see Section 2) and the seat (see Section 3).
2 Remove the two retaining screws from the underside of the seat cowling, on each side of the tail light (see illustration).
3 Unscrew the bolts securing the grab rail to the frame and remove the grab rail (see illustration).
4 Remove the two bolts and washers securing each side of the seat cowling to the shock absorber upper mountings (see illustrations). Note that on some R models, the left-hand bolt is shorter than the right-hand one.
5 Carefully pull each side of the front of the cowling out away from the frame and draw the cowling rearwards and off the motorcycle, taking care not to bend the sides excessively (see illustration). Note the bushes in the bolt holes in the sides of the cowling (see illustration).

Installation

6 Installation is the reverse of removal. Take care not to pull the sides apart excessively when locating the cowling on the frame. Ensure the bushes are in place before installing the side securing bolts and tighten the bolts to the torque setting specified at the beginning of this Chapter.

CBF models

Removal

7 Remove the seats (see Section 3).
8 Remove the blanking caps from the passenger grab-rail bolts (see illustration). Unscrew the bolts and remove the grab-rails (see illustration).
9 Unscrew the centre panel bolts and remove the panel, noting how the tab on each side at the front engages with each side section (see illustrations).

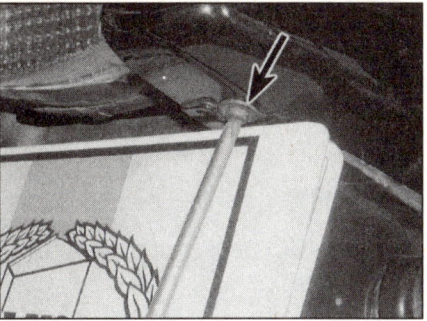

4.2 Remove the screws on each side of the tail light (arrowed)

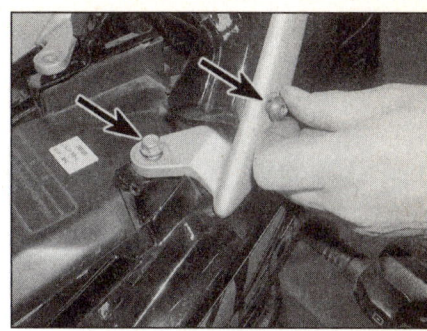

4.3 Grab rail is retained by two bolts (arrowed) on each side

4.4 Remove the bolts from the shock absorber upper mountings

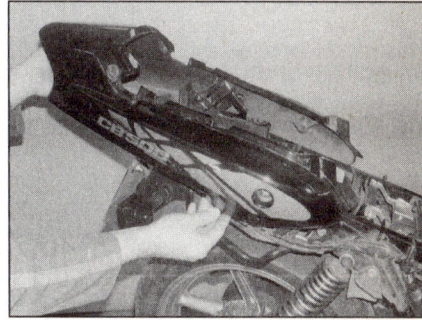

4.5a Ease the cowling rearwards off the bike . . .

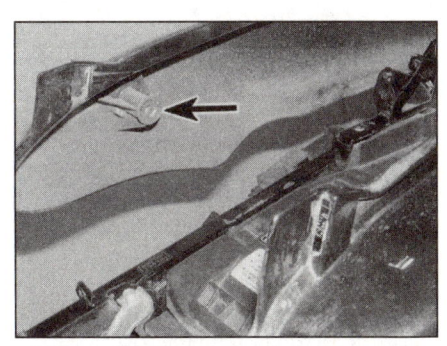

4.5b . . . taking care to retain the bushes (arrowed) in the bolt holes

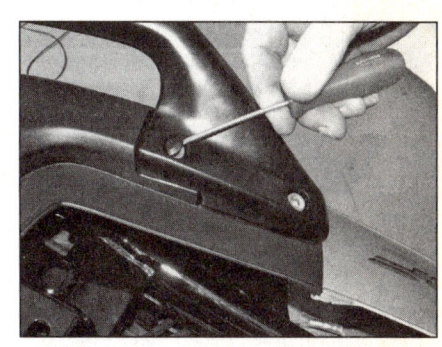

4.8a Remove the blanking caps . . .

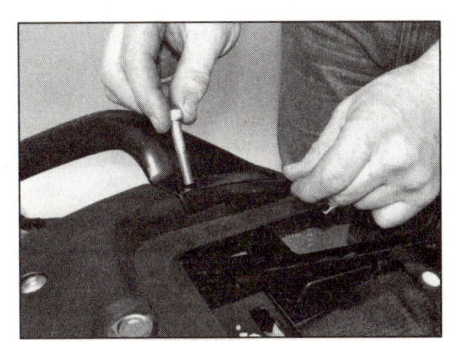

4.8b . . . then unscrew the bolts and remove the grab rails

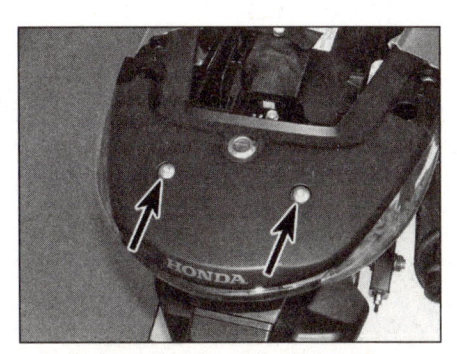

4.9a Unscrew the bolts (arrowed) . . .

4.9b . . . and remove the centre section

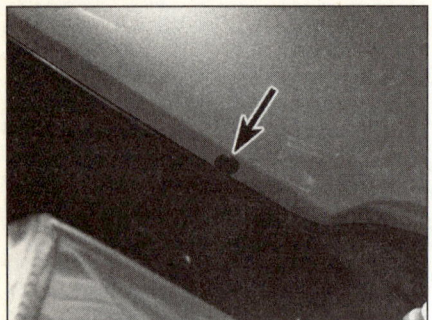

8•4 Bodywork

4.10a Release the trim clip (arrowed) . . .

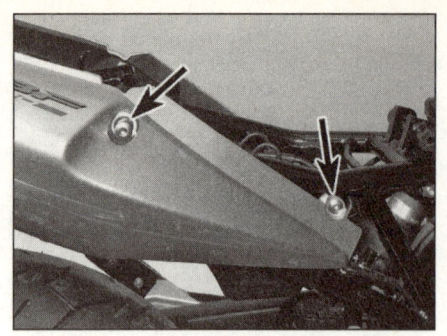

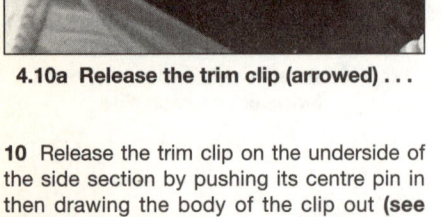

4.10b . . . then unscrew the bolts (arrowed) . . .

4.11 . . . and remove the cowling

10 Release the trim clip on the underside of the side section by pushing its centre pin in then drawing the body of the clip out **(see illustration)**. Unscrew the bolt at the front and the stud bolt in the middle, noting the collar fitted with it **(see illustration)**.

11 Release the tab at the back and lift the top off the grab-rail bolt post and remove the cowling **(see illustration)**.

Installation

12 Installation is the reverse of removal. Reset the trim clip by pushing the pin out of the body. Fit the body into the hole, then push the pin in to lock it.

5 Radiator side panels (R, T, V, W, X, Y and 2 models) – removal and installation

1 Remove the upper and lower retaining screws and their washers **(see illustration)**.
2 Pull the rear edge of the panel away from the motorcycle to release the peg on the back of the panel from the grommet on the lower edge of the fuel tank **(see illustration)**.
3 Installation is the reverse of removal.

6 Rear view mirrors – removal and installation

1 Lift the cover, where fitted, on the rear view mirror mounting lock nut, slacken the locknut and unscrew the mirror from the handlebar bracket.
2 Installation is the reverse of removal. Sit on the motorcycle and adjust the mirror position before tightening the locknut. Final adjustment can be made by tilting the mirror.

7 Fairing (SW, SX, SY and S2 models) – removal and installation

Removal

1 Disconnect the front turn signal wiring connectors on each side inside the fairing.

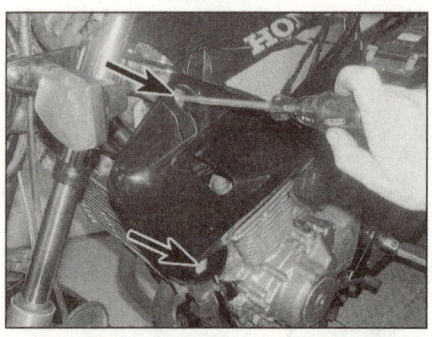

5.1 Upper and lower retaining screws (arrowed)

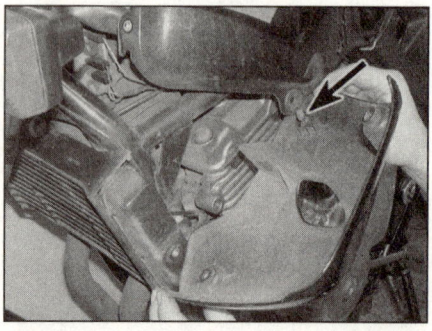

5.2 Peg (arrowed) locates in lower edge of fuel tank

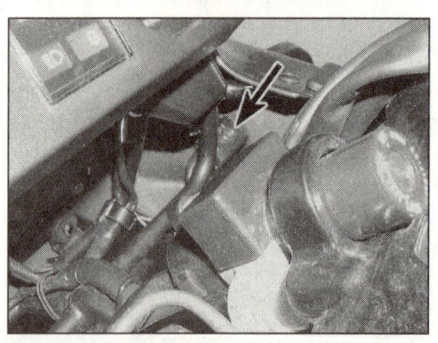

7.1a Unscrew the nut (arrowed) . . .

7.1b . . . and withdraw the turn signal through the fairing

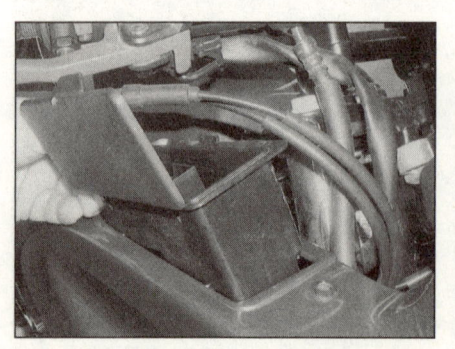

Unscrew the nuts securing the turn signal assembly brackets to the fairing frame and withdraw the signal assemblies through the fairing **(see illustrations)**.
2 Pull the storage box out of the recess in the right-hand side of the fairing **(see illustration)**.
3 Remove the screws securing the sides of the fairing to the mounting brackets **(see illustration)**. Pull the lower rear edge of the

7.2 Remove the storage box

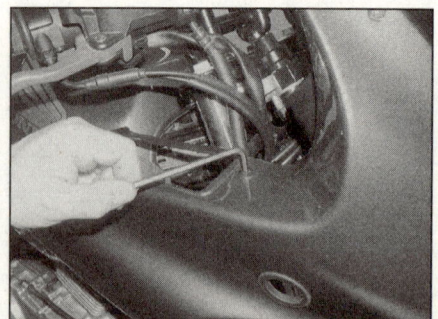

7.3a Remove the screws from the mounting bracket . . .

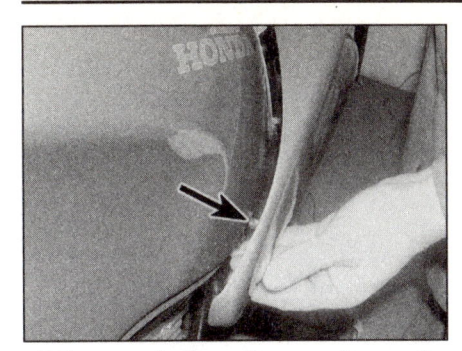

Bodywork 8•5

7.3b ... and release the pegs (arrowed) from the lower edge of the fuel tank

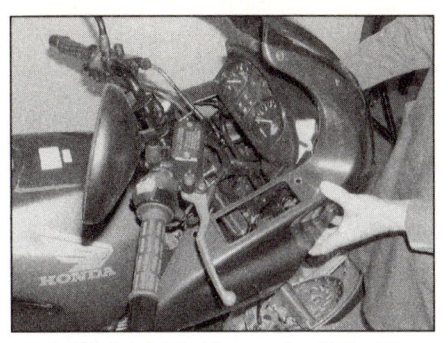

7.4 Lift the fairing forwards off the bike

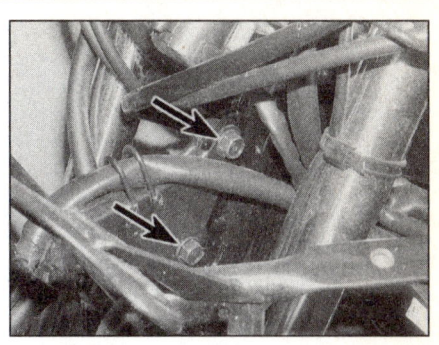

7.6 Fairing stay mounting bolts (arrowed)

fairing away from the motorcycle to release the pegs on the back of the fairing from the grommets in the lower edge of the fuel tank **(see illustration)**.

4 Disconnect the wiring connector for the sidelight and pull the headlight connector off the back of the headlight unit, then carefully pull the fairing forward and off the motorcycle **(see illustration)**.

5 The fairing is constructed in four sections; left and right-hand panels, front panel and windshield. The headlight unit is bolted onto the back of the front panel (see Chapter 9). The individual sections of the fairing can be separated for renewal or repair by unscrewing the connecting screws; the windshield screws are retained by nuts, the panel screws are retained by metal plates. Note the position of all screws on disassembly.

6 The fairing stay is mounted to the steering head by two bolts **(see illustration)**. To remove the frame, first remove the instrument cluster (see Chapter 9) and unclip the wiring loom for the instruments and the headlight. Unscrew the frame mounting bolts and remove the frame.

7 The two fairing mounting brackets on the front edge of the fuel tank can be removed by unscrewing the mounting bolts **(see illustration)**.

Installation

8 Installation is the reverse of removal, noting the following:
a) If removed, do not overtighten the fairing panel screws, especially the windshield screws.

b) Make sure the wiring looms are secured to the fairing stay.
c) When installing the fairing, ensure the two pins on the back of the headlight unit bracket locate in the grommets on the front lower edge of the fairing stay.
d) Install the glove box only after the fairing has been installed.
e) Make sure the wiring connectors are securely connected.

9 Turn the handlebars from lock to lock to ensure their movement is not restricted by the fairing.

10 Test the operation of the lights before riding the motorcycle

8 Front mudguard – removal and installation

CB models

Removal

1 Remove the front wheel (see Chapter 7) and pull the speedometer cable out of the cable guide on the front mudguard.

2 Remove the four bolts securing the mudguard to the fork sliders and remove the mudguard by lifting it up between the fork legs and withdrawing it forwards **(see illustration)**.

3 A metal plate inside the mudguard retains four mounting nuts; only remove the plate if it is damaged and must be renewed.

4 The speedometer cable guide is a push fit in the left-hand side of the mudguard **(see illustration 8.2)**.

Installation

5 Installation is the reverse of removal, noting the following:
a) Route the speedometer cable through the cable guide.
b) Secure the front brake hose clamp with the rear right-hand mudguard mounting bolt.
c) Tighten the mudguard mounting bolts to the torque setting specified at the beginning of this Chapter.

CBF models

Removal

6 Remove the front wheel (see Chapter 7).

7 Unscrew the two hex bolts, noting that they thread into plate nuts on the inside of the mudguard which could come loose and turn or drop off **(see illustration)**. On the right-hand side note how the bolt secures the brake hose, and on ABS models the wheel sensor wiring.

8 Unscrew the two Allen bolts and remove the mudguard by lifting it up between the fork legs and drawing it forwards.

Installation

9 Installation is the reverse of removal, noting the following:
a) Secure the front brake hose and on ABS models the sensor wiring with the rear right-hand bolt.
b) Tighten the bolts to the torque setting specified at the beginning of this Chapter.

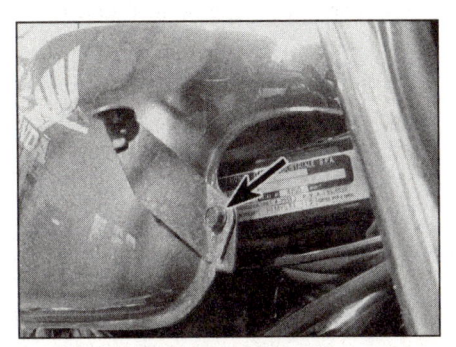

7.7 Fairing mounting bracket is retained by single bolt (arrowed)

8.2 Lift the mudguard clear of the fork sliders. Speedometer cable guide arrowed – CB models

8.7 Unscrew the hex bolt (A) on each side, then the Allen bolt (B) on each side

Notes

Chapter 9
Electrical system

Contents

Alternator – removal, inspection and installation 32
Battery – charging 4
Battery – removal, installation, inspection and maintenance 3
Brake light switches – check and replacement 11
Brake/tail light/licence plate light bulbs – test and renewal 9
Charging system – leakage and output tests 31
Charging system testing – general information and precautions 30
Clutch switch – check and replacement 22
Diode – check and replacement 23
Electrical system – fault finding 2
Fuses – check and renewal 5
General information 1
Handlebar switches – check 24
Handlebar switches – removal and installation 25
Headlight aim – check and adjustment . . . . see Chapter 1
Headlight and sidelight bulbs – test and renewal 7
Headlight assembly – removal and installation 8
Horn – check and replacement 26
Ignition (main) switch – check, removal and installation 18
Ignition system components . . . . see Chapter 5
Instrument and warning light bulbs – renewal 17
Instrument cluster and speedometer cable/speed sensor –
  removal and installation 15
Instruments – check and replacement 16
Lighting system – check 6
Neutral switch – check, removal and installation 20
Oil pressure switch – check, removal and installation 19
Regulator/rectifier – check and replacement 33
Sidestand switch – check and replacement 21
Starter motor – disassembly, inspection and reassembly 29
Starter motor – removal and installation 28
Starter relay – check and replacement 27
Tail light assembly – removal and installation 10
Turn signal bulbs – test and renewal 13
Turn signal assemblies – removal and installation 14
Turn signal circuit – check 12

Degrees of difficulty

| **Easy,** suitable for novice with little experience | **Fairly easy,** suitable for beginner with some experience | **Fairly difficult,** suitable for competent DIY mechanic | **Difficult,** suitable for experienced DIY mechanic | **Very difficult,** suitable for expert DIY or professional |

Specifications

Battery
Capacity
  CB models 12V, 8Ah
  CBF models 12V, 8.6Ah
Voltage
  Fully charged @ 20°C . . . . over 13.0V
  Discharged @ 20°C . . . . below 12.3V
Charging rate
  Normal . . . . 0.9A for 5 to 10 hrs
  Quick . . . . 4.0A for 1 hr

Charging system
Current leakage . . . . 1mA (max)
Regulated voltage output . . . . 14.0 to 15.5V @ 5000 rpm
Charging coil resistance
  CB models . . . . 0.18 to 0.20 ohm
  CBF models . . . . 0.10 to 1.0 ohm

Starter motor
Brush length
  Standard . . . . 12.0 to 13.0 mm
  Service limit (min) . . . . 8.5 mm

Fuses

CB models
Main	30A
Headlight	10A
Tail light, signal, brake light, horn	15A
Ignition, starter	10A
Fan	10A

CBF models
Main	30A
Headlight	10A
Instruments, tail light, side light, brake light, horn	10A
Ignition, starter	10A
Fan	10A
Instrument back-up	10A
Turn signal	10A
ABS system (where fitted)	
Main	10A
Pump motor	30A
Fail-safe relay	30A

Bulbs

CB models
Headlight	60/55W H4 halogen
Sidelight	4.0W
Brake/tail light	21/5W
Turn signal lights	21W
Warning lights	3.0W
Instrument lights	3.4W and 1.7W

CBF models
Headlight	60/55W H4 halogen
Sidelight	5.0W
Brake/tail light	21/5W
Licence plate light	5.0W
Turn signal lights	21W
Warning lights (HI beam, neutral, oil, fuel, ABS)	LEDs
Turn signal warning lights	1.7W
Instrument illumination	1.7W

Torque settings

Headlight mounting bolts (R, T, V, W, X, Y and 2 only)	25 Nm
Oil pressure switch	12 Nm
Ignition (main) switch bolts	
CB models	27 Nm
CBF models	25 Nm
Steering stem nut	103 Nm
Fork clamp bolts	23 Nm
Neutral switch	12 Nm
Sidestand switch bolt	10 Nm
Alternator stator bolts	12 Nm
Alternator rotor bolt	95 Nm
Engine cover bolts	12 Nm

1 General information

All models have a 12-volt electrical system charged by a three-phase alternator with a separate regulator/rectifier. The regulator maintains the charging system output within the specified range to prevent overcharging, and the rectifier converts the ac (alternating current) output of the alternator to dc (direct current) to power the lights and other components and to charge the battery. The alternator is mounted inside the left-hand engine cover

The starting system includes the starter motor, the relay, the battery and the various wires and switches. If the engine kill switch is in the RUN position and the ignition (main) switch is ON, the starting system will allow the starter motor to operate only if the transmission is in neutral (neutral light on) or, if the transmission is in gear, if the clutch lever is pulled into the handlebar and the sidestand is up. The starter motor is mounted on the engine unit behind the cylinders.

Note: *Keep in mind that electrical parts, once purchased, cannot be returned. To avoid unnecessary expense, make very sure the faulty component has been positively identified before buying a new part.*

2 Electrical system – fault finding

Warning: *To prevent the risk of short circuits, the ignition (main) switch must always be OFF and the battery negative (-ve) terminal should be disconnected before any of the bike's other electrical components are disturbed. Don't forget to reconnect the terminal securely once work is finished or if battery power is needed for circuit testing.*

1 A typical electrical circuit consists of an

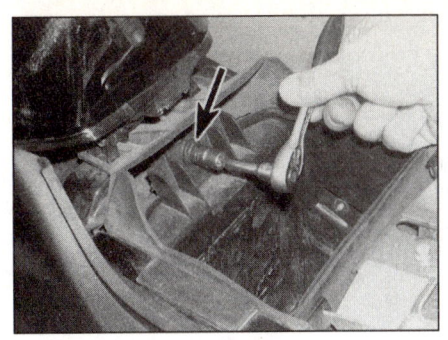

Electrical system 9•3

3.2 Battery cover is retained by a single bolt (arrowed)

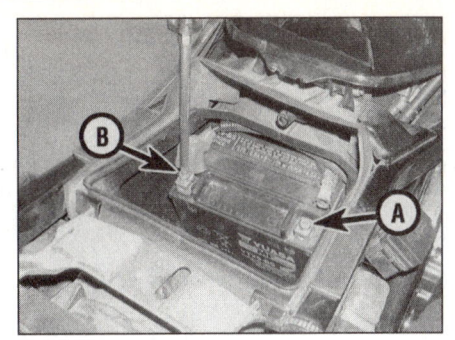

3.3 First disconnect negative (-ve) (A) then positive (+ve) (B) terminals

3 Battery – removal, installation, inspection and maintenance

Caution: *Be extremely careful when handling or working around the battery. The electrolyte is very caustic and an explosive gas (hydrogen) is given off when the battery is charging.*

Removal and installation – CB models

1 Remove the seat (see Chapter 8).
2 Unbolt the battery cover and pull the cover back to gain access to the battery **(see illustration)**.
3 First disconnect the battery negative (-ve) lead from the battery terminal, then lift the red insulating cover and disconnect the positive (+ve) lead **(see illustration)**. Lift the battery from the bike, taking care not to lose the nuts from inside the terminal blocks when the battery is moved.
4 Before installation, clean the battery terminals, terminal screws, nuts and lead ends with a wire brush, knife or wire wool to ensure a good electrical connection. Reconnect the leads, connecting the positive (+ve) lead first, and replace the red insulating cover.

electrical component, the switches, relays, etc. related to that component and the wiring and connectors that hook the component to both the battery and the frame. To aid in locating a problem in any electrical circuit, refer to the wiring diagrams at the end of this Chapter.

2 Before tackling any faulty electrical circuit, first study the wiring diagram (see end of this Chapter) thoroughly to get a complete picture of what makes up that individual circuit. Trouble spots, for instance, can often be narrowed down by noting if other components related to that circuit are operating properly or not. If several components or circuits fail at one time, chances are the fault lies in the fuse or earth (ground) connection, as several circuits are often routed through the same fuse and earth (ground) connections.

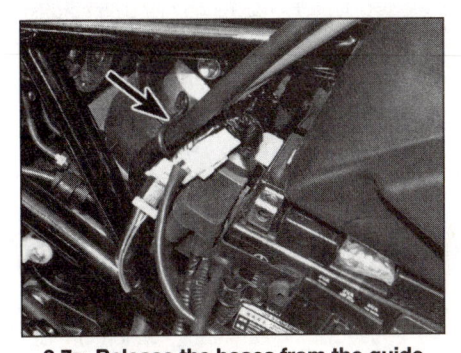

3 Electrical problems often stem from simple causes, such as loose or corroded connections or a blown fuse. Prior to any electrical fault finding, always visually check the condition of the fuse, wires and connections in the problem circuit. Intermittent failures can be especially frustrating, since you can't always duplicate the failure when it's convenient to test. In such situations, a good practice is to clean all connections in the affected circuit, whether or not they appear to be good. All of the connections and wires should also be wiggled to check for looseness which can cause intermittent failure.

4 If testing instruments are going to be utilised, use the wiring diagram to plan where you will make the necessary connections in order to accurately pinpoint the trouble spot.

5 The basic tools needed for electrical fault finding include a battery and bulb test circuit, a continuity tester, a test light, and a jumper wire. A multimeter capable of reading volts, ohms and amps is also very useful as an alternative to the above, and is necessary for performing more extensive tests and checks.

> **HAYNES HiNT** Refer to Fault Finding Equipment in the Reference section for details of how to use electrical test equipment.

> **HAYNES HiNT** Battery corrosion can be kept to a minimum by applying a layer of petroleum jelly to the terminals after the cables have been connected.

5 Replace the battery cover then install the bolt retaining the cover and tighten it securely. Install the seat (see Chapter 8).

Removal and installation – CBF models

6 Remove the right-hand frame side panel (see Chapter 8).
7 Release the hoses from the hose guide **(see illustration)**. Unscrew the battery holder bolts and displace the holder.

3.7a Release the hoses from the guide (arrowed)...

3.7b ...then unscrew the bolts (arrowed)...

3.7c ...and displace the holder

9•4 Electrical system

3.8a Disconnect the negative lead, then the positive lead ...

3.8b ... and remove the battery

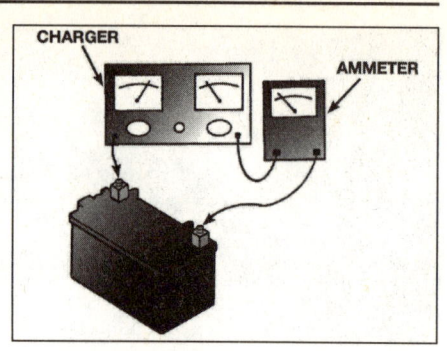

4.3 If the charger has no built-in ammeter, connect one in series as shown. DO NOT connect the ammeter between the battery terminals or it will be ruined

8 First disconnect the battery negative (-ve) lead from the battery terminal, then lift the red insulating cover and disconnect the positive (+ve) lead **(see illustration)**. Remove the battery from the bike **(see illustration)**.

9 Before installation, clean the battery terminals, terminal screws and lead ends with a wire brush, knife or wire wool to ensure a good electrical connection. Reconnect the leads, connecting the positive (+ve) lead first, and replace the red insulating cover.

10 Replace the battery holder and hoses. Install the side panel (see Chapter 8).

Inspection and maintenance

11 The battery fitted to the models covered in this manual is of the maintenance free (sealed) type and therefore does not require topping up. However, the following checks should still be regularly performed.

12 Check the battery terminals and leads for tightness and corrosion. If corrosion is evident, unscrew the terminal screws and disconnect the leads from the battery, disconnecting the negative (-ve) terminal first. Wash the terminals and lead ends in a solution of baking soda and hot water and dry them thoroughly. If necessary, further clean the terminals and lead ends with a wire brush, knife or wire wool. Reconnect the leads, connecting the positive (+ve) terminal first, and apply a thin coat of petroleum jelly to the connections to slow further corrosion.

13 The battery case should be kept clean to prevent current leakage, which can discharge the battery over a period of time (especially when it sits unused). Wash the outside of the case with a solution of baking soda and water. Rinse the battery thoroughly, then dry it.

14 Look for cracks in the case and renew the battery if any are found. If acid has been spilled on the frame or battery box, neutralise it with a baking soda and water solution, dry it thoroughly, then touch up any damaged paint.

15 If the motorcycle sits unused for long periods of time, disconnect the battery negative (-ve) terminal. Honda recommend that the battery is recharged once every two weeks when the motorcycle is not being used (see Section 4).

16 The condition of the battery can be assessed by measuring the voltage present at the battery terminals with a multimeter. Connect the multimeter positive (+ve) probe to the battery positive (+ve) terminal, and the negative (-ve) probe to the battery negative (-ve) terminal. When fully charged there should be more than 13 volts present. If the voltage falls to 12.3 volts the battery must be removed, disconnecting the negative (-ve) terminal first, and recharged (see Section 4).

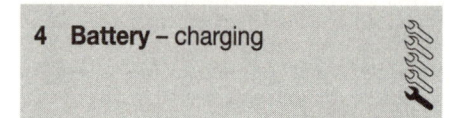

4 Battery – charging

Caution: Be extremely careful when handling or working around the battery. The electrolyte is very caustic and an explosive gas (hydrogen) is given off when the battery is charging.

1 Ensure the charger is suitable for charging a 12V battery.

2 Remove the battery from the motorcycle (see Section 3). Connect the charger to the battery **BEFORE** switching the charger **ON**. Make sure that the positive (+ve) lead on the charger is connected to the positive (+ve) terminal on the battery, and the negative (-ve) lead is connected to the negative (-ve) terminal.

3 Few owners will have access to an expensive current controlled charger, so if a normal domestic charger is used check that after a possible initial peak, the charge rate falls to a safe level **(see illustration)**. If the battery becomes hot during charging **STOP**.

Further charging will cause damage. **Note:** *In emergencies the battery can be charged at a high rate of around 4.0 amps for a period of 1 hour. However, this is not recommended and the low amp (trickle) charge is by far the safer method of charging the battery.*

4 If the recharged battery discharges rapidly when left disconnected, it is likely that an internal short caused by physical damage or sulphation has occurred and a new battery will be required. A sound battery will tend to lose its charge at about 1% per day.

5 Install the battery (see Section 3).

6 If the motorcycle sits unused for long periods of time, Honda recommend recharging the battery once every two weeks. If the battery is left on the motorcycle, disconnect the negative (-ve) lead, otherwise store the battery in a cool, dry place.

5 Fuses – check and renewal

1 The electrical system is protected by fuses of different ratings. All except the main fuse are housed in the fusebox, which on CB models is located behind the right-hand frame side panel and on CBF models is under the passenger's seat **(see illustrations)**. The main fuse is integral with the starter relay, which is to the rear of the fusebox on CB models and behind the right-hand frame side panel next to

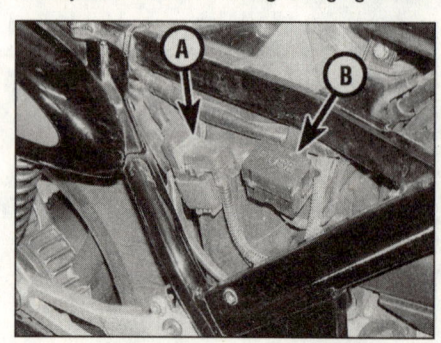

5.1a Main fuse (A), fusebox (B) – CB models

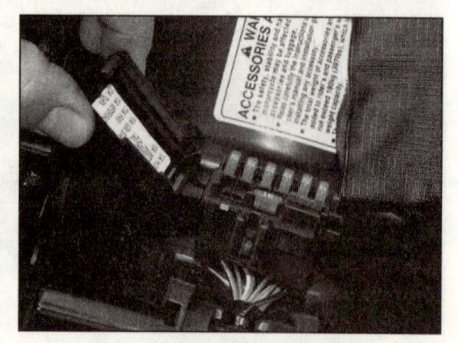

5.1b Fusebox (arrowed) ...

Electrical system 9•5

5.1c ... and main fuse in starter relay (arrowed) – CBF models

5.1d ABS main fuse (arrowed) ...

5.1e ... and other fuses (arrowed) – CBF models with ABS

the battery on CBF models **(see illustration)**. On CBF models with ABS the ABS main fuse is in its own holder, and the other ABS fuses along with a spare are in a separate holder, both behind the right-hand frame side panel next to the battery **(see illustrations)**.

2 To access the fuses, remove the right-hand side panel or passenger seat according to model (see Chapter 8) and unclip the fusebox lid **(see illustration)**. To access the main fuse, remove the side panel (see Chapter 8) and disconnect the red starter relay wiring connector **(see illustration)**.

3 The fuses can be removed and checked visually. If you can't pull the fuse out with your fingertips, use a pair of suitable pliers. A blown fuse is easily identified by a break in the element **(see illustration)**. Each fuse is clearly marked with its rating and must only be replaced by a fuse of the correct rating. A spare fuse of each rating is housed in the fusebox, and a spare main fuse is housed in the bottom of the starter relay. If a spare fuse is used, always renew it so that a spare of each rating is carried on the motorcycle at all times.

⚠ **Warning: Never put in a fuse of a higher rating or bridge the terminals with any other substitute, however temporary it may be. Serious damage may be done to the circuit, or a fire may start.**

4 If a fuse blows, be sure to check the wiring circuit very carefully for evidence of a short-circuit. Look for bare wires and chafed, melted or burned insulation. If the fuse is renewed before the cause is located, the new fuse will blow immediately.

5 Occasionally a fuse will blow or cause an open-circuit for no obvious reason. Corrosion of the fuse ends and fusebox terminals may occur and cause poor fuse contact. If this happens, remove the corrosion with a wire brush or wire wool, then spray the fuse end and terminals with electrical contact cleaner.

6 Lighting system – check

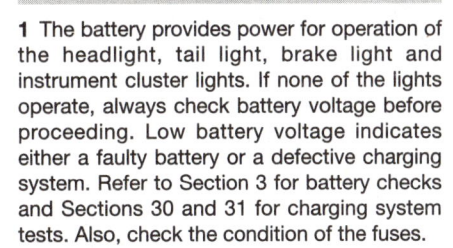

1 The battery provides power for operation of the headlight, tail light, brake light and instrument cluster lights. If none of the lights operate, always check battery voltage before proceeding. Low battery voltage indicates either a faulty battery or a defective charging system. Refer to Section 3 for battery checks and Sections 30 and 31 for charging system tests. Also, check the condition of the fuses.

Headlight

2 If the headlight fails to work, first check the bulb and bulb terminals (see Section 7) and then the headlight fuse (see Section 5). If they are both good, the problem lies in the wiring or one of the switches in the circuit. Refer to Section 24 for the switch testing procedures, and also the wiring diagrams at the end of this Chapter.

3 German market models and some Northern European market models are fitted with headlight relays for HI (blue wiring) and LO (white wiring) beams (see wiring diagrams). If there appears to be a fault with the headlight but the headlight beams come on when the bulb is connected directly to the battery with jumper wires, the problem could be in one of the relays. To gain access to the relays, first remove the fuel tank (see Chapter 4). The relays are located on brackets on either side of the frame above the carburettors; the LO beam relay is on the left-hand side, the HI beam relay is on the right-hand side.

4 If a relay is suspected of being faulty, the easiest way to determine this is to substitute it with another relay by simply swapping the relays. If the beam in question then works, the faulty relay must be renewed.

5 To test the HI beam relay if another relay is not available for testing purposes, disconnect the relay wiring connector then switch the ignition ON and the light switch ON and turn the HI/LO beam switch to HI. Check for voltage at the wiring connector blue/red terminal. If there is voltage at the terminal, check the wiring between the relay and the headlight. If the wiring is good, renew the relay. If there is no voltage, check the wiring between the connector and the fusebox (see wiring diagrams at the end of this Chapter). Turn the ignition OFF when the check is complete.

6 To test the LO beam relay, disconnect the relay wiring connector then switch the ignition ON and the light switch ON and turn the HI/LO beam switch to LO. Check for voltage at the wiring connector white/red terminal. If

5.2a Unclip fusebox lid to access fuses

5.2b Unclip relay wiring connector to access main fuse

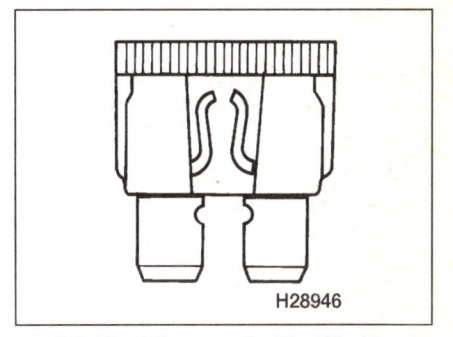

5.3 A blown fuse can be identified by a break in its element

9•6 Electrical system

7.1a Remove the headlight rim screws (arrowed)...

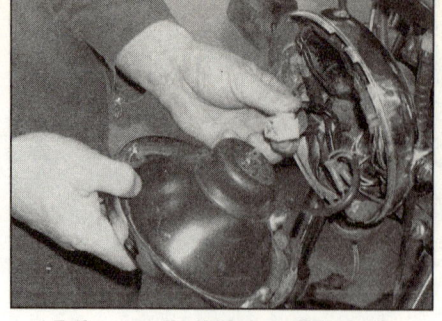

7.1b ...pull off the headlight bulb connector...

7.1c ...and pull the sidelight bulb out of its socket

there is voltage at the terminal, check the wiring between the relay and the headlight. If the wiring is good, renew the relay. If there is no voltage, check the wiring between the connector and the fusebox (see wiring diagrams at the end of this Chapter). Turn the ignition OFF when the check is complete.

Tail light

7 If the tail light fails to work, first check the bulb and the bulb terminals (see Section 9), then the fuse (see Section 5), then check for voltage at the brown terminal on the supply side of the tail light wiring connector. If voltage is present, check the earth (ground) circuit for an open or poor connection.

8 If no voltage is indicated, check the wiring between the tail light, ignition switch and fuse. Also check the lighting switch (see Section 24).

Brake light

9 If the brake light fails to work, first check the bulbs and the bulb terminals (see Section 9), then the fuse (see Section 5). Check for voltage at the green/yellow terminal on the supply side of the tail light wiring connector, with the brake lever pulled in or the brake pedal depressed. If voltage is present, check the earth (ground) circuit for an open or poor connection.

10 If no voltage is indicated, check the wiring between the brake light and the switches, then check the brake light switches (see Section 11).

7 Headlight and sidelight bulbs – test and renewal

Note: *The headlight bulb is of the quartz-halogen type. Do not touch the bulb glass as skin acids will shorten the bulb's service life. If the bulb is accidentally touched, it should be wiped carefully when cold with a rag soaked in methylated spirit and dried before fitting. Use a paper towel or dry cloth when handling the bulb.*

> **Warning:** Allow the bulb time to cool before removing it if the headlight has just been on.

Headlight bulb – CB500-R, T, V, W, X, Y and 2 models, and CBF models

1 Unscrew the three headlight rim retaining screws and remove the screws from the headlight shell. Pull the rim and headlight out of the shell, disconnect the wiring connector from the back of the headlight bulb and pull the sidelight bulbholder out of its socket in the headlight **(see illustrations)**.

2 To test the bulb, first ensure that the bulb terminals are clean and free of corrosion. Clean the terminals with a knife or wire wool, then use jumper wires to connect the bulb earth (ground) terminal to the battery negative (-ve) terminal, then alternately connect the bulb HI and LO beam terminals to the battery positive (+ve) terminal. If either of the bulb elements fails to illuminate, renew the bulb.

3 Remove the rubber cover from the back of the headlight bulb, noting how it fits **(see illustration)**.

4 Release the bulb retaining clip, noting how it fits, then remove the bulb **(see illustrations)**.

5 Fit the new bulb, bearing in mind the information in the **Note** above. Make sure the tabs on the bulb fit correctly in the slots in the bulb housing, and secure the bulb in position with the retaining clip.

6 Install the rubber cover, making sure it is correctly seated and with the 'TOP' mark at the top. Check that the contacts inside the wiring connector are clean and free from corrosion, then connect the wiring connector and press the sidelight bulbholder into its socket.

7 Install the rim and headlight in the headlight shell, install the three retaining screws and tighten the screws securely.

8 Check the operation of the headlight.

> **HAYNES HINT:** *Always use a paper towel or dry cloth when handling new bulbs to prevent injury if the bulb should break and to increase bulb life.*

Sidelight bulb – CB500-R, T, V, W, X, Y and 2 models, and CBF models

9 Remove the rim and headlight from its shell (see Step 1).

7.3 Remove the rubber cover

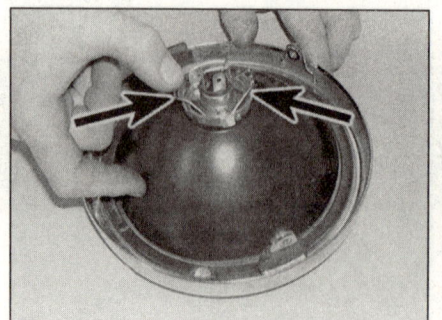

7.4a Release the retaining clips (arrowed)...

7.4b ...and withdraw the bulb

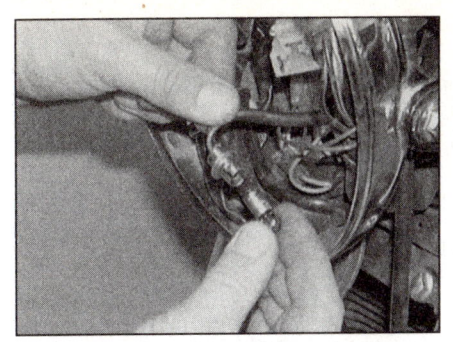

Electrical system 9•7

7.10 Twist anticlockwise and pull to remove the sidelight bulb

7.17 Remove the rubber cover, noting the vent (arrowed)

7.18a Twist the retaining ring anti-clockwise and remove

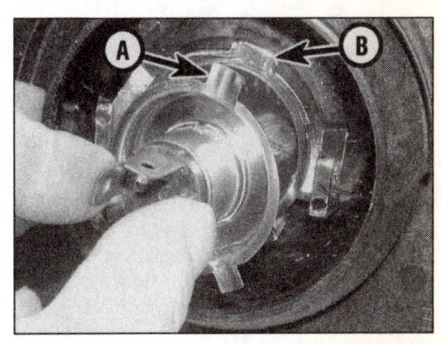

7.18b Tab (A) on bulb locates in recess (B)

10 Carefully twist anticlockwise and pull the sidelight bulb out of the bulbholder **(see illustration)**.
11 To test the bulb, first ensure that the metal bulb body and the bottom terminal are clean and free of corrosion; clean them with a knife or wire wool. Use jumper wires to connect the bulb body to the battery negative (-ve) terminal and the bulb bottom terminal to the battery positive (+ve) terminal. If the bulb fails to illuminate, renew the bulb.
12 Check that the contacts inside the bulbholder are clean and free from corrosion. Install the new bulb in the bulbholder, then press the bulbholder into its socket.
13 Reconnect the headlight bulb and install the rim and headlight (see Step 7).
14 Check the operation of the sidelight.

Headlight bulb – CB500-SW, SX, SY and S2 models

15 Disconnect the wiring connector from the back of the headlight bulb.
16 To test the bulb see Step 2.
17 Remove the rubber cover from the back of the headlight unit, noting how it fits **(see illustration)**.
18 Turn the bulb retaining ring anti-clockwise, remove the ring, then remove the bulb **(see illustrations)**.
19 Fit the new bulb, bearing in mind the information in the **Note** above. Make sure the tabs on the bulb fit correctly in the slots in the bulb housing, and secure the bulb in position with the retaining ring **(see illustrations 7.18a and b)**.
20 Install the rubber cover, making sure it is correctly seated and with the vent at the bottom **(see illustration 7.17)**.
21 Check that the contacts inside the wiring connector are clean and free from corrosion, then connect the wiring connector to the back of the headlight bulb.
22 Check the operation of the headlight.

HAYNES HiNT *Always use a paper towel or dry cloth when handling new bulbs to prevent injury if the bulb should break and to increase bulb life.*

Sidelight bulb – CB500-SW, SX, SY and S2 models

23 Remove the rubber plug from the underside of the fairing to gain access to the sidelight **(see illustration)**.
24 Pull the sidelight bulbholder out of its socket in the underside of the headlight and pull the sidelight bulb out of the bulbholder **(see illustration)**.
25 To test the bulb, use jumper wires to connect one of the bulb terminal wires to the battery positive (+ve) terminal and the other bulb terminal wire to the battery negative (-ve) terminal. If the bulb fails to illuminate, renew the bulb.

26 Check that the contacts inside the bulbholder are clean and free from corrosion. Install the new bulb in the bulbholder, then press the bulbholder into its socket and install the rubber plug in the fairing.
27 Check the operation of the sidelight.

8 Headlight assembly – removal and installation

CB500-R, T, V, W, X, Y and 2 models, and CBF models

1 Unscrew the three headlight rim retaining screws and remove the screws from the headlight shell **(see illustration 7.1a)**. Pull the

7.23 Remove plug to access the sidelight

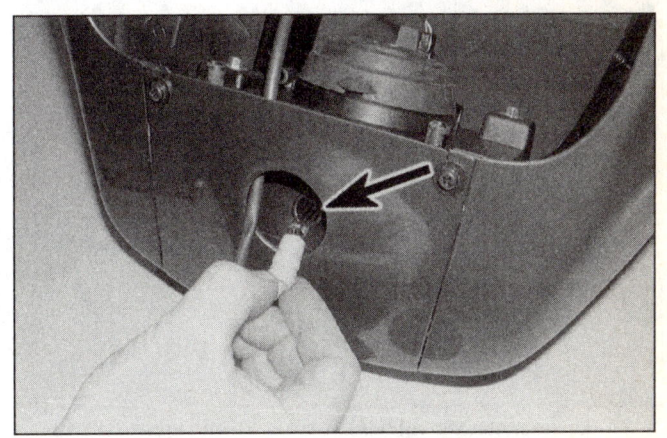

7.24 Bulbholder is a push fit in the headlight shell

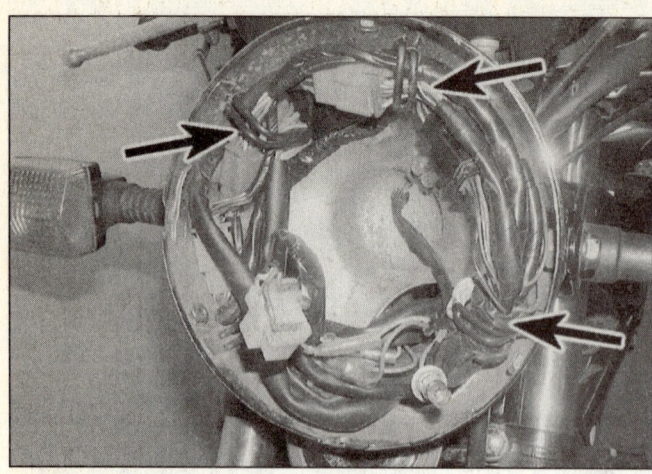

8.2 Note the arrangement of the wiring and the clamps (arrowed)

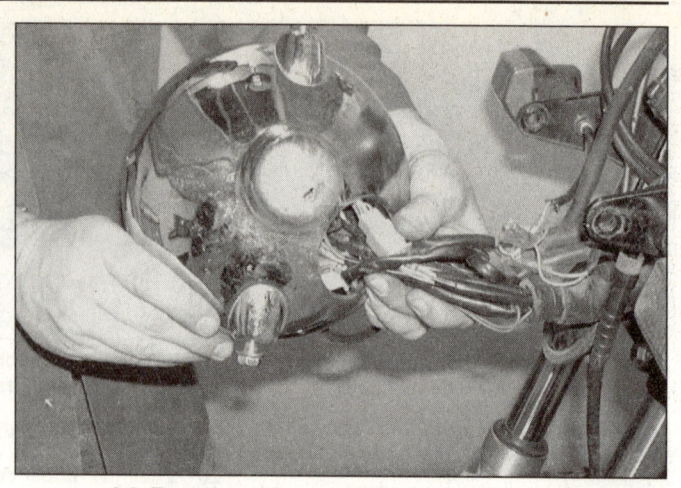

8.3 Ease the wiring out of the back of the shell

rim and headlight out of the shell, disconnect the wiring connector from the back of the headlight bulb and pull the sidelight bulbholder free **(see illustrations 7.1b and c)**.

2 To remove the headlight shell, first note the arrangement of the wiring inside the shell, then free the wiring from any clamps and disconnect any wiring connectors as necessary **(see illustration)**. Label connectors, where necessary, to aid correct reconnection.

3 Unscrew the bolts securing the shell to the headlight brackets, noting the alignment marks on the shell and the brackets, and remove the shell, carefully easing the wiring out of the back of the shell **(see illustration)**.

4 The headlight brackets are held in place between the top and bottom fork yokes and can only be removed by displacing the top yoke (see Chapter 6, Section 8).

5 Installation is the reverse of removal. Make that all the wiring is correctly connected and secured and that the marks on the shell and the brackets align **(see illustration)**. Check the operation of the headlight and sidelight. Check the headlight aim (see Chapter 1).

CB500-SW, SX, SY and S2 models

6 Remove the fairing (see Chapter 8).
7 The headlight is secured to the fairing by three bolts **(see illustration)**. Unscrew the two lower bolts and remove the bolts and the bracket, noting how it fits. Unscrew the top bolt and remove the bolt and the spacer between the headlight unit and the fairing. Remove the headlight.

8 Installation is the reverse of removal. Make sure the spacer is installed between the headlight unit and the fairing on the top mounting bolt **(see illustration)** and that the bracket retained by the lower headlight mounting bolts is installed flat edge uppermost. Check the operation of the headlight and sidelight. Check the headlight aim (see Chapter 1).

9 Brake/tail light/licence plate light bulbs – test and renewal

Brake and tail light bulb – all models

1 Remove the two screws securing the tail

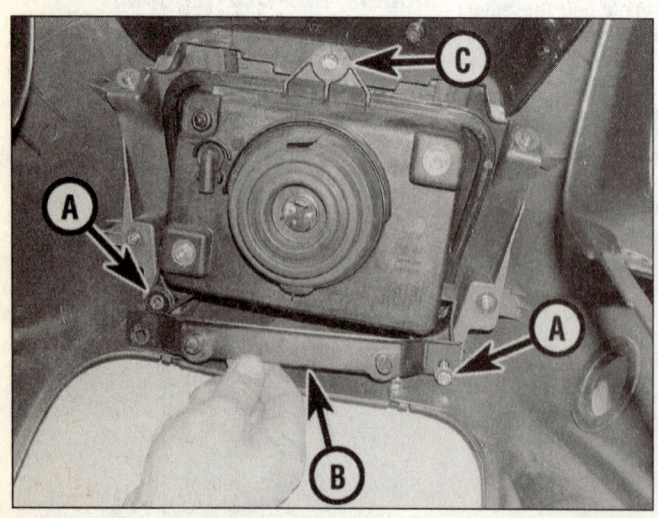

8.5 Align the marks (arrowed) on the headlight shell and brackets

8.7 Lower mounting bolts (A), bracket (B) and upper bolt (C)

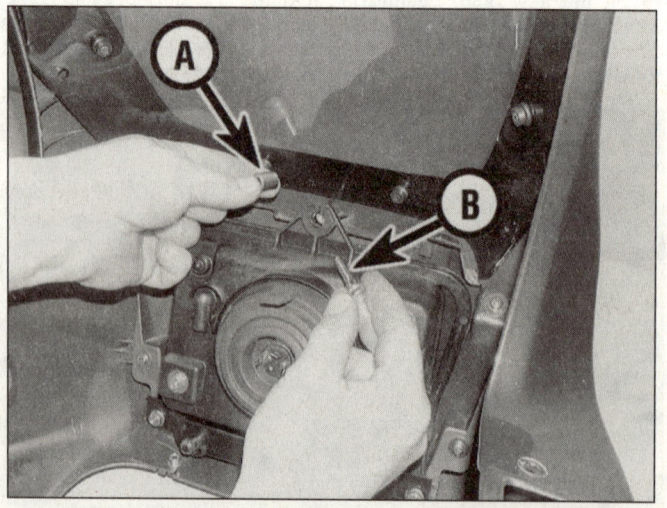

8.8 Don't forget the spacer (A) when installing the upper mounting bolt (B)

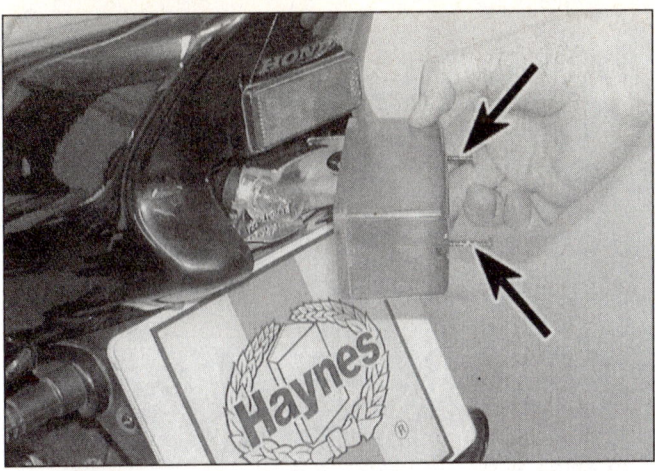

Electrical system 9•9

9.1a Tail light lens is retained by screws (arrowed) – CB models

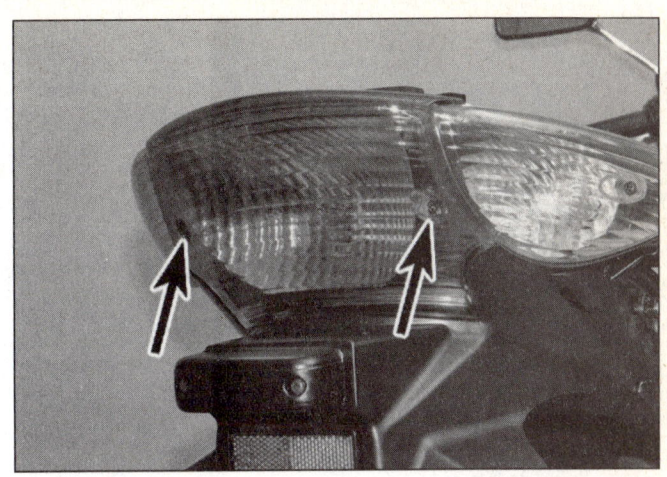

9.1b Tail light lens screws (arrowed) – CBF models

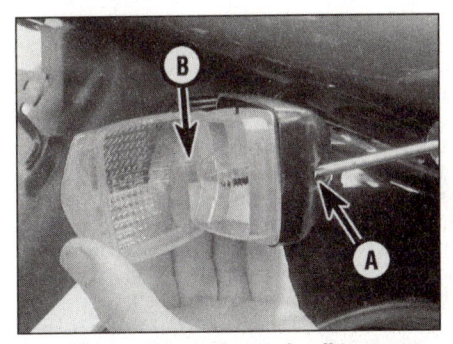

9.2 Twist anticlockwise and pull to remove the bulb

9.7 Undo the screws and remove the lens . . .

9.8 . . . then pull the bulb out

light lens and remove the lens, noting how it fits **(see illustrations)**.

2 Push the bulb into the socket and twist it anti-clockwise to remove it **(see illustration)**.

3 To test the bulb, first ensure that the metal bulb body and the two bottom terminals are clean and free of corrosion; clean them with a knife or wire wool. Use jumper wires to connect the bulb body to the battery negative (-ve) terminal and then alternately connect the bottom terminals (brake light and tail light) to the battery positive (+ve) terminal. If either of the bulb elements fails to illuminate, renew the bulb.

4 Check that the contacts inside the socket are clean and free from corrosion. Line up the pins on the bulb body with the slots in the socket, then push the bulb in and turn it clockwise until it locks into place. **Note:** *The pins on the bulb are offset so it can only be installed one way. It is a good idea to use a paper towel or dry cloth when handling the new bulb to prevent injury if the bulb should break and to increase bulb life.*

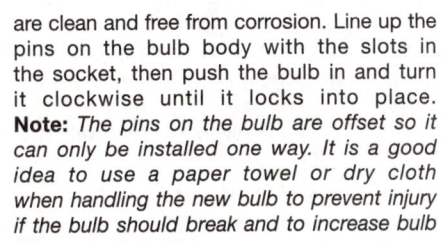

5 Check the condition of the lens sealing gasket and renew it if it is damaged, then install the lens and tighten the screws securely.

6 Check the operation of the brake/tail light.

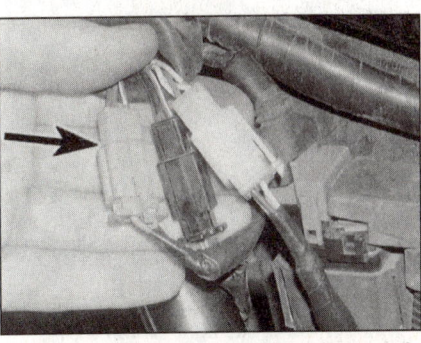

Licence plate light bulb – CBF models

7 Undo the two screws and remove the lens **(see illustration)**.

8 Pull the bulb out of the socket and replace it with a new one **(see illustration)**.

9 Check the condition of the lens sealing gasket and renew it if it is damaged, then install the lens and tighten the screws securely.

10 Check the operation of the light.

10 Tail light assembly – removal and installation

CB models
Removal

1 Remove the frame side panels and the seat cowling (see Chapter 8).

2 Draw back the rubber boot on the right-hand side of the frame and disconnect the green wiring connector **(see illustration)** Free the tail light wiring from the clips which secure it to the frame.

3 Unscrew the two nuts and bolts securing the tail light assembly to the rear mudguard, remove the nuts, bolts and spacers, noting how they fit through the bushes in the mudguard assembly, and remove the tail light **(see illustration)**.

10.2 Tail light wiring connector (arrowed)

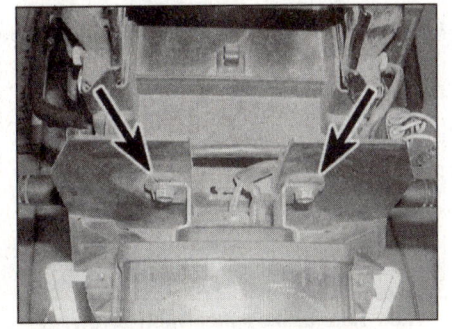

10.3 Tail light assembly is retained by nuts, bolts and spacers (arrowed)

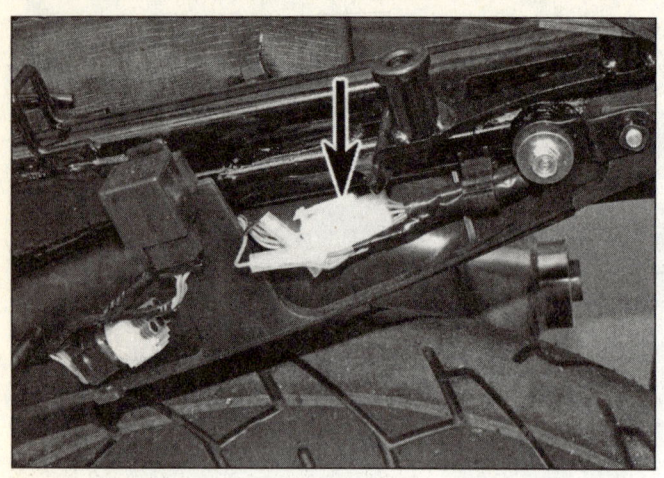

9•10 Electrical system

10.8 Disconnect the wiring connector (arrowed) . . .

10.9 . . . then undo the screws and bolts

4 If necessary, unscrew the bolts securing the mounting brackets to the back of the tail light assembly and remove the brackets.

Installation

5 Installation is the reverse of removal. Ensure that the spacers for the tail light assembly bolts (see Step 3) are installed from the forward facing side of the mudguard assembly.

6 Check the operation of the tail light and the brake light.

CBF models

7 Remove the seat cowling (see Chapter 8).
8 Disconnect the tail light and turn signal wiring connector **(see illustration)**. Free the wiring from the clip.
9 Undo the screws and bolts securing the tail light assembly and remove the tail light **(see illustration)**.
10 Installation is the reverse of removal.
11 Check the operation of the tail light and the brake light.

11 Brake light switches – check and replacement

Circuit check

1 Before checking any electrical circuit, check the bulb (see Section 9) and fuse which feeds the tail light, signal, brake light and horn circuit (see Section 5).
2 Using a multimeter or test light connected to a good earth (ground), with the ignition ON check for voltage at the black/brown wire terminal on the brake light switch wiring connector with the lever/pedal at rest – see Step 8 for access to the rear switch connector. If there's no voltage present, check the wiring between the switch and the fuse (see wiring diagrams at the end of this Chapter).
3 If there is voltage at the black/brown wire terminal of the switch, touch the test probe to the green/yellow wire terminal, then pull the brake lever in or depress the brake pedal. If no reading is obtained or the test light doesn't light up, renew the switch.
4 If a reading is obtained or the test light does light up, yet the bulb still does not come on, check the wiring between the switch and the brake light bulb (see the wiring diagrams at the end of this Chapter).

Switch replacement

Front brake lever switch

5 The switch is mounted on the underside of the brake master cylinder. Disconnect the wiring connectors from the switch **(see illustration 5.4 in Chapter 7)**.
6 Remove the single screw securing the switch to the bottom of the master cylinder and remove the switch.
7 Installation is the reverse of removal. The switch isn't adjustable.

Rear brake pedal switch

8 The switch is mounted on the inside of the right-hand footrest bracket **(see illustration)**. On CB models remove the right-hand frame side panel (see Chapter 8). On CBF models remove the left-hand frame side panel (see Chapter 8). Trace the wiring from the switch and disconnect it at the black (CB) or white (CBF) 2-pin connector **(see illustration)**.
9 Detach the lower end of the switch spring from the brake pedal. On CB models unscrew and remove the switch. On CBF models lift the switch out of its holder.
10 Installation is the reverse of removal. Make sure the brake light is activated just before the rear brake pedal takes effect. If adjustment is necessary, hold the switch and turn the adjusting ring on the switch body until the brake light is activated as required.

12 Turn signal circuit – check

1 The battery provides power for operation of the turn signal lights, so if they do not operate, always check the battery voltage first. Low battery voltage indicates either a faulty battery or a defective charging system. Refer to Section 3 for battery checks and Sections 30 and 31 for charging system tests. Also, check the fuse (see Section 5) and the switch (see Section 24).
2 Most turn signal problems are the result of a failed bulb or corroded socket. This is especially true when the turn signals function

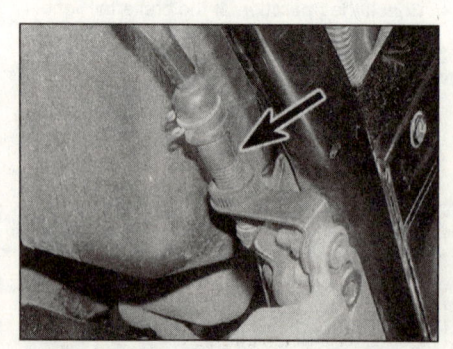

11.8a Rear brake light switch (arrowed)

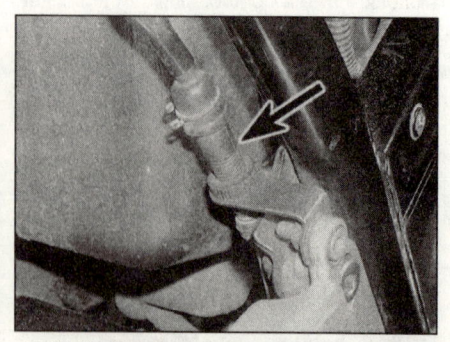

11.8b Rear brake light switch wiring connector is one of many in the boot – CBF models

Electrical system 9•11

12.3a Turn signal relay (arrowed)

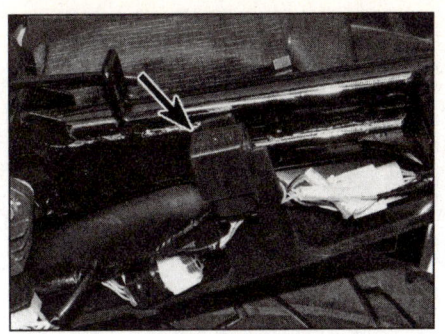

12.3b Turn signal relay (arrowed) – CBF models

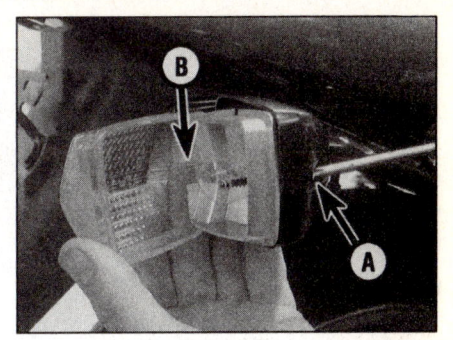

13.1 Signal lens is retained by screw (A) and tab (B)

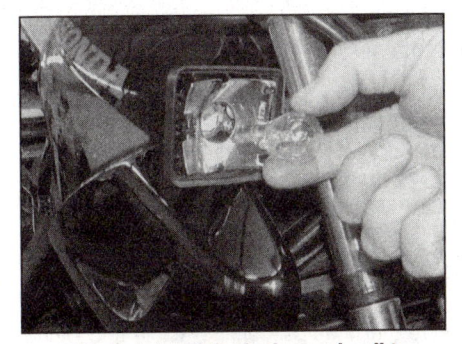

13.2 Twist anticlockwise and pull to remove the bulb

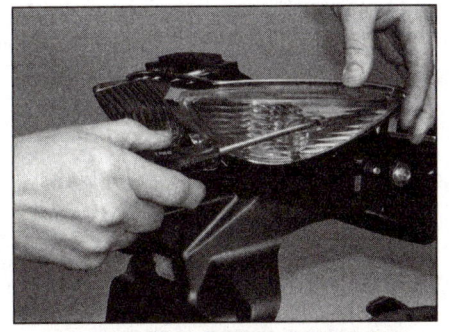

13.8 Undo the screw and remove the lens . . .

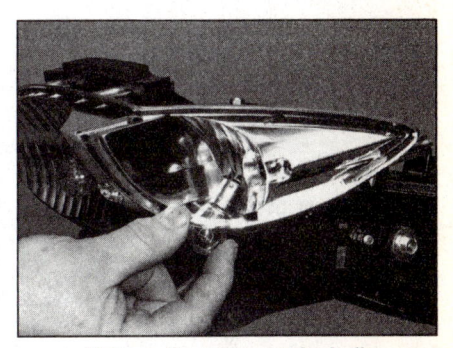

13.9 . . . then remove the bulb

properly in one direction, but not in the other. Check the bulbs and the sockets (see Section 13).
3 If the bulbs and sockets are good, on CB models remove the seat cowling and on CBF models remove the passenger's seat, and for better access if required the left-hand section of the seat cowling (see Chapter 8) for access to the turn signal relay which is mounted on the right-hand side of the frame on CB models and the left on CBF models **(see illustrations)**. Disconnect the wiring connector then switch the ignition ON and check for voltage at the wiring connector black/brown (CB models) or white/green (CBF models) terminal.
4 If there is voltage at the terminal, check the grey wire between the relay and the turn signal switch for continuity and the green earth wire for continuity. If the wiring is good, renew the relay.
5 If there is no voltage at the relay black/brown wire terminal, check the wiring between the connector and the fuse (see wiring diagrams at the end of this Chapter).

13 Turn signal bulbs – test and renewal

Front (all models) and rear (CB models)

1 Remove the screw on the back of the turn signal body securing the signal lens and remove the lens, noting how it fits **(see illustration)**.

2 Push the bulb into the socket and twist it anti-clockwise to remove it **(see illustration)**.
3 To test the bulb, first ensure that the metal bulb body and the bottom terminal is clean and free of corrosion; clean them with a knife or wire wool. Use jumper wires to connect the bulb body to the battery negative (-ve) terminal and the bottom terminal to the battery positive (+ve) terminal. If the bulb fails to illuminate, renew the bulb.
4 Check that the contacts inside the socket are clean and free from corrosion. Line up the pins on the bulb body with the slots in the socket, then push the bulb in and turn it clockwise until it locks into place. **Note:** *It is a good idea to use a paper towel or dry cloth when handling the new bulb to prevent injury if the bulb should break and to increase bulb life.*
5 Check the condition of the lens sealing gasket (where fitted) and renew it if it is damaged. Install the lens, locating the tab on the side of the lens in the turn signal body, and tighten the screw securely.
6 Check the operation of the turn signal light.

Rear (CBF models)

7 Undo the tail light lens screws and remove the lens **(see illustration 9.1b)**.
8 Undo the turn signal lens screw and remove the lens **(see illustration)**.
9 Push the bulb into the socket and twist it anti-clockwise to remove it **(see illustration)**.
10 Test the bulb as in Step 3

11 Check that the contacts inside the socket are clean and free from corrosion. Line up the pins on the bulb body with the slots in the socket, then push the bulb in and turn it clockwise until it locks into place.
12 Install the turn signal lens then the tail light lens. Do not overtighten the screws.
13 Check the operation of the turn signal light.

14 Turn signal assemblies – removal and installation

Front

1 Disconnect the turn signal wiring connectors **(see illustration)** – on CB500-R, T, V, W, X, Y and 2 models and CBF models they are inside the headlight shell (see Section 8). Ease the

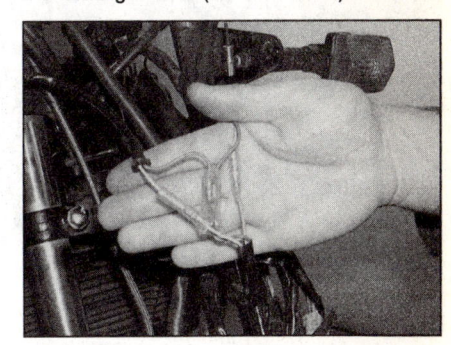

14.1 Wiring connectors for the front turn signals

14.5 Wiring connectors for the rear turn signals

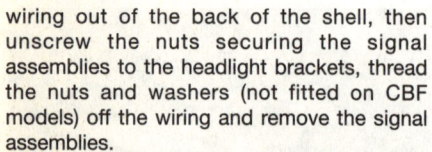

15.3 Unscrew the ring (arrowed) and detach the cable

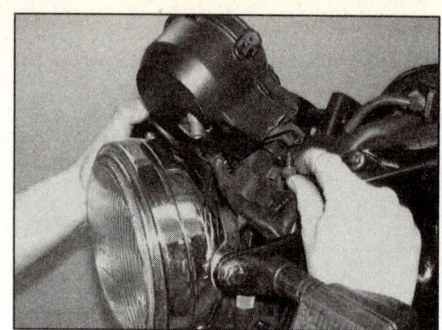

15.4 Detach the instrument cluster from the top yoke

wiring out of the back of the shell, then unscrew the nuts securing the signal assemblies to the headlight brackets, thread the nuts and washers (not fitted on CBF models) off the wiring and remove the signal assemblies.

2 On SW, SX, SY and S2 models the wiring connectors are inside the fairing. Disconnect the connectors, then unscrew the nuts securing the signal assembly brackets to the fairing stay and withdraw the signal assemblies through the fairing **(see illustrations 7.1a and b in Chapter 8)**.

3 Installation is the reverse of removal. On all models, ensure that the wiring connections are firm and that the wiring is securely clamped in place.

4 Check the operation of the turn signals.

Rear – CB models

5 Remove the seat cowling (see Chapter 8) and locate the wiring connectors inside a rubber boot secured to the right-hand side of the frame **(see illustration)**. Disconnect the connectors, then unscrew the nuts securing the signal assemblies to the frame, thread the nuts and washers off the wiring and remove the signal assemblies.

6 Installation is the reverse of removal. Ensure that the wiring connections are firm and protected by the rubber boot, and that the wiring is securely clamped in place.

7 Check the operation of the turn signals.

Rear – CBF models

8 The turn signal is part of the tail light – see Section 10.

15 Instrument cluster and speedometer cable/speed sensor – removal and installation

Instrument cluster – CB500-R, T, V, W, X, Y and 2 models

1 Remove the headlight from its shell (see Section 7)

2 Trace the wiring from the instrument cluster to the white 6-pin and red 6-pin connectors inside the headlight shell and disconnect the connectors. Note the position of the wiring

inside the shell then free the instrument wiring from the clamps and carefully ease it out of the back of the shell.

3 Unscrew the knurled ring securing the speedometer cable to the back of the speedometer and detach the cable **(see illustration)**.

4 Unscrew the two bolts securing the instrument cluster to the underside of the top yoke and carefully remove it **(see illustration)**.

5 Installation is the reverse of removal. Make sure that the speedometer cable and wiring are correctly routed and secured.

Instrument cluster – CB500-SW, SX, SY and S2 models

6 Remove the fairing (see Chapter 8).

7 Unscrew the knurled ring securing the speedometer cable to the back of the speedometer and detach the cable **(see illustration)**.

8 Pull back the rubber boot on the instrument wiring and disconnect the red and black wiring connectors and the earth (ground) wire **(see illustration)**.

9 Unscrew the three nuts securing the instrument cluster to the mounting bracket and carefully remove it **(see illustration)**.

10 Installation is the reverse of removal. Make sure that the speedometer cable and wiring are correctly routed and secured.

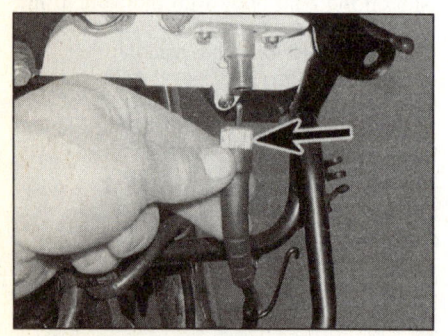

15.7 Unscrew the ring (arrowed) and detach the cable

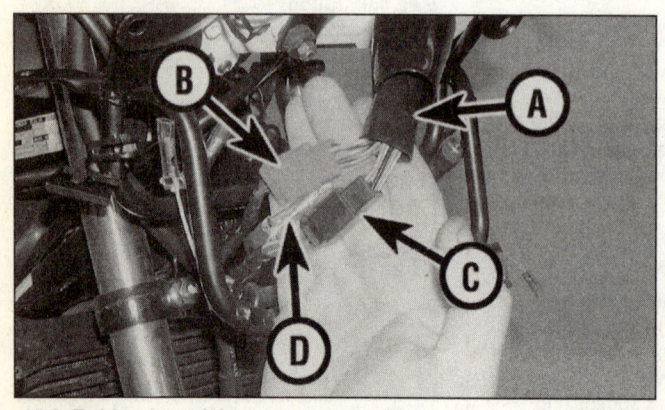

15.8 Rubber boot (A), red connector (B), black connector (C) and earth (ground) (D)

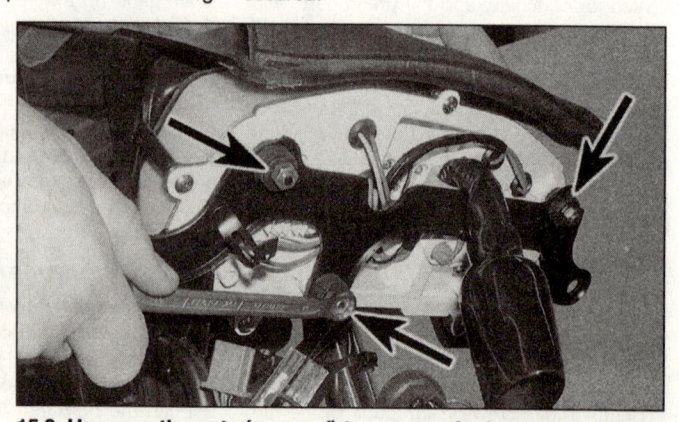

15.9 Unscrew the nuts (arrowed) to remove the instrument cluster

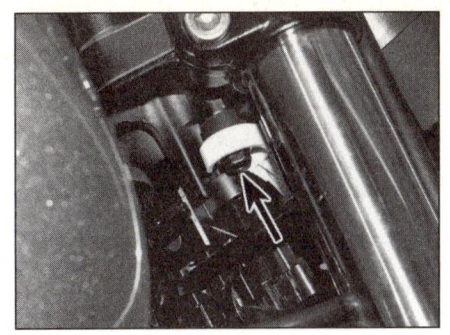

Electrical system 9•13

15.13a Unscrew the bolt (arrowed) on each side . . .

15.13b . . . and the two nuts (arrowed)

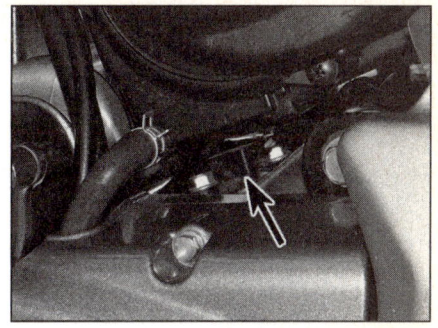

15.24 Speed sensor (arrowed)

Instrument cluster – CBF models

11 Remove the headlight from its shell (see Section 7). Trace the wiring from the instrument cluster to the white 6-pin and black 9-pin connectors inside the shell and disconnect them. Note the position of the wiring inside the shell then free it from the clamps and carefully ease it out of the back.
12 Displace the handlebars (see Chapter 6).
13 Unscrew the two bolts securing the instrument cluster to the underside of the top yoke, noting the collars **(see illustration)**. Unscrew the two nuts securing the instrument cluster to the bracket **(see illustration)**.
14 Slacken the fork clamp bolts in the top yoke. Unscrew the steering stem nut. Ease the yoke up off the forks until there is enough clearance for the instrument cluster to be removed.
15 Installation is the reverse of removal. Tighten the steering stem nut and the fork clamp bolts to the torque settings specified at the beginning of the Chapter.

Speedometer cable – CB models

16 On SW, SX, SY and S2 models, remove the fairing (see Chapter 8)
17 Unscrew the knurled ring securing the speedometer cable to the back of the speedometer and detach the cable **(see illustrations 15.3 or 15.7)**.
18 Remove the screw that retains the lower end of the cable in the speedometer drive unit on the left-hand side of the front wheel hub and withdraw the cable **(see illustration)**.
19 Remove the cable, noting its routing through the guide on the front mudguard.

20 Installation is the reverse of removal. Engage the forked lower end of the cable on the spade drive from the speedometer gearbox and tighten the retaining screw securely **(see illustration)**.
21 Route the cable through the guide on the mudguard, and the fairing bracket on SW, SX and SY models **(see illustration)**, and connect the squared upper end of the cable to the speedometer, then tighten the retaining ring securely.
22 Check that the cable doesn't restrict steering movement or interfere with any other components.
23 On SW, SX and SY models, install the fairing (see Chapter 8).

Speed sensor – CBF models

24 The speed sensor is mounted on the crankcase behind the starter motor **(see illustration)**. To check the sensor refer to Section 16.
25 Remove the frame side panels (see Chapter 8), and the starter motor (see Section 28).
26 Trace the wiring from the sensor and disconnect it at the 3-pin black connector inside the rubber boot **(see illustration 11.8b)**. Release the cable-tie and feed the wiring back to the sensor, noting its routing.
27 Clean around the sensor so no dirt can fall into the engine. Unscrew the bolts and remove the sensor. Remove and discard its O-ring as a new one must be used. Plug the sensor orifice with clean rag to prevent anything falling into the engine.
28 Installation is the reverse of removal – use a new O-ring and smear it with oil.

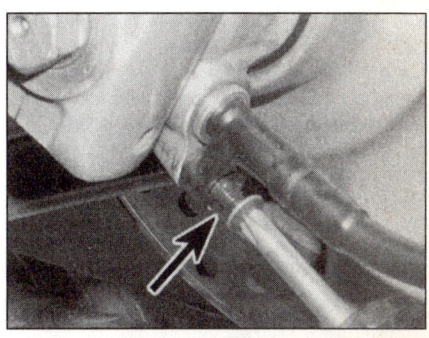

15.18 Remove the screw (arrowed) and withdraw the cable

16 Instruments – check and replacement

Speedometer

Check

1 Special equipment is required to check the operation of this meter. If it is believed to be faulty, take the speedometer to a Honda dealer for assessment.

Replacement – R, T, V, W, X, Y and 2 models

2 Remove the instrument cluster (see Section 15).
3 Remove the screws securing the instrument cluster back and remove the back **(see illustrations)**.

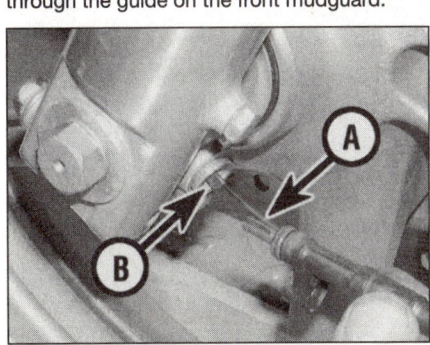

15.20 Forked end (A) engages on spade drive (B)

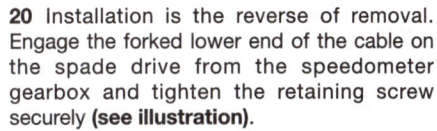

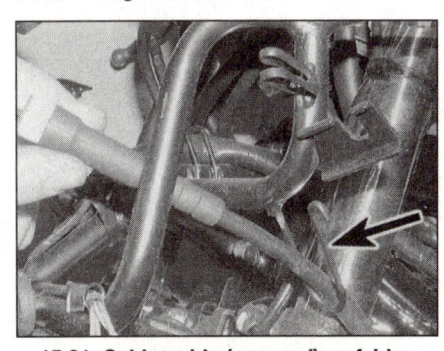

15.21 Cable guide (arrowed) on fairing bracket – SW, SX and SY models

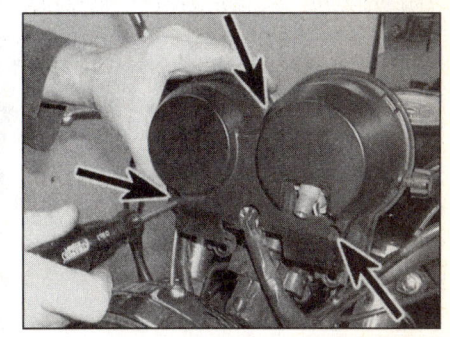

16.3a Remove the screws (arrowed) . . .

9•14 Electrical system

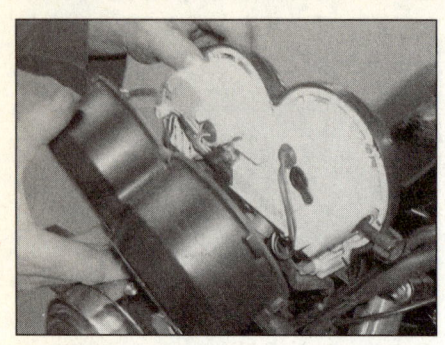

16.3b ... and remove the back

6 Remove the two screws securing the speedometer to the casing and carefully lift the speedometer out of the casing **(see illustration 16.5)**.
7 Installation is the reverse of removal.

Replacement –
SW, SX, SY and S2 models

8 Remove the instrument cluster (see Section 15).
9 Pull the odometer trip knob off the trip spindle.
10 Remove the five screws securing the instrument cluster top cover to the casing and separate the cover, plate, lens and casing, noting the order of assembly **(see illustration)**.
11 Remove the two screws securing the speedometer to the casing and carefully lift the speedometer out of the casing **(see illustration 16.10)**.
12 Installation is the reverse of removal.

Tachometer – CB models

Check
13 Special equipment is required to check the operation of this meter. If it is believed to be faulty, take the tachometer to a Honda dealer or automotive electrician for assessment.

Replacement –
R, T, V, W, X, Y and 2 models

Note: *On these models, the tachometer and temperature gauge are a composite unit.*

14 Remove the instrument cluster (see Section 15) and follow Steps 3 and 4.
15 Unscrew the two tachometer and three temperature gauge wiring connectors from the casing **(see illustration)**.
16 Remove the three screws securing the instrument cluster top cover and remove the cover **(see illustration 16.5)**.
17 Remove the two screws securing the tachometer and temperature gauge to the casing and carefully lift the meter and gauge out of the casing **(see illustration)**.
18 Installation is the reverse of removal. Make sure the wires are correctly and securely connected (see wiring diagrams at the end of this Chapter).

Replacement –
SW, SX, SY and S2 models

19 Remove the instrument cluster (see Section 15).

4 Unscrew the nuts securing the instrument cluster to the mounting bracket and remove the nuts, washers and bracket, noting the position of the bushes on the bracket **(see illustrations)**.
5 Using a very small Phillips screwdriver, remove the screw in the centre of the odometer trip knob and remove the knob. Remove the three screws securing the instrument cluster top cover and remove the cover **(see illustration)**.

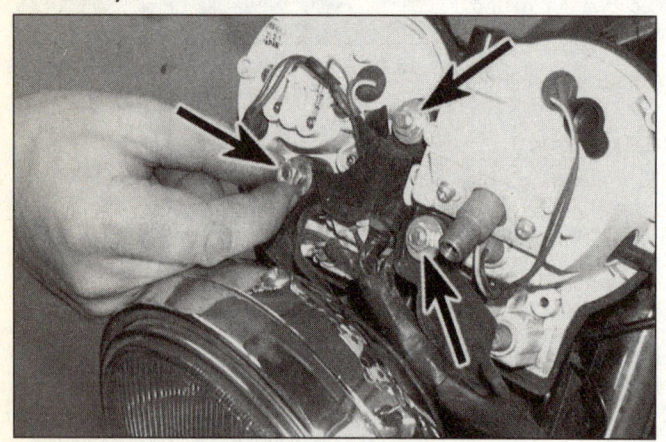

16.4a Remove the nuts and washers (arrowed) ...

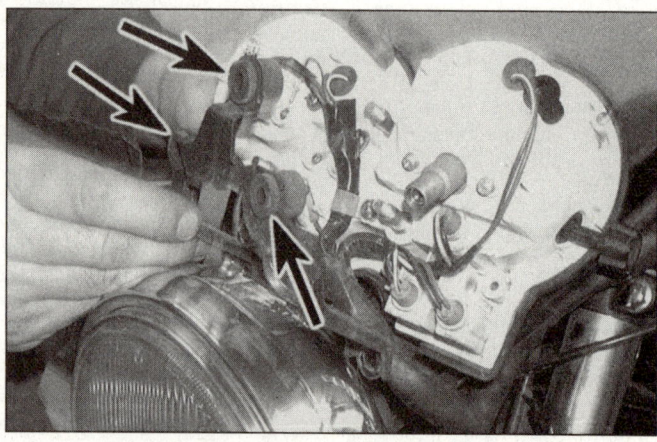

16.4b ... noting position of the bushes (arrowed)

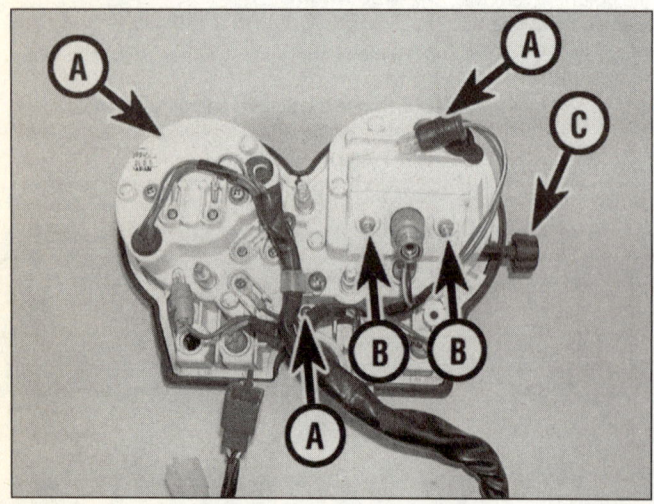

16.5 Top cover is retained by screws (A), speedometer is retained by screws (B). Odometer trip knob (C)

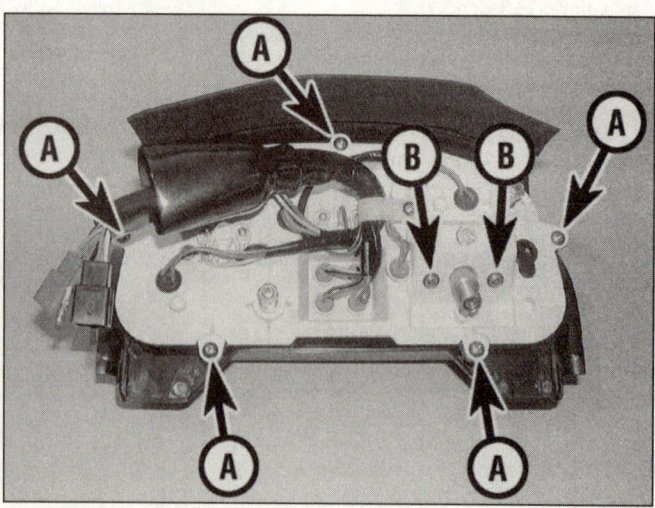

16.10 Top cover is retained by screws (A), speedometer is retained by screws (B)

Electrical system 9•15

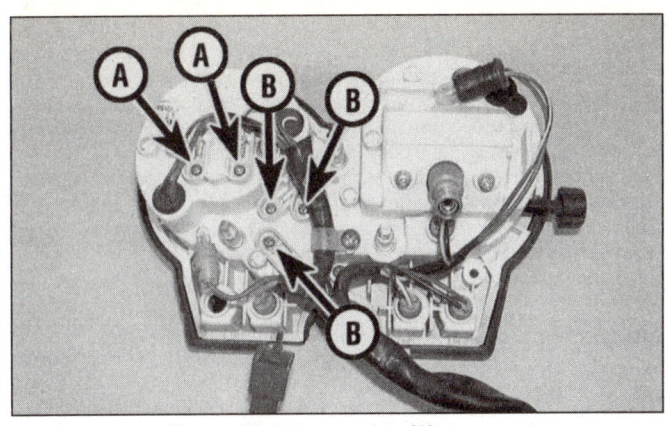

16.15 Tachometer wiring connectors (A), temperature gauge wiring connectors (B)

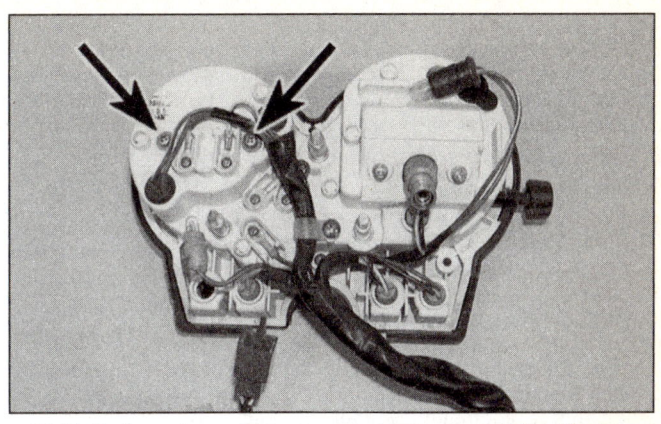

16.17 Tachometer and temperature gauge are retained by two screws (arrowed)

20 Unscrew the two tachometer wiring connectors from the casing **(see illustration)**. Pull the odometer trip knob off the trip spindle.
21 Remove the five screws securing the instrument cluster top cover to the casing and separate the cover, plate, lens and casing, noting the order of assembly **(see illustration 16.10)**.
22 Remove the two screws securing the tachometer to the casing and carefully lift the tachometer out of the casing **(see illustration 16.20)**.
23 Installation is the reverse of removal. Make sure the wires are correctly and securely connected (see wiring diagrams at the end of this Chapter).

Coolant temperature gauge – CB models

Check
24 See Chapter 3.

Replacement – R, T, V, W, X, Y and 2 models
Note: On these models, the tachometer and temperature gauge are a composite unit. Follow Steps 14 to 18.

Replacement – SW, SX, SY and S2 models
25 Remove the instrument cluster (see Section 15).
26 Pull the odometer trip knob off the trip spindle. Remove the five screws securing the instrument cluster top cover to the casing and separate the cover, plate, lens and casing, noting the order of assembly **(see illus-tration 16.10)**.
27 Remove the three screws securing the temperature gauge wiring connectors and the gauge from the bottom of the casing and carefully lift the gauge out of the casing **(see illustration)**.
28 Installation is the reverse of removal. Make sure the wires are correctly and securely connected (see wiring diagrams at the end of this Chapter).

Power check – CBF models
29 If none of the instruments or displays are working, first check the instrument fuses (see Section 5).
30 If the fuse is good, remove the headlight from its shell (see Section 7). Trace the wiring from the instrument cluster to the white 6-pin and black 9-pin connectors inside the headlight shell and disconnect them. Check the connectors for loose or broken connections.
31 To check the power input line, check for battery voltage between the black/brown wire terminal on the loom side of the 9-pin connector and a good earth (ground) with the ignition switch ON. There should be battery voltage. If there is no voltage, refer to the wiring diagrams and check the black/brown circuit for loose or broken connections or a damaged wire.
32 To check the back-up power line, check for battery voltage between the red/green wire terminal on the loom side of the 6-pin connector and a good earth (ground) with the ignition switch OFF. There should be battery voltage. If there is no voltage, refer to the wiring diagrams and check the red/green circuit for loose or broken connections or a damaged wire.
33 If there is voltage, and to check the earth (ground) line, check for continuity between the green/black wire terminal on the loom side of the 9-pin connector and earth (ground). If there is no continuity, check the circuit for loose or broken connections or a damaged wire.
34 If no faults can be found, also check the

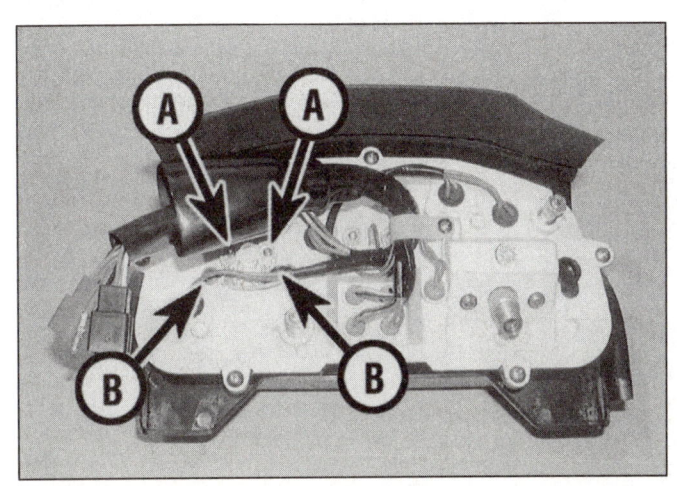

16.20 Tachometer wiring connectors (A), tachometer retaining screws (B)

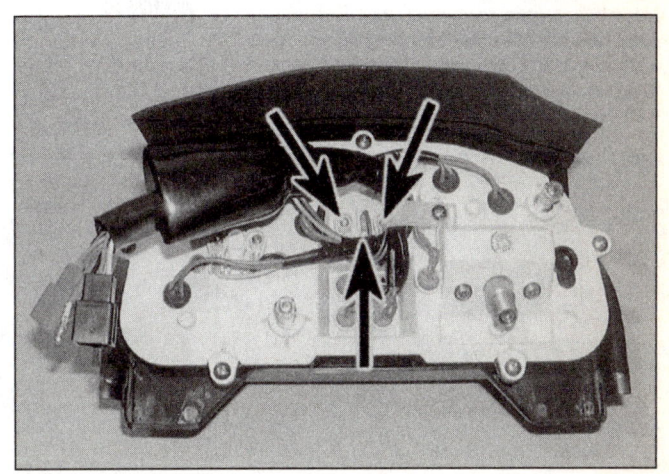

16.27 Temperature gauge wiring connectors and retaining screws (arrowed)

9•16 Electrical system

wiring between the cluster connectors and the cluster itself for continuity – there is a panel on the back of the cluster that can be removed to access the internal connectors. Next check the individual instruments as follows.

Speedometer and speed sensor check

35 First do the power check (see above).
36 If the power to the instruments is good, connect a voltmeter between the pink/green and green/black wire terminals on the 9-pin connector, with it still connected (insert the probes in the back of the connector). With the machine on its centrestand and the ignition switch ON, turn the rear wheel by hand and check that a fluctuating voltage reading between 0 – 5 volts is obtained.
37 If no reading is obtained, remove the left-hand frame side panel (see Chapter 8). Trace the wiring from the speed sensor, which is mounted on the crankcase behind the starter motor, and disconnect it at the 3-pin black connector inside the rubber boot **(see illustration 11.8b)**. Check the connectors for loose or broken terminals or a damaged wire. Check for continuity in the wiring between the instrument cluster wiring connector and the speed sensor wiring connector. If the wiring is good, then the speed sensor is faulty.
38 If the wiring and sensor are good, the speedometer is faulty – remove the instrument assembly from the housing and replace it with a new one (see Steps 43 to 46).

Tachometer check

39 First do the power check (see above).
40 If the power to the instruments is good, check the tachometer input peak voltage. Honda specify their own Imrie diagnostic tester (model 625), or the peak voltage adapter (Pt. No. 07HGJ-0020100) with an aftermarket digital multimeter having an impedance of 10 M-ohm/DCV minimum, for this test. Connect the positive (+) lead of the voltmeter and peak voltage adapter arrangement to the yellow/green wire terminal on the loom side of the connector and the negative (–) lead to the green/black wire terminal. Start the engine and measure the tachometer input peak voltage, which should be at least 10.5 volts. If the peak voltage is normal then the tachometer is faulty – remove the instrument assembly from the housing and replace it with a new one (see Steps 43 to 46).
41 If there is no reading, check for continuity in the yellow/green wire between the instrument cluster wiring connector and the ignition control unit wiring connector (see Chapter 5 for access). If there is no continuity there is a break in the wire or faulty connector. If that is good check for continuity to earth in the green/black wire. Refer to the wiring diagrams and trace and rectify any fault. If the wiring is good the ICU could be faulty.

Coolant temperature gauge check

42 See Chapter 3.

Instrument replacement

43 Remove the instrument cluster (see Section 15).
44 Undo the rear cover screws and lift off the cover.
45 Undo the wiring clamp screw. Carefully pull the bulbholders out. Remove the instrument wiring covers and disconnect the connectors.
46 Undo the ten screws on the back of the instrument housing, noting where the ones with washers fit. Turn the instrument cluster over and remove the front cover. Release the switch bases from the locating pins and lift the instrument assembly from the housing.
47 Installation is the reverse of removal.

17 Instrument and warning light bulbs – renewal

Note: *The neutral light is part of the safety circuit which prevents the engine from starting with the transmission in gear unless the clutch lever is pulled in and the sidestand is up, and prevents the engine from running with the sidestand down unless the transmission is in neutral. An apparent fault with the neutral light may be due to a defective diode (see Section 23).*

CB500-R, T, V, W, X, Y and 2 models

1 Unscrew the knurled ring securing the speedometer cable to the back of the speedometer and detach the cable **(see illustration 15.3)**.
2 Unscrew the two bolts securing the instrument cluster to the underside of the top yoke and remove the bolts **(see illustration 15.4)**. Turn the instruments to gain access to the screws securing the instrument cluster back and remove the back **(see illustrations 16.3a and b)**.
3 Some of the bulbs are accessible with the cluster mounting bracket in place, but access to others requires removal of the mounting bracket **(see illustration 16.4a)**.
4 Gently pull the bulbholder out of the back of the instrument casing, then pull the bulb out of the bulbholder **(see illustrations)**.
5 If the socket contacts are dirty or corroded, scrape them clean and spray with electrical contact cleaner before a new bulb is installed.
6 Carefully push the new bulb into the holder, then push the holder into the casing.
7 Installation of the instrument cluster is the reverse of removal.

CB500-SW, SX, SY and S2 models

8 Remove the fairing (see Chapter 8).
9 Some of the bulbs are accessible with the instrument cluster in place, but access to others requires removing the cluster from the mounting bracket **(see illustration 15.9)**.
10 Gently pull the bulbholder out of the back of the instrument casing, then pull the bulb out of the bulbholder **(see illustration)**.
11 If the socket contacts are dirty or corroded, scrape them clean and spray with electrical contact cleaner before a new bulb is installed.
12 Carefully push the new bulb into the holder, then push the holder into the casing.
13 Installation of the instrument cluster is the reverse of removal.

CBF models

14 Remove the instrument cluster (see Section 15).
15 Undo the rear cover screws and lift off the cover.
16 Carefully pull the bulbholder out of the back of the instrument casing, then pull the bulb out of the bulbholder.
17 Carefully push the new bulb into the holder, then push the holder into the casing.
18 Fit the rear cover. Install the instrument cluster (see Section 15).

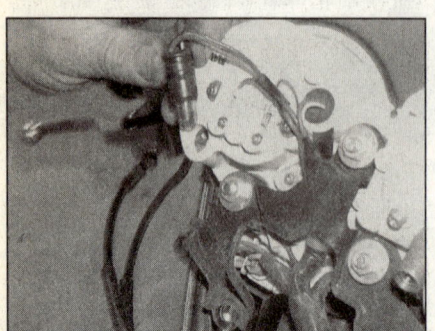

17.4a Pull the bulbholder out of the casing . . .

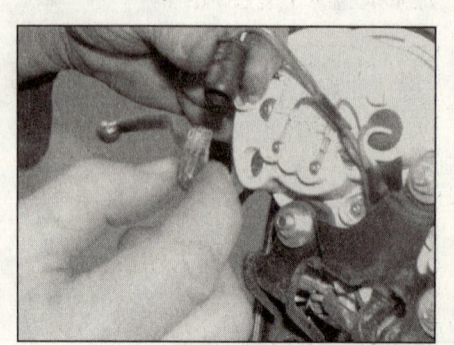

17.4b . . . and pull the bulb out of the holder

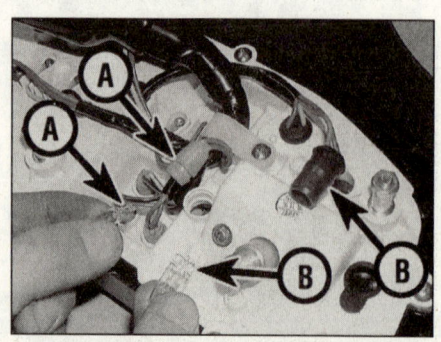

17.10 Both sizes of bulbs (A) and (B) are a push fit in the bulbholders

Electrical system 9•17

18 Ignition (main) switch – check, removal and installation

⚠ **Warning: To prevent the risk of short circuits, remove the rider's seat (see Chapter 8) and disconnect the battery negative (-ve) lead before making any ignition (main) switch checks.**

Check

1 On CB500-R, T, V, W, X, Y and 2 models, and on CBF models, remove the headlight from its shell (see Section 7). Trace the ignition (main) switch wiring back from the base of the switch and disconnect it at the black 3-pin (CB) or white 4-pin (CBF) connector inside the headlight shell.
2 On CB500-SW, SX, SY and S2 models, remove the fairing (see Chapter 8). Trace the ignition (main) switch wiring back from the base of the switch and disconnect it at the black 3-pin connector.
3 Using an ohmmeter or a continuity tester, check the continuity of the connector terminal pairs (see the *wiring diagrams* at the end of this Chapter). Having already disconnected the battery negative (-ve) lead, insert the key in the switch. Continuity should exist between the terminals connected by a solid line on the diagram when the switch is in the indicated position.
4 If the switch fails any of the tests, renew it.

Removal

5 The ignition switch is secured to the top yoke with two Torx bolts on CB models and by two one-way security bolts (i.e. they can be done up but not undone) on CBF models **(see illustration)**. To gain access to these bolts, disconnect the ignition wiring connector (see Step 1 or 2), then remove the handlebars and fork top yoke (see Chapter 6).
6 Remove the bolts and withdraw the switch from the top yoke – on CBF models you must drift the bolts round using a suitable punch or chisel, taking care to support the top yoke securely and to protect it with some rag.

Installation

7 Installation is the reverse of removal. On CBF models Honda specify to use new bolts. Tighten the bolts to the torque setting specified at the beginning of this Chapter. Make sure the wiring is securely connected and correctly routed.

19 Oil pressure switch – check, removal and installation

Check

1 The oil pressure switch is screwed into the crankcase in front of the engine left-hand cover. The oil pressure warning light should come on when the ignition is first turned on, then extinguish once the engine is started. If it does not come on, yet the bulb and fuse (tail light, signal, brake light, horn circuit fuse) are good, pull the rubber boot off the switch and remove the screw securing the wiring connector **(see illustration)**.
2 With the ignition switched ON, earth (ground) the wire on the crankcase and check to see if the warning light comes on. If the light comes on, the switch is defective and must be renewed.
3 If the warning light comes on whilst the engine is running, yet the oil pressure is known to be good (see Chapter 1), remove the wire from the oil pressure switch with the engine running. If the light goes out, the switch is defective and must be renewed. If the light remains illuminated, the wire between the switch and instrument cluster must be earthed (grounded) at some point (see the *wiring diagrams* at the end of this Chapter).

⚠ **Warning: Avoid working on a hot engine – the exhaust pipes, crankcases and engine oil can cause severe burns. Do not allow exhaust gases to build up in the work area; either perform the check outside or use an exhaust gas extraction system**

Removal

4 Pull the rubber boot off the switch and remove the screw securing the wiring connector.
5 Unscrew the oil pressure switch and withdraw it from the crankcase.

Installation

6 Apply a suitable sealant to the upper portion of the switch threads near the switch body, leaving the bottom 3 to 4 mm of thread clean. Install the switch in the crankcase and tighten it to the torque setting specified at the beginning of this Chapter.
7 Attach the wiring connector and secure it with the screw, then fit the rubber boot.
8 Run the engine and check that the warning light operates correctly.

20 Neutral switch – check, removal and installation

Note: *The neutral switch is part of the safety circuit which prevents the engine from starting with the transmission in gear unless the clutch lever is pulled in and the sidestand is up, and prevents the engine from running with the sidestand down unless the transmission is in neutral. An apparent fault with the switch may be due to a defective diode (see Section 23).*

Check

1 The neutral switch is screwed into the left-hand side of the engine behind the front sprocket cover. To check the switch, first remove the sprocket cover and chain guide (see Chapter 6, Section 14).
2 If the neutral warning light does not come on when the ignition is turned ON with the transmission in neutral, detach the wiring connector and earth (ground) the wire on the crankcase **(see illustration)**. If the light comes on, the switch is defective and must be renewed.
3 If the light does not come on, check for voltage at the connector and check the wire for continuity between the connector, diode (see Section 23), instrument cluster and fusebox (see the *wiring diagrams* at the end of this Chapter). Also check the warning light bulb (see Section 17), and the tail light, signal, brake light, horn circuit fuse (see Section 5).
4 If the warning light does not extinguish when a gear is selected and the ignition is ON, detach the wiring connector. If the light goes out, the switch is defective and must be renewed. If the light remains on, the wire

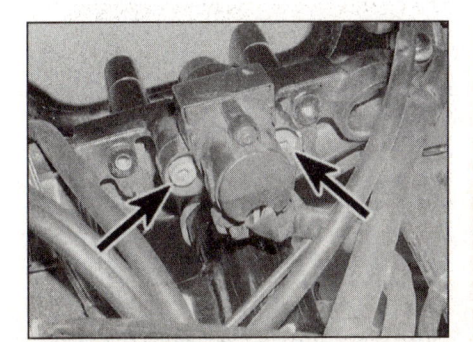

18.5 Torx bolts (arrowed) secure the ignition switch

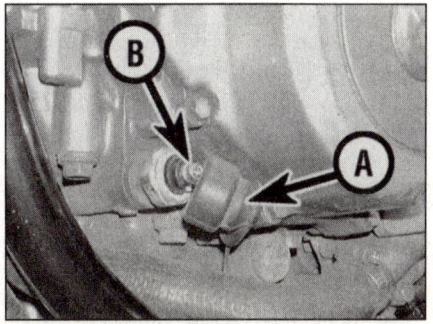

19.1 Pull back the boot (A) and remove the screw (B)

20.2 Neutral switch wiring connector (arrowed)

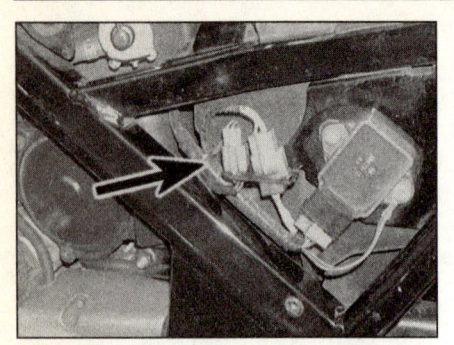

9•18 Electrical system

21.3 Sidestand switch wiring connector (arrowed) – CB models

21.7 Sidestand switch bolt (arrowed)

22.2 Disconnect the wiring connectors (arrowed)

between the connector and instrument cluster must be earthed (grounded) at some point.

Removal

5 Remove the sprocket cover and chain guide and detach the wiring connector from the switch **(see illustration 20.2)**.
6 Unscrew the switch and withdraw it from the engine casing.

Installation

7 Apply a smear of sealant to the threads of the switch, taking care not to cover the contact point.
8 Install the switch, tighten it to the torque setting specified at the beginning of this Chapter, then reconnect the switch wire.
9 Check the operation of the neutral light.
10 Install the chain guide and sprocket cover.

21 Sidestand switch – check and replacement

Note: *The sidestand switch is part of the safety circuit which prevents the engine from starting with the transmission in gear unless the clutch lever is pulled in and the sidestand is up, and prevents the engine from running with the sidestand down unless the transmission is in neutral.*

Check

1 The sidestand switch is mounted on the back of the sidestand.
2 Sit astride the motorcycle and start the engine with the sidestand up and the transmission in neutral. Pull the clutch lever in and engage first gear, then lower the sidestand; the engine should stop.
3 To check the operation of the switch using an ohmmeter or continuity test light, place the motorcycle on its centre stand, remove the left-hand frame side panel (see Chapter 8) and trace the wiring back from the switch to the green connector and disconnect it **(see illustration)**. Connect the meter to the green/ white and green wires on the switch side of the connector. With the sidestand up there should be continuity (zero resistance) between the terminals, and with the stand down there should

be no continuity (infinite resistance). On CB models now connect the meter to the yellow/ black and green wires on the switch side of the connector. With the sidestand down there should be continuity (zero resistance) between the terminals, and with the stand up there should be no continuity (infinite resistance).
4 If the switch fails either or both of these tests it is defective and must be renewed.
5 If the switch is good, check the wiring between the various components in the starter safety circuit (see the *wiring diagrams* at the end of this Chapter).

Replacement

6 Place the motorcycle on its centre stand. Remove the front sprocket cover and chain guide and the left-hand frame side panel (see Chapter 8). Trace the wiring back from the switch to the green connector and disconnect it **(see illustration 21.3)**. Work back along the switch wiring, freeing it from any relevant retaining clips and ties, noting its correct routing.
7 Unscrew the switch bolt and remove the switch from the stand, noting how it fits **(see illustration)**.
8 Locate the tab on the inside of the new switch in the hole in the sidestand and align the switch body with the stand spring post **(see illustration 21.7)**. Tighten the bolt to the torque setting specified at the beginning of this Chapter.
9 Make sure the wiring is correctly routed up to the connector and retained by all the necessary clips and ties.
10 Reconnect the wiring connector and check the operation of the sidestand switch (see Step 2).
11 Install the chain guide and sprocket cover, and frame side panel (see Chapter 8).

22 Clutch switch – check and replacement

Note: *The clutch switch is part of the safety circuit which prevents the engine from starting with the transmission in gear unless the clutch lever is pulled in and the sidestand is up, and prevents the engine from running with the sidestand down unless the transmission is in neutral.*

Check

1 The clutch switch is housed in the clutch lever bracket.
2 To check the switch, disconnect the wiring connectors from the switch **(see illustration)**. Connect the probes of an ohmmeter or a continuity test light to the two switch terminals. With the clutch lever pulled in, continuity should be indicated (zero resistance). With the clutch lever out, no continuity (infinite resistance) should be indicated. Reconnect the wiring connectors.
3 If the switch is good, check for voltage at one of the terminals on the clutch switch wiring connectors with the ignition ON – there will be voltage on one terminal and zero on the other. If voltage is indicated, check the other components in the starter circuit as described in the relevant sections of this Chapter. If no voltage is indicated, or if all components are good, check the wiring between the various components (see the *wiring diagrams* at the end of this Chapter).

Replacement

4 Remove the clutch lever (see Chapter 6, Section 5).
5 Disconnect the wiring connectors from the clutch switch **(see illustration 22.2)**. Using a small screwdriver, push the switch from the connector end and withdraw it from inside the bracket.
6 Installation is the reverse of removal. Make sure the ridge on the top of the switch locates in the cutout in the lever bracket, and push the switch fully home.

23 Diode – check and replacement

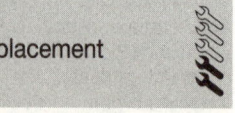

Note: *The diode switch is part of the safety circuit which prevents the engine from starting with the transmission in gear unless the clutch lever is pulled in and the sidestand is up, and prevents the engine from running with the sidestand down unless the transmission is in neutral.*

Check

1 On CB models remove the fuel tank (see Chapter 4). The diode is a small block that

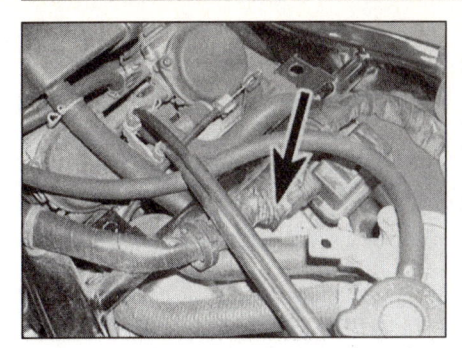

Electrical system 9•19

23.1 Locate the diode on the main wiring harness – CB models

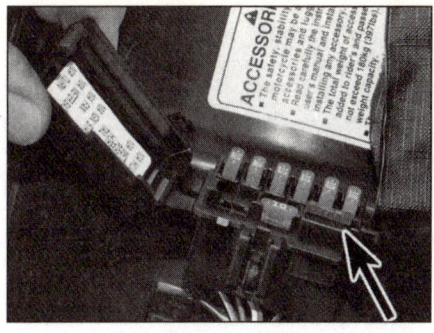

23.2 The diode (arrowed) is in the fusebox on CBF models

plugs into the main wiring harness **(see illustration)**. Disconnect the diode from the harness, removing any tape used to bind it.
2 On CBF models remove the passenger seat (see Chapter 8). The diode is in the fusebox – unclip the lid to access it, then pull the diode out of its socket **(see illustration)**.
3 Using an ohmmeter or continuity tester, connect one probe to one of the outer terminals of the diode and the other probe to the middle terminal of the diode and test for continuity. Now reverse the probes. The diode should only show continuity in one direction (as indicated by the symbol on the body of the diode). If the diode shows continuity or no continuity in both directions, it should be renewed.
4 If the diode is good, repeat the test between the other outer terminal and the middle terminal. Again, the diode should only show continuity in one direction. If it doesn't it should be renewed.
5 If the diode is good, check the other components in the starter circuit as described in the relevant sections of this Chapter. If all components are good, check the wiring between the various components (see the *wiring diagrams* at the end of this Chapter).

Replacement

6 Follow Steps 1 or 2 and connect the new diode to the main wiring harness. Use insulating tape to bind it in place on CB models.

24 Handlebar switches – check

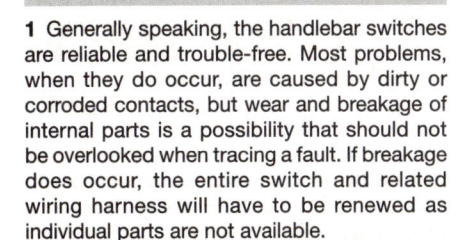

1 Generally speaking, the handlebar switches are reliable and trouble-free. Most problems, when they do occur, are caused by dirty or corroded contacts, but wear and breakage of internal parts is a possibility that should not be overlooked when tracing a fault. If breakage does occur, the entire switch and related wiring harness will have to be renewed as individual parts are not available.
2 The switches can be checked for continuity using an ohmmeter or a continuity test light. Always disconnect the battery negative (-ve) lead, which will prevent the possibility of a short circuit, before making the checks.
3 Trace the wiring from the switch in question

back to its connector and disconnect it. On CB500-R, T, V, W, X, Y and 2 models and CBF models the connectors are inside the headlight shell (see Section 7). On CB500-SW, SX, SY and S2 models the connectors are inside the fairing (see Chapter 8).
4 The right-hand switch connector is the red 9-pin connector and the left-hand switch connectors are the green 6-pin and black 4-pin connectors.
5 Check for continuity between the terminals of the switch wiring with the switch in the various positions (ie switch off – no continuity, switch on – continuity) – see the *wiring diagrams* at the end of this Chapter.
6 If the continuity check indicates a problem exists, refer to Section 25, remove the switch and spray the switch contacts with electrical contact cleaner. If they are accessible, the contacts can be scraped clean carefully with a knife or polished with crocus cloth. If switch

components are damaged or broken, it will be obvious when the switch is disassembled.
7 Clean the inside of the switch body thoroughly and smear the contacts with silicone grease before reassembly

25 Handlebar switches – removal and installation

Removal

1 If the switch unit is to be removed from the motorcycle, rather than just displaced from the handlebar, it will be necessary to disconnect the switch wiring from the main wiring harness.
2 Trace the wiring from the switch in question back to its connector and disconnect it. On CB500-R, T, V, W, X, Y and 2 models and CBF models the connectors are inside the headlight shell (see Section 7). On CB500-SW, SX, SY and S2 models the connectors are inside the fairing (see Chapter 8).
3 The right-hand switch connector is the red 9-pin connector and the left-hand switch connectors are the green 6-pin and black 4-pin connectors. Work back along the wiring, freeing it from all the relevant clips and ties and noting its correct routing.
4 In addition, the right-hand switch wiring is connected to the front brake light switch (two wires) and the left-hand switch wiring is connected to the clutch switch (two wires), horn (two wires) and turn signals (four wires).
5 Unscrew the two handlebar switch screws and free the switch from the handlebar by separating the halves **(see illustrations)**. Note

25.5a Unscrew the two right-hand switch screws (rear screw arrowed) . . .

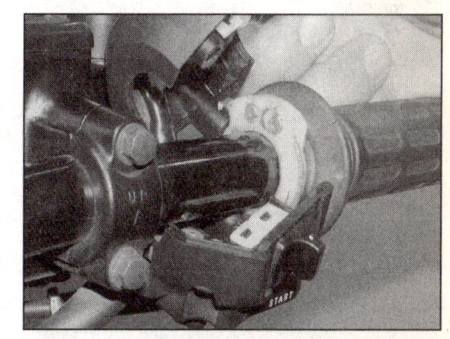

25.5b . . . and detach the switch from the handlebar

25.5c Unscrew the two left-hand switch screws (rear screw arrowed) . . .

25.5d . . . and detach the switch from the handlebar

9•20 Electrical system

26.1a Horn mounting bracket – R, T, V, W, X, Y and 2 models

26.1b Horn mounting bracket – SW, SX, SY and S2 models

26.1c Horn mounting bracket – CBF models

that the choke (left side) and throttle (right side) cables must be detached to fully remove the switches (see Chapter 4).

Installation

6 Installation is the reverse of removal. Make sure the locating pin in the lower half of the switch locates in the hole in the underside of the handlebar and that all the wiring is correctly connected and secured. Check the operation of the switches and related components.

26 Horn – check and replacement

Check

1 The horn is mounted to a bracket on the bottom yoke on CB models, and below the radiator on the left-hand side on CBF models **(see illustrations)**.
2 Disconnect the wiring connectors from the horn and ensure that the contacts are clean and free from corrosion **(see illustration)**.
3 To test the horn, use jumper wires to connect one of the horn terminals to the battery positive (+ve) terminal and the other horn terminal to the battery negative (-ve) terminal. If the horn sounds, check the switch (see Section 24), the fuse (see Section 5) and the wiring between the switch and the horn

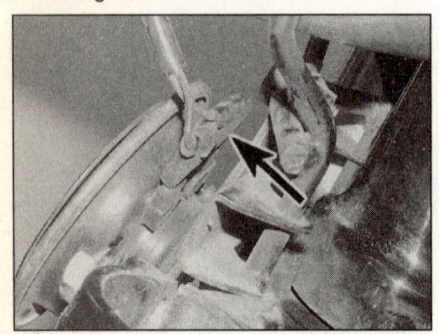

26.2 Horn wiring connectors (arrowed)

(see the *wiring diagrams* at the end of this Chapter).
4 If the horn doesn't sound, renew it.

Replacement

5 If required, remove the fairing for improved access on SW, SX, SY and S2 models (see Chapter 8).
6 Disconnect the wiring connectors from the horn, then unscrew the bolt securing the horn to the mounting bracket and remove it from the bike **(see illustrations 26.1a and b)**.
7 Install the horn and tighten the bolt securely. Connect the wiring connectors to the horn and test the horn.

27 Starter relay – check and replacement

Check

1 The starter relay is located on the right-hand side of the bike behind the right-hand frame side panel **(see illustration 5.1)**.
2 If the starter circuit is faulty, first check the main fuse and ignition/starter fuse (see Section 5).
3 To check the relay, lift the rubber terminal cover and unscrew the bolt securing the starter motor lead **(see illustration)**; position the lead well away from the relay terminal. With the ignition switch ON, the engine kill switch in the RUN position

27.3 Disconnect the starter motor lead (arrowed)

and the transmission in neutral, press the starter switch. The relay should be heard to click.
4 If the relay doesn't click, switch the ignition OFF and remove the relay (see Steps 8 and 9), then test it as follows.
5 Set a multimeter to the ohms x 1 scale and connect it across the relay's starter motor and battery lead terminals **(see illustration 27.3)**. Using a fully-charged 12 volt battery and two insulated jumper wires, connect the positive (+ve) terminal of the battery to the yellow/red wire terminal of the relay, and the negative (-ve) terminal to the green/red wire terminal of the relay. At this point the relay should be heard to click and the multimeter read 0 ohms (continuity). If the relay does not click when voltage is applied and indicates no continuity (infinite resistance) across its terminals, it is faulty and must be renewed.
6 If the relay is good, check for voltage between the yellow/red wire and the green/red wire terminals in the connector when the starter button is pressed with the ignition switch ON, the engine kill switch in the RUN position and the transmission in neutral.
7 Check the other components in the starter circuit as described in the relevant sections of this Chapter. If all components are good, check the wiring between the various components (see the *wiring diagrams* at the end of this Chapter).

Replacement

8 Disconnect the battery terminals, remembering to disconnect the negative (-ve) terminal first.
9 Disconnect the relay wiring connector, then unscrew the two bolts securing the starter motor and battery leads to the relay and detach the leads **(see illustrations 27.3)**. Remove the relay with its rubber sleeve from its mounting lug on the frame.
10 Installation is the reverse of removal. Make sure the connector terminals, battery and starter motor leads are clean and securely tightened.
11 Connect the negative (-ve) lead last when reconnecting the battery.

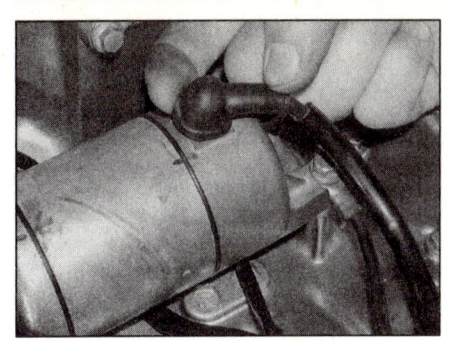

Electrical system 9•21

28.3 Starter motor terminal cover

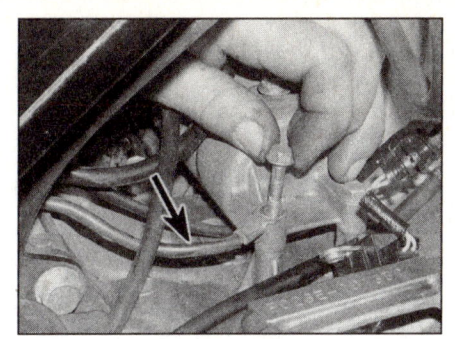

28.4 Starter motor earth lead (arrowed)

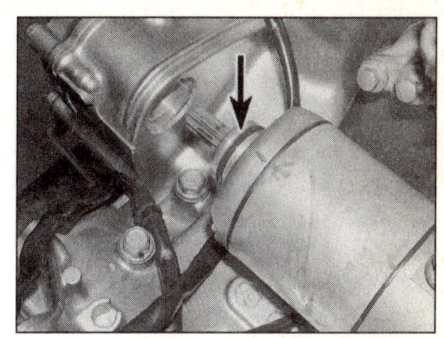

28.5 Withdraw the starter motor and discard the O-ring (arrowed)

28 Starter motor – removal and installation

Removal

1 The starter motor is mounted on the crankcase behind the cylinders.
2 Disconnect the battery negative (-ve) lead (see Section 3).
3 Peel back the rubber terminal cover on the starter motor, then remove the nut securing the starter lead to the motor terminal and detach the lead **(see illustration)**.
4 Unscrew the two bolts securing the starter motor to the crankcase, noting the earth lead attached to the rear bolt **(see illustration)**.
5 Slide the starter motor out from the crankcase and remove it from the machine **(see illustration)**. Remove the O-ring on the end of the starter motor body and discard it as a new one must be used.

Installation

6 Install a new O-ring on the end of the starter motor body and ensure it is seated in its groove **(see illustration)**. Apply a smear of engine oil to the O-ring to aid installation.
7 Manoeuvre the motor into position and slide it into the crankcase. Ensure that the starter motor teeth mesh correctly with those of the starter idle gear. Install the mounting bolts, not forgetting to fit the earth lead with the rear bolt, and tighten them securely **(see illustration 28.4)**.
8 Connect the starter lead to the motor terminal and secure it with the nut. Hold the lower nut on the terminal with an open-ended spanner while tightening the top nut **(see illustration)**. Make sure the rubber cover is correctly seated over the terminal.
9 Connect the battery negative (-ve) lead.

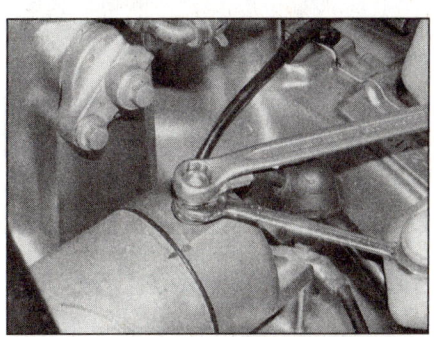

29 Starter motor – disassembly, inspection and reassembly

Disassembly

1 Remove the starter motor (see Section 28).
2 Note the alignment marks between the main housing and the front and rear covers, or make your own if they aren't clear **(see illustration)**.
3 Unscrew the two long bolts, noting how the D-shaped washers locate, and withdraw them from the starter motor **(see illustration)**. Discard their O-rings as new ones must be used.
4 Wrap some insulating tape around the teeth on the end of the starter motor shaft – this will protect the oil seal from damage as the front cover is removed. Remove the front cover from the motor and remove the tabbed thrust washer from inside the front cover **(see illustration)**.

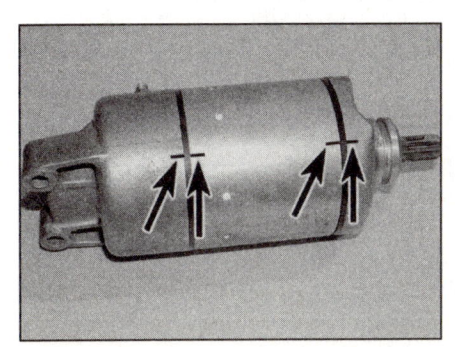

28.6 Ensure O-ring seats in its groove (arrowed)

28.8 Use two spanners when tightening the terminal nut

29.2 Note the alignment marks (arrowed) between the housing and the covers

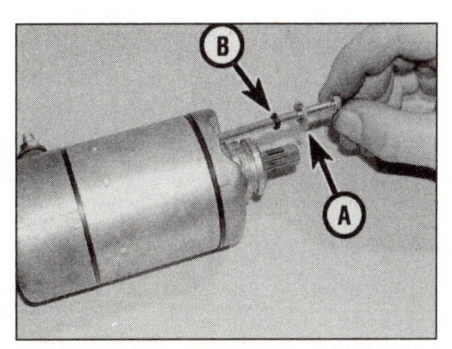

29.3 D-shaped washer (A) and O-ring (B)

29.4 Remove the front cover and the tabbed thrust washer (arrowed)

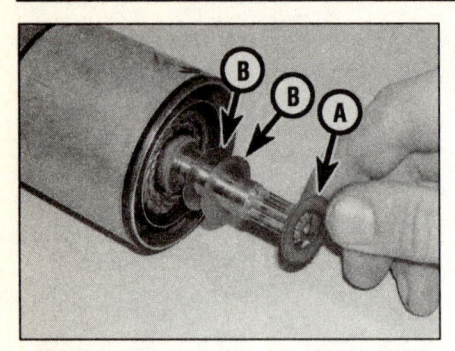

9•22 Electrical system

29.5 Insulating washer (A) and shims (B)

29.6 Remove the main housing and discard the O-rings

29.7a Separate the armature commutator (arrowed) from the brushplate . . .

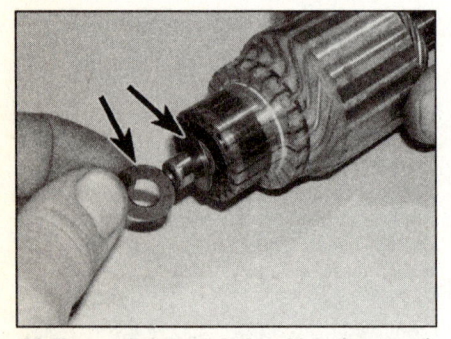

29.7b . . . and remove the shims (arrowed) from the end of the shaft

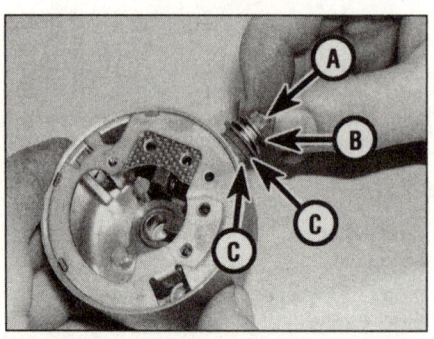

29.8a Remove the terminal nut (A), washer (B) and insulating washers (C) . . .

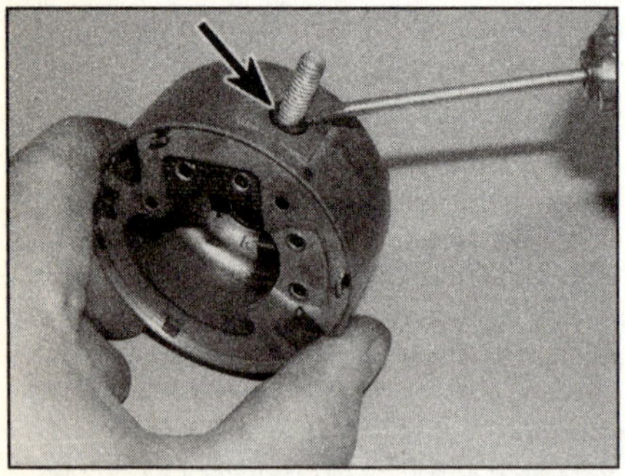

29.8b . . . and then the O-ring (arrowed)

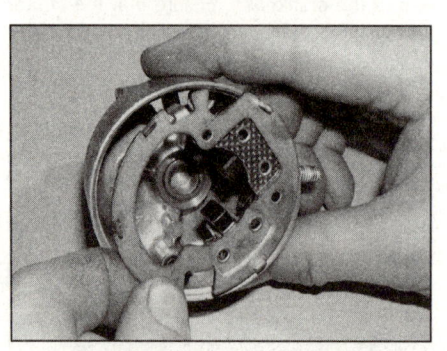

29.9a Withdraw the brushplate assembly . . .

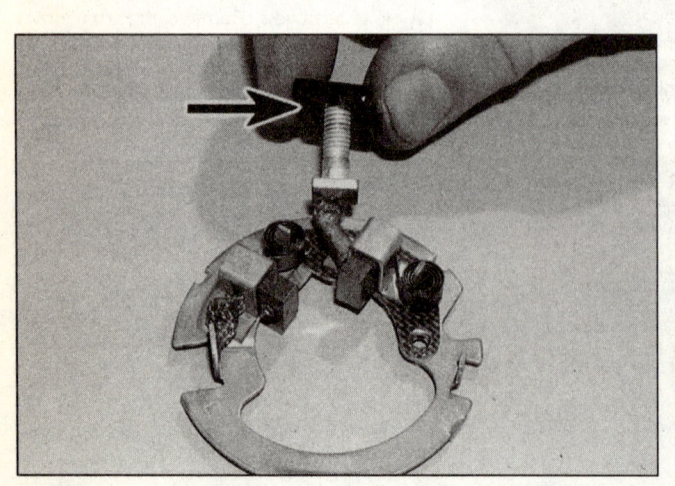

29.9b . . . and remove the square insulating washer (arrowed)

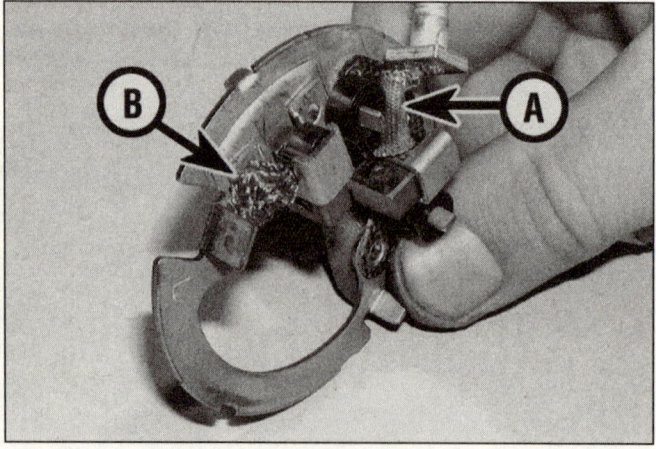

29.10 Brush (A) is attached to the terminal, brush (B) is attached to the brushplate

5 Remove the insulating washer and shims from the front end of the armature shaft, noting the order in which they are fitted **(see illustration)**.

6 Remove the main housing and remove the cover O-rings from the housing and discard them as new ones must be fitted **(see illustration)**.

7 Remove the rear cover and brushplate assembly from the armature commutator **(see illustration)**. Remove the shims from the rear end of the armature shaft **(see illustration)**.

8 Noting the order in which they are fitted, unscrew the terminal nut and remove it along with its washer, the insulating washers and O-ring **(see illustrations)**.

9 Withdraw the terminal and brushplate assembly from the rear cover and remove the square insulating washer from the terminal **(see illustrations)**.

10 Lift the brush springs and slide the brushes out from their holders, noting that one brush is attached to the terminal and the other is attached to the brushplate **(see illustration)**.

Inspection

11 Check the general condition of all the starter motor components **(see**

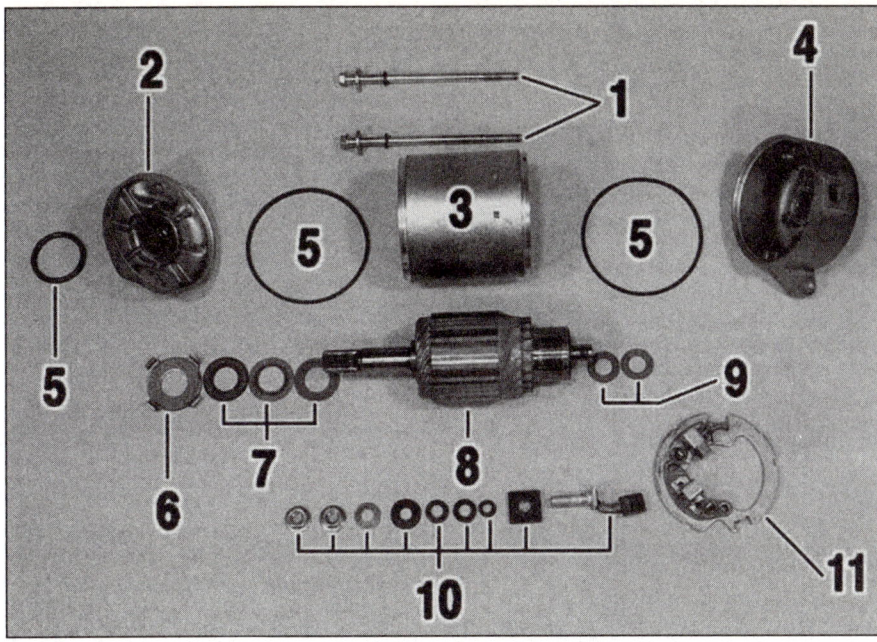

Electrical system 9•23

29.11a Starter motor components

1 Long bolts
2 Front cover
3 Main housing
4 Rear cover
5 O-rings
6 Tabbed washer
7 Insulating washer and front shims
8 Armature
9 Rear shims
10 Terminal bolt assembly
11 Brushplate assembly

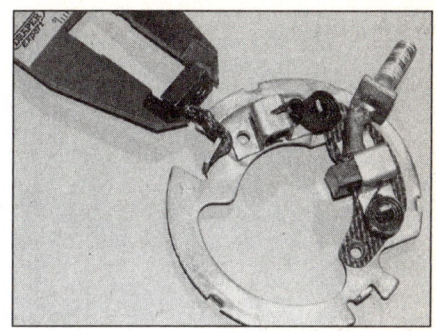

29.11b Measure the brush length

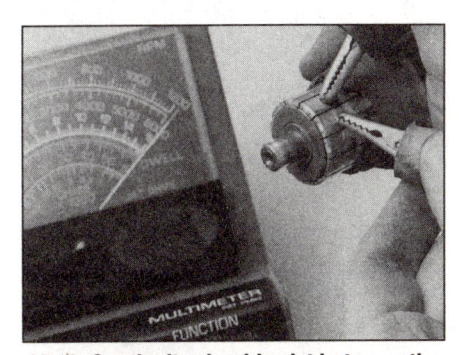

29.13 Continuity should exist between the commutator bars

29.14 There should be no continuity between the commutator bars and the armature shaft

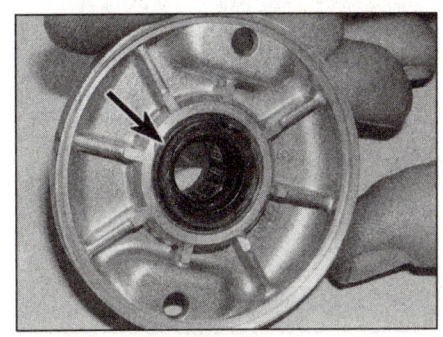

29.18 Inspect the front cover oil seal (arrowed)

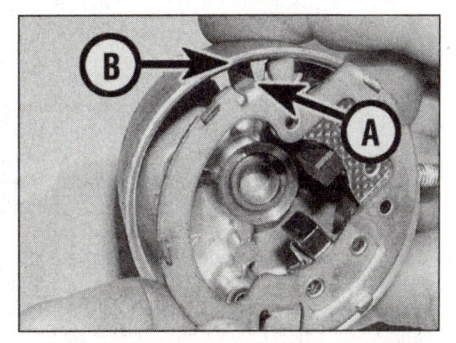

29.21 Fit brushplate tab (A) into slot (B)

illustration). The parts that are most likely to require attention are the brushes. Measure the length of the brushes and compare the results to the brush length listed in this Chapter's Specifications (see illustration). If either of the brushes are worn beyond the service limit, renew the brush assembly. If the brushes are not worn excessively, cracked, chipped, or otherwise damaged, they may be re-used.

12 Inspect the commutator bars on the armature for scoring, scratches and discoloration. The commutator can be cleaned and polished with crocus cloth, but do not use sandpaper or emery paper. After cleaning, wipe away any residue with a cloth soaked in electrical system cleaner or denatured alcohol.

13 Using an ohmmeter or a continuity test light, check for continuity between the commutator bars (see illustration). Continuity (zero resistance) should exist between each bar and all of the others.

14 Also, check for continuity between the commutator bars and the armature shaft (see illustration). There should be no continuity (infinite resistance) between the commutator and the shaft. If the checks indicate otherwise, the armature is defective.

15 Check for continuity between each brush and the brush terminal. There should be continuity (zero resistance). Check for continuity between the terminal bolt and the housing (when assembled). There should be no continuity (infinite resistance).

16 Check the front end of the armature shaft for worn, cracked, chipped and broken teeth. If the shaft is damaged or worn, renew the armature.

17 Inspect the end covers for signs of cracks or wear. Inspect the magnets in the main housing and the housing itself for cracks.

18 Inspect the insulating washers and front cover oil seal for signs of damage and renew them if necessary (see illustration).

Reassembly

19 Slide the brushes back into position in their holders and place the brush spring ends onto the brushes.

20 Ensure that the square insulating washer is in place on the terminal (see illustration 29.9b), then insert the terminal through the rear cover from the inside. Fit the O-ring, insulating washers and standard washer on the terminal and then fit and tighten the nut (see illustrations 29.8b and a).

21 Fit the brushplate assembly into the cover, making sure its tab is correctly located in the slot in the cover (see illustration).

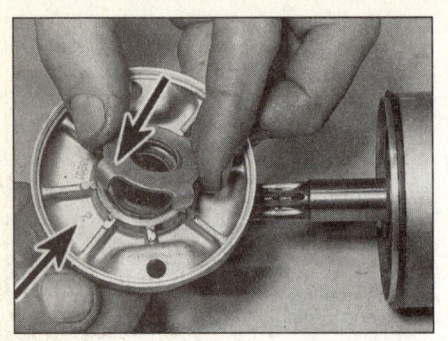

29.24 Align washer tabs with inside of front cover

29.25a Ensure the marks on the housing and end covers are aligned . . .

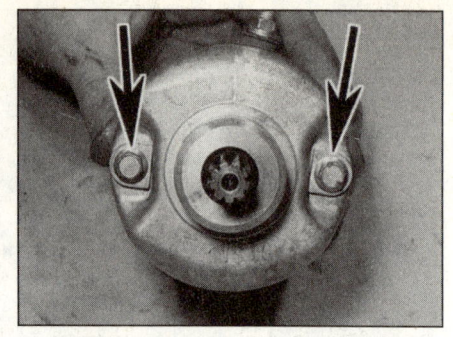

29.25b . . . and the flat edges of the D-washers are correctly located

22 Slide the shims onto the rear end of the armature shaft **(see illustration 29.7b)**. Lubricate the shaft with a smear of grease, then insert the shaft into the rear cover, locating the brushes on the commutator as you do, taking care not to damage the brushes. Check that each brush is securely pressed against the commutator by its spring and is free to move easily in its holder.
23 Fit new O-rings onto the main housing, then fit the housing over the armature and onto the rear cover, aligning the marks made on removal (see Step 2).
24 Slide the shims and then the insulating washer onto the front end of the armature shaft **(see illustration 29.5)** and lubricate the shaft with a smear of grease. Apply a smear of grease to the inside of the front cover oil seal and fit the tabbed washer into the cover, making sure the tabs locate correctly **(see illustration)**. Install the cover onto the main housing, aligning the marks made on removal (see Step 2).
25 Ensure that the D-shaped washers are in place on the long bolts and slide a new O-ring onto each of the bolts. Check the marks on the rear cover, main housing and front cover are correctly aligned, then install the long bolts and tighten them securely, making sure the flat edge on each D-washer is correctly fitted against the front cover **(see illustrations)**.
26 Remove the protective tape from the shaft end and install the starter motor (see Section 28).

30 Charging system testing – general information and precautions

1 If the performance of the charging system is suspect, the system as a whole should be checked first, followed by testing of the individual components. **Note:** *Before beginning the checks, make sure the battery is fully charged and that the voltage between its terminals is greater than 13 volts, and that all system connections are clean and tight (see Section 3).*

2 Checking the output of the charging system and the performance of the various components within the charging system requires the use of a multimeter (with voltage, current and resistance checking facilities). Ensure that the multimeter battery is in good condition before commencing tests.
3 When making the tests, follow the procedures carefully to prevent incorrect connections or short circuits, as irreparable damage to electrical system components may result if short circuits occur.
4 If a multimeter is not available, the job of checking the charging system should be left to a Honda dealer or automotive electrician.

31 Charging system – leakage and output tests

1 If the charging system of the machine is thought to be faulty carry-out the following tests with a multimeter.

Leakage test
2 Ensure that the ignition switch is OFF and disconnect the lead from the battery negative (-ve) terminal (see Section 3).
3 Set the multimeter to the Amps function and connect its negative (-ve) probe to the battery negative (-ve) terminal, and positive (+ve) probe to the disconnected negative (-ve) lead **(see illustration)**. Always set the meter to a high amps range initially and then bring it down to the mA (milli Amps) range; if there is a high current flow in the circuit it may blow the meter's fuse.
Caution: *Always connect an ammeter (multimeter set to Amps) in series, never in parallel with the battery, otherwise it will be damaged. Do not turn the ignition ON or operate the starter motor when the ammeter is connected – a sudden surge in current will blow the meter's fuse.*
4 If the current leakage indicated exceeds the amount specified at the beginning of this Chapter, there is probably a short circuit in the wiring. Disconnect the meter and reconnect the negative (-ve) lead to the battery, tightening it securely.

5 If leakage is indicated, use the wiring diagrams at the end of this book to systematically disconnect individual electrical components and repeat the test until the source is identified.

Output test
6 Start the engine and warm it up to normal operating temperature, then stop the engine. Remove the battery cover (see Section 3).
7 Connect the multimeter set to the 0 – 20 volts DC scale (voltmeter) across the terminals of the battery (positive (+ve) probe to battery positive (+ve) terminal, negative (-ve) probe to battery negative (-ve) terminal). Start the engine and turn the headlight HI beam on. Slowly increase the engine speed to 5,000 rpm and note the reading obtained, then stop the engine and turn the ignition OFF; do not allow the engine to overheat. The regulated voltage output should be as specified at the beginning of this Chapter. If the voltage output is outside these limits, check the alternator and the regulator (see Sections 32 and 33).

⚠ **Warning:** *Do not allow exhaust gases to build up in the work area; either perform the check outside or use an exhaust gas extraction system.*

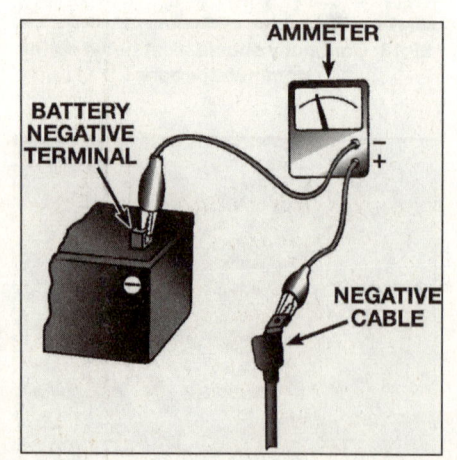

31.3 Checking the charging system leakage rate – connect the meter as shown

Electrical system 9•25

32.2 Disconnect the alternator wiring connector

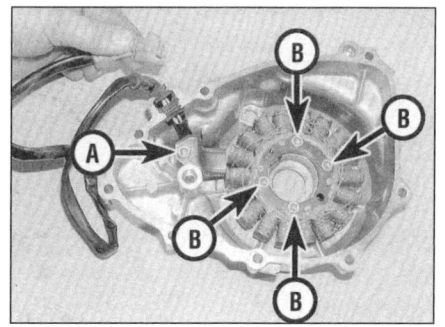

32.8 Alternator wiring clamp bolt (A) and stator bolts (B)

32.9a Unscrew the rotor bolt using a strap to hold the rotor . . .

32 Alternator – check, removal and installation

Check

1 Remove the left-hand frame side panel (see Chapter 8).
2 Trace the alternator wiring back from the top of the left-hand engine cover and disconnect it at the white 3-pin connector containing the three yellow wires **(see illustration)**.
3 Using the multimeter set to the resistance (ohms) scale, measure the resistance between each of the yellow wire terminals on the alternator side of the connector, taking a total of three readings. If the stator coil windings are in good condition the three readings should be within the range shown in the Specifications at the start of this Chapter.
4 Check for continuity between each terminal and earth (ground); there should be no continuity (infinite resistance) between any of the terminals and earth. If not, the alternator stator coil assembly is at fault and should be renewed. **Note:** *Before condemning the stator coils, check the fault is not due to damaged wiring between the connector and coils.*

Removal

5 Disconnect the wiring as described in Steps 1 and 2). Free the alternator wiring from any clips or guides and feed it through to the left-hand engine cover.
6 Drain the engine oil (see Chapter 1).
7 Remove the left-hand engine cover (see Chapter 2, Section 13).
8 To remove the alternator stator, first unscrew the bolt securing the alternator wiring clamp to the cover and remove the bolt, clamp and the wiring grommets from the engine cover. Unscrew the bolts securing the alternator stator to the cover and remove the stator **(see illustration)**.
9 To remove the rotor bolt from the end of the crankshaft it is necessary to stop the rotor from turning; use either the Honda service tool (pt. no. 07725-0040000) or a rotor holding strap **(see illustration)**. Alternatively, if the engine is in the frame, place the transmission in gear and have an assistant apply the rear brake to stop the rotor turning, then unscrew the bolt and remove the bolt and washer **(see illustration)**.
10 To remove the rotor from the shaft it is necessary to use a centre-bolt type rotor puller (Honda service tool pt. no. 07733-0020001 or a pattern alternative) – do not use a legged puller on the rotor. Thread the rotor puller into the centre of the rotor and turn it until the rotor is displaced from the shaft **(see illustration)**. Remove the Woodruff key from its slot in the end of the crankshaft for safekeeping if it is loose **(see illustration)**.

Installation

11 Install the stator in the cover, aligning the wiring with the outlet hole **(see illustration 32.8)**. Apply a suitable non-permanent thread-locking compound to the stator bolt threads, then install the bolts and tighten them to the torque setting specified at the beginning of this Chapter. Apply a suitable sealant to the wiring grommets, then install them in the seats in the cover. Secure the wiring with its clamp and tighten the clamp bolt.
12 Clean the tapered end of the crankshaft and the corresponding mating surface on the inside of the rotor with a suitable solvent. If removed, fit the Woodruff key into its slot in the crankshaft **(see illustration)**. Make sure that no metal objects have attached themselves to the magnets on the inside of the rotor, then install the rotor onto the shaft, aligning the slot

32.9b . . . then remove the bolt and washer

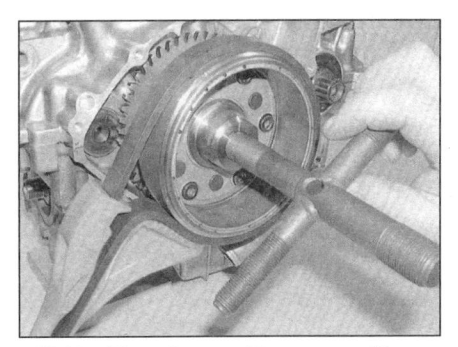

32.10a The rotor must be removed from the shaft with a puller as shown . . .

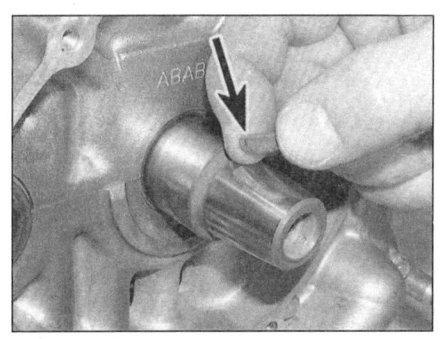

32.10b . . . then remove the Woodruff key from the shaft

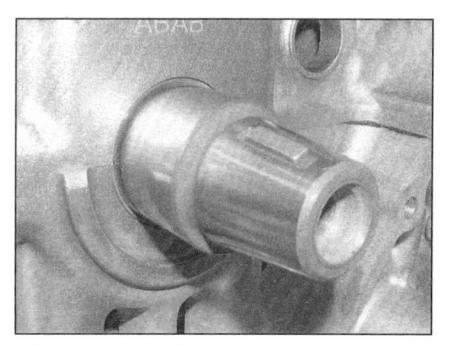

32.12a Fit the Woodruff key in its slot . . .

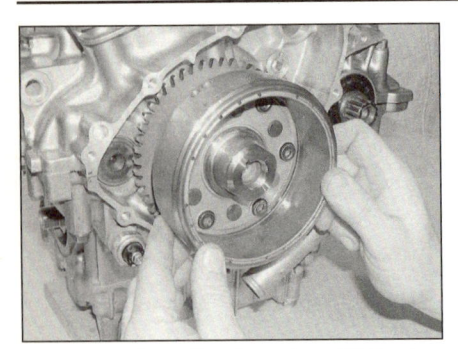

9•26 Electrical system

32.12b ... then align the rotor with the key and install it on the shaft

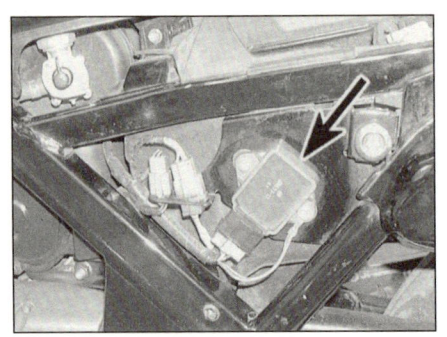

33.1a The regulator/rectifier – CB models

33.1b Regulator/rectifier – CBF models

in the rotor with the Woodruff key **(see illustration)**.

13 Apply a drop of clean engine oil to the threads of the rotor bolt, then install the bolt and washer and tighten the bolt to the torque setting specified at the beginning of this Chapter, using the method employed on removal to prevent the rotor from turning **(see illustration 32.9a)**.

14 Install the left-hand engine cover (see Chapter 2, Section 13).

15 Refill the engine with oil to the correct level (see *Daily (pre-ride) checks*).

16 Feed the alternator wiring back to its connector, making sure it is correctly routed, and reconnect it (see Steps 5 and 2).

33 Regulator/rectifier – check and replacement

> **HAYNES HINT** *Clues to a faulty regulator are constantly blowing bulbs, with brightness varying considerably with engine speed, and battery overheating.*

Check

1 On CB models remove the left-hand frame side panel (see Chapter 8) – the regulator/rectifier is mounted on the left-hand side of the frame **(see illustration)**. On CBF models remove the left-hand section of the seat cowling (see Chapter 8) – the regulator/rectifier is mounted on the sub-frame **(see illustration)**.

2 Disconnect the wiring connector, then connect the multimeter positive (+ve) probe to the red/white terminal and the negative (-ve) probe to the green terminal on the wiring connector and check for voltage with the ignition switched ON. Full battery voltage should be present. Switch the ignition OFF.

3 Switch the multimeter to the resistance (ohms) scale. Measure the resistance between each of the yellow wire terminals on the wiring connector, taking a total of three readings. If the stator coil windings are in good condition (see Section 32) and the wiring between the alternator and the rectifier is good, the three readings should be within the range shown for stator coil resistance in the Specifications at the start of this Chapter.

4 Check for continuity between the green terminal of the wiring connector and earth (ground) on the frame. There should be continuity.

5 If the above checks do not provide the specified results, check the wiring between the battery, regulator/rectifier and alternator (see the *wiring diagrams* at the end of this Chapter).

6 On CB models, if the wiring is good, test the regulator/rectifier unit. Remove the unit from the motorcycle (see Step 7) and use a multimeter set to the appropriate scale to check the resistance between the various terminals on the regulator/rectifier **(see illustration)**. If the readings do not compare closely with those shown in the accompanying table the regulator/rectifier unit must be renewed, although you are advised to have your findings confirmed by a Honda dealer before doing so. No test data is given for CBF models.

Note: *The use of certain multimeters could lead to false readings being obtained, as could a low battery in the meter and contact between the meter probes and your fingers when the meter is in use. Honda recommend the use of Sanwa or Kowa analogue type multimeters.*

Replacement

7 Disconnect the battery negative (-ve) lead (see Section 3). Refer to Step 1 for access to the regulator/rectifier. Unscrew the two bolts securing the regulator/rectifier and remove it, on CB models noting the earth wire secured by the bottom bolt. Disconnect the wiring connector from the regulator/rectifier.

8 To install the regulator/rectifier, ensure that the terminals in the connector are clean and connect it to the unit. Install the mounting bolts, on CB models with the earth wire on the bottom bolt, and tighten the bolts.

9 On CB models install the left-hand frame side panel (see Chapter 8). On CBF models install the left-hand section of the seat cowling (see Chapter 8). Reconnect the battery negative lead.

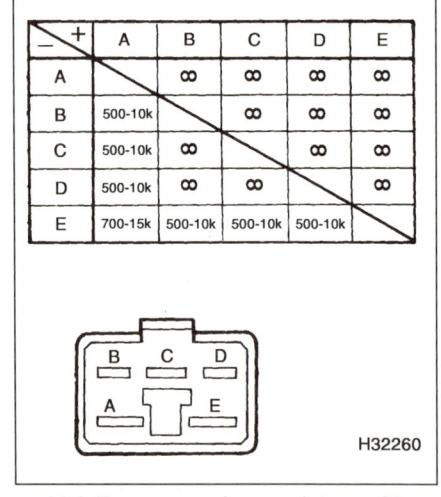

– / +	A	B	C	D	E
A		∞	∞	∞	∞
B	500-10k		∞	∞	∞
C	500-10k	∞		∞	∞
D	500-10k	∞	∞		∞
E	700-15k	500-10k	500-10k	500-10k	

33.6 Test connections and data – CB models

Sanwa tester – K-ohms range
Kowa tester – 100 ohms range
∞ – infinite resistance (no continuity)

Electrical system 9•27

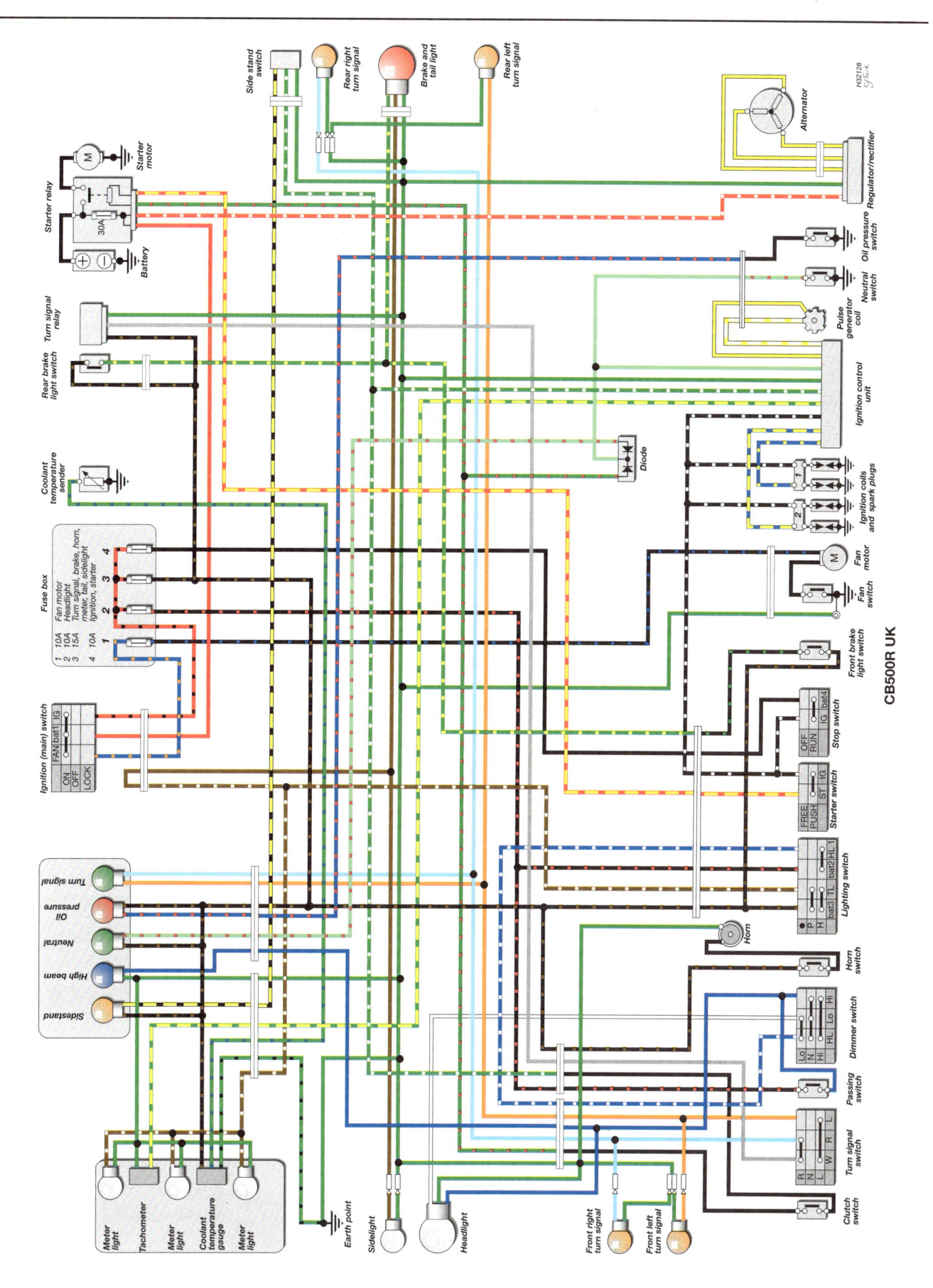

9•28 Wiring diagrams

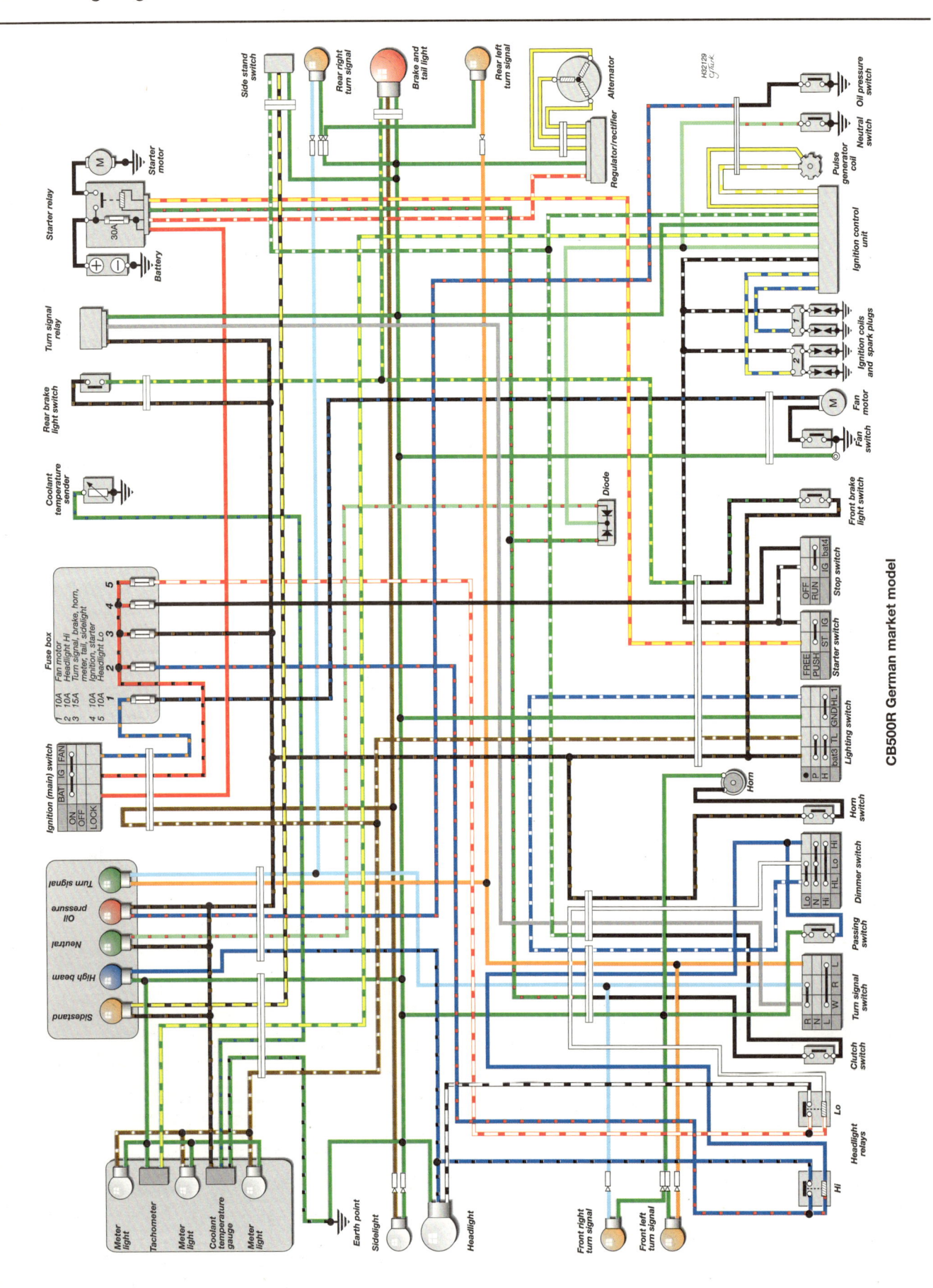

Electrical system 9•29

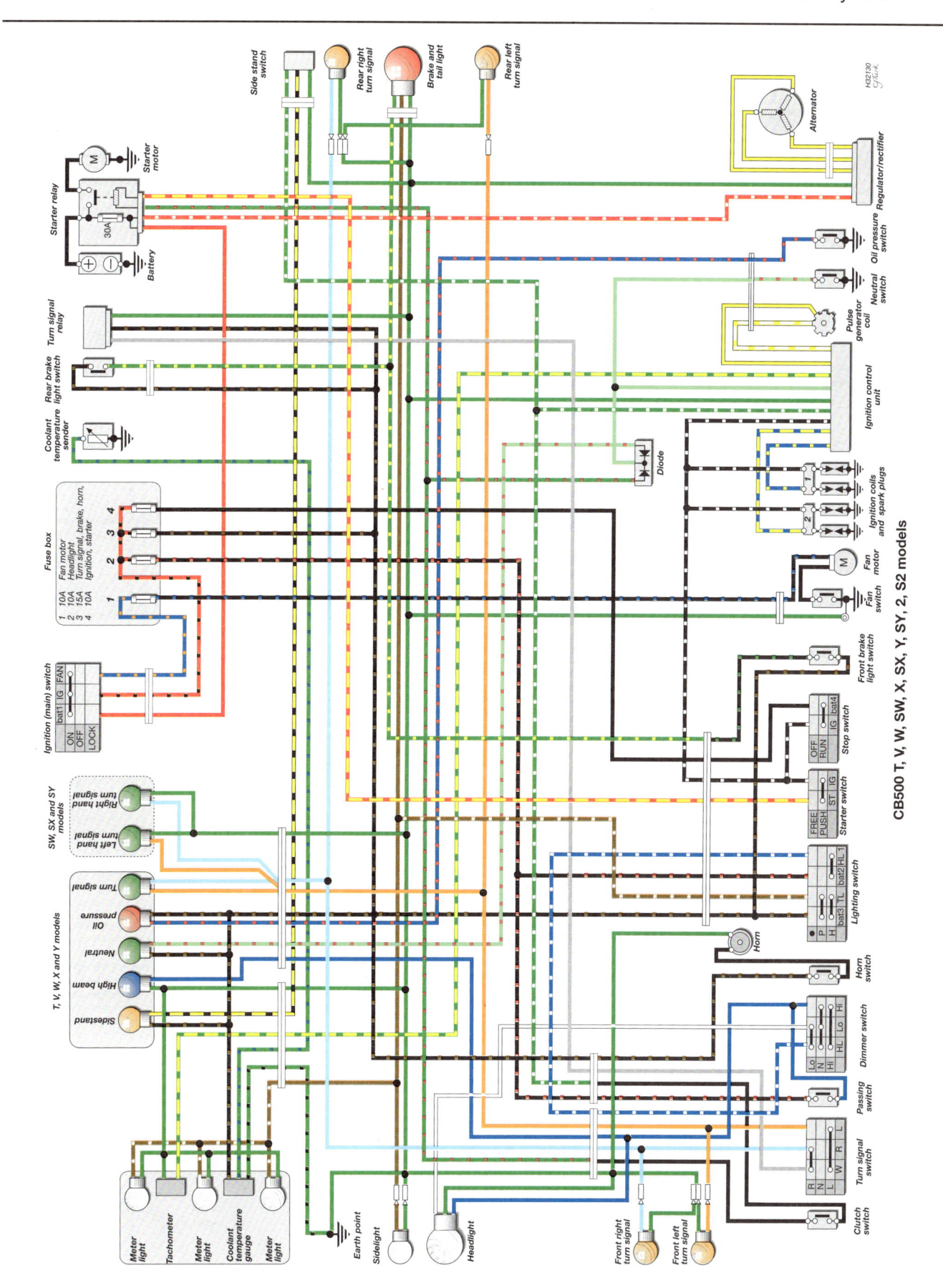

CB500 T, V, W, SW, X, SX, Y, SY, 2, S2 models

9•30 Wiring diagrams

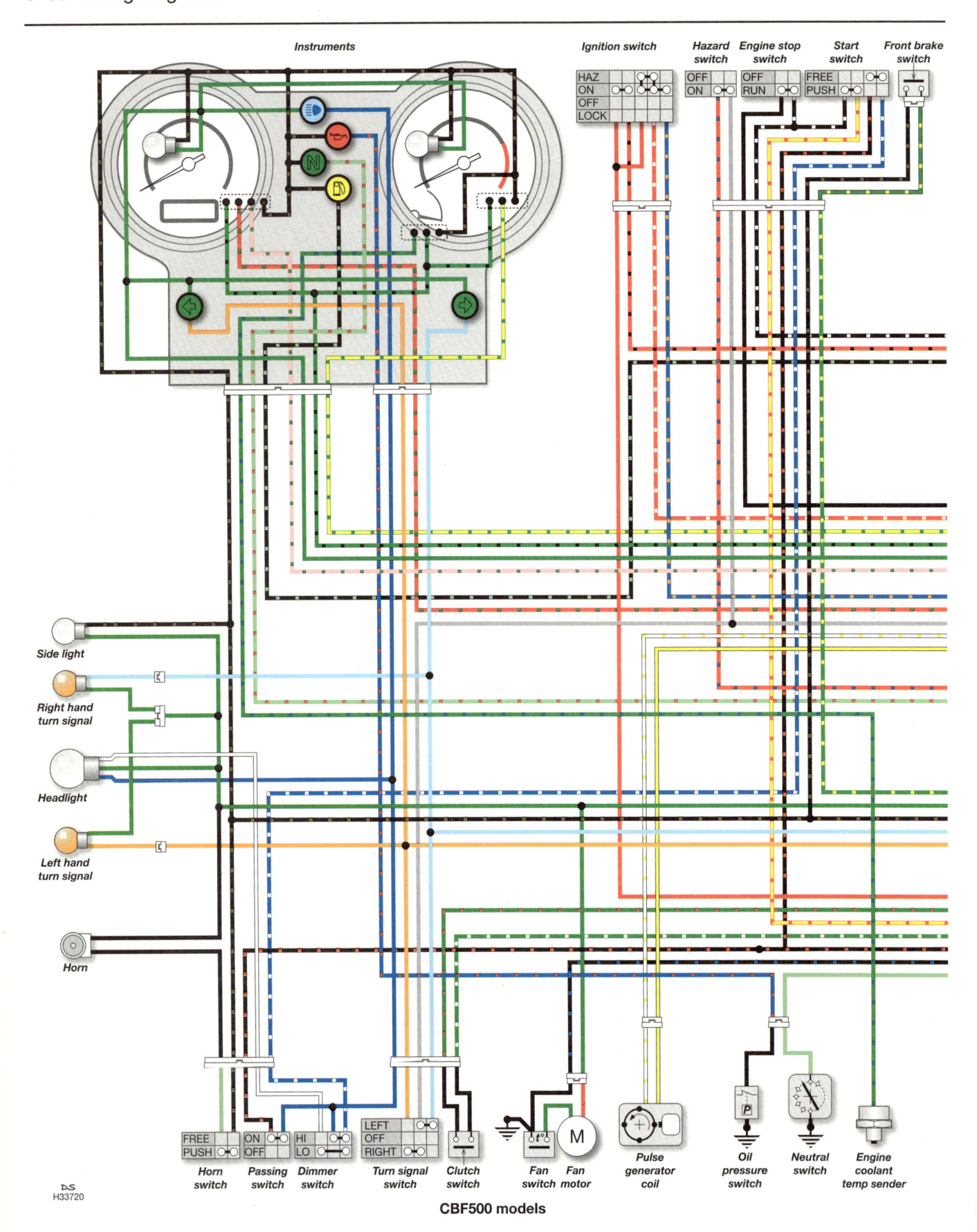

CBF500 models

Electrical system 9•31

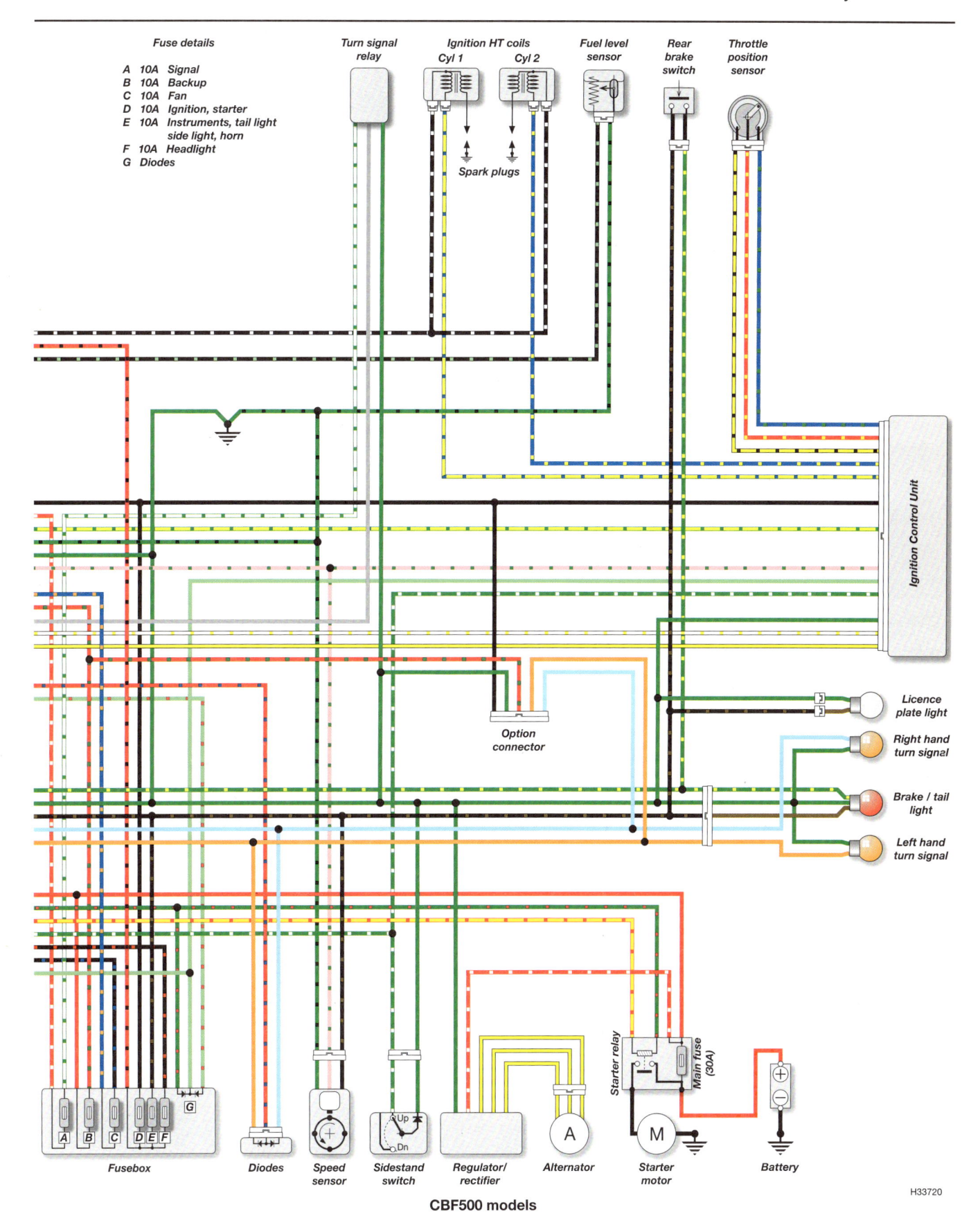

Fuse details

A 10A Signal
B 10A Backup
C 10A Fan
D 10A Ignition, starter
E 10A Instruments, tail light side light, horn
F 10A Headlight
G Diodes

CBF500 models

H33720

9•32 Wiring diagrams

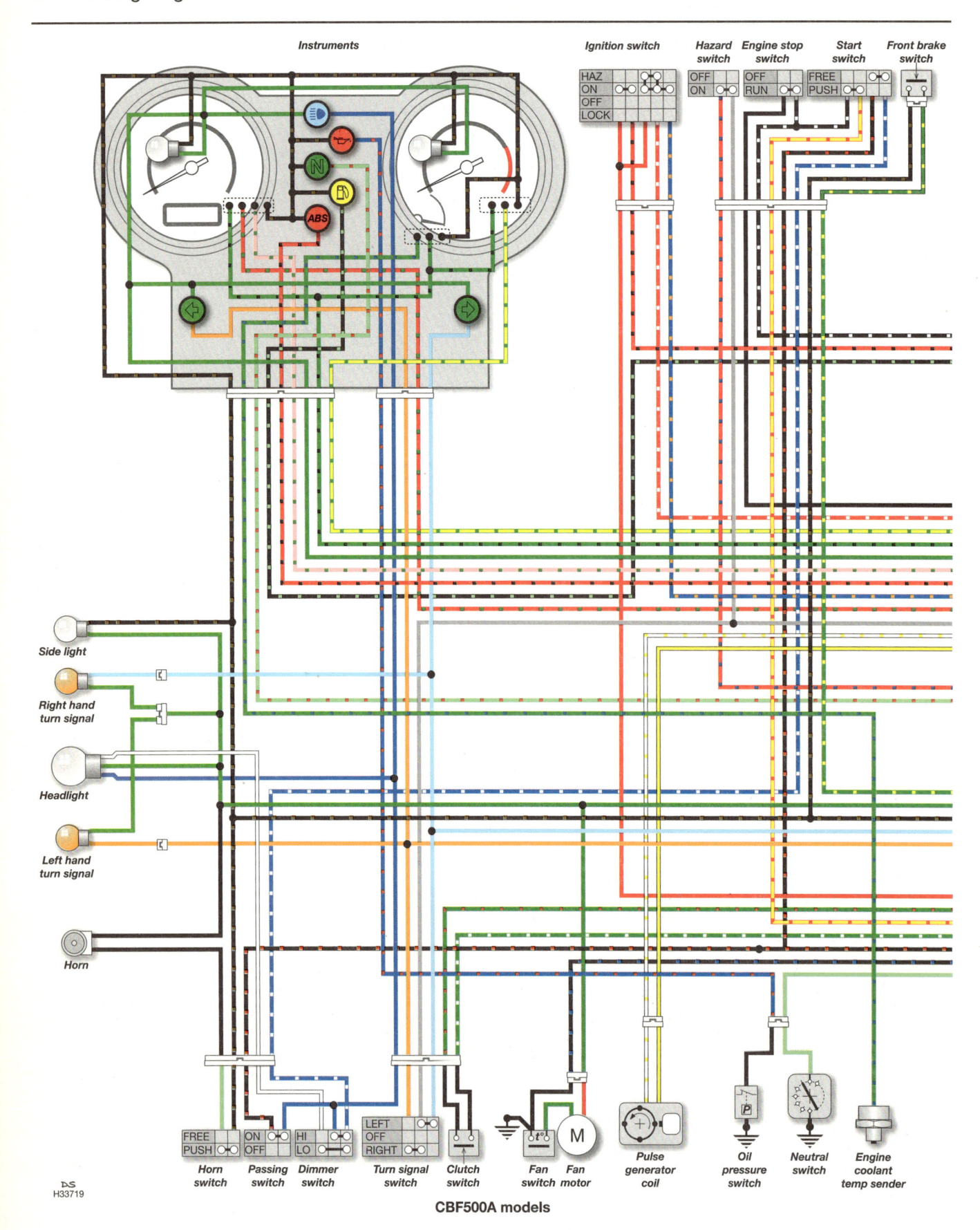

CBF500A models

Electrical system 9•33

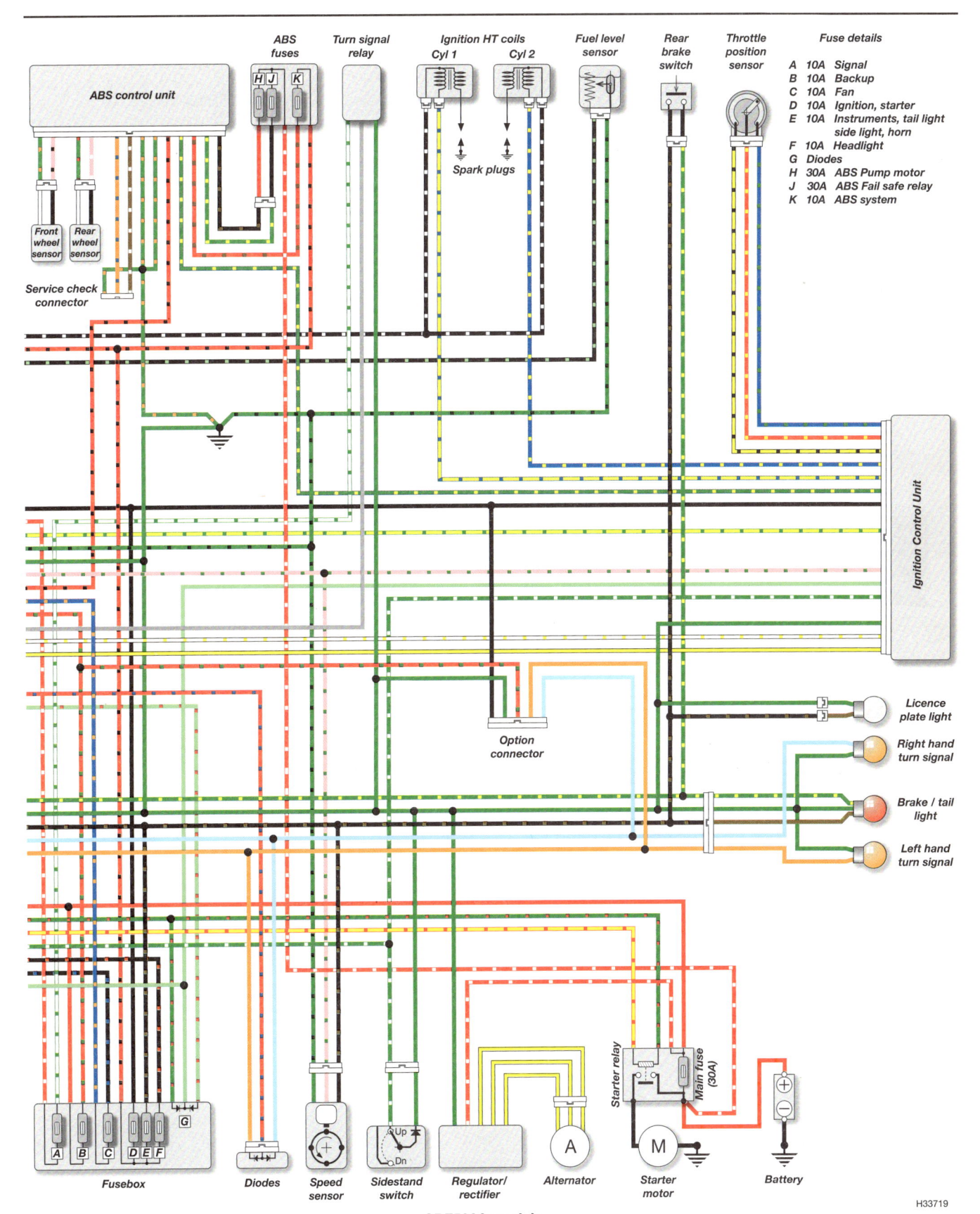

ABS control unit

Front wheel sensor
Rear wheel sensor
Service check connector

ABS fuses H J K
Turn signal relay
Ignition HT coils Cyl 1 Cyl 2
Spark plugs
Fuel level sensor
Rear brake switch
Throttle position sensor

Fuse details
A 10A Signal
B 10A Backup
C 10A Fan
D 10A Ignition, starter
E 10A Instruments, tail light side light, horn
F 10A Headlight
G Diodes
H 30A ABS Pump motor
J 30A ABS Fail safe relay
K 10A ABS system

Ignition Control Unit

Option connector

Licence plate light
Right hand turn signal
Brake / tail light
Left hand turn signal

Fusebox A B C D E F G
Diodes
Speed sensor
Sidestand switch Up Dn
Regulator/ rectifier
Alternator A
Starter relay
Main fuse (30A)
Starter motor M
Battery

CBF500A models

H33719

Notes

Reference

Tools and Workshop Tips — REF•2
- Building up a tool kit and equipping your workshop ● Using tools ● Understanding bearing, seal, fastener and chain sizes and markings ● Repair techniques

Security — REF•20
- Locks and chains ● U-locks ● Disc locks ● Alarms and immobilisers ● Security marking systems ● Tips on how to prevent bike theft

Lubricants and fluids — REF•23
- Engine oils ● Transmission (gear) oils ● Coolant/anti-freeze ● Fork oils and suspension fluids ● Brake/clutch fluids ● Spray lubes, degreasers and solvents

Conversion Factors — REF•26

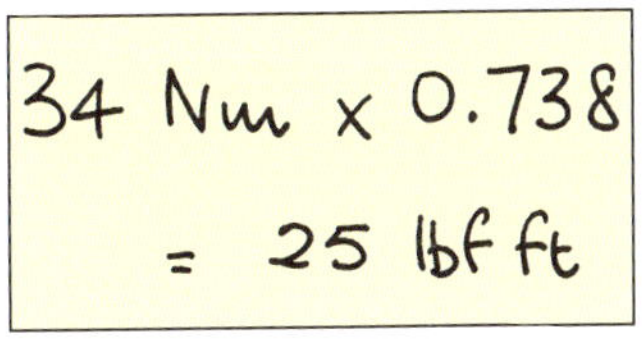

34 Nm × 0.738 = 25 lbf ft

- Formulae for conversion of the metric (SI) units used throughout the manual into Imperial measures

MOT Test Checks — REF•27
- A guide to the UK MOT test ● Which items are tested ● How to prepare your motorcycle for the test and perform a pre-test check

Storage — REF•32
- How to prepare your motorcycle for going into storage and protect essential systems ● How to get the motorcycle back on the road

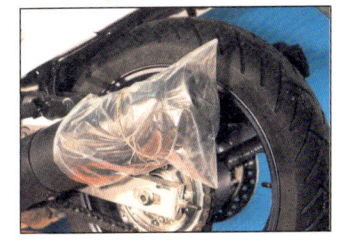

Fault Finding — REF•35
- Common faults and their likely causes ● How to check engine cylinder compression ● How to make electrical tests and use test meters

Index — REF•49

REF•2 Tools and Workshop Tips

Buying tools

A toolkit is a fundamental requirement for servicing and repairing a motorcycle. Although there will be an initial expense in building up enough tools for servicing, this will soon be offset by the savings made by doing the job yourself. As experience and confidence grow, additional tools can be added to enable the repair and overhaul of the motorcycle. Many of the specialist tools are expensive and not often used so it may be preferable to hire them, or for a group of friends or motorcycle club to join in the purchase.

As a rule, it is better to buy more expensive, good quality tools. Cheaper tools are likely to wear out faster and need to be renewed more often, nullifying the original saving.

> **Warning:** To avoid the risk of a poor quality tool breaking in use, causing injury or damage to the component being worked on, always aim to purchase tools which meet the relevant national safety standards.

The following lists of tools do not represent the manufacturer's service tools, but serve as a guide to help the owner decide which tools are needed for this level of work. In addition, items such as an electric drill, hacksaw, files, soldering iron and a workbench equipped with a vice, may be needed. Although not classed as tools, a selection of bolts, screws, nuts, washers and pieces of tubing always come in useful.

For more information about tools, refer to the Haynes *Motorcycle Workshop Practice Techbook* (Bk. No. 3470).

Manufacturer's service tools

Inevitably certain tasks require the use of a service tool. Where possible an alternative tool or method of approach is recommended, but sometimes there is no option if personal injury or damage to the component is to be avoided. Where required, service tools are referred to in the relevant procedure.

Service tools can usually only be purchased from a motorcycle dealer and are identified by a part number. Some of the commonly-used tools, such as rotor pullers, are available in aftermarket form from mail-order motorcycle tool and accessory suppliers.

Maintenance and minor repair tools

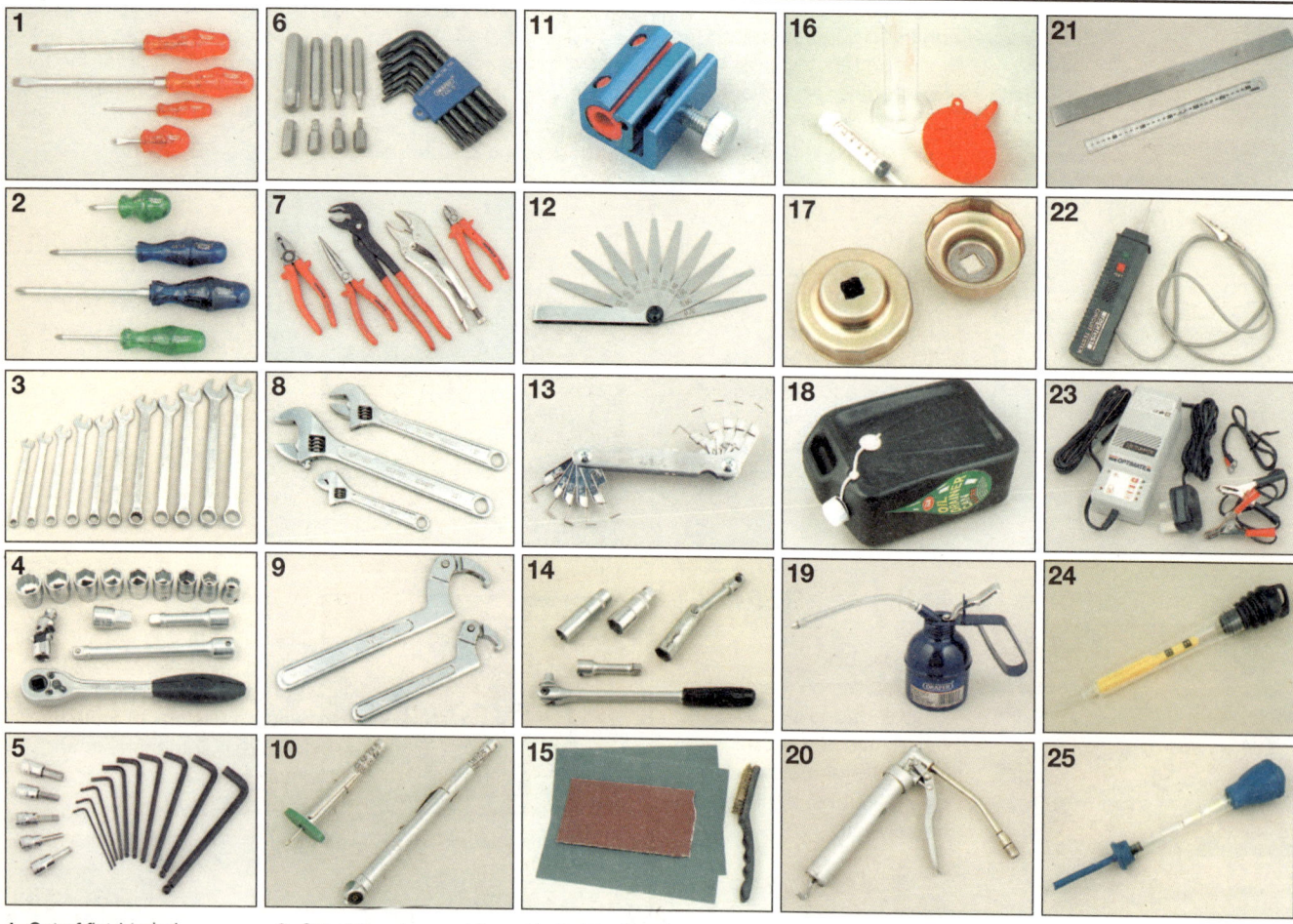

1. Set of flat-bladed screwdrivers
2. Set of Phillips head screwdrivers
3. Combination open-end and ring spanners
4. Socket set (3/8 inch or 1/2 inch drive)
5. Set of Allen keys or bits
6. Set of Torx keys or bits
7. Pliers, cutters and self-locking grips (Mole grips)
8. Adjustable spanners
9. C-spanners
10. Tread depth gauge and tyre pressure gauge
11. Cable oiler clamp
12. Feeler gauges
13. Spark plug gap measuring tool
14. Spark plug spanner or deep plug sockets
15. Wire brush and emery paper
16. Calibrated syringe, measuring vessel and funnel
17. Oil filter adapters
18. Oil drainer can or tray
19. Pump type oil can
20. Grease gun
21. Straight-edge and steel rule
22. Continuity tester
23. Battery charger
24. Hydrometer (for battery specific gravity check)
25. Anti-freeze tester (for liquid-cooled engines)

Tools and Workshop Tips REF•3

Repair and overhaul tools

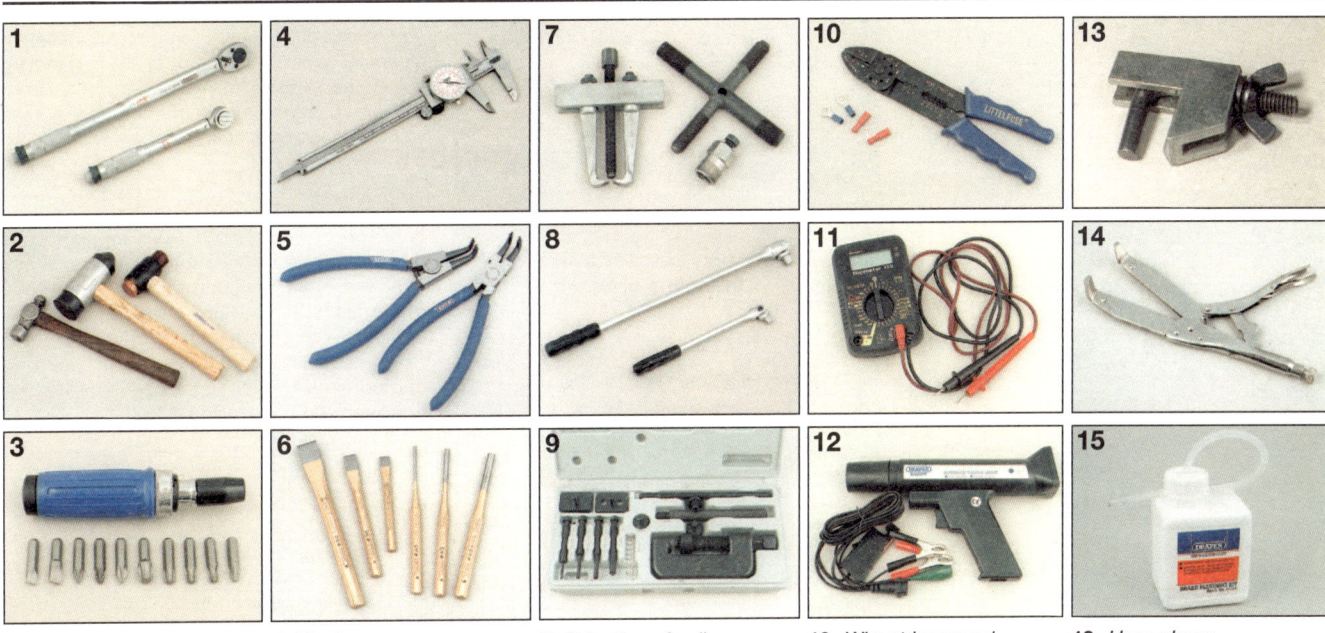

1. Torque wrench (small and mid-ranges)
2. Conventional, plastic or soft-faced hammers
3. Impact driver set
4. Vernier gauge
5. Circlip pliers (internal and external, or combination)
6. Set of cold chisels and punches
7. Selection of pullers
8. Breaker bars
9. Chain breaking/riveting tool set
10. Wire stripper and crimper tool
11. Multimeter (measures amps, volts and ohms)
12. Stroboscope (for dynamic timing checks)
13. Hose clamp (wingnut type shown)
14. Clutch holding tool
15. One-man brake/clutch bleeder kit

Specialist tools

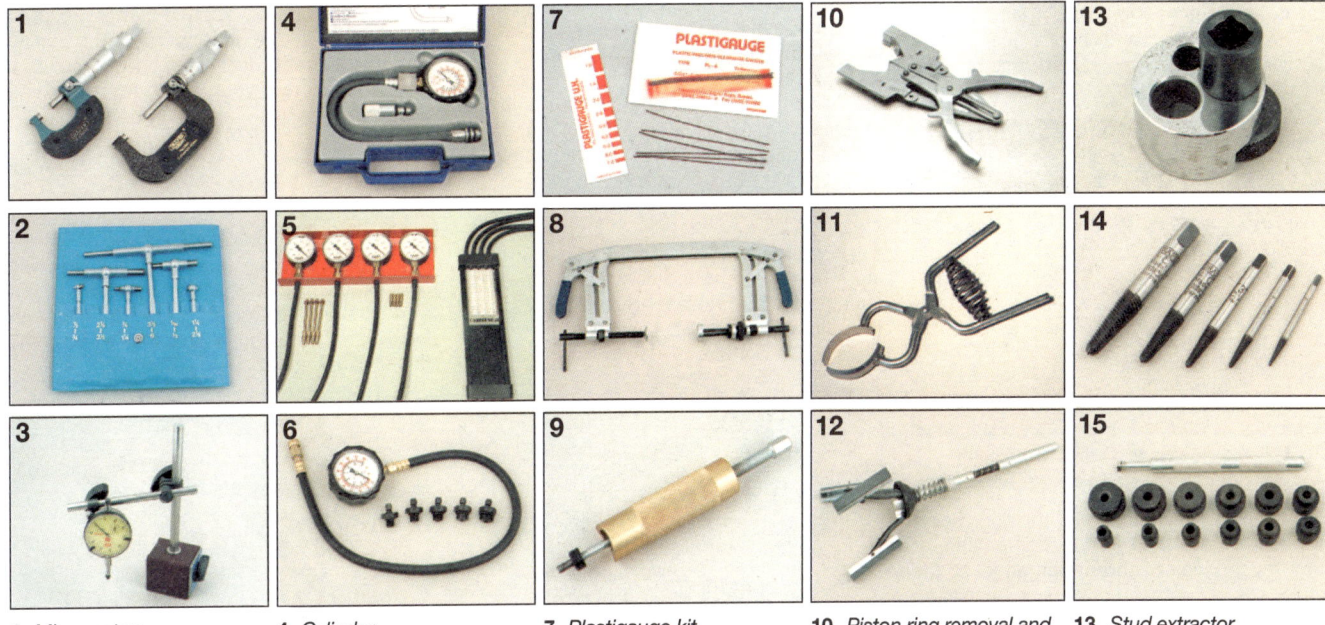

1. Micrometers (external type)
2. Telescoping gauges
3. Dial gauge
4. Cylinder compression gauge
5. Vacuum gauges (left) or manometer (right)
6. Oil pressure gauge
7. Plastigauge kit
8. Valve spring compressor (4-stroke engines)
9. Piston pin drawbolt tool
10. Piston ring removal and installation tool
11. Piston ring clamp
12. Cylinder bore hone (stone type shown)
13. Stud extractor
14. Screw extractor set
15. Bearing driver set

REF•4 Tools and Workshop Tips

1 Workshop equipment and facilities

The workbench

● Work is made much easier by raising the bike up on a ramp - components are much more accessible if raised to waist level. The hydraulic or pneumatic types seen in the dealer's workshop are a sound investment if you undertake a lot of repairs or overhauls **(see illustration 1.1)**.

1.1 Hydraulic motorcycle ramp

● If raised off ground level, the bike must be supported on the ramp to avoid it falling. Most ramps incorporate a front wheel locating clamp which can be adjusted to suit different diameter wheels. When tightening the clamp, take care not to mark the wheel rim or damage the tyre - use wood blocks on each side to prevent this.
● Secure the bike to the ramp using tie-downs **(see illustration 1.2)**. If the bike has only a sidestand, and hence leans at a dangerous angle when raised, support the bike on an auxiliary stand.

1.2 Tie-downs are used around the passenger footrests to secure the bike

● Auxiliary (paddock) stands are widely available from mail order companies or motorcycle dealers and attach either to the wheel axle or swingarm pivot **(see illustration 1.3)**. If the motorcycle has a centrestand, you can support it under the crankcase to prevent it toppling whilst either wheel is removed **(see illustration 1.4)**.

1.3 This auxiliary stand attaches to the swingarm pivot

1.4 Always use a block of wood between the engine and jack head when supporting the engine in this way

Fumes and fire

● Refer to the Safety first! page at the beginning of the manual for full details. Make sure your workshop is equipped with a fire extinguisher suitable for fuel-related fires (Class B fire - flammable liquids) - it is not sufficient to have a water-filled extinguisher.
● Always ensure adequate ventilation is available. Unless an exhaust gas extraction system is available for use, ensure that the engine is run outside of the workshop.
● If working on the fuel system, make sure the workshop is ventilated to avoid a build-up of fumes. This applies equally to fume build-up when charging a battery. Do not smoke or allow anyone else to smoke in the workshop.

Fluids

● If you need to drain fuel from the tank, store it in an approved container marked as suitable for the storage of petrol (gasoline) **(see illustration 1.5)**. Do not store fuel in glass jars or bottles.

1.5 Use an approved can only for storing petrol (gasoline)

● Use proprietary engine degreasers or solvents which have a high flash-point, such as paraffin (kerosene), for cleaning off oil, grease and dirt - never use petrol (gasoline) for cleaning. Wear rubber gloves when handling solvent and engine degreaser. The fumes from certain solvents can be dangerous - always work in a well-ventilated area.

Dust, eye and hand protection

● Protect your lungs from inhalation of dust particles by wearing a filtering mask over the nose and mouth. Many frictional materials still contain asbestos which is dangerous to your health. Protect your eyes from spouts of liquid and sprung components by wearing a pair of protective goggles **(see illustration 1.6)**.

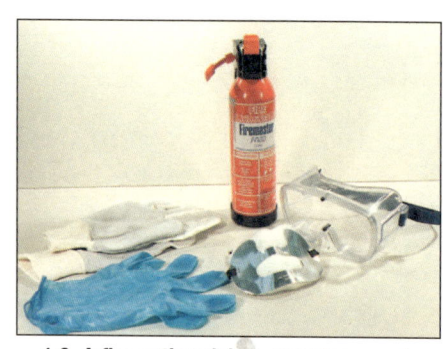

1.6 A fire extinguisher, goggles, mask and protective gloves should be at hand in the workshop

● Protect your hands from contact with solvents, fuel and oils by wearing rubber gloves. Alternatively apply a barrier cream to your hands before starting work. If handling hot components or fluids, wear suitable gloves to protect your hands from scalding and burns.

What to do with old fluids

● Old cleaning solvent, fuel, coolant and oils should not be poured down domestic drains or onto the ground. Package the fluid up in old oil containers, label it accordingly, and take it to a garage or disposal facility. Contact your local authority for location of such sites or ring the oil care hotline.

OIL CARE
0800 66 33 66
www.oilbankline.org.uk

Note: It is antisocial and illegal to dump oil down the drain. To find the location of your local oil recycling bank, call this number free.

In the USA, note that any oil supplier must accept used oil for recycling.

Tools and Workshop Tips REF•5

2 Fasteners - screws, bolts and nuts

Fastener types and applications

Bolts and screws

● Fastener head types are either of hexagonal, Torx or splined design, with internal and external versions of each type **(see illustrations 2.1 and 2.2)**; splined head fasteners are not in common use on motorcycles. The conventional slotted or Phillips head design is used for certain screws. Bolt or screw length is always measured from the underside of the head to the end of the item **(see illustration 2.11)**.

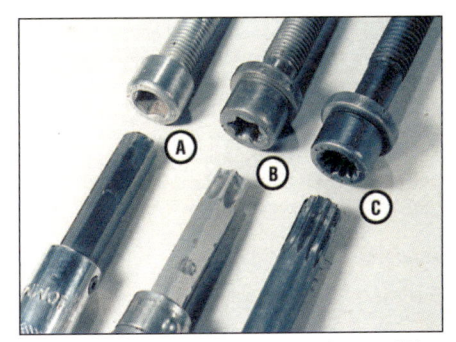

2.1 Internal hexagon/Allen (A), Torx (B) and splined (C) fasteners, with corresponding bits

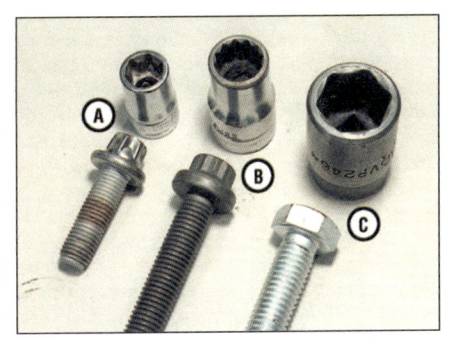

2.2 External Torx (A), splined (B) and hexagon (C) fasteners, with corresponding sockets

● Certain fasteners on the motorcycle have a tensile marking on their heads, the higher the marking the stronger the fastener. High tensile fasteners generally carry a 10 or higher marking. Never replace a high tensile fastener with one of a lower tensile strength.

Washers (see illustration 2.3)

● Plain washers are used between a fastener head and a component to prevent damage to the component or to spread the load when torque is applied. Plain washers can also be used as spacers or shims in certain assemblies. Copper or aluminium plain washers are often used as sealing washers on drain plugs.

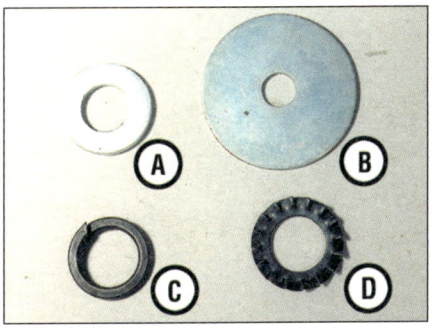

2.3 Plain washer (A), penny washer (B), spring washer (C) and serrated washer (D)

● The split-ring spring washer works by applying axial tension between the fastener head and component. If flattened, it is fatigued and must be renewed. If a plain (flat) washer is used on the fastener, position the spring washer between the fastener and the plain washer.

● Serrated star type washers dig into the fastener and component faces, preventing loosening. They are often used on electrical earth (ground) connections to the frame.

● Cone type washers (sometimes called Belleville) are conical and when tightened apply axial tension between the fastener head and component. They must be installed with the dished side against the component and often carry an OUTSIDE marking on their outer face. If flattened, they are fatigued and must be renewed.

● Tab washers are used to lock plain nuts or bolts on a shaft. A portion of the tab washer is bent up hard against one flat of the nut or bolt to prevent it loosening. Due to the tab washer being deformed in use, a new tab washer should be used every time it is disturbed.

● Wave washers are used to take up endfloat on a shaft. They provide light springing and prevent excessive side-to-side play of a component. Can be found on rocker arm shafts.

Nuts and split pins

● Conventional plain nuts are usually six-sided **(see illustration 2.4)**. They are sized by thread diameter and pitch. High tensile nuts carry a number on one end to denote their tensile strength.

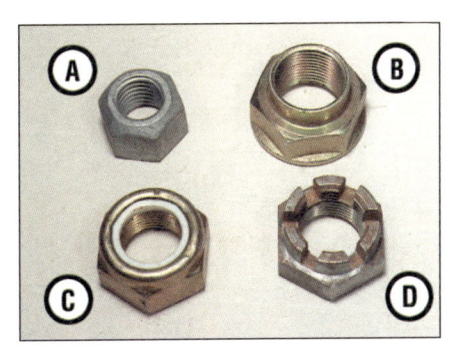

2.4 Plain nut (A), shouldered locknut (B), nylon insert nut (C) and castellated nut (D)

● Self-locking nuts either have a nylon insert, or two spring metal tabs, or a shoulder which is staked into a groove in the shaft - their advantage over conventional plain nuts is a resistance to loosening due to vibration. The nylon insert type can be used a number of times, but must be renewed when the friction of the nylon insert is reduced, ie when the nut spins freely on the shaft. The spring tab type can be reused unless the tabs are damaged. The shouldered type must be renewed every time it is disturbed.

● Split pins (cotter pins) are used to lock a castellated nut to a shaft or to prevent slackening of a plain nut. Common applications are wheel axles and brake torque arms. Because the split pin arms are deformed to lock around the nut a new split pin must always be used on installation - always fit the correct size split pin which will fit snugly in the shaft hole. Make sure the split pin arms are correctly located around the nut **(see illustrations 2.5 and 2.6)**.

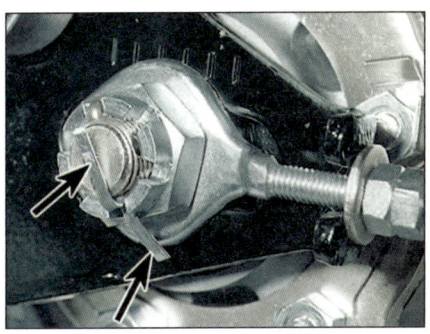

2.5 Bend split pin (cotter pin) arms as shown (arrows) to secure a castellated nut

2.6 Bend split pin (cotter pin) arms as shown to secure a plain nut

Caution: If the castellated nut slots do not align with the shaft hole after tightening to the torque setting, tighten the nut until the next slot aligns with the hole - never slacken the nut to align its slot.

● R-pins (shaped like the letter R), or slip pins as they are sometimes called, are sprung and can be reused if they are otherwise in good condition. Always install R-pins with their closed end facing forwards **(see illustration 2.7)**.

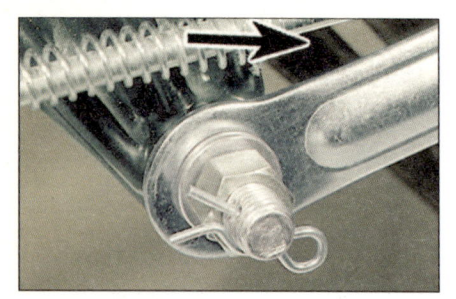

Tools and Workshop Tips

2.7 Correct fitting of R-pin. Arrow indicates forward direction

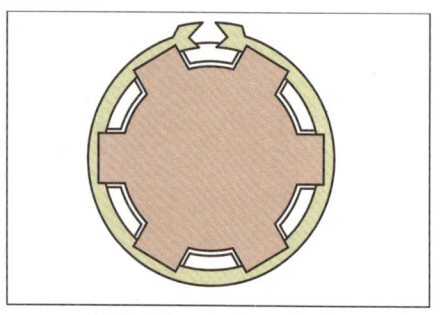

2.10 Align circlip opening with shaft channel

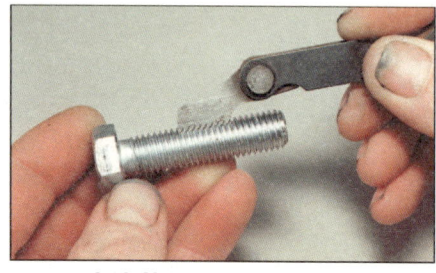

2.12 Using a thread gauge to measure pitch

Circlips (see illustration 2.8)

- Circlips (sometimes called snap-rings) are used to retain components on a shaft or in a housing and have corresponding external or internal ears to permit removal. Parallel-sided (machined) circlips can be installed either way round in their groove, whereas stamped circlips (which have a chamfered edge on one face) must be installed with the chamfer facing away from the direction of thrust load **(see illustration 2.9)**.

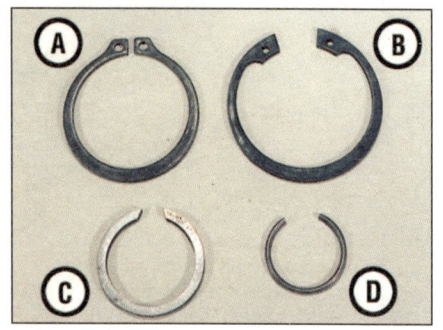

- Circlips can wear due to the thrust of components and become loose in their grooves, with the subsequent danger of becoming dislodged in operation. For this reason, renewal is advised every time a circlip is disturbed.

- Wire circlips are commonly used as piston pin retaining clips. If a removal tang is provided, long-nosed pliers can be used to dislodge them, otherwise careful use of a small flat-bladed screwdriver is necessary. Wire circlips should be renewed every time they are disturbed.

Thread diameter and pitch

- Diameter of a male thread (screw, bolt or stud) is the outside diameter of the threaded portion **(see illustration 2.11)**. Most motorcycle manufacturers use the ISO (International Standards Organisation) metric system expressed in millimetres, eg M6 refers to a 6 mm diameter thread. Sizing is the same for nuts, except that the thread diameter is measured across the valleys of the nut.

- Pitch is the distance between the peaks of the thread **(see illustration 2.11)**. It is expressed in millimetres, thus a common bolt size may be expressed as 6.0 x 1.0 mm (6 mm thread diameter and 1 mm pitch). Generally pitch increases in proportion to thread diameter, although there are always exceptions.

- Thread diameter and pitch are related for conventional fastener applications and the accompanying table can be used as a guide. Additionally, the AF (Across Flats), spanner or socket size dimension of the bolt or nut **(see illustration 2.11)** is linked to thread and pitch specification. Thread pitch can be measured with a thread gauge **(see illustration 2.12)**.

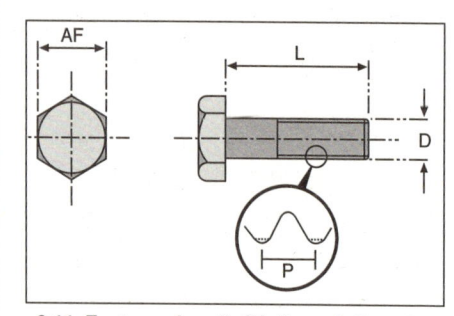

2.8 External stamped circlip (A), internal stamped circlip (B), machined circlip (C) and wire circlip (D)

- Always use circlip pliers to remove and install circlips; expand or compress them just enough to remove them. After installation, rotate the circlip in its groove to ensure it is securely seated. If installing a circlip on a splined shaft, always align its opening with a shaft channel to ensure the circlip ends are well supported and unlikely to catch **(see illustration 2.10)**.

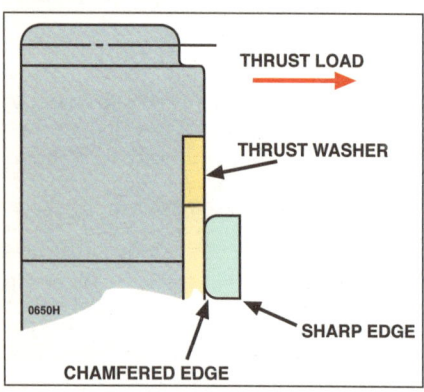

2.9 Correct fitting of a stamped circlip

2.11 Fastener length (L), thread diameter (D), thread pitch (P) and head size (AF)

AF size	Thread diameter x pitch (mm)
8 mm	M5 x 0.8
8 mm	M6 x 1.0
10 mm	M6 x 1.0
12 mm	M8 x 1.25
14 mm	M10 x 1.25
17 mm	M12 x 1.25

- The threads of most fasteners are of the right-hand type, ie they are turned clockwise to tighten and anti-clockwise to loosen. The reverse situation applies to left-hand thread fasteners, which are turned anti-clockwise to tighten and clockwise to loosen. Left-hand threads are used where rotation of a component might loosen a conventional right-hand thread fastener.

Seized fasteners

- Corrosion of external fasteners due to water or reaction between two dissimilar metals can occur over a period of time. It will build up sooner in wet conditions or in countries where salt is used on the roads during the winter. If a fastener is severely corroded it is likely that normal methods of removal will fail and result in its head being ruined. When you attempt removal, the fastener thread should be heard to crack free and unscrew easily - if it doesn't, stop there before damaging something.

- A smart tap on the head of the fastener will often succeed in breaking free corrosion which has occurred in the threads **(see illustration 2.13)**.

- An aerosol penetrating fluid (such as WD-40) applied the night beforehand may work its way down into the thread and ease removal. Depending on the location, you may be able to make up a Plasticine well around the fastener head and fill it with penetrating fluid.

2.13 A sharp tap on the head of a fastener will often break free a corroded thread

Tools and Workshop Tips REF•7

• If you are working on an engine internal component, corrosion will most likely not be a problem due to the well lubricated environment. However, components can be very tight and an impact driver is a useful tool in freeing them **(see illustration 2.14)**.

2.14 Using an impact driver to free a fastener

• Where corrosion has occurred between dissimilar metals (eg steel and aluminium alloy), the application of heat to the fastener head will create a disproportionate expansion rate between the two metals and break the seizure caused by the corrosion. Whether heat can be applied depends on the location of the fastener - any surrounding components likely to be damaged must first be removed **(see illustration 2.15)**. Heat can be applied using a paint stripper heat gun or clothes iron, or by immersing the component in boiling water - wear protective gloves to prevent scalding or burns to the hands.

2.15 Using heat to free a seized fastener

• As a last resort, it is possible to use a hammer and cold chisel to work the fastener head unscrewed **(see illustration 2.16)**. This will damage the fastener, but more importantly extreme care must be taken not to damage the surrounding component.

Caution: Remember that the component being secured is generally of more value than the bolt, nut or screw - when the fastener is freed, do not unscrew it with force, instead work the fastener back and forth when resistance is felt to prevent thread damage.

2.16 Using a hammer and chisel to free a seized fastener

Broken fasteners and damaged heads

• If the shank of a broken bolt or screw is accessible you can grip it with self-locking grips. The knurled wheel type stud extractor tool or self-gripping stud puller tool is particularly useful for removing the long studs which screw into the cylinder mouth surface of the crankcase or bolts and screws from which the head has broken off **(see illustration 2.17)**. Studs can also be removed by locking two nuts together on the threaded end of the stud and using a spanner on the lower nut **(see illustration 2.18)**.

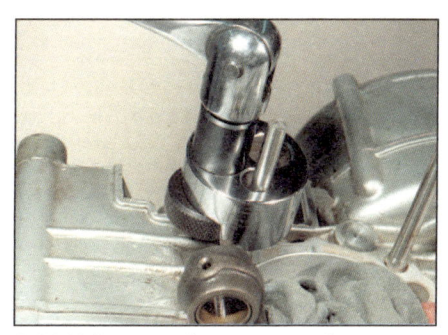

2.17 Using a stud extractor tool to remove a broken crankcase stud

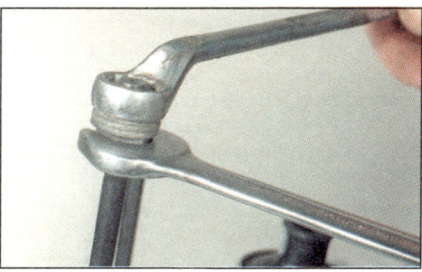

2.18 Two nuts can be locked together to unscrew a stud from a component

• A bolt or screw which has broken off below or level with the casing must be extracted using a screw extractor set. Centre punch the fastener to centralise the drill bit, then drill a hole in the fastener **(see illustration 2.19)**. Select a drill bit which is approximately half to three-quarters the diameter of the fastener and drill to a depth which will accommodate the extractor. Use the largest size extractor possible, but avoid leaving too small a wall thickness otherwise the extractor will merely force the fastener walls outwards wedging it in the casing thread.

• If a spiral type extractor is used, thread it anti-clockwise into the fastener. As it is screwed in, it will grip the fastener and unscrew it from the casing **(see illustration 2.20)**.

2.19 When using a screw extractor, first drill a hole in the fastener . . .

2.20 . . . then thread the extractor anti-clockwise into the fastener

• If a taper type extractor is used, tap it into the fastener so that it is firmly wedged in place. Unscrew the extractor (anti-clockwise) to draw the fastener out.

Warning: Stud extractors are very hard and may break off in the fastener if care is not taken - ask an engineer about spark erosion if this happens.

• Alternatively, the broken bolt/screw can be drilled out and the hole retapped for an oversize bolt/screw or a diamond-section thread insert. It is essential that the drilling is carried out squarely and to the correct depth, otherwise the casing may be ruined - if in doubt, entrust the work to an engineer.

• Bolts and nuts with rounded corners cause the correct size spanner or socket to slip when force is applied. Of the types of spanner/socket available always use a six-point type rather than an eight or twelve-point type - better grip

REF•8 Tools and Workshop Tips

2.21 Comparison of surface drive ring spanner (left) with 12-point type (right)

is obtained. Surface drive spanners grip the middle of the hex flats, rather than the corners, and are thus good in cases of damaged heads **(see illustration 2.21)**.

● Slotted-head or Phillips-head screws are often damaged by the use of the wrong size screwdriver. Allen-head and Torx-head screws are much less likely to sustain damage. If enough of the screw head is exposed you can use a hacksaw to cut a slot in its head and then use a conventional flat-bladed screwdriver to remove it. Alternatively use a hammer and cold chisel to tap the head of the fastener around to slacken it. Always replace damaged fasteners with new ones, preferably Torx or Allen-head type.

> **HAYNES HiNT**
>
> *A dab of valve grinding compound between the screw head and screwdriver tip will often give a good grip.*

Thread repair

● Threads (particularly those in aluminium alloy components) can be damaged by overtightening, being assembled with dirt in the threads, or from a component working loose and vibrating. Eventually the thread will fail completely, and it will be impossible to tighten the fastener.

● If a thread is damaged or clogged with old locking compound it can be renovated with a thread repair tool (thread chaser) **(see illustrations 2.22 and 2.23)**; special thread

2.22 A thread repair tool being used to correct an internal thread

2.23 A thread repair tool being used to correct an external thread

chasers are available for spark plug hole threads. The tool will not cut a new thread, but clean and true the original thread. Make sure that you use the correct diameter and pitch tool. Similarly, external threads can be cleaned up with a die or a thread restorer file **(see illustration 2.24)**.

2.24 Using a thread restorer file

● It is possible to drill out the old thread and retap the component to the next thread size. This will work where there is enough surrounding material and a new bolt or screw can be obtained. Sometimes, however, this is not possible - such as where the bolt/screw passes through another component which must also be suitably modified, also in cases where a spark plug or oil drain plug cannot be obtained in a larger diameter thread size.

● The diamond-section thread insert (often known by its popular trade name of Heli-Coil) is a simple and effective method of renewing the thread and retaining the original size. A kit can be purchased which contains the tap, insert and installing tool **(see illustration 2.25)**. Drill out the damaged thread with the size drill specified **(see illustration 2.26)**. Carefully retap the thread **(see illustration 2.27)**. Install the

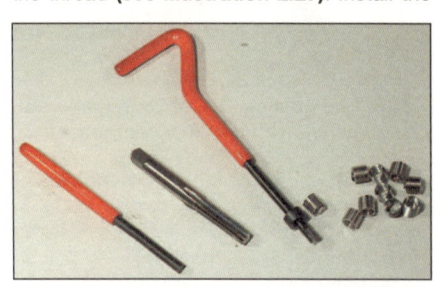

2.25 Obtain a thread insert kit to suit the thread diameter and pitch required

2.26 To install a thread insert, first drill out the original thread . . .

2.27 . . . tap a new thread . . .

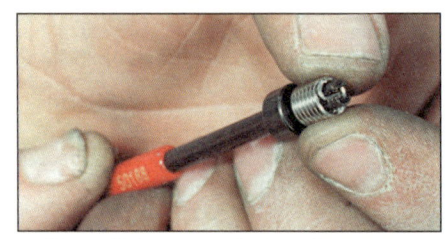

2.28 . . . fit insert on the installing tool . . .

2.29 . . . and thread into the component . . .

2.30 . . . break off the tang when complete

insert on the installing tool and thread it slowly into place using a light downward pressure **(see illustrations 2.28 and 2.29)**. When positioned between a 1/4 and 1/2 turn below the surface withdraw the installing tool and use the break-off tool to press down on the tang, breaking it off **(see illustration 2.30)**.

● There are epoxy thread repair kits on the market which can rebuild stripped internal threads, although this repair should not be used on high load-bearing components.

Tools and Workshop Tips

Thread locking and sealing compounds

● Locking compounds are used in locations where the fastener is prone to loosening due to vibration or on important safety-related items which might cause loss of control of the motorcycle if they fail. It is also used where important fasteners cannot be secured by other means such as lockwashers or split pins.

● Before applying locking compound, make sure that the threads (internal and external) are clean and dry with all old compound removed. Select a compound to suit the component being secured - a non-permanent general locking and sealing type is suitable for most applications, but a high strength type is needed for permanent fixing of studs in castings. Apply a drop or two of the compound to the first few threads of the fastener, then thread it into place and tighten to the specified torque. Do not apply excessive thread locking compound otherwise the thread may be damaged on subsequent removal.

● Certain fasteners are impregnated with a dry film type coating of locking compound on their threads. Always renew this type of fastener if disturbed.

● Anti-seize compounds, such as copper-based greases, can be applied to protect threads from seizure due to extreme heat and corrosion. A common instance is spark plug threads and exhaust system fasteners.

3 Measuring tools and gauges

Feeler gauges

● Feeler gauges (or blades) are used for measuring small gaps and clearances **(see illustration 3.1)**. They can also be used to measure endfloat (sideplay) of a component on a shaft where access is not possible with a dial gauge.

● Feeler gauge sets should be treated with care and not be bent or damaged. They are etched with their size on one face. Keep them clean and very lightly oiled to prevent corrosion build-up.

3.1 Feeler gauges are used for measuring small gaps and clearances - thickness is marked on one face of gauge

● When measuring a clearance, select a gauge which is a light sliding fit between the two components. You may need to use two gauges together to measure the clearance accurately.

Micrometers

● A micrometer is a precision tool capable of measuring to 0.01 or 0.001 of a millimetre. It should always be stored in its case and not in the general toolbox. It must be kept clean and never dropped, otherwise its frame or measuring anvils could be distorted resulting in inaccurate readings.

● External micrometers are used for measuring outside diameters of components and have many more applications than internal micrometers. Micrometers are available in different size ranges, eg 0 to 25 mm, 25 to 50 mm, and upwards in 25 mm steps; some large micrometers have interchangeable anvils to allow a range of measurements to be taken. Generally the largest precision measurement you are likely to take on a motorcycle is the piston diameter.

● Internal micrometers (or bore micrometers) are used for measuring inside diameters, such as valve guides and cylinder bores. Telescoping gauges and small hole gauges are used in conjunction with an external micrometer, whereas the more expensive internal micrometers have their own measuring device.

External micrometer

Note: *The conventional analogue type instrument is described. Although much easier to read, digital micrometers are considerably more expensive.*

● Always check the calibration of the micrometer before use. With the anvils closed (0 to 25 mm type) or set over a test gauge (for the larger types) the scale should read zero **(see illustration 3.2)**; make sure that the anvils (and test piece) are clean first. Any discrepancy can be adjusted by referring to the instructions supplied with the tool. Remember that the micrometer is a precision measuring tool - don't force the anvils closed, use the ratchet (4) on the end of the micrometer to close it. In this way, a measured force is always applied.

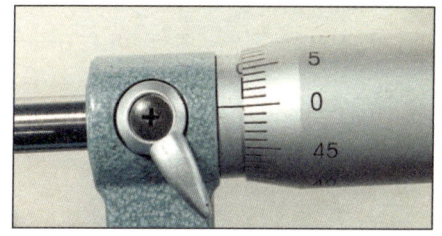

3.2 Check micrometer calibration before use

● To use, first make sure that the item being measured is clean. Place the anvil of the micrometer (1) against the item and use the thimble (2) to bring the spindle (3) lightly into contact with the other side of the item **(see illustration 3.3)**. Don't tighten the thimble down because this will damage the micrometer - instead use the ratchet (4) on the end of the micrometer. The ratchet mechanism applies a measured force preventing damage to the instrument.

● The micrometer is read by referring to the linear scale on the sleeve and the annular scale on the thimble. Read off the sleeve first to obtain the base measurement, then add the fine measurement from the thimble to obtain the overall reading. The linear scale on the sleeve represents the measuring range of the micrometer (eg 0 to 25 mm). The annular scale

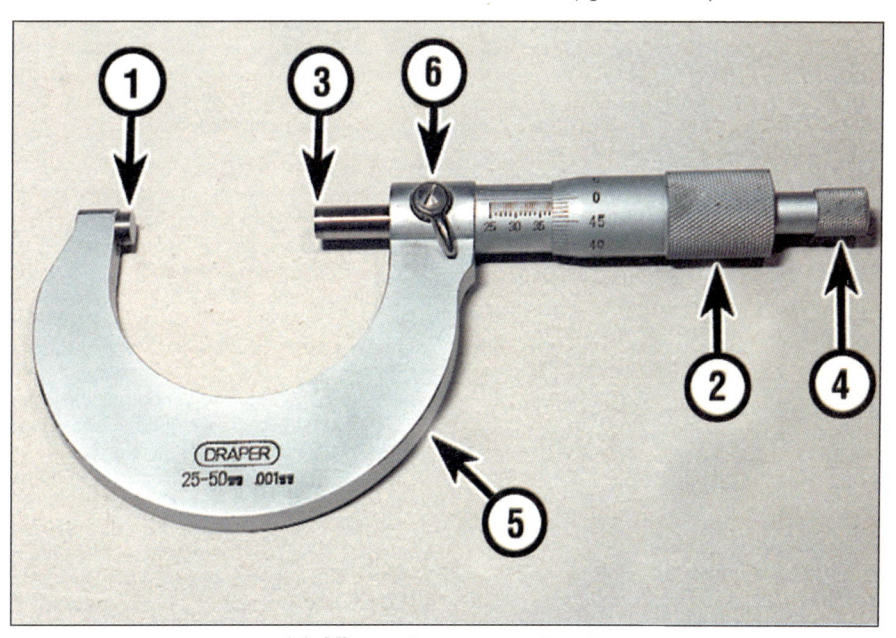

3.3 Micrometer component parts

1 Anvil
2 Thimble
3 Spindle
4 Ratchet
5 Frame
6 Locking lever

Tools and Workshop Tips

on the thimble will be in graduations of 0.01 mm (or as marked on the frame) - one full revolution of the thimble will move 0.5 mm on the linear scale. Take the reading where the datum line on the sleeve intersects the thimble's scale. Always position the eye directly above the scale otherwise an inaccurate reading will result.

In the example shown the item measures 2.95 mm **(see illustration 3.4)**:

Linear scale	2.00 mm
Linear scale	0.50 mm
Annular scale	0.45 mm
Total figure	**2.95 mm**

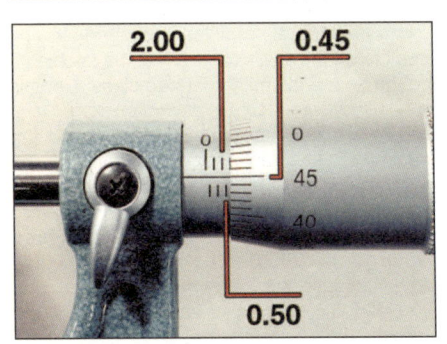

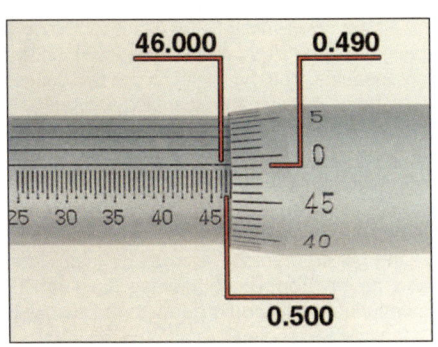

3.5 Micrometer reading of 46.99 mm on linear and annular scales . . .

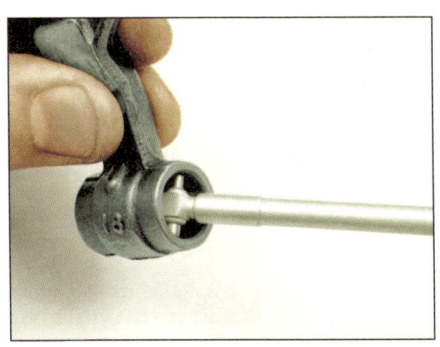

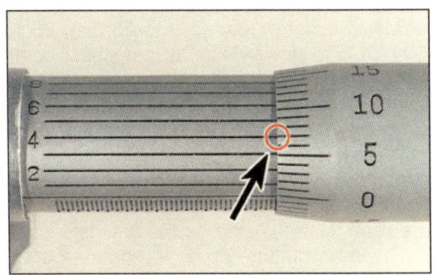

3.7 Expand the telescoping gauge in the bore, lock its position . . .

3.4 Micrometer reading of 2.95 mm

3.6 . . . and 0.004 mm on vernier scale

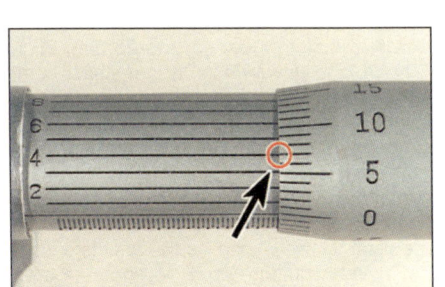

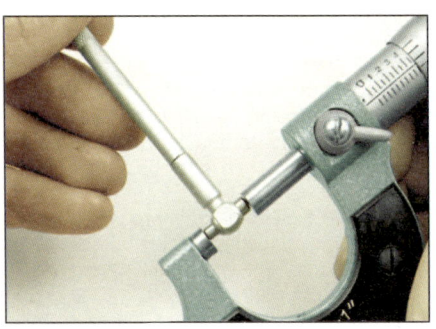

3.8 . . . then measure the gauge with a micrometer

Most micrometers have a locking lever (6) on the frame to hold the setting in place, allowing the item to be removed from the micrometer.
● Some micrometers have a vernier scale on their sleeve, providing an even finer measurement to be taken, in 0.001 increments of a millimetre. Take the sleeve and thimble measurement as described above, then check which graduation on the vernier scale aligns with that of the annular scale on the thimble **Note:** *The eye must be perpendicular to the scale when taking the vernier reading - if necessary rotate the body of the micrometer to ensure this.* Multiply the vernier scale figure by 0.001 and add it to the base and fine measurement figures.

In the example shown the item measures 46.994 mm **(see illustrations 3.5 and 3.6)**:

Linear scale (base)	46.000 mm
Linear scale (base)	00.500 mm
Annular scale (fine)	00.490 mm
Vernier scale	00.004 mm
Total figure	**46.994 mm**

Internal micrometer

● Internal micrometers are available for measuring bore diameters, but are expensive and unlikely to be available for home use. It is suggested that a set of telescoping gauges and small hole gauges, both of which must be used with an external micrometer, will suffice for taking internal measurements on a motorcycle.
● Telescoping gauges can be used to measure internal diameters of components. Select a gauge with the correct size range, make sure its ends are clean and insert it into the bore. Expand the gauge, then lock its position and withdraw it from the bore **(see illustration 3.7)**. Measure across the gauge ends with a micrometer **(see illustration 3.8)**.
● Very small diameter bores (such as valve guides) are measured with a small hole gauge. Once adjusted to a slip-fit inside the component, its position is locked and the gauge withdrawn for measurement with a micrometer **(see illustrations 3.9 and 3.10)**.

Vernier caliper

Note: *The conventional linear and dial gauge type instruments are described. Digital types are easier to read, but are far more expensive.*
● The vernier caliper does not provide the precision of a micrometer, but is versatile in being able to measure internal and external diameters. Some types also incorporate a depth gauge. It is ideal for measuring clutch plate friction material and spring free lengths.
● To use the conventional linear scale vernier, slacken off the vernier clamp screws (1) and set its jaws over (2), or inside (3), the item to be measured **(see illustration 3.11)**. Slide the jaw into contact, using the thumbwheel (4) for fine movement of the sliding scale (5) then tighten the clamp screws (1). Read off the main scale (6) where the zero on the sliding scale (5) intersects it, taking the whole number to the left of the zero; this provides the base measurement. View along the sliding scale and select the division which

3.9 Expand the small hole gauge in the bore, lock its position . . .

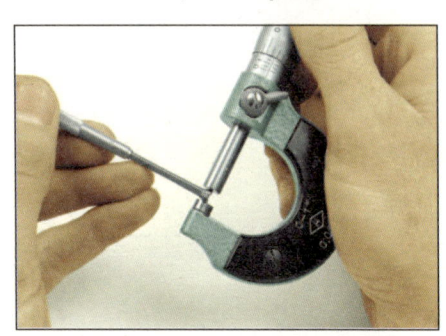

3.10 . . . then measure the gauge with a micrometer

lines up exactly with any of the divisions on the main scale, noting that the divisions usually represents 0.02 of a millimetre. Add this fine measurement to the base measurement to obtain the total reading.

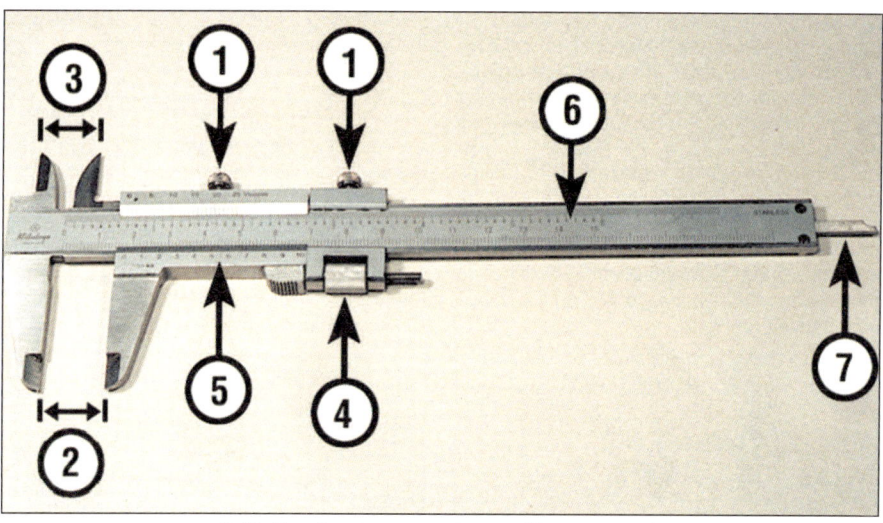

Tools and Workshop Tips   REF•11

3.11 Vernier component parts (linear gauge)

1. Clamp screws
2. External jaws
3. Internal jaws
4. Thumbwheel
5. Sliding scale
6. Main scale
7. Depth gauge

In the example shown the item measures 55.92 mm **(see illustration 3.12)**:

Base measurement	55.00 mm
Fine measurement	00.92 mm
Total figure	**55.92 mm**

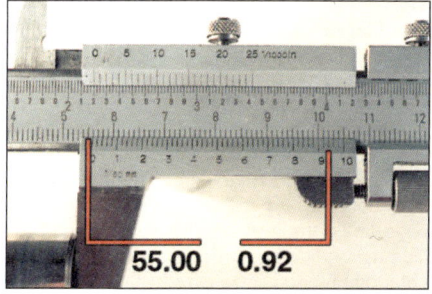

● Some vernier calipers are equipped with a dial gauge for fine measurement. Before use, check that the jaws are clean, then close them fully and check that the dial gauge reads zero. If necessary adjust the gauge ring accordingly. Slacken the vernier clamp screw (1) and set its jaws over (2), or inside (3), the item to be measured **(see illustration 3.13)**. Slide the jaws into contact, using the thumbwheel (4) for fine movement. Read off the main scale (5) where the edge of the sliding scale (6) intersects it, taking the whole number to the left of the zero; this provides the base measurement. Read off the needle position on the dial gauge (7) scale to provide the fine measurement; each division represents 0.05 of a millimetre. Add this fine measurement to the base measurement to obtain the total reading.

In the example shown the item measures 55.95 mm **(see illustration 3.14)**:

Base measurement	55.00 mm
Fine measurement	00.95 mm
Total figure	**55.95 mm**

3.12 Vernier gauge reading of 55.92 mm

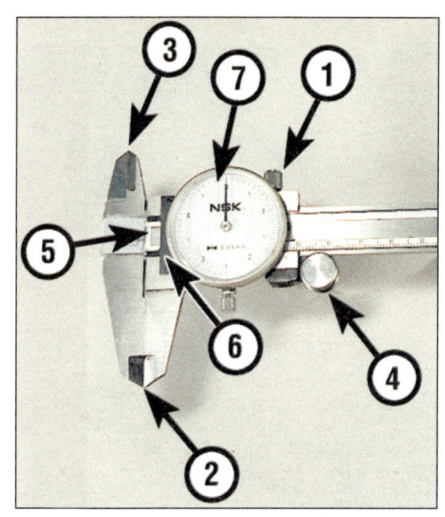

3.13 Vernier component parts (dial gauge)

1. Clamp screw
2. External jaws
3. Internal jaws
4. Thumbwheel
5. Main scale
6. Sliding scale
7. Dial gauge

3.14 Vernier gauge reading of 55.95 mm

Plastigauge

● Plastigauge is a plastic material which can be compressed between two surfaces to measure the oil clearance between them. The width of the compressed Plastigauge is measured against a calibrated scale to determine the clearance.

● Common uses of Plastigauge are for measuring the clearance between crankshaft journal and main bearing inserts, between crankshaft journal and big-end bearing inserts, and between camshaft and bearing surfaces. The following example describes big-end oil clearance measurement.

● Handle the Plastigauge material carefully to prevent distortion. Using a sharp knife, cut a length which corresponds with the width of the bearing being measured and place it carefully across the journal so that it is parallel with the shaft **(see illustration 3.15)**. Carefully install both bearing shells and the connecting rod. Without rotating the rod on the journal tighten its bolts or nuts (as applicable) to the specified torque. The connecting rod and bearings are then disassembled and the crushed Plastigauge examined.

3.15 Plastigauge placed across shaft journal

● Using the scale provided in the Plastigauge kit, measure the width of the material to determine the oil clearance **(see illustration 3.16)**. Always remove all traces of Plastigauge after use using your fingernails.

Caution: Arriving at the correct clearance demands that the assembly is torqued correctly, according to the settings and sequence (where applicable) provided by the motorcycle manufacturer.

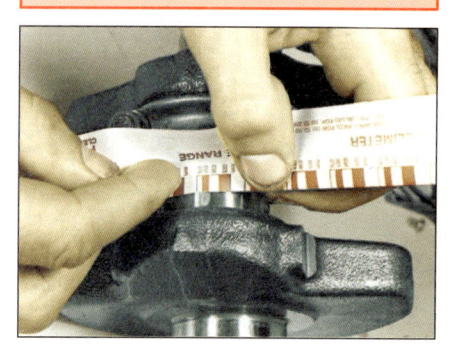

3.16 Measuring the width of the crushed Plastigauge

Tools and Workshop Tips

Dial gauge or DTI (Dial Test Indicator)

● A dial gauge can be used to accurately measure small amounts of movement. Typical uses are measuring shaft runout or shaft endfloat (sideplay) and setting piston position for ignition timing on two-strokes. A dial gauge set usually comes with a range of different probes and adapters and mounting equipment.

● The gauge needle must point to zero when at rest. Rotate the ring around its periphery to zero the gauge.

● Check that the gauge is capable of reading the extent of movement in the work. Most gauges have a small dial set in the face which records whole millimetres of movement as well as the fine scale around the face periphery which is calibrated in 0.01 mm divisions. Read off the small dial first to obtain the base measurement, then add the measurement from the fine scale to obtain the total reading.

In the example shown the gauge reads 1.48 mm **(see illustration 3.17)**:

Base measurement	1.00 mm
Fine measurement	0.48 mm
Total figure	**1.48 mm**

3.17 Dial gauge reading of 1.48 mm

● If measuring shaft runout, the shaft must be supported in vee-blocks and the gauge mounted on a stand perpendicular to the shaft. Rest the tip of the gauge against the centre of the shaft and rotate the shaft slowly whilst watching the gauge reading **(see illustration 3.18)**. Take several measurements along the length of the shaft and record the maximum gauge reading as the amount of runout in the shaft. **Note:** *The reading obtained will be total runout at that point - some manufacturers specify that the runout figure is halved to compare with their specified runout limit.*

● Endfloat (sideplay) measurement requires that the gauge is mounted securely to the surrounding component with its probe touching the end of the shaft. Using hand pressure, push and pull on the shaft noting the maximum endfloat recorded on the gauge **(see illustration 3.19)**.

3.19 Using a dial gauge to measure shaft endfloat

● A dial gauge with suitable adapters can be used to determine piston position BTDC on two-stroke engines for the purposes of ignition timing. The gauge, adapter and suitable length probe are installed in the place of the spark plug and the gauge zeroed at TDC. If the piston position is specified as 1.14 mm BTDC, rotate the engine back to 2.00 mm BTDC, then slowly forwards to 1.14 mm BTDC.

Cylinder compression gauges

● A compression gauge is used for measuring cylinder compression. Either the rubber-cone type or the threaded adapter type can be used. The latter is preferred to ensure a perfect seal against the cylinder head. A 0 to 300 psi (0 to 20 Bar) type gauge (for petrol/gasoline engines) will be suitable for motorcycles.

● The spark plug is removed and the gauge either held hard against the cylinder head (cone type) or the gauge adapter screwed into the cylinder head (threaded type) **(see illustration 3.20)**. Cylinder compression is measured with the engine turning over, but not running - carry out the compression test as described in *Fault Finding Equipment*. The gauge will hold the reading until manually released.

Oil pressure gauge

● An oil pressure gauge is used for measuring engine oil pressure. Most gauges come with a set of adapters to fit the thread of the take-off point **(see illustration 3.21)**. If the take-off point specified by the motorcycle manufacturer is an external oil pipe union, make sure that the specified replacement union is used to prevent oil starvation.

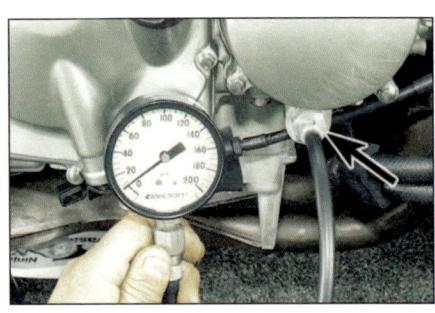

3.21 Oil pressure gauge and take-off point adapter (arrow)

● Oil pressure is measured with the engine running (at a specific rpm) and often the manufacturer will specify pressure limits for a cold and hot engine.

Straight-edge and surface plate

● If checking the gasket face of a component for warpage, place a steel rule or precision straight-edge across the gasket face and measure any gap between the straight-edge and component with feeler gauges **(see illustration 3.22)**. Check diagonally across the component and between mounting holes **(see illustration 3.23)**.

3.22 Use a straight-edge and feeler gauges to check for warpage

3.18 Using a dial gauge to measure shaft runout

3.20 Using a rubber-cone type cylinder compression gauge

3.23 Check for warpage in these directions

Tools and Workshop Tips

- Checking individual components for warpage, such as clutch plain (metal) plates, requires a perfectly flat plate or piece or plate glass and feeler gauges.

4 Torque and leverage

What is torque?

- Torque describes the twisting force about a shaft. The amount of torque applied is determined by the distance from the centre of the shaft to the end of the lever and the amount of force being applied to the end of the lever; distance multiplied by force equals torque.
- The manufacturer applies a measured torque to a bolt or nut to ensure that it will not slacken in use and to hold two components securely together without movement in the joint. The actual torque setting depends on the thread size, bolt or nut material and the composition of the components being held.
- Too little torque may cause the fastener to loosen due to vibration, whereas too much torque will distort the joint faces of the component or cause the fastener to shear off. Always stick to the specified torque setting.

Using a torque wrench

- Check the calibration of the torque wrench and make sure it has a suitable range for the job. Torque wrenches are available in Nm (Newton-metres), kgf m (kilograms-force metre), lbf ft (pounds-feet), lbf in (inch-pounds). Do not confuse lbf ft with lbf in.
- Adjust the tool to the desired torque on the scale (see illustration 4.1). If your torque wrench is not calibrated in the units specified, carefully convert the figure (see *Conversion Factors*). A manufacturer sometimes gives a torque setting as a range (8 to 10 Nm) rather than a single figure - in this case set the tool midway between the two settings. The same torque may be expressed as 9 Nm ± 1 Nm. Some torque wrenches have a method of locking the setting so that it isn't inadvertently altered during use.

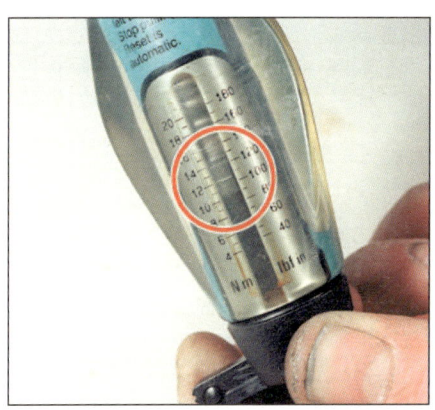

4.1 Set the torque wrench index mark to the setting required, in this case 12 Nm

- Install the bolts/nuts in their correct location and secure them lightly. Their threads must be clean and free of any old locking compound. Unless specified the threads and flange should be dry - oiled threads are necessary in certain circumstances and the manufacturer will take this into account in the specified torque figure. Similarly, the manufacturer may also specify the application of thread-locking compound.
- Tighten the fasteners in the specified sequence until the torque wrench clicks, indicating that the torque setting has been reached. Apply the torque again to double-check the setting. Where different thread diameter fasteners secure the component, as a rule tighten the larger diameter ones first.
- When the torque wrench has been finished with, release the lock (where applicable) and fully back off its setting to zero - do not leave the torque wrench tensioned. Also, do not use a torque wrench for slackening a fastener.

Angle-tightening

- Manufacturers often specify a figure in degrees for final tightening of a fastener. This usually follows tightening to a specific torque setting.
- A degree disc can be set and attached to the socket (see illustration 4.2) or a protractor can be used to mark the angle of movement on the bolt/nut head and the surrounding casting (see illustration 4.3).

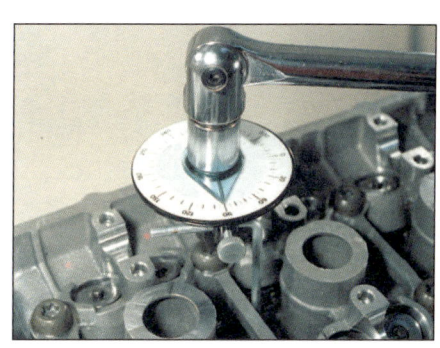

4.2 Angle tightening can be accomplished with a torque-angle gauge . . .

4.3 . . . or by marking the angle on the surrounding component

Loosening sequences

- Where more than one bolt/nut secures a component, loosen each fastener evenly a little at a time. In this way, not all the stress of the joint is held by one fastener and the components are not likely to distort.
- If a tightening sequence is provided, work in the REVERSE of this, but if not, work from the outside in, in a criss-cross sequence (see illustration 4.4).

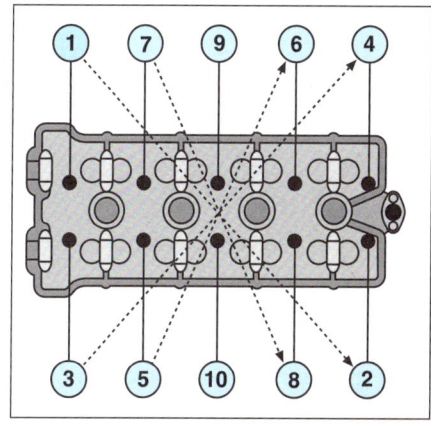

4.4 When slackening, work from the outside inwards

Tightening sequences

- If a component is held by more than one fastener it is important that the retaining bolts/nuts are tightened evenly to prevent uneven stress build-up and distortion of sealing faces. This is especially important on high-compression joints such as the cylinder head.
- A sequence is usually provided by the manufacturer, either in a diagram or actually marked in the casting. If not, always start in the centre and work outwards in a criss-cross pattern (see illustration 4.5). Start off by securing all bolts/nuts finger-tight, then set the torque wrench and tighten each fastener by a small amount in sequence until the final torque is reached. By following this practice,

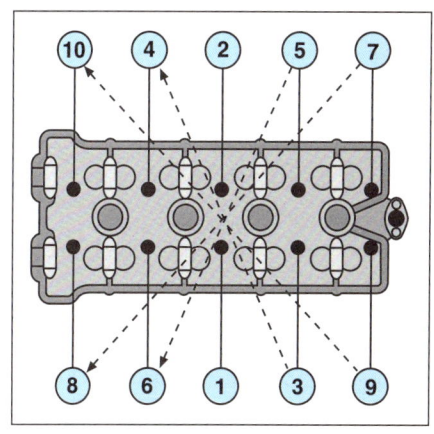

4.5 When tightening, work from the inside outwards

Tools and Workshop Tips

the joint will be held evenly and will not be distorted. Important joints, such as the cylinder head and big-end fasteners often have two- or three-stage torque settings.

Applying leverage

● Use tools at the correct angle. Position a socket wrench or spanner on the bolt/nut so that you pull it towards you when loosening. If this can't be done, push the spanner without curling your fingers around it **(see illustration 4.6)** - the spanner may slip or the fastener loosen suddenly, resulting in your fingers being crushed against a component.

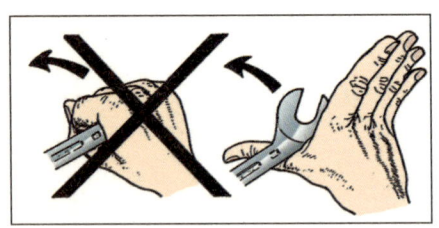

4.6 If you can't pull on the spanner to loosen a fastener, push with your hand open

● Additional leverage is gained by extending the length of the lever. The best way to do this is to use a breaker bar instead of the regular length tool, or to slip a length of tubing over the end of the spanner or socket wrench.
● If additional leverage will not work, the fastener head is either damaged or firmly corroded in place (see *Fasteners*).

5 Bearings

Bearing removal and installation

Drivers and sockets

● Before removing a bearing, always inspect the casing to see which way it must be driven out - some casings will have retaining plates or a cast step. Also check for any identifying markings on the bearing and if installed to a certain depth, measure this at this stage. Some roller bearings are sealed on one side - take note of the original fitted position.
● Bearings can be driven out of a casing using a bearing driver tool (with the correct size head) or a socket of the correct diameter. Select the driver head or socket so that it contacts the outer race of the bearing, not the balls/rollers or inner race. Always support the casing around the bearing housing with wood blocks, otherwise there is a risk of fracture. The bearing is driven out with a few blows on the driver or socket from a heavy mallet. Unless access is severely restricted (as with wheel bearings), a pin-punch is not recommended unless it is moved around the bearing to keep it square in its housing.

● The same equipment can be used to install bearings. Make sure the bearing housing is supported on wood blocks and line up the bearing in its housing. Fit the bearing as noted on removal - generally they are installed with their marked side facing outwards. Tap the bearing squarely into its housing using a driver or socket which bears only on the bearing's outer race - contact with the bearing balls/rollers or inner race will destroy it **(see illustrations 5.1 and 5.2)**.
● Check that the bearing inner race and balls/rollers rotate freely.

5.1 Using a bearing driver against the bearing's outer race

5.2 Using a large socket against the bearing's outer race

Pullers and slide-hammers

● Where a bearing is pressed on a shaft a puller will be required to extract it **(see illustration 5.3)**. Make sure that the puller clamp or legs fit securely behind the bearing and are unlikely to slip out. If pulling a bearing off a gear shaft for example, you may have to locate the puller behind a gear pinion if there is no access to the race and draw the gear pinion off the shaft as well **(see illustration 5.4)**.

Caution: Ensure that the puller's centre bolt locates securely against the end of the shaft and will not slip when pressure is applied. Also ensure that puller does not damage the shaft end.

5.4 Where no access is available to the rear of the bearing, it is sometimes possible to draw off the adjacent component

● Operate the puller so that its centre bolt exerts pressure on the shaft end and draws the bearing off the shaft.
● When installing the bearing on the shaft, tap only on the bearing's inner race - contact with the balls/rollers or outer race with destroy the bearing. Use a socket or length of tubing as a drift which fits over the shaft end **(see illustration 5.5)**.

5.5 When installing a bearing on a shaft use a piece of tubing which bears only on the bearing's inner race

● Where a bearing locates in a blind hole in a casing, it cannot be driven or pulled out as described above. A slide-hammer with knife-edged bearing puller attachment will be required. The puller attachment passes through the bearing and when tightened expands to fit firmly behind the bearing **(see illustration 5.6)**. By operating the slide-hammer part of the tool the bearing is jarred out of its housing **(see illustration 5.7)**.
● It is possible, if the bearing is of reasonable weight, for it to drop out of its housing if the casing is heated as described opposite. If this

5.3 This bearing puller clamps behind the bearing and pressure is applied to the shaft end to draw the bearing off

Tools and Workshop Tips REF•15

5.6 Expand the bearing puller so that it locks behind the bearing . . .

5.7 . . . attach the slide hammer to the bearing puller

method is attempted, first prepare a work surface which will enable the casing to be tapped face down to help dislodge the bearing - a wood surface is ideal since it will not damage the casing's gasket surface. Wearing protective gloves, tap the heated casing several times against the work surface to dislodge the bearing under its own weight **(see illustration 5.8)**.

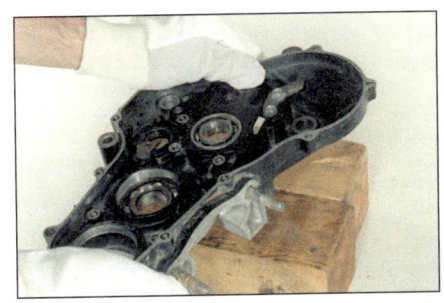

5.8 Tapping a casing face down on wood blocks can often dislodge a bearing

● Bearings can be installed in blind holes using the driver or socket method described above.

Drawbolts

● Where a bearing or bush is set in the eye of a component, such as a suspension linkage arm or connecting rod small-end, removal by drift may damage the component. Furthermore, a rubber bushing in a shock absorber eye cannot successfully be driven out of position. If access is available to a engineering press, the task is straightforward. If not, a drawbolt can be fabricated to extract the bearing or bush.

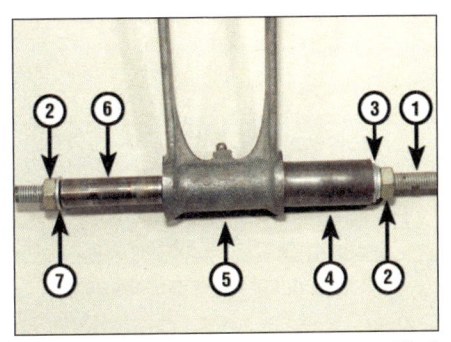

5.9 Drawbolt component parts assembled on a suspension arm

1 Bolt or length of threaded bar
2 Nuts
3 Washer (external diameter greater than tubing internal diameter)
4 Tubing (internal diameter sufficient to accommodate bearing)
5 Suspension arm with bearing
6 Tubing (external diameter slightly smaller than bearing)
7 Washer (external diameter slightly smaller than bearing)

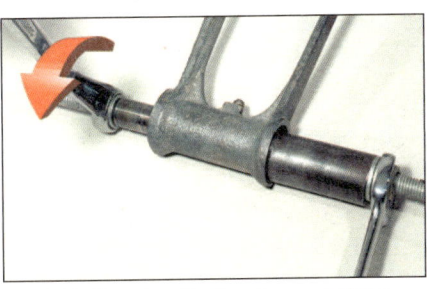

5.10 Drawing the bearing out of the suspension arm

● To extract the bearing/bush you will need a long bolt with nut (or piece of threaded bar with two nuts), a piece of tubing which has an internal diameter larger than the bearing/bush, another piece of tubing which has an external diameter slightly smaller than the bearing/bush, and a selection of washers **(see illustrations 5.9 and 5.10)**. Note that the pieces of tubing must be of the same length, or longer, than the bearing/bush.

● The same kit (without the pieces of tubing) can be used to draw the new bearing/bush back into place **(see illustration 5.11)**.

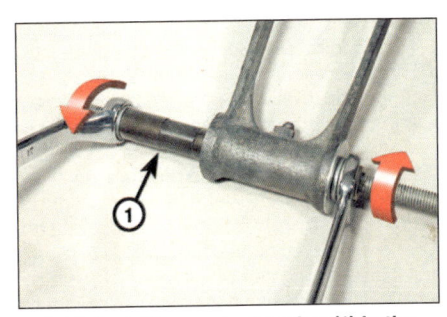

5.11 Installing a new bearing (1) in the suspension arm

Temperature change

● If the bearing's outer race is a tight fit in the casing, the aluminium casing can be heated to release its grip on the bearing. Aluminium will expand at a greater rate than the steel bearing outer race. There are several ways to do this, but avoid any localised extreme heat (such as a blow torch) - aluminium alloy has a low melting point.

● Approved methods of heating a casing are using a domestic oven (heated to 100°C) or immersing the casing in boiling water **(see illustration 5.12)**. Low temperature range localised heat sources such as a paint stripper heat gun or clothes iron can also be used **(see illustration 5.13)**. Alternatively, soak a rag in boiling water, wring it out and wrap it around the bearing housing.

> ⚠ **Warning: All of these methods require care in use to prevent scalding and burns to the hands. Wear protective gloves when handling hot components.**

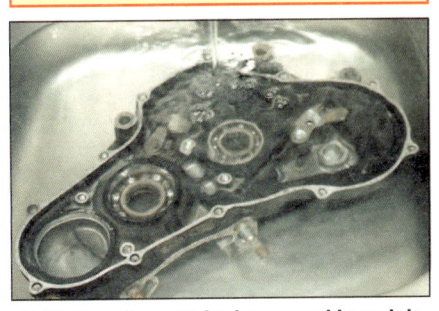

5.12 A casing can be immersed in a sink of boiling water to aid bearing removal

5.13 Using a localised heat source to aid bearing removal

● If heating the whole casing note that plastic components, such as the neutral switch, may suffer - remove them beforehand.

● After heating, remove the bearing as described above. You may find that the expansion is sufficient for the bearing to fall out of the casing under its own weight or with a light tap on the driver or socket.

● If necessary, the casing can be heated to aid bearing installation, and this is sometimes the recommended procedure if the motorcycle manufacturer has designed the housing and bearing fit with this intention.

REF•16 Tools and Workshop Tips

- Installation of bearings can be eased by placing them in a freezer the night before installation. The steel bearing will contract slightly, allowing easy insertion in its housing. This is often useful when installing steering head outer races in the frame.

Bearing types and markings

- Plain shell bearings, ball bearings, needle roller bearings and tapered roller bearings will all be found on motorcycles (see illustrations 5.14 and 5.15). The ball and roller types are usually caged between an inner and outer race, but uncaged variations may be found.

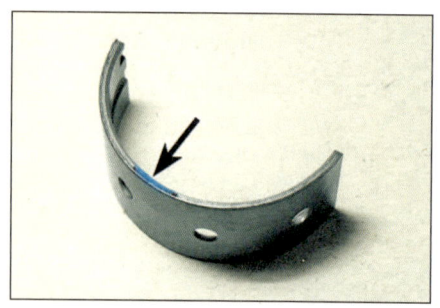

5.14 Shell bearings are either plain or grooved. They are usually identified by colour code (arrow)

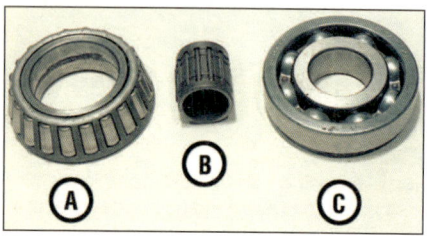

5.15 Tapered roller bearing (A), needle roller bearing (B) and ball journal bearing (C)

- Shell bearings (often called inserts) are usually found at the crankshaft main and connecting rod big-end where they are good at coping with high loads. They are made of a phosphor-bronze material and are impregnated with self-lubricating properties.
- Ball bearings and needle roller bearings consist of a steel inner and outer race with the balls or rollers between the races. They require constant lubrication by oil or grease and are good at coping with axial loads. Taper roller bearings consist of rollers set in a tapered cage set on the inner race; the outer race is separate. They are good at coping with axial loads and prevent movement along the shaft - a typical application is in the steering head.
- Bearing manufacturers produce bearings to ISO size standards and stamp one face of the bearing to indicate its internal and external diameter, load capacity and type (see illustration 5.16).
- Metal bushes are usually of phosphor-bronze material. Rubber bushes are used in suspension mounting eyes. Fibre bushes have also been used in suspension pivots.

5.16 Typical bearing marking

Bearing fault finding

- If a bearing outer race has spun in its housing, the housing material will be damaged. You can use a bearing locking compound to bond the outer race in place if damage is not too severe.
- Shell bearings will fail due to damage of their working surface, as a result of lack of lubrication, corrosion or abrasive particles in the oil (see illustration 5.17). Small particles of dirt in the oil may embed in the bearing material whereas larger particles will score the bearing and shaft journal. If a number of short journeys are made, insufficient heat will be generated to drive off condensation which has built up on the bearings.

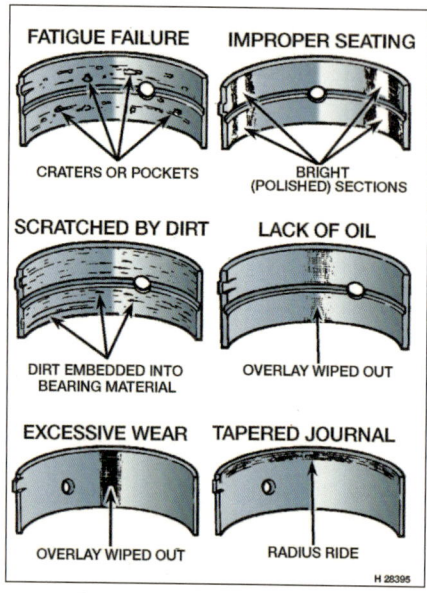

5.17 Typical bearing failures

- Ball and roller bearings will fail due to lack of lubrication or damage to the balls or rollers. Tapered-roller bearings can be damaged by overloading them. Unless the bearing is sealed on both sides, wash it in paraffin (kerosene) to remove all old grease then allow it to dry. Make a visual inspection looking to dented balls or rollers, damaged cages and worn or pitted races (see illustration 5.18).
- A ball bearing can be checked for wear by listening to it when spun. Apply a film of light oil to the bearing and hold it close to the ear - hold the outer race with one hand and spin the inner

5.18 Example of ball journal bearing with damaged balls and cages

5.19 Hold outer race and listen to inner race when spun

race with the other hand (see illustration 5.19). The bearing should be almost silent when spun; if it grates or rattles it is worn.

6 Oil seals

Oil seal removal and installation

- Oil seals should be renewed every time a component is dismantled. This is because the seal lips will become set to the sealing surface and will not necessarily reseal.
- Oil seals can be prised out of position using a large flat-bladed screwdriver (see illustration 6.1). In the case of crankcase seals, check first that the seal is not lipped on the inside, preventing its removal with the crankcases joined.

6.1 Prise out oil seals with a large flat-bladed screwdriver

- New seals are usually installed with their marked face (containing the seal reference code) outwards and the spring side towards the fluid being retained. In certain cases, such as a two-stroke engine crankshaft seal, a double lipped seal may be used due to there being fluid or gas on each side of the joint.

Tools and Workshop Tips REF•17

● Use a bearing driver or socket which bears only on the outer hard edge of the seal to install it in the casing - tapping on the inner edge will damage the sealing lip.

Oil seal types and markings

● Oil seals are usually of the single-lipped type. Double-lipped seals are found where a liquid or gas is on both sides of the joint.
● Oil seals can harden and lose their sealing ability if the motorcycle has been in storage for a long period - renewal is the only solution.
● Oil seal manufacturers also conform to the ISO markings for seal size - these are moulded into the outer face of the seal (see illustration 6.2).

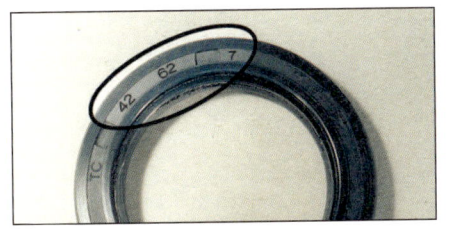

6.2 These oil seal markings indicate inside diameter, outside diameter and seal thickness

7 Gaskets and sealants

Types of gasket and sealant

● Gaskets are used to seal the mating surfaces between components and keep lubricants, fluids, vacuum or pressure contained within the assembly. Aluminium gaskets are sometimes found at the cylinder joints, but most gaskets are paper-based. If the mating surfaces of the components being joined are undamaged the gasket can be installed dry, although a dab of sealant or grease will be useful to hold it in place during assembly.
● RTV (Room Temperature Vulcanising) silicone rubber sealants cure when exposed to moisture in the atmosphere. These sealants are good at filling pits or irregular gasket faces, but will tend to be forced out of the joint under very high torque. They can be used to replace a paper gasket, but first make sure that the width of the paper gasket is not essential to the shimming of internal components. RTV sealants should not be used on components containing petrol (gasoline).
● Non-hardening, semi-hardening and hard setting liquid gasket compounds can be used with a gasket or between a metal-to-metal joint. Select the sealant to suit the application: universal non-hardening sealant can be used on virtually all joints; semi-hardening on joint faces which are rough or damaged; hard setting sealant on joints which require a permanent bond and are subjected to high temperature and pressure. **Note:** *Check first if the paper gasket has a bead of sealant impregnated in its surface before applying additional sealant.*
● When choosing a sealant, make sure it is suitable for the application, particularly if being applied in a high-temperature area or in the vicinity of fuel. Certain manufacturers produce sealants in either clear, silver or black colours to match the finish of the engine. This has a particular application on motorcycles where much of the engine is exposed.
● Do not over-apply sealant. That which is squeezed out on the outside of the joint can be wiped off, whereas an excess of sealant on the inside can break off and clog oilways.

Breaking a sealed joint

● Age, heat, pressure and the use of hard setting sealant can cause two components to stick together so tightly that they are difficult to separate using finger pressure alone. Do not resort to using levers unless there is a pry point provided for this purpose (see illustration 7.1) or else the gasket surfaces will be damaged.
● Use a soft-faced hammer (see illustration 7.2) or a wood block and conventional hammer to strike the component near the mating surface. Avoid hammering against cast extremities since they may break off. If this method fails, try using a wood wedge between the two components.

Caution: If the joint will not separate, double-check that you have removed all the fasteners.

7.1 If a pry point is provided, apply gently pressure with a flat-bladed screwdriver

7.2 Tap around the joint with a soft-faced mallet if necessary - don't strike cooling fins

Removal of old gasket and sealant

● Paper gaskets will most likely come away complete, leaving only a few traces stuck on the sealing faces of the components. It is imperative that all traces are removed to ensure correct sealing of the new gasket.
● Very carefully scrape all traces of gasket away making sure that the sealing surfaces are not gouged or scored by the scraper (see illustrations 7.3, 7.4 and 7.5). Stubborn deposits can be removed by spraying with an aerosol gasket remover. Final preparation of

HAYNES HiNT

Most components have one or two hollow locating dowels between the two gasket faces. If a dowel cannot be removed, do not resort to gripping it with pliers - it will almost certainly be distorted. Install a close-fitting socket or Phillips screwdriver into the dowel and then grip the outer edge of the dowel to free it.

7.3 Paper gaskets can be scraped off with a gasket scraper tool . . .

7.4 . . . a knife blade . . .

7.5 . . . or a household scraper

REF•18 Tools and Workshop Tips

7.6 Fine abrasive paper is wrapped around a flat file to clean up the gasket face

7.7 A kitchen scourer can be used on stubborn deposits

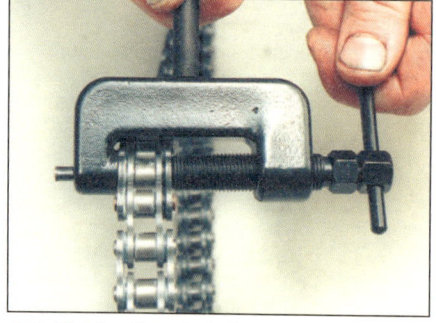

8.1 Tighten the chain breaker to push the pin out of the link . . .

8.2 . . . withdraw the pin, remove the tool . . .

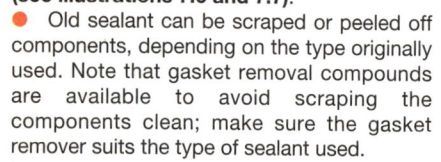

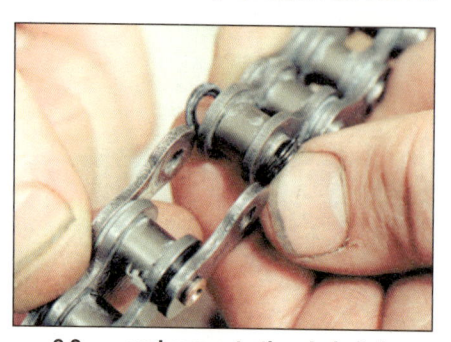

8.3 . . . and separate the chain link

8.4 Insert the new soft link, with O-rings, through the chain ends . . .

8.5 . . . install the O-rings over the pin ends . . .

8.6 . . . followed by the sideplate

8.7 Push the sideplate into position using a clamp

the gasket surface can be made with very fine abrasive paper or a plastic kitchen scourer **(see illustrations 7.6 and 7.7)**.

● Old sealant can be scraped or peeled off components, depending on the type originally used. Note that gasket removal compounds are available to avoid scraping the components clean; make sure the gasket remover suits the type of sealant used.

8 Chains

Breaking and joining final drive chains

● Drive chains for all but small bikes are continuous and do not have a clip-type connecting link. The chain must be broken using a chain breaker tool and the new chain securely riveted together using a new soft rivet-type link. Never use a clip-type connecting link instead of a rivet-type link, except in an emergency. Various chain breaking and riveting tools are available, either as separate tools or combined as illustrated in the accompanying photographs - read the instructions supplied with the tool carefully.

> **Warning: The need to rivet the new link pins correctly cannot be overstressed - loss of control of the motorcycle is very likely to result if the chain breaks in use.**

● Rotate the chain and look for the soft link. The soft link pins look like they have been deeply centre-punched instead of peened over like all the other pins **(see illustration 8.9)** and its sideplate may be a different colour. Position the soft link midway between the sprockets and assemble the chain breaker tool over one of the soft link pins **(see illustration 8.1)**. Operate the tool to push the pin out through the chain **(see illustration 8.2)**. On an O-ring chain, remove the O-rings **(see illustration 8.3)**. Carry out the same procedure on the other soft link pin.

> **Caution: Certain soft link pins (particularly on the larger chains) may require their ends to be filed or ground off before they can be pressed out using the tool.**

● Check that you have the correct size and strength (standard or heavy duty) new soft link - do not reuse the old link. Look for the size marking on the chain sideplates **(see illustration 8.10)**.

● Position the chain ends so that they are engaged over the rear sprocket. On an O-ring chain, install a new O-ring over each pin of the link and insert the link through the two chain ends **(see illustration 8.4)**. Install a new O-ring over the end of each pin, followed by the sideplate (with the chain manufacturer's marking facing outwards) **(see illustrations 8.5 and 8.6)**. On an unsealed chain, insert the link through the two chain ends, then install the sideplate with the chain manufacturer's marking facing outwards.

● Note that it may not be possible to install the sideplate using finger pressure alone. If using a joining tool, assemble it so that the plates of the tool clamp the link and press the sideplate over the pins **(see illustration 8.7)**. Otherwise, use two small sockets placed over

Tools and Workshop Tips REF•19

8.8 Assemble the chain riveting tool over one pin at a time and tighten it fully

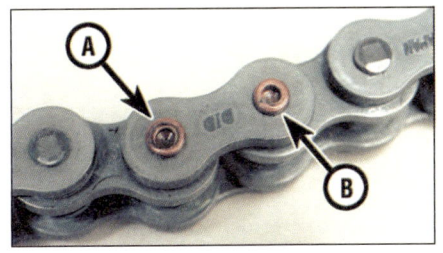

8.9 Pin end correctly riveted (A), pin end unriveted (B)

the rivet ends and two pieces of the wood between a G-clamp. Operate the clamp to press the sideplate over the pins.
● Assemble the joining tool over one pin (following the maker's instructions) and tighten the tool down to spread the pin end securely **(see illustrations 8.8 and 8.9)**. Do the same on the other pin.

> **Warning:** Check that the pin ends are secure and that there is no danger of the sideplate coming loose. If the pin ends are cracked the soft link must be renewed.

Final drive chain sizing

● Chains are sized using a three digit number, followed by a suffix to denote the chain type **(see illustration 8.10)**. Chain type is either standard or heavy duty (thicker sideplates), and also unsealed or O-ring/X-ring type.
● The first digit of the number relates to the pitch of the chain, ie the distance from the centre of one pin to the centre of the next pin **(see illustration 8.11)**. Pitch is expressed in eighths of an inch, as follows:

8.10 Typical chain size and type marking

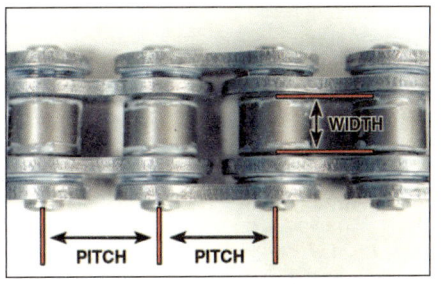

8.11 Chain dimensions

Sizes commencing with a 4 (eg 428) have a pitch of 1/2 inch (12.7 mm)

Sizes commencing with a 5 (eg 520) have a pitch of 5/8 inch (15.9 mm)

Sizes commencing with a 6 (eg 630) have a pitch of 3/4 inch (19.1 mm)

● The second and third digits of the chain size relate to the width of the rollers, again in imperial units, eg the 525 shown has 5/16 inch (7.94 mm) rollers **(see illustration 8.11)**.

9 Hoses

Clamping to prevent flow

● Small-bore flexible hoses can be clamped to prevent fluid flow whilst a component is worked on. Whichever method is used, ensure that the hose material is not permanently distorted or damaged by the clamp.
a) A brake hose clamp available from auto accessory shops **(see illustration 9.1)**.
b) A wingnut type hose clamp **(see illustration 9.2)**.

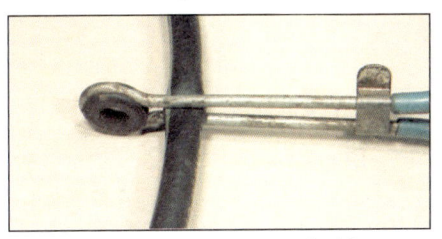

9.1 Hoses can be clamped with an automotive brake hose clamp . . .

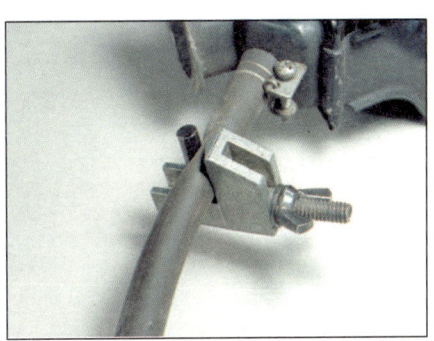

9.2 . . . a wingnut type hose clamp . . .

c) Two sockets placed each side of the hose and held with straight-jawed self-locking grips **(see illustration 9.3)**.
d) Thick card each side of the hose held between straight-jawed self-locking grips **(see illustration 9.4)**.

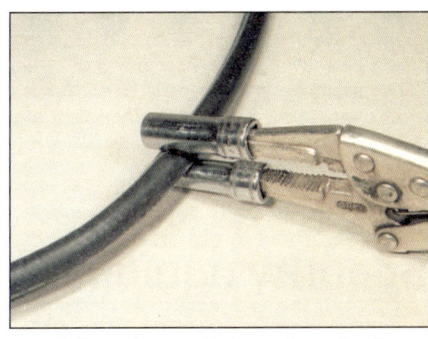

9.3 . . . two sockets and a pair of self-locking grips . . .

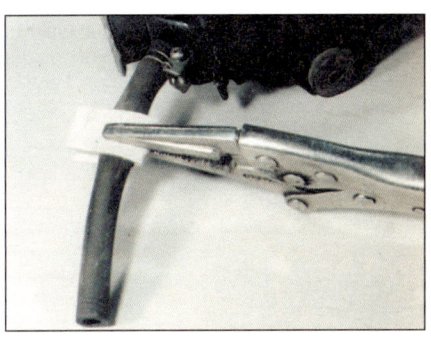

9.4 . . . or thick card and self-locking grips

Freeing and fitting hoses

● Always make sure the hose clamp is moved well clear of the hose end. Grip the hose with your hand and rotate it whilst pulling it off the union. If the hose has hardened due to age and will not move, slit it with a sharp knife and peel its ends off the union **(see illustration 9.5)**.
● Resist the temptation to use grease or soap on the unions to aid installation; although it helps the hose slip over the union it will equally aid the escape of fluid from the joint. It is preferable to soften the hose ends in hot water and wet the inside surface of the hose with water or a fluid which will evaporate.

9.5 Cutting a coolant hose free with a sharp knife

Security

Introduction

In less time than it takes to read this introduction, a thief could steal your motorcycle. Returning only to find your bike has gone is one of the worst feelings in the world. Even if the motorcycle is insured against theft, once you've got over the initial shock, you will have the inconvenience of dealing with the police and your insurance company.

The motorcycle is an easy target for the professional thief and the joyrider alike and the official figures on motorcycle theft make for depressing reading; on average a motorcycle is stolen every 16 minutes in the UK!

Motorcycle thefts fall into two categories, those stolen 'to order' and those taken by opportunists. The thief stealing to order will be on the look out for a specific make and model and will go to extraordinary lengths to obtain that motorcycle. The opportunist thief on the other hand will look for easy targets which can be stolen with the minimum of effort and risk.

Whilst it is never going to be possible to make your machine 100% secure, it is estimated that around half of all stolen motorcycles are taken by opportunist thieves. Remember that the opportunist thief is always on the look out for the easy option: if there are two similar motorcycles parked side-by-side, they will target the one with the lowest level of security. By taking a few precautions, you can reduce the chances of your motorcycle being stolen.

Security equipment

There are many specialised motorcycle security devices available and the following text summarises their applications and their good and bad points.

Once you have decided on the type of security equipment which best suits your needs, we recommended that you read one of the many equipment tests regularly carried out by the motorcycle press. These tests compare the products from all the major manufacturers and give impartial ratings on their effectiveness, value-for-money and ease of use.

No one item of security equipment can provide complete protection. It is highly recommended that two or more of the items described below are combined to increase the security of your motorcycle (a lock and chain plus an alarm system is just about ideal). The more security measures fitted to the bike, the less likely it is to be stolen.

Lock and chain

Pros: *Very flexible to use; can be used to secure the motorcycle to almost any immovable object. On some locks and chains, the lock can be used on its own as a disc lock (see below).*

Cons: *Can be very heavy and awkward to carry on the motorcycle, although some types will be supplied with a carry bag which can be strapped to the pillion seat.*

● Heavy-duty chains and locks are an excellent security measure **(see illustration 1)**. Whenever the motorcycle is parked, use the lock and chain to secure the machine to a solid, immovable object such as a post or railings. This will prevent the machine from being ridden away or being lifted into the back of a van.

● When fitting the chain, always ensure the chain is routed around the motorcycle frame or swingarm **(see illustrations 2 and 3)**. Never merely pass the chain around one of the wheel rims; a thief may unbolt the wheel and lift the rest of the machine into a van, leaving you with just the wheel! Try to avoid having excess chain free, thus making it difficult to use cutting tools, and keep the chain and lock off the ground to prevent thieves attacking it with a cold chisel. Position the lock so that its cold barrel is facing downwards; this will make it harder for the thief to attack the lock mechanism.

1 Ensure the lock and chain you buy is of good quality and long enough to shackle your bike to a solid object

2 Pass the chain through the bike's frame, rather than just through a wheel . . .

3 . . . and loop it around a solid object

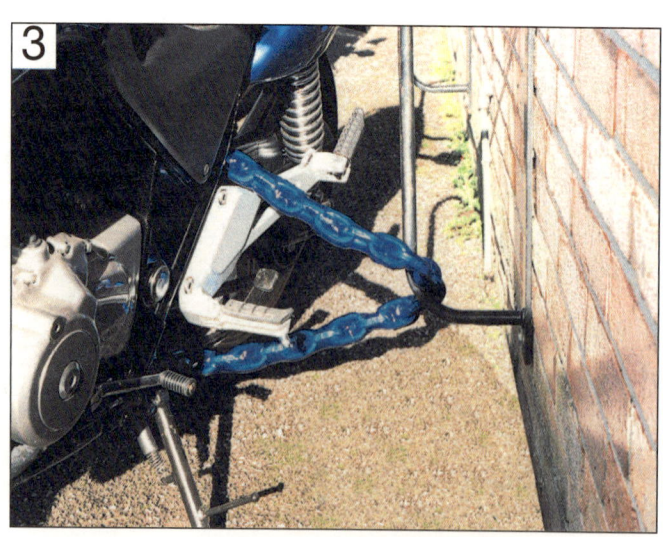

Security REF•21

U-locks

Pros: *Highly effective deterrent which can be used to secure the bike to a post or railings. Most U-locks come with a carrier which allows the lock to be easily carried on the bike.*

Cons: *Not as flexible to use as a lock and chain.*

● These are solid locks which are similar in use to a lock and chain. U-locks are lighter than a lock and chain but not so flexible to use. The length and shape of the lock shackle limit the objects to which the bike can be secured **(see illustration 4)**.

U-locks can be used to secure the bike to a solid object – ensure you purchase one which is long enough

Disc locks

Pros: *Small, light and very easy to carry; most can be stored underneath the seat.*

Cons: *Does not prevent the motorcycle being lifted into a van. Can be very embarrassing if you forget to remove the lock before attempting to ride off!*

● Disc locks are designed to be attached to the front brake disc. The lock passes through one of the holes in the disc and prevents the wheel rotating by jamming against the fork/brake caliper **(see illustration 5)**. Some are equipped with an alarm siren which sounds if the disc lock is moved; this not only acts as a theft deterrent but also as a handy reminder if you try to move the bike with the lock still fitted.

● Combining the disc lock with a length of cable which can be looped around a post or railings provides an additional measure of security **(see illustration 6)**.

A typical disc lock attached through one of the holes in the disc

Alarms and immobilisers

Pros: *Once installed it is completely hassle-free to use. If the system is 'Thatcham' or 'Sold Secure-approved', insurance companies may give you a discount.*

Cons: *Can be expensive to buy and complex to install. No system will prevent the motorcycle from being lifted into a van and taken away.*

● Electronic alarms and immobilisers are available to suit a variety of budgets. There are three different types of system available: pure alarms, pure immobilisers, and the more expensive systems which are combined alarm/immobilisers **(see illustration 7)**.

● An alarm system is designed to emit an audible warning if the motorcycle is being tampered with.

● An immobiliser prevents the motorcycle being started and ridden away by disabling its electrical systems.

● When purchasing an alarm/immobiliser system, check the cost of installing the system unless you are able to do it yourself. If the motorcycle is not used regularly, another consideration is the current drain of the system. All alarm/immobiliser systems are powered by the motorcycle's battery; purchasing a system with a very low current drain could prevent the battery losing its charge whilst the motorcycle is not being used.

A disc lock combined with a security cable provides additional protection

A typical alarm/immobiliser system

REF•22 Security

Indelible markings can be applied to most areas of the bike – always apply the manufacturer's sticker to warn off thieves

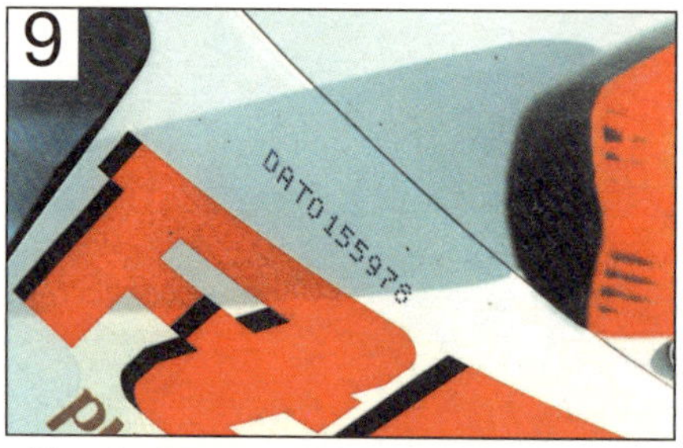

Chemically-etched code numbers can be applied to main body panels . . .

. . . again, always ensure that the kit manufacturer's sticker is applied in a prominent position

Security marking kits

Pros: *Very cheap and effective deterrent. Many insurance companies will give you a discount on your insurance premium if a recognised security marking kit is used on your motorcycle.*

Cons: *Does not prevent the motorcycle being stolen by joyriders.*

● There are many different types of security marking kits available. The idea is to mark as many parts of the motorcycle as possible with a unique security number **(see illustrations 8, 9 and 10)**. A form will be included with the kit to register your personal details and those of the motorcycle with the kit manufacturer. This register is made available to the police to help them trace the rightful owner of any motorcycle or components which they recover should all other forms of identification have been removed. Always apply the warning stickers provided with the kit to deter thieves.

Ground anchors, wheel clamps and security posts

Pros: *An excellent form of security which will deter all but the most determined of thieves.*

Cons: *Awkward to install and can be expensive.*

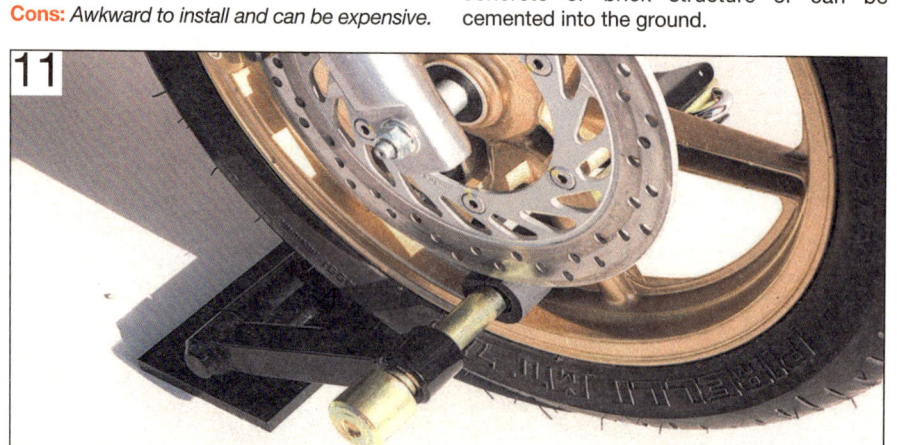

Permanent ground anchors provide an excellent level of security when the bike is at home

● Whilst the motorcycle is at home, it is a good idea to attach it securely to the floor or a solid wall, even if it is kept in a securely locked garage. Various types of ground anchors, security posts and wheel clamps are available for this purpose **(see illustration 11)**. These security devices are either bolted to a solid concrete or brick structure or can be cemented into the ground.

Security at home

A high percentage of motorcycle thefts are from the owner's home. Here are some things to consider whenever your motorcycle is at home:

✔ Where possible, always keep the motorcycle in a securely locked garage. Never rely solely on the standard lock on the garage door, these are usual hopelessly inadequate. Fit an additional locking mechanism to the door and consider having the garage alarmed. A security light, activated by a movement sensor, is also a good investment.

✔ Always secure the motorcycle to the ground or a wall, even if it is inside a securely locked garage.

✔ Do not regularly leave the motorcycle outside your home, try to keep it out of sight wherever possible. If a garage is not available, fit a motorcycle cover over the bike to disguise its true identity.

✔ It is not uncommon for thieves to follow a motorcyclist home to find out where the bike is kept. They will then return at a later date. Be aware of this whenever you are returning home on your motorcycle. If you suspect you are being followed, do not return home, instead ride to a garage or shop and stop as a precaution.

✔ When selling a motorcycle, do not provide your home address or the location where the bike is normally kept. Arrange to meet the buyer at a location away from your home. Thieves have been known to pose as potential buyers to find out where motorcycles are kept and then return later to steal them.

Security away from the home

As well as fitting security equipment to your motorcycle here are a few general rules to follow whenever you park your motorcycle.
✔ Park in a busy, public place.
✔ Use car parks which incorporate security features, such as CCTV.

✔ At night, park in a well-lit area, preferably directly underneath a street light.
✔ Engage the steering lock.
✔ Secure the motorcycle to a solid, immovable object such as a post or railings with an additional lock. If this is not possible, secure the bike to a friend's motorcycle. Some public parking places provide security loops for motorcycles.
✔ Never leave your helmet or luggage attached to the motorcycle. Take them with you at all times.

Lubricants and fluids

A wide range of lubricants, fluids and cleaning agents is available for motor-cycles. This is a guide as to what is available, its applications and properties.

Four-stroke engine oil

- Engine oil is without doubt the most important component of any four-stroke engine. Modern motorcycle engines place a lot of demands on their oil and choosing the right type is essential. Using an unsuitable oil will lead to an increased rate of engine wear and could result in serious engine damage. Before purchasing oil, always check the recommended oil specification given by the manufacturer. The manufacturer will state a recommended 'type or classification' and also a specific 'viscosity' range for engine oil.
- The oil 'type or classification' is identified by its API (American Petroleum Institute) rating. The API rating will be in the form of two letters, e.g. SG. The S identifies the oil as being suitable for use in a petrol (gasoline) engine (S stands for spark ignition) and the second letter, ranging from A to J, identifies the oil's performance rating. The later this letter, the higher the specification of the oil; for example API SG oil exceeds the requirements of API SF oil. **Note:** *On some oils there may also be a second rating consisting of another two letters, the first letter being C, e.g. API SF/CD. This rating indicates the oil is also suitable for use in a diesel engines (the C stands for compression ignition) and is thus of no relevance for motorcycle use.*
- The 'viscosity' of the oil is identified by its SAE (Society of Automotive Engineers) rating. All modern engines require multigrade oils and the SAE rating will consist of two numbers, the first followed by a W, e.g. 10W/40. The first number indicates the viscosity rating of the oil at low temperatures (W stands for winter – tested at –20ºC) and the second number represents the viscosity of the oil at high temperatures (tested at 100ºC). The lower the number, the thinner the oil. For example an oil with an SAE 10W/40 rating will give better cold starting and running than an SAE 15W/40 oil.
- As well as ensuring the 'type' and 'viscosity' of the oil match the recommendations, another consideration to make when buying engine oil is whether to purchase a standard mineral-based oil, a semi-synthetic oil (also known as a synthetic blend or synthetic-based oil) or a fully-synthetic oil. Although all oils will have a similar rating and viscosity, their cost will vary considerably; mineral-based oils are the cheapest, the fully-synthetic oils the most expensive with the semi-synthetic oils falling somewhere in-between. This decision is very much up to the owner, but it should be noted that modern synthetic oils have far better lubricating and cleaning qualities than traditional mineral-based oils and tend to retain these properties for far longer. Bearing in mind the operating conditions inside a modern, high-revving motorcycle engine it is highly recommended that a fully synthetic oil is used. The extra expense at each service could save you money in the long term by preventing premature engine wear.
- As a final note always ensure that the oil is specifically designed for use in motorcycle engines. Engine oils designed primarily for use in car engines sometimes contain additives or friction modifiers which could cause clutch slip on a motorcycle fitted with a wet-clutch.

Two-stroke engine oil

- Modern two-stroke engines, with their high power outputs, place high demands on their oil. If engine seizure is to be avoided it is essential that a high-quality oil is used. Two-stroke oils differ hugely from four-stroke oils. The oil lubricates only the crankshaft and piston(s) (the transmission has its own lubricating oil) and is used on a total-loss basis where it is burnt completely during the combustion process.
- The Japanese have recently introduced a classification system for two-stroke oils, the JASO rating. This rating is in the form of two letters, either FA, FB or FC – FA is the lowest classification and FC the highest. Ensure the oil being used meets or exceeds the recommended rating specified by the manufacturer.
- As well as ensuring the oil rating matches the recommendation, another consideration to make when buying engine oil is whether to purchase a standard mineral-based oil, a semi-synthetic oil (also known as a synthetic blend or synthetic-based oil) or a fully-synthetic oil. The cost of each type of oil varies considerably; mineral-based oils are the cheapest, the fully-synthetic oils the most expensive with the semi-synthetic oils falling somewhere in-between. This decision is very much up to the owner, but it should be noted that modern synthetic oils have far better lubricating properties and burn cleaner than traditional mineral-based oils. It is therefore recommended that a fully synthetic oil is used. The extra expense could save you money in the long term by preventing premature engine wear, engine performance will be improved, carbon deposits and exhaust smoke will be reduced.

Lubricants and fluids

● Always ensure that the oil is specifically designed for use in an injector system. Many high quality two-stroke oils are designed for competition use and need to be pre-mixed with fuel. These oils are of a much higher viscosity and are not designed to flow through the injector pumps used on road-going two-stroke motorcycles.

Transmission (gear) oil

● On a two-stroke engine, the transmission and clutch are lubricated by their own separate oil bath which must be changed in accordance with the Maintenance Schedule.
● Although the engine and transmission units of most four-strokes use a common lubrication supply, there are some exceptions where the engine and gearbox have separate oil reservoirs and a dry clutch is used.
● Motorcycle manufacturers will either recommend a monograde transmission oil or a four-stroke multigrade engine oil to lubricate the transmission.
● Transmission oils, or gear oils as they are often called, are designed specifically for use in transmission systems. The viscosity of these oils is represented by an SAE number, but the scale of measurement applied is different to that used to grade engine oils. As a rough guide a SAE90 gear oil will be of the same viscosity as an SAE50 engine oil.

Shaft drive oil

● On models equipped with shaft final drive, the shaft drive gears are will have their own oil supply. The manufacturer will state a recommended 'type or classification' and also a specific 'viscosity' range in the same manner as for four-stroke engine oil.
● Gear oil classification is given by the number which follows the API GL (GL standing for gear lubricant) rating, the higher the number, the higher the specification of the oil, e.g. API GL5 oil is a higher specification than API GL4 oil. Ensure the oil meets or exceeds the classification specified and is of the correct viscosity. The viscosity of gear oils is also represented by an SAE number but the scale of measurement used is different to that used to grade engine oils. As a rough guide an SAE90 gear oil will be of the same viscosity as an SAE50 engine oil.
● If the use of an EP (Extreme Pressure) gear oil is specified, ensure the oil purchased is suitable.

Fork oil and suspension fluid

● Conventional telescopic front forks are hydraulic and require fork oil to work. To ensure the forks function correctly, the fork oil must be changed in accordance with the Maintenance Schedule.
● Fork oil is available in a variety of viscosities, identified by their SAE rating; fork oil ratings vary from light (SAE 5) to heavy (SAE 30). When purchasing fork oil, ensure the viscosity rating matches that specified by the manufacturer.
● Some lubricant manufacturers also produce a range of high-quality suspension fluids which are very similar to fork oil but are designed mainly for competition use. These fluids may have a different viscosity rating system which is not to be confused with the SAE rating of normal fork oil. Refer to the manufacturer's instructions if in any doubt.

Brake and clutch fluid

● All disc brake systems and some clutch systems are hydraulically operated. To ensure correct operation, the hydraulic fluid must be changed in accordance with the Maintenance Schedule.
● Brake and clutch fluid is classified by its DOT rating with most motorcycle manufacturers specifying DOT 3 or 4 fluid. Both fluid types are glycol-based and can be mixed together without adverse effect; DOT 4 fluid exceeds the requirements of DOT 3 fluid. Although it is safe to use DOT 4 fluid in a system designed for use with DOT 3 fluid, never use DOT 3 fluid in a system which specifies the use of DOT 4 as this will adversely affect the system's performance. The type required for the system will be marked on the fluid reservoir cap.
● Some manufacturers also produce a DOT 5 hydraulic fluid. DOT 5 hydraulic fluid is silicone-based and is not compatible with the glycol-based DOT 3 and 4 fluids. Never mix DOT 5 fluid with DOT 3 or 4 fluid as this will seriously affect the performance of the hydraulic system.

Coolant/antifreeze

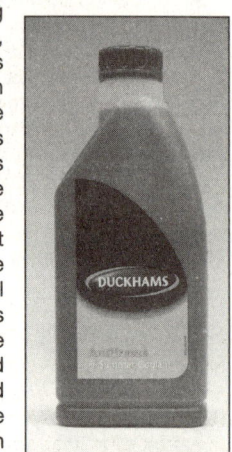

● When purchasing coolant/antifreeze, always ensure it is suitable for use in an aluminium engine and contains corrosion inhibitors to prevent possible blockages of the internal coolant passages of the system. As a general rule, most coolants are designed to be used neat and should not be diluted whereas antifreeze can be mixed with distilled water to provide a coolant solution of the required strength. Refer to the manufacturer's instructions on the bottle.
● Ensure the coolant is changed in accordance with the Maintenance Schedule.

Chain lube

● Chain lube is an aerosol-type spray lubricant specifically designed for use on motorcycle final drive chains. Chain lube has two functions, to minimise friction between the final drive chain and sprockets and to prevent corrosion of the chain. Regular use of a good-quality chain lube will extend the life of the drive chain and sprockets and thus maximise the power being transmitted from the transmission to the rear wheel.
● When using chain lube, always allow some time for the solvents in the lube to evaporate before riding the motorcycle. This will minimise the amount of lube which will

Lubricants and fluids REF•25

'fling' off from the chain when the motorcycle is used. If the motorcycle is equipped with an 'O-ring' chain, ensure the chain lube is labelled as being suitable for use on 'O-ring' chains.

Degreasers and solvents

● There are many different types of solvents and degreasers available to remove the grime and grease which accumulate around the motorcycle during normal use. Degreasers and solvents are usually available as an aerosol-type spray or as a liquid which you apply with a brush. Always closely follow the manufacturer's instructions and wear eye protection during use. Be aware that many solvents are flammable and may give off noxious fumes; take adequate precautions when using them (see Safety First!).

● For general cleaning, use one of the many solvents or degreasers available from most motorcycle accessory shops. These solvents are usually applied then left for a certain time before being washed off with water.

Brake cleaner is a solvent specifically designed to remove all traces of oil, grease and dust from braking system components. Brake cleaner is designed to evaporate quickly and leaves behind no residue.

Carburettor cleaner is an aerosol-type solvent specifically designed to clear carburettor blockages and break down the hard deposits and gum often found inside carburettors during overhaul.

Contact cleaner is an aerosol-type solvent designed for cleaning electrical components. The cleaner will remove all traces of oil and dirt from components such as switch contacts or fouled spark plugs and then dry, leaving behind no residue.

Gasket remover is an aerosol-type solvent designed for removing stubborn gaskets from engine components during overhaul. Gasket remover will minimise the amount of scraping required to remove the gasket and therefore reduce the risk of damage to the mating surface.

Spray lubricants

● Aerosol-based spray lubricants are widely available and are excellent for lubricating lever pivots and exposed cables and switches. Try to use a lubricant which is of the dry-film type as the fluid evaporates, leaving behind a dry-film of lubricant. Lubricants which leave behind an oily residue will attract dust and dirt which will increase the rate of wear of the cable/lever.

● Most lubricants also act as a moisture dispersant and a penetrating fluid. This means they can also be used to 'dry out' electrical components such as wiring connectors or switches as well as helping to free seized fasteners.

Greases

● Grease is used to lubricate many of the pivot-points. A good-quality multi-purpose grease is suitable for most applications but some manufacturers will specify the use of specialist greases for use on components such as swingarm and suspension linkage bushes. These specialist greases can be purchased from most motorcycle (or car) accessory shops; commonly specified types include molybdenum disulphide grease, lithium-based grease, graphite-based grease, silicone-based grease and high-temperature copper-based grease.

Gasket sealing compounds

● Gasket sealing compounds can be used in conjunction with gaskets, to improve their sealing capabilities, or on their own to seal metal-to-metal joints. Depending on their type, sealing compounds either set hard or stay relatively soft and pliable.

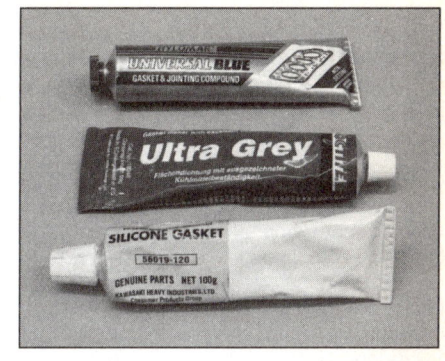

● When purchasing a gasket sealing compound, ensure that it is designed specifically for use on an internal combustion engine. General multi-purpose sealants available from DIY stores may appear visibly similar but they are not designed to withstand the extreme heat or contact with fuel and oil encountered when used on an engine (see 'Tools and Workshop Tips' for further information).

Thread locking compound

● Thread locking compounds are used to secure certain threaded fasteners in position to prevent them from loosening due to vibration. Thread locking compounds can be purchased from most motorcycle (and car) accessory shops. Ensure the threads of the both components are completely clean and dry before sparingly applying the locking compound (see 'Tools and Workshop Tips' for further information).

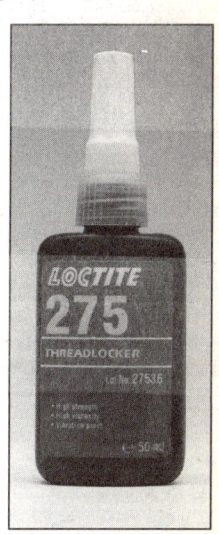

Fuel additives

● Fuel additives which protect and clean the fuel system components are widely available. These additives are designed to remove all traces of deposits that build up on the carburettors/injectors and prevent wear, helping the fuel system to operate more efficiently. If a fuel additive is being used, check that it is suitable for use with your motorcycle, especially if your motorcycle is equipped with a catalytic converter.

● Octane boosters are also available. These additives are designed to improve the performance of highly-tuned engines being run on normal pump-fuel and are of no real use on standard motorcycles.

Conversion Factors

Length (distance)
Inches (in)	x 25.4	= Millimetres (mm)	x 0.0394	=	Inches (in)
Feet (ft)	x 0.305	= Metres (m)	x 3.281	=	Feet (ft)
Miles	x 1.609	= Kilometres (km)	x 0.621	=	Miles

Volume (capacity)
Cubic inches (cu in; in³)	x 16.387	= Cubic centimetres (cc; cm³)	x 0.061	=	Cubic inches (cu in; in³)
Imperial pints (Imp pt)	x 0.568	= Litres (l)	x 1.76	=	Imperial pints (Imp pt)
Imperial quarts (Imp qt)	x 1.137	= Litres (l)	x 0.88	=	Imperial quarts (Imp qt)
Imperial quarts (Imp qt)	x 1.201	= US quarts (US qt)	x 0.833	=	Imperial quarts (Imp qt)
US quarts (US qt)	x 0.946	= Litres (l)	x 1.057	=	US quarts (US qt)
Imperial gallons (Imp gal)	x 4.546	= Litres (l)	x 0.22	=	Imperial gallons (Imp gal)
Imperial gallons (Imp gal)	x 1.201	= US gallons (US gal)	x 0.833	=	Imperial gallons (Imp gal)
US gallons (US gal)	x 3.785	= Litres (l)	x 0.264	=	US gallons (US gal)

Mass (weight)
Ounces (oz)	x 28.35	= Grams (g)	x 0.035	=	Ounces (oz)
Pounds (lb)	x 0.454	= Kilograms (kg)	x 2.205	=	Pounds (lb)

Force
Ounces-force (ozf; oz)	x 0.278	= Newtons (N)	x 3.6	=	Ounces-force (ozf; oz)
Pounds-force (lbf; lb)	x 4.448	= Newtons (N)	x 0.225	=	Pounds-force (lbf; lb)
Newtons (N)	x 0.1	= Kilograms-force (kgf; kg)	x 9.81	=	Newtons (N)

Pressure
Pounds-force per square inch (psi; lbf/in²; lb/in²)	x 0.070	= Kilograms-force per square centimetre (kgf/cm²; kg/cm²)	x 14.223	=	Pounds-force per square inch (psi; lbf/in²; lb/in²)
Pounds-force per square inch (psi; lbf/in²; lb/in²)	x 0.068	= Atmospheres (atm)	x 14.696	=	Pounds-force per square inch (psi; lbf/in²; lb/in²)
Pounds-force per square inch (psi; lbf/in²; lb/in²)	x 0.069	= Bars	x 14.5	=	Pounds-force per square inch (psi; lbf/in²; lb/in²)
Pounds-force per square inch (psi; lbf/in²; lb/in²)	x 6.895	= Kilopascals (kPa)	x 0.145	=	Pounds-force per square inch (psi; lbf/in²; lb/in²)
Kilopascals (kPa)	x 0.01	= Kilograms-force per square centimetre (kgf/cm²; kg/cm²)	x 98.1	=	Kilopascals (kPa)
Millibar (mbar)	x 100	= Pascals (Pa)	x 0.01	=	Millibar (mbar)
Millibar (mbar)	x 0.0145	= Pounds-force per square inch (psi; lbf/in²; lb/in²)	x 68.947	=	Millibar (mbar)
Millibar (mbar)	x 0.75	= Millimetres of mercury (mmHg)	x 1.333	=	Millibar (mbar)
Millibar (mbar)	x 0.401	= Inches of water (inH₂O)	x 2.491	=	Millibar (mbar)
Millimetres of mercury (mmHg)	x 0.535	= Inches of water (inH₂O)	x 1.868	=	Millimetres of mercury (mmHg)
Inches of water (inH₂O)	x 0.036	= Pounds-force per square inch (psi; lbf/in²; lb/in²)	x 27.68	=	Inches of water (inH₂O)

Torque (moment of force)
Pounds-force inches (lbf in; lb in)	x 1.152	= Kilograms-force centimetre (kgf cm; kg cm)	x 0.868	=	Pounds-force inches (lbf in; lb in)
Pounds-force inches (lbf in; lb in)	x 0.113	= Newton metres (Nm)	x 8.85	=	Pounds-force inches (lbf in; lb in)
Pounds-force inches (lbf in; lb in)	x 0.083	= Pounds-force feet (lbf ft; lb ft)	x 12	=	Pounds-force inches (lbf in; lb in)
Pounds-force feet (lbf ft; lb ft)	x 0.138	= Kilograms-force metres (kgf m; kg m)	x 7.233	=	Pounds-force feet (lbf ft; lb ft)
Pounds-force feet (lbf ft; lb ft)	x 1.356	= Newton metres (Nm)	x 0.738	=	Pounds-force feet (lbf ft; lb ft)
Newton metres (Nm)	x 0.102	= Kilograms-force metres (kgf m; kg m)	x 9.804	=	Newton metres (Nm)

Power
Horsepower (hp)	x 745.7	= Watts (W)	x 0.0013	=	Horsepower (hp)

Velocity (speed)
Miles per hour (miles/hr; mph)	x 1.609	= Kilometres per hour (km/hr; kph)	x 0.621	=	Miles per hour (miles/hr; mph)

Fuel consumption*
Miles per gallon (mpg)	x 0.354	= Kilometres per litre (km/l)	x 2.825	=	Miles per gallon (mpg)

Temperature
Degrees Fahrenheit = (°C x 1.8) + 32   Degrees Celsius (Degrees Centigrade; °C) = (°F - 32) x 0.56

* It is common practice to convert from miles per gallon (mpg) to litres/100 kilometres (l/100km), where mpg x l/100 km = 282

MOT Test Checks   REF•27

About the MOT Test

In the UK, all vehicles more than three years old are subject to an annual test to ensure that they meet minimum safety requirements. A current test certificate must be issued before a machine can be used on public roads, and is required before a road fund licence can be issued. Riding without a current test certificate will also invalidate your insurance.

For most owners, the MOT test is an annual cause for anxiety, and this is largely due to owners not being sure what needs to be checked prior to submitting the motorcycle for testing. The simple answer is that a fully roadworthy motorcycle will have no difficulty in passing the test.

This is a guide to getting your motorcycle through the MOT test. Obviously it will not be possible to examine the motorcycle to the same standard as the professional MOT tester, particularly in view of the equipment required for some of the checks. However, working through the following procedures will enable you to identify any problem areas before submitting the motorcycle for the test.

It has only been possible to summarise the test requirements here, based on the regulations in force at the time of printing. Test standards are becoming increasingly stringent, although there are some exemptions for older vehicles. More information about the MOT test can be obtained from the TSO publications, *How Safe is your Motorcycle* and *The MOT Inspection Manual for Motorcycle Testing*.

Many of the checks require that one of the wheels is raised off the ground. If the motorcycle doesn't have a centre stand, note that an auxiliary stand will be required. Additionally, the help of an assistant may prove useful.

Certain exceptions apply to machines under 50 cc, machines without a lighting system, and Classic bikes - if in doubt about any of the requirements listed below seek confirmation from an MOT tester prior to submitting the motorcycle for the test.

Check that the frame number is clearly visible.

Electrical System

Lights, turn signals, horn and reflector

✔ With the ignition on, check the operation of the following electrical components. **Note:** *The electrical components on certain small-capacity machines are powered by the generator, requiring that the engine is run for this check.*

a) *Headlight and tail light. Check that both illuminate in the low and high beam switch positions.*

b) *Position lights. Check that the front position (or sidelight) and tail light illuminate in this switch position.*

c) *Turn signals. Check that all flash at the correct rate, and that the warning light(s) function correctly. Check that the turn signal switch works correctly.*

d) *Hazard warning system (where fitted). Check that all four turn signals flash in this switch position.*

e) *Brake stop light. Check that the light comes on when the front and rear brakes are independently applied. Models first used on or after 1st April 1986 must have a brake light switch on each brake.*

f) *Horn. Check that the sound is continuous and of reasonable volume.*

✔ Check that there is a red reflector on the rear of the machine, either mounted separately or as part of the tail light lens.

✔ Check the condition of the headlight, tail light and turn signal lenses.

Headlight beam height

✔ The MOT tester will perform a headlight beam height check using specialised beam setting equipment **(see illustration 1)**. This equipment will not be available to the home mechanic, but if you suspect that the headlight is incorrectly set or may have been maladjusted in the past, you can perform a rough test as follows.

✔ Position the bike in a straight line facing a brick wall. The bike must be off its stand, upright and with a rider seated. Measure the height from the ground to the centre of the headlight and mark a horizontal line on the wall at this height. Position the motorcycle 3.8 metres from the wall and draw a vertical line up the wall central to the centreline of the motorcycle. Switch to dipped beam and check that the beam pattern falls slightly lower than the horizontal line and to the left of the vertical line **(see illustration 2)**.

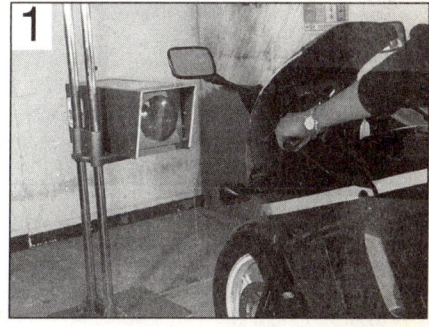

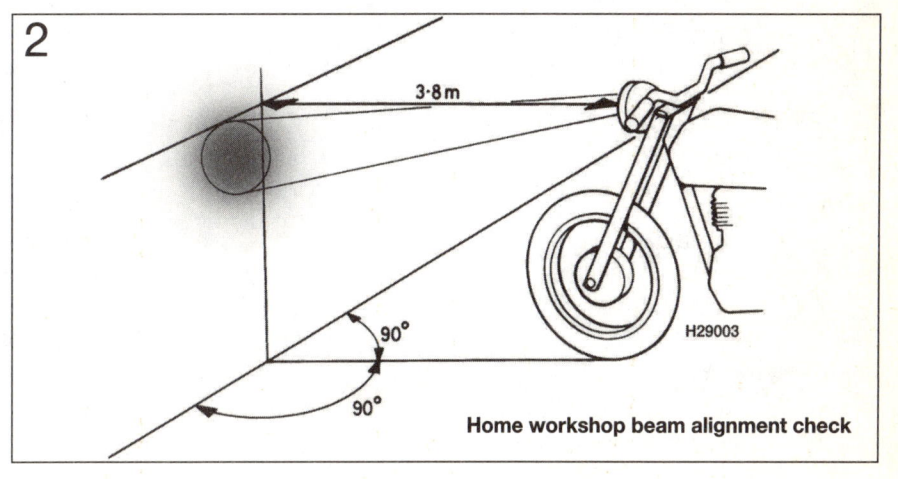

1 Headlight beam height checking equipment

2 Home workshop beam alignment check

REF•28 MOT Test Checks

Exhaust System and Final Drive

Exhaust

✔ Check that the exhaust mountings are secure and that the system does not foul any of the rear suspension components.
✔ Start the motorcycle. When the revs are increased, check that the exhaust is neither holed nor leaking from any of its joints. On a linked system, check that the collector box is not leaking due to corrosion.
✔ Note that the exhaust decibel level ("loudness" of the exhaust) is assessed at the discretion of the tester. If the motorcycle was first used on or after 1st January 1985 the silencer must carry the BSAU 193 stamp, or a marking relating to its make and model, or be of OE (original equipment) manufacture. If the silencer is marked NOT FOR ROAD USE, RACING USE ONLY or similar, it will fail the MOT.

Final drive

✔ On chain or belt drive machines, check that the chain/belt is in good condition and does not have excessive slack. Also check that the sprocket is securely mounted on the rear wheel hub. Check that the chain/belt guard is in place.
✔ On shaft drive bikes, check for oil leaking from the drive unit and fouling the rear tyre.

Steering and Suspension

Steering

✔ With the front wheel raised off the ground, rotate the steering from lock to lock. The handlebar or switches must not contact the fuel tank or be close enough to trap the rider's hand. Problems can be caused by damaged lock stops on the lower yoke and frame, or by the fitting of non-standard handlebars.
✔ When performing the lock to lock check, also ensure that the steering moves freely without drag or notchiness. Steering movement can be impaired by poorly routed cables, or by overtight head bearings or worn bearings. The tester will perform a check of the steering head bearing lower race by mounting the front wheel on a surface plate, then performing a lock to lock check with the weight of the machine on the lower bearing (see illustration 3).
✔ Grasp the fork sliders (lower legs) and attempt to push and pull on the forks (see illustration 4). Any play in the steering head bearings will be felt. Note that in extreme cases, wear of the front fork bushes can be misinterpreted for head bearing play.
✔ Check that the handlebars are securely mounted.
✔ Check that the handlebar grip rubbers are secure. They should by bonded to the bar left end and to the throttle cable pulley on the right end.

Front wheel mounted on a surface plate for steering head bearing lower race check

Front suspension

✔ With the motorcycle off the stand, hold the front brake on and pump the front forks up and down (see illustration 5). Check that they are adequately damped.

Checking the steering head bearings for freeplay

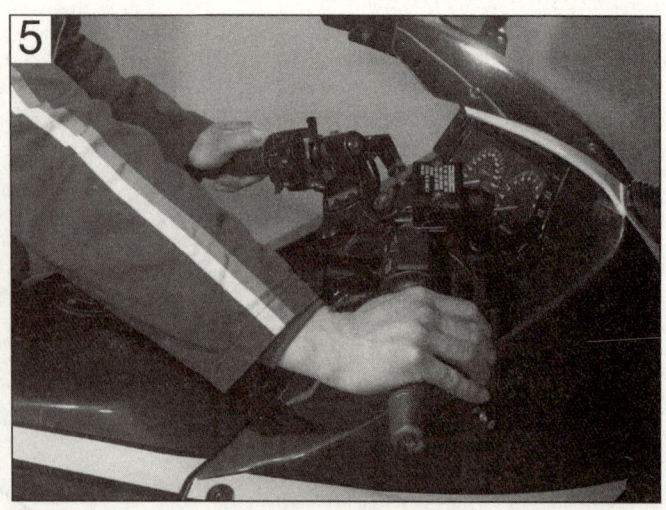

Hold the front brake on and pump the front forks up and down to check operation

MOT Test Checks REF•29

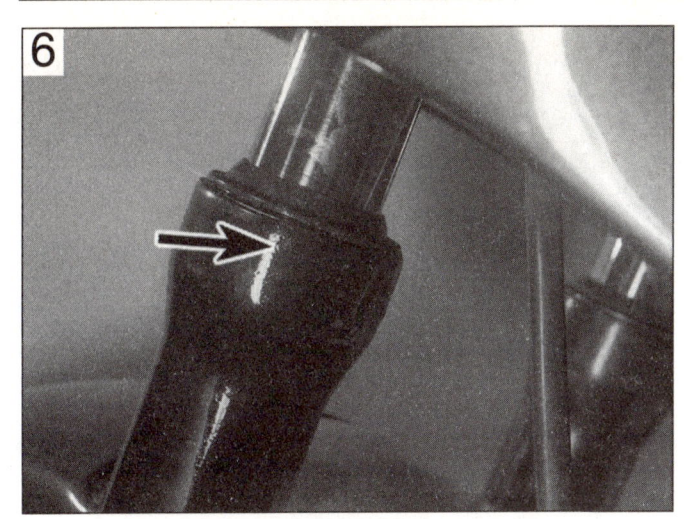

6 Inspect the area around the fork dust seal for oil leakage (arrow)

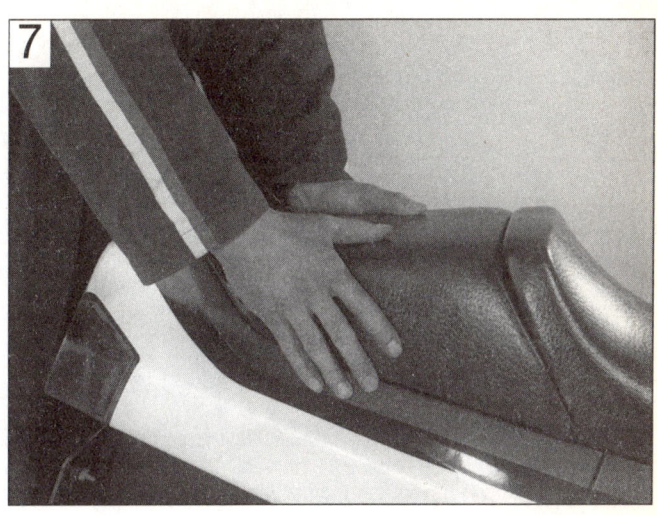

7 Bounce the rear of the motorcycle to check rear suspension operation

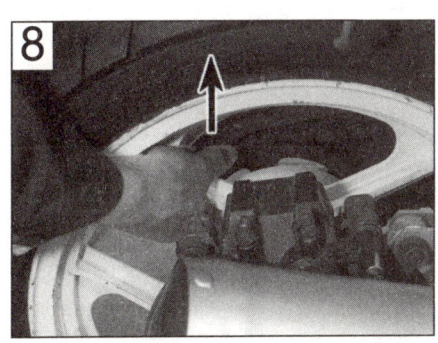

8 Checking for rear suspension linkage play

✔ Inspect the area above and around the front fork oil seals **(see illustration 6)**. There should be no sign of oil on the fork tube (stanchion) nor leaking down the slider (lower leg). On models so equipped, check that there is no oil leaking from the anti-dive units.
✔ On models with swingarm front suspension, check that there is no freeplay in the linkage when moved from side to side.

Rear suspension

✔ With the motorcycle off the stand and an assistant supporting the motorcycle by its handlebars, bounce the rear suspension **(see illustration 7)**. Check that the suspension components do not foul on any of the cycle parts and check that the shock absorber(s) provide adequate damping.
✔ Visually inspect the shock absorber(s) and check that there is no sign of oil leakage from its damper. This is somewhat restricted on certain single shock models due to the location of the shock absorber.
✔ With the rear wheel raised off the ground, grasp the wheel at the highest point and attempt to pull it up **(see illustration 8)**. Any play in the swingarm pivot or suspension linkage bearings will be felt as movement.
Note: *Do not confuse play with actual suspension movement.* Failure to lubricate suspension linkage bearings can lead to bearing failure **(see illustration 9)**.
✔ With the rear wheel raised off the ground, grasp the swingarm ends and attempt to move the swingarm from side to side and forwards and backwards - any play indicates wear of the swingarm pivot bearings **(see illustration 10)**.

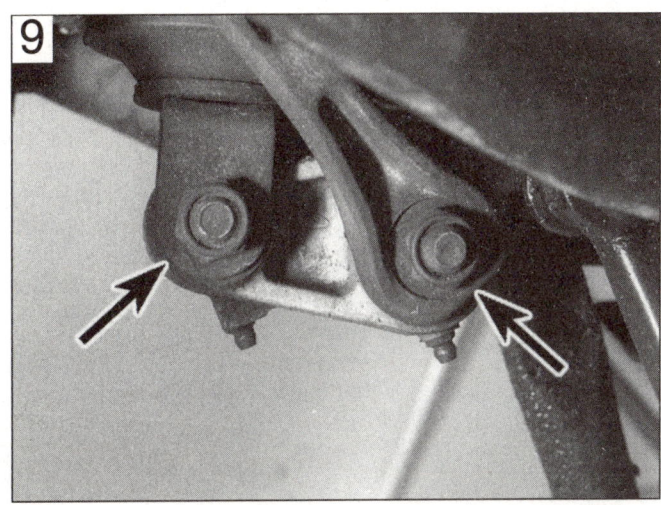

9 Worn suspension linkage pivots (arrows) are usually the cause of play in the rear suspension

10 Grasp the swingarm at the ends to check for play in its pivot bearings

REF•30 MOT Test Checks

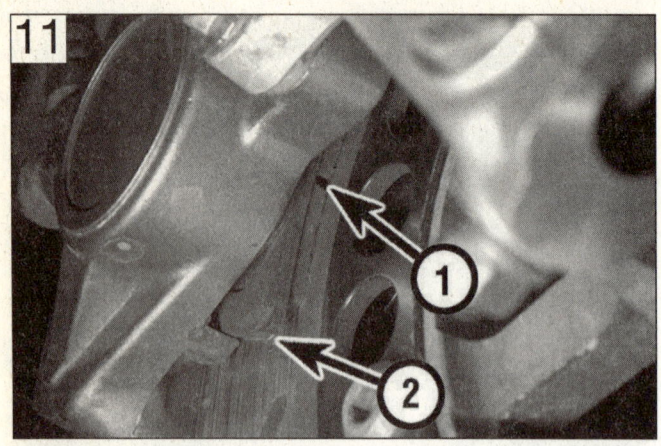

Brake pad wear can usually be viewed without removing the caliper. Most pads have wear indicator grooves (1) and some also have indicator tangs (2)

On drum brakes, check the angle of the operating lever with the brake fully applied. Most drum brakes have a wear indicator pointer and scale.

Brakes, Wheels and Tyres

Brakes

✔ With the wheel raised off the ground, apply the brake then free it off, and check that the wheel is about to revolve freely without brake drag.
✔ On disc brakes, examine the disc itself. Check that it is securely mounted and not cracked.
✔ On disc brakes, view the pad material through the caliper mouth and check that the pads are not worn down beyond the limit **(see illustration 11)**.
✔ On drum brakes, check that when the brake is applied the angle between the operating lever and cable or rod is not too great **(see illustration 12)**. Check also that the operating lever doesn't foul any other components.
✔ On disc brakes, examine the flexible hoses from top to bottom. Have an assistant hold the brake on so that the fluid in the hose is under pressure, and check that there is no sign of fluid leakage, bulges or cracking. If there are any metal brake pipes or unions, check that these are free from corrosion and damage. Where a brake-linked anti-dive system is fitted, check the hoses to the anti-dive in a similar manner.
✔ Check that the rear brake torque arm is secure and that its fasteners are secured by self-locking nuts or castellated nuts with split-pins or R-pins **(see illustration 13)**.
✔ On models with ABS, check that the self-check warning light in the instrument panel works.
✔ The MOT tester will perform a test of the motorcycle's braking efficiency based on a calculation of rider and motorcycle weight. Although this cannot be carried out at home, you can at least ensure that the braking systems are properly maintained. For hydraulic disc brakes, check the fluid level, lever/pedal feel (bleed of air if its spongy) and pad material. For drum brakes, check adjustment, cable or rod operation and shoe lining thickness.

Wheels and tyres

✔ Check the wheel condition. Cast wheels should be free from cracks and if of the built-up design, all fasteners should be secure. Spoked wheels should be checked for broken, corroded, loose or bent spokes.
✔ With the wheel raised off the ground, spin the wheel and visually check that the tyre and wheel run true. Check that the tyre does not foul the suspension or mudguards.
✔ With the wheel raised off the ground, grasp the wheel and attempt to move it about the axle (spindle) **(see illustration 14)**. Any play felt here indicates wheel bearing failure.

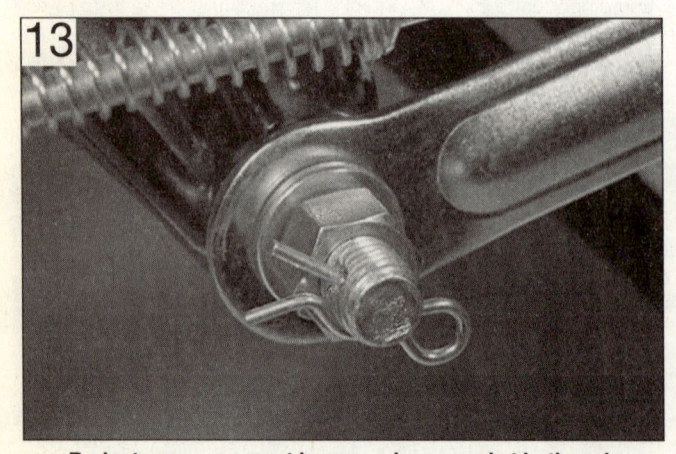

Brake torque arm must be properly secured at both ends

Check for wheel bearing play by trying to move the wheel about the axle (spindle)

MOT Test Checks REF•31

15

Checking the tyre tread depth

16

Tyre direction of rotation arrow can be found on tyre sidewall

17

Castellated type wheel axle (spindle) nut must be secured by a split pin or R-pin

18

Two straightedges are used to check wheel alignment

✔ Check the tyre tread depth, tread condition and sidewall condition **(see illustration 15)**.
✔ Check the tyre type. Front and rear tyre types must be compatible and be suitable for road use. Tyres marked NOT FOR ROAD USE, COMPETITION USE ONLY or similar, will fail the MOT.

✔ If the tyre sidewall carries a direction of rotation arrow, this must be pointing in the direction of normal wheel rotation **(see illustration 16)**.
✔ Check that the wheel axle (spindle) nuts (where applicable) are properly secured. A self-locking nut or castellated nut with a split-pin or R-pin can be used **(see illustration 17)**.
✔ Wheel alignment is checked with the motorcycle off the stand and a rider seated. With the front wheel pointing straight ahead, two perfectly straight lengths of metal or wood and placed against the sidewalls of both tyres **(see illustration 18)**. The gap each side of the front tyre must be equidistant on both sides. Incorrect wheel alignment may be due to a cocked rear wheel (often as the result of poor chain adjustment) or in extreme cases, a bent frame.

General checks and condition

✔ Check the security of all major fasteners, bodypanels, seat, fairings (where fitted) and mudguards.

✔ Check that the rider and pillion footrests, handlebar levers and brake pedal are securely mounted.

✔ Check for corrosion on the frame or any load-bearing components. If severe, this may affect the structure, particularly under stress.

Sidecars

A motorcycle fitted with a sidecar requires additional checks relating to the stability of the machine and security of attachment and swivel joints, plus specific wheel alignment (toe-in) requirements. Additionally, tyre and lighting requirements differ from conventional motorcycle use. Owners are advised to check MOT test requirements with an official test centre.

REF•32 Storage

Preparing for storage

Before you start

If repairs or an overhaul is needed, see that this is carried out now rather than left until you want to ride the bike again.

Give the bike a good wash and scrub all dirt from its underside. Make sure the bike dries completely before preparing for storage.

Engine

- Remove the spark plug(s) and lubricate the cylinder bores with approximately a teaspoon of motor oil using a spout-type oil can **(see illustration 1)**. Reinstall the spark plug(s). Crank the engine over a couple of times to coat the piston rings and bores with oil. If the bike has a kickstart, use this to turn the engine over. If not, flick the kill switch to the OFF position and crank the engine over on the starter **(see illustration 2)**. If the nature on the ignition system prevents the starter operating with the kill switch in the OFF position, remove the spark plugs and fit them back in their caps; ensure that the plugs are earthed (grounded) against the cylinder head when the starter is operated **(see illustration 3)**.

⚠️ **Warning: It is important that the plugs are earthed (grounded) away from the spark plug holes otherwise there is a risk of atomised fuel from the cylinders igniting.**

HAYNES HINT: *On a single cylinder four-stroke engine, you can seal the combustion chamber completely by positioning the piston at TDC on the compression stroke.*

- Drain the carburettor(s) otherwise there is a risk of jets becoming blocked by gum deposits from the fuel **(see illustration 4)**.

- If the bike is going into long-term storage, consider adding a fuel stabiliser to the fuel in the tank. If the tank is drained completely, corrosion of its internal surfaces may occur if left unprotected for a long period. The tank can be treated with a rust preventative especially for this purpose. Alternatively, remove the tank and pour half a litre of motor oil into it, install the filler cap and shake the tank to coat its internals with oil before draining off the excess. The same effect can also be achieved by spraying WD40 or a similar water-dispersant around the inside of the tank via its flexible nozzle.

- Make sure the cooling system contains the correct mix of antifreeze. Antifreeze also contains important corrosion inhibitors.

- The air intakes and exhaust can be sealed off by covering or plugging the openings. Ensure that you do not seal in any condensation; run the engine until it is hot,

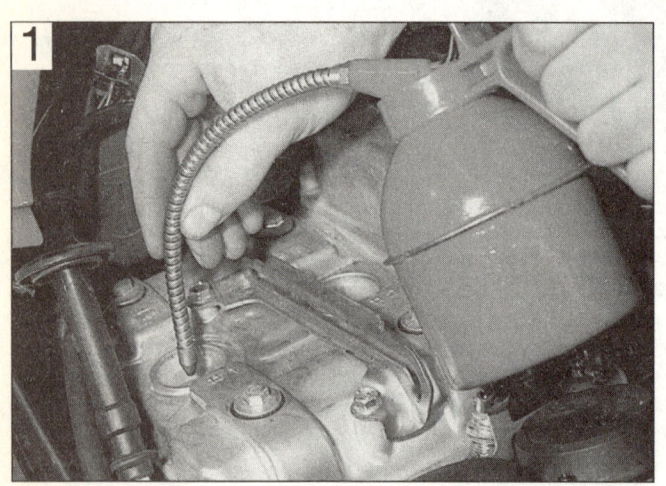

1 Squirt a drop of motor oil into each cylinder

2 Flick the kill switch to OFF . . .

3 . . . and ensure that the metal bodies of the plugs (arrows) are earthed against the cylinder head

4 Connect a hose to the carburettor float chamber drain stub (arrow) and unscrew the drain screw

Storage   REF•33

Exhausts can be sealed off with a plastic bag

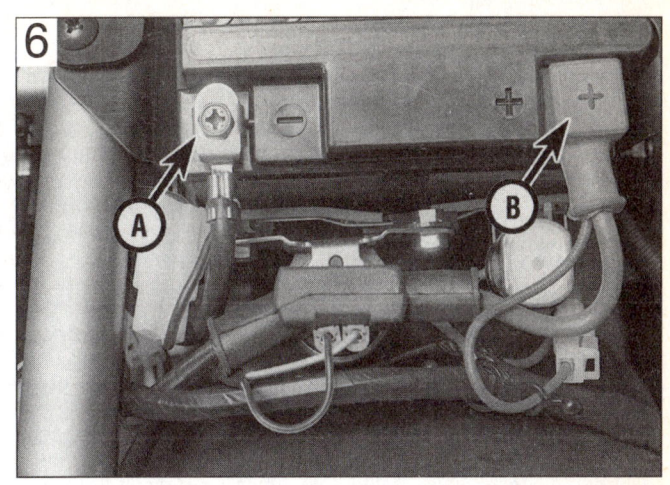

Disconnect the negative lead (A) first, followed by the positive lead (B)

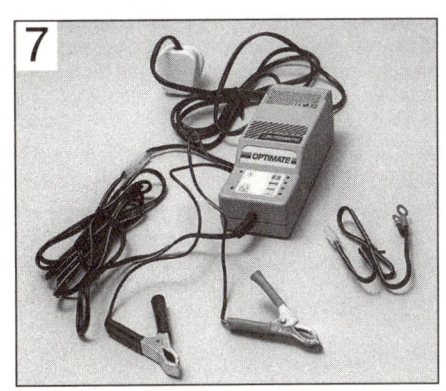

Use a suitable battery charger - this kit also assess battery condition

then switch off and allow to cool. Tape a piece of thick plastic over the silencer end(s) **(see illustration 5)**. Note that some advocate pouring a tablespoon of motor oil into the silencer(s) before sealing them off.

Battery

● Remove it from the bike - in extreme cases of cold the battery may freeze and crack its case **(see illustration 6)**.

● Check the electrolyte level and top up if necessary (conventional refillable batteries). Clean the terminals.
● Store the battery off the motorcycle and away from any sources of fire. Position a wooden block under the battery if it is to sit on the ground.
● Give the battery a trickle charge for a few hours every month **(see illustration 7)**.

Tyres

● Place the bike on its centrestand or an auxiliary stand which will support the motorcycle in an upright position. Position wood blocks under the tyres to keep them off the ground and to provide insulation from damp. If the bike is being put into long-term storage, ideally both tyres should be off the ground; not only will this protect the tyres, but will also ensure that no load is placed on the steering head or wheel bearings.
● Deflate each tyre by 5 to 10 psi, no more or the beads may unseat from the rim, making subsequent inflation difficult on tubeless tyres.

Pivots and controls

● Lubricate all lever, pedal, stand and footrest pivot points. If grease nipples are fitted to the rear suspension components, apply lubricant to the pivots.
● Lubricate all control cables.

Cycle components

● Apply a wax protectant to all painted and plastic components. Wipe off any excess, but don't polish to a shine. Where fitted, clean the screen with soap and water.
● Coat metal parts with Vaseline (petroleum jelly). When applying this to the fork tubes, do not compress the forks otherwise the seals will rot from contact with the Vaseline.
● Apply a vinyl cleaner to the seat.

Storage conditions

● Aim to store the bike in a shed or garage which does not leak and is free from damp.
● Drape an old blanket or bedspread over the bike to protect it from dust and direct contact with sunlight (which will fade paint). This also hides the bike from prying eyes. Beware of tight-fitting plastic covers which may allow condensation to form and settle on the bike.

Getting back on the road

Engine and transmission

● Change the oil and replace the oil filter. If this was done prior to storage, check that the oil hasn't emulsified - a thick whitish substance which occurs through condensation.
● Remove the spark plugs. Using a spout-type oil can, squirt a few drops of oil into the cylinder(s). This will provide initial lubrication as the piston rings and bores comes back into contact. Service the spark plugs, or fit new ones, and install them in the engine.

● Check that the clutch isn't stuck on. The plates can stick together if left standing for some time, preventing clutch operation. Engage a gear and try rocking the bike back and forth with the clutch lever held against the handlebar. If this doesn't work on cable-operated clutches, hold the clutch lever back against the handlebar with a strong elastic band or cable tie for a couple of hours **(see illustration 8)**.
● If the air intakes or silencer end(s) were blocked off, remove the bung or cover used.
● If the fuel tank was coated with a rust

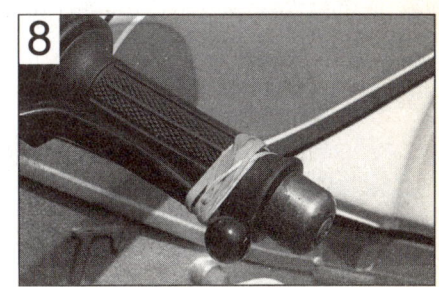

Hold clutch lever back against the handlebar with elastic bands or a cable tie

Storage

preventative, oil or a stabiliser added to the fuel, drain and flush the tank and dispose of the fuel sensibly. If no action was taken with the fuel tank prior to storage, it is advised that the old fuel is disposed of since it will go off over a period of time. Refill the fuel tank with fresh fuel.

Frame and running gear

- Oil all pivot points and cables.
- Check the tyre pressures. They will definitely need inflating if pressures were reduced for storage.
- Lubricate the final drive chain (where applicable).
- Remove any protective coating applied to the fork tubes (stanchions) since this may well destroy the fork seals. If the fork tubes weren't protected and have picked up rust spots, remove them with very fine abrasive paper and refinish with metal polish.
- Check that both brakes operate correctly. Apply each brake hard and check that it's not possible to move the motorcycle forwards, then check that the brake frees off again once released. Brake caliper pistons can stick due to corrosion around the piston head, or on the sliding caliper types, due to corrosion of the slider pins. If the brake doesn't free after repeated operation, take the caliper off for examination. Similarly drum brakes can stick due to a seized operating cam, cable or rod linkage.
- If the motorcycle has been in long-term storage, renew the brake fluid and clutch fluid (where applicable).
- Depending on where the bike has been stored, the wiring, cables and hoses may have been nibbled by rodents. Make a visual check and investigate disturbed wiring loom tape.

Battery

- If the battery has been previously removal and given top up charges it can simply be reconnected. Remember to connect the positive cable first and the negative cable last.
- On conventional refillable batteries, if the battery has not received any attention, remove it from the motorcycle and check its electrolyte level. Top up if necessary then charge the battery. If the battery fails to hold a charge and a visual checks show heavy white sulphation of the plates, the battery is probably defective and must be renewed. This is particularly likely if the battery is old. Confirm battery condition with a specific gravity check.
- On sealed (MF) batteries, if the battery has not received any attention, remove it from the motorcycle and charge it according to the information on the battery case - if the battery fails to hold a charge it must be renewed.

Starting procedure

- If a kickstart is fitted, turn the engine over a couple of times with the ignition OFF to distribute oil around the engine. If no kickstart is fitted, flick the engine kill switch OFF and the ignition ON and crank the engine over a couple of times to work oil around the upper cylinder components. If the nature of the ignition system is such that the starter won't work with the kill switch OFF, remove the spark plugs, fit them back into their caps and earth (ground) their bodies on the cylinder head. Reinstall the spark plugs afterwards.
- Switch the kill switch to RUN, operate the choke and start the engine. If the engine won't start don't continue cranking the engine - not only will this flatten the battery, but the starter motor will overheat. Switch the ignition off and try again later. If the engine refuses to start, go through the fault finding procedures in this manual. **Note:** *If the bike has been in storage for a long time, old fuel or a carburettor blockage may be the problem. Gum deposits in carburettors can block jets - if a carburettor cleaner doesn't prove successful the carburettors must be dismantled for cleaning.*
- Once the engine has started, check that the lights, turn signals and horn work properly.
- Treat the bike gently for the first ride and check all fluid levels on completion. Settle the bike back into the maintenance schedule.

Fault Finding REF•35

This Section provides an easy reference-guide to the more common faults that are likely to afflict your machine. Obviously, the opportunities are almost limitless for faults to occur as a result of obscure failures, and to try and cover all eventualities would require a book. Indeed, a number have been written on the subject.

Successful troubleshooting is not a mysterious 'black art' but the application of a bit of knowledge combined with a systematic and logical approach to the problem. Approach any troubleshooting by first accurately identifying the symptom and then checking through the list of possible causes, starting with the simplest or most obvious and progressing in stages to the most complex.

Take nothing for granted, but above all apply liberal quantities of common sense.

The main symptom of a fault is given in the text as a major heading below which are listed the various systems or areas which may contain the fault. Details of each possible cause for a fault and the remedial action to be taken are given, in brief, in the paragraphs below each heading. Further information should be sought in the relevant Chapter.

1 Engine doesn't start or is difficult to start
- [ ] Starter motor doesn't rotate
- [ ] Starter motor rotates but engine does not turn over
- [ ] Starter works but engine won't turn over (seized)
- [ ] No fuel flow
- [ ] Engine flooded
- [ ] No spark or weak spark
- [ ] Compression low
- [ ] Stalls after starting
- [ ] Rough idle

2 Poor running at low speed
- [ ] Spark weak
- [ ] Fuel/air mixture incorrect
- [ ] Compression low
- [ ] Poor acceleration

3 Poor running or no power at high speed
- [ ] Firing incorrect
- [ ] Fuel/air mixture incorrect
- [ ] Compression low
- [ ] Knocking or pinging
- [ ] Miscellaneous causes

4 Overheating
- [ ] Engine overheats
- [ ] Firing incorrect
- [ ] Fuel/air mixture incorrect
- [ ] Compression too high
- [ ] Engine load excessive
- [ ] Lubrication inadequate
- [ ] Miscellaneous causes

5 Clutch problems
- [ ] Clutch slipping
- [ ] Clutch not disengaging completely

6 Gear shifting problems
- [ ] Doesn't go into gear, or lever doesn't return
- [ ] Jumps out of gear
- [ ] Overshifts

7 Abnormal engine noise
- [ ] Knocking or pinging
- [ ] Piston slap or rattling
- [ ] Valve noise
- [ ] Other noise

8 Abnormal driveline noise
- [ ] Clutch noise
- [ ] Transmission noise
- [ ] Final drive noise

9 Abnormal frame and suspension noise
- [ ] Front end noise
- [ ] Shock absorber noise
- [ ] Brake noise

10 Oil pressure warning light comes on
- [ ] Engine lubrication system
- [ ] Electrical system

11 Excessive exhaust smoke
- [ ] White smoke
- [ ] Black smoke
- [ ] Brown smoke

12 Poor handling or stability
- [ ] Handlebar hard to turn
- [ ] Handlebar shakes or vibrates excessively
- [ ] Handlebar pulls to one side
- [ ] Poor shock absorbing qualities

13 Braking problems
- [ ] Brakes are spongy, don't hold
- [ ] Brake lever or pedal pulsates
- [ ] Brakes drag

14 Electrical problems
- [ ] Battery dead or weak
- [ ] Battery overcharged

Fault Finding

1 Engine doesn't start or is difficult to start

Starter motor doesn't rotate
- [ ] Engine kill switch OFF.
- [ ] Fuse blown. Check main fuse and starter circuit fuse (Chapter 9).
- [ ] Battery voltage low. Check and recharge battery (Chapter 9).
- [ ] Starter motor defective. Make sure the wiring to the starter is secure. Make sure the starter relay clicks when the start button is pushed. If the relay clicks, then the fault is in the wiring or motor.
- [ ] Starter relay faulty. Check it according to the procedure in Chapter 9.
- [ ] Starter switch not contacting. The contacts could be wet, corroded or dirty. Disassemble and clean the switch (Chapter 9).
- [ ] Wiring open or shorted. Check all wiring connections and harnesses to make sure that they are dry, tight and not corroded. Also check for broken or frayed wires that can cause a short to ground (earth) (see wiring diagram, Chapter 9).
- [ ] Ignition (main) switch defective. Check the switch according to the procedure in Chapter 9. Replace the switch with a new one if it is defective.
- [ ] Engine kill switch defective. Check for wet, dirty or corroded contacts. Clean or replace the switch as necessary (Chapter 9).
- [ ] Faulty neutral, side stand or clutch switch. Check the wiring to each switch and the switch itself according to the procedures in Chapter 9.

Starter motor rotates but engine does not turn over
- [ ] Starter motor clutch defective. Inspect and repair or replace (Chapter 2).
- [ ] Damaged idler or starter gears. Inspect and replace the damaged parts (Chapter 2).

Starter works but engine won't turn over (seized)
- [ ] Seized engine caused by one or more internally damaged components. Failure due to wear, abuse or lack of lubrication. Damage can include seized valves, followers, camshafts, pistons, crankshaft, connecting rod bearings, or transmission gears or bearings. Refer to Chapter 2 for engine disassembly.

No fuel flow
- [ ] No fuel in tank.
- [ ] Fuel tank breather hose obstructed.
- [ ] Fuel tap filter clogged. Remove the tap and clean the filter (see Chapter 4).
- [ ] Fuel line clogged. Pull the fuel line loose and carefully blow through it.
- [ ] Float needle valve clogged. For both valves to be clogged, either a very bad batch of fuel with an unusual additive has been used, or some other foreign material has entered the tank. Many times after a machine has been stored for many months without running, the fuel turns to a varnish-like liquid and forms deposits on the inlet needle valves and jets. The carburettors should be removed and overhauled if draining the float chambers doesn't solve the problem.

Engine flooded
- [ ] Float height too high. Check as described in Chapter 4.
- [ ] Float needle valve worn or stuck open (carburettor models). A piece of dirt, rust or other debris can cause the valve to seat improperly, causing excess fuel to be admitted to the float chamber. In this case, the float chamber should be cleaned and the needle valve and seat inspected. If the needle and seat are worn, then the leaking will persist and the parts should be replaced with new ones (Chapter 4).
- [ ] Starting technique incorrect. Under normal circumstances (i.e., if all the carburettor functions are sound) the machine should start with little or no throttle. When the engine is cold, the choke should be operated and the engine started without opening the throttle. When the engine is at operating temperature, only a very slight amount of throttle should be necessary.

No spark or weak spark
- [ ] Ignition switch OFF.
- [ ] Engine kill switch turned to the OFF position.
- [ ] Battery voltage low. Check and recharge the battery as necessary (Chapter 9).
- [ ] Spark plugs dirty, defective or worn out. Locate reason for fouled plugs using spark plug condition chart and follow the plug maintenance procedures (Chapter 1).
- [ ] Spark plug caps or secondary (HT) wiring faulty. Check condition. Replace either or both components if cracks or deterioration are evident (Chapter 5).
- [ ] Spark plug caps not making good contact. Make sure that the plug caps fit snugly over the plug ends.
- [ ] Ignition control unit defective. Check the unit, referring to Chapter 5 for details.
- [ ] Pulse generator defective. Check the unit, referring to Chapter 5 for details.
- [ ] Ignition HT coils defective. Check the coils, referring to Chapter 5.
- [ ] Ignition or kill switch shorted. This is usually caused by water, corrosion, damage or excessive wear. The switches can be disassembled and cleaned with electrical contact cleaner. If cleaning does not help, replace the switches (Chapter 9).
- [ ] Wiring shorted or broken between:
    a) Ignition (main) switch and engine kill switch (or blown fuse)
    b) Ignition control unit and engine kill switch
    c) Ignition control unit and ignition HT coils
    d) Ignition HT coils and spark plugs
    e) Ignition control unit and pulse generator
- [ ] Make sure that all wiring connections are clean, dry and tight. Look for chafed and broken wires (Chapters 5 and 9).

Compression low
- [ ] Spark plugs loose. Remove the plugs and inspect their threads. Reinstall and tighten to the specified torque (Chapter 1).
- [ ] Cylinder head not sufficiently tightened down. If the cylinder head is suspected of being loose, then there's a chance that the gasket or head is damaged if the problem has persisted for any length of time. The head bolts should be tightened to the proper torque in the correct sequence (Chapter 2).
- [ ] Improper valve clearance. This means that the valve is not closing completely and compression pressure is leaking past the valve. Check and adjust the valve clearances (Chapter 1).
- [ ] Cylinder and/or piston worn. Excessive wear will cause compression pressure to leak past the rings. This is usually accompanied by worn rings as well. A top-end overhaul is necessary (Chapter 2).
- [ ] Piston rings worn, weak, broken, or sticking. Broken or sticking piston rings usually indicate a lubrication or carburation problem that causes excess carbon deposits or seizures to form on the pistons and rings. Top-end overhaul is necessary (Chapter 2).
- [ ] Piston ring-to-groove clearance excessive. This is caused by excessive wear of the piston ring lands. Piston replacement is necessary (Chapter 2).
- [ ] Cylinder head gasket damaged. If a head is allowed to become loose, or if excessive carbon build-up on the piston crown and combustion chamber causes extremely high compression, the head gasket may leak. Retorquing the head is not always sufficient to restore the seal, so gasket replacement is necessary (Chapter 2).
- [ ] Cylinder head warped. This is caused by overheating or improperly tightened head bolts. Machine shop resurfacing or head replacement is necessary (Chapter 2).
- [ ] Valve spring broken or weak. Caused by component failure or wear; the springs must be replaced (Chapter 2).

Fault Finding REF•37

1 Engine doesn't start or is difficult to start (continued)

Compression low (continued)

☐ Valve not seating properly. This is caused by a bent valve (from over-revving or improper valve adjustment), burned valve or seat (improper carburation) or an accumulation of carbon deposits on the seat (from carburation or lubrication problems). The valves must be cleaned and/or replaced and the seats serviced if possible (Chapter 2).

Stalls after starting

☐ Improper choke action (carburettor models). Make sure the choke linkage shaft is getting a full stroke and staying in the out position (Chapter 4).
☐ Ignition malfunction. See Chapter 5.
☐ Carburettor malfunction. See Chapter 4.
☐ Fuel contaminated. The fuel can be contaminated with either dirt or water, or can change chemically if the machine is allowed to sit for several months or more. Drain the tank and float chambers (Chapter 4).
☐ Intake air leak. Check for loose carburettor-to-intake manifold connections, loose or missing vacuum gauge adapter screws, caps or hoses, or loose carburettor tops (Chapter 4).
☐ Engine idle speed incorrect. Turn idle adjusting screw until the engine idles at the specified rpm (Chapter 1).

Rough idle

☐ Ignition malfunction. See Chapter 5.
☐ Idle speed incorrect. See Chapter 1.
☐ Carburettors not synchronised. Adjust them with vacuum gauge or manometer set as described in Chapter 1.
☐ Carburettor malfunction. See Chapter 4.
☐ Fuel contaminated. The fuel can be contaminated with either dirt or water, or can change chemically if the machine is allowed to sit for several months or more. Drain the tank and float chambers or fuel rail (Chapter 4).
☐ Intake air leak. Check for loose carburettor-to-intake manifold connections, loose or missing vacuum gauge adapter screws, caps or hoses, or loose carburettor tops (Chapter 4).
☐ Air filter clogged. Replace the air filter element (Chapter 1).

2 Poor running at low speeds

Spark weak

☐ Battery voltage low. Check and recharge battery (Chapter 9).
☐ Spark plugs fouled, defective or worn out. Refer to Chapter 1 for spark plug maintenance.
☐ Spark plug cap or HT wiring defective. Refer to Chapters 1 and 5 for details on the ignition system.
☐ Spark plug caps not making contact.
☐ Incorrect spark plugs. Wrong type, heat range or cap configuration. Check and install correct plugs listed in Chapter 1.
☐ Ignition control unit defective. See Chapter 5.
☐ Pulse generator defective. See Chapter 5.
☐ Ignition HT coils defective. See Chapter 5.

Fuel/air mixture incorrect

☐ Pilot screws out of adjustment (Chapter 4). Pilot jet or air passage clogged. Remove and overhaul the carburettors (Chapter 4).
☐ Air bleed holes clogged. Remove carburettor and blow out all passages (Chapter 4).
☐ Air filter clogged, poorly sealed or missing (Chapter 1).
☐ Air filter housing poorly sealed. Look for cracks, holes or loose clamps and replace or repair defective parts.
☐ Fuel level too high or too low. Check the float height (Chapter 4).
☐ Fuel tank breather hose obstructed.
☐ Carburettor intake manifolds loose. Check for cracks, breaks, tears or loose clamps. Replace the rubber intake manifold joints if split or perished.

Compression low

☐ Spark plugs loose. Remove the plugs and inspect their threads. Reinstall and tighten to the specified torque (Chapter 1).
☐ Cylinder head not sufficiently tightened down. If the cylinder head is suspected of being loose, then there's a chance that the gasket and head are damaged if the problem has persisted for any length of time. The head bolts should be tightened to the proper torque in the correct sequence (Chapter 2).
☐ Improper valve clearance. This means that the valve is not closing completely and compression pressure is leaking past the valve. Check and adjust the valve clearances (Chapter 1).
☐ Cylinder and/or piston worn. Excessive wear will cause compression pressure to leak past the rings. This is usually accompanied by worn rings as well. A top end overhaul is necessary (Chapter 2).
☐ Piston rings worn, weak, broken, or sticking. Broken or sticking piston rings usually indicate a lubrication or carburation problem that causes excess carbon deposits or seizures to form on the pistons and rings. Top-end overhaul is necessary (Chapter 2).
☐ Piston ring-to-groove clearance excessive. This is caused by excessive wear of the piston ring lands. Piston replacement is necessary (Chapter 2).
☐ Cylinder head gasket damaged. If a head is allowed to become loose, or if excessive carbon build-up on the piston crown and combustion chamber causes extremely high compression, the head gasket may leak. Retorquing the head is not always sufficient to restore the seal, so gasket replacement is necessary (Chapter 2).
☐ Cylinder head warped. This is caused by overheating or improperly tightened head bolts. Machine shop resurfacing or head replacement is necessary (Chapter 2).
☐ Valve spring broken or weak. Caused by component failure or wear; the springs must be replaced (Chapter 2).
☐ Valve not seating properly. This is caused by a bent valve (from over-revving or improper valve adjustment), burned valve or seat (improper carburation) or an accumulation of carbon deposits on the seat (from carburation, lubrication problems). The valves must be cleaned and/or replaced and the seats serviced if possible (Chapter 2).

Poor acceleration

☐ Carburettors leaking or dirty. Overhaul them (Chapter 4).
☐ Timing not advancing. The pulse generator or the ignition control module may be defective. If so, they must be replaced with new ones, as they can't be repaired.
☐ Carburettors not synchronised. Adjust them with a vacuum gauge set or manometer (Chapter 1).
☐ Engine oil viscosity too high. Using a heavier oil than that recommended in Chapter 1 can damage the oil pump or lubrication system and cause drag on the engine.
☐ Brakes dragging. Usually caused by debris which has entered the brake piston seals, or from a warped disc or bent axle. Repair as necessary (Chapter 7).

Fault Finding

3 Poor running or no power at high speed

Firing incorrect
- [ ] Air filter restricted. Clean or replace filter (Chapter 1).
- [ ] Spark plugs fouled, defective or worn out. See Chapter 1 for spark plug maintenance.
- [ ] Spark plug caps or HT wiring defective. See Chapters 1 and 5 for details of the ignition system.
- [ ] Spark plug caps not in good contact. See Chapter 5.
- [ ] Incorrect spark plugs. Wrong type, heat range or cap configuration. Check and install correct plugs listed in Chapter 1.
- [ ] Ignition control unit defective. See Chapter 5.
- [ ] Ignition coils defective. See Chapter 5.

Fuel/air mixture incorrect
- [ ] Main jet or fuel injector clogged or wrong size. Dirt, water or other contaminants can clog them. Clean the fuel filter, and the carburettors (Chapter 4). The standard jetting is for sea level atmospheric pressure and oxygen content.
- [ ] Air bleed holes clogged. Remove and overhaul carburettors (Chapter 4).
- [ ] Air filter clogged, poorly sealed, or missing (Chapter 1).
- [ ] Air filter housing poorly sealed. Look for cracks, holes or loose clamps, and replace or repair defective parts.
- [ ] Fuel level too high or too low. Check the float height (Chapter 4).
- [ ] Fuel tank breather hose obstructed.
- [ ] Carburettor intake manifolds loose. Check for cracks, breaks, tears or loose clamps. Replace the rubber intake manifolds if they are split or perished (Chapter 4).

Compression low
- [ ] Spark plugs loose. Remove the plugs and inspect their threads. Reinstall and tighten to the specified torque (Chapter 1).
- [ ] Cylinder head not sufficiently tightened down. If the cylinder head is suspected of being loose, then there's a chance that the gasket and head are damaged if the problem has persisted for any length of time. The head bolts should be tightened to the proper torque in the correct sequence (Chapter 2).
- [ ] Improper valve clearance. This means that the valve is not closing completely and compression pressure is leaking past the valve. Check and adjust the valve clearances (Chapter 1).
- [ ] Cylinder and/or piston worn. Excessive wear will cause compression pressure to leak past the rings. This is usually accompanied by worn rings as well. A top-end overhaul is necessary (Chapter 2).
- [ ] Piston rings worn, weak, broken, or sticking. Broken or sticking piston rings usually indicate a lubrication or carburation problem that causes excess carbon deposits or seizures to form on the pistons and rings. Top-end overhaul is necessary (Chapter 2).
- [ ] Piston ring-to-groove clearance excessive. This is caused by excessive wear of the piston ring lands. Piston replacement is necessary (Chapter 2).
- [ ] Cylinder head gasket damaged. If a head is allowed to become loose, or if excessive carbon build-up on the piston crown and combustion chamber causes extremely high compression, the head gasket may leak. Retorquing the head is not always sufficient to restore the seal, so gasket replacement is necessary (Chapter 2).
- [ ] Cylinder head warped. This is caused by overheating or improperly tightened head bolts. Machine shop resurfacing or head replacement is necessary (Chapter 2).
- [ ] Valve spring broken or weak. Caused by component failure or wear; the springs must be replaced (Chapter 2).
- [ ] Valve not seating properly. This is caused by a bent valve (from over-revving or improper valve adjustment), burned valve or seat (improper carburation) or an accumulation of carbon deposits on the seat (from carburation or lubrication problems). The valves must be cleaned and/or replaced and the seats serviced if possible (Chapter 2).

Knocking or pinking
- [ ] Carbon build-up in combustion chamber. Use of a fuel additive that will dissolve the adhesive bonding the carbon particles to the crown and chamber is the easiest way to remove the build-up. Otherwise, the cylinder head will have to be removed and decarbonised (Chapter 2).
- [ ] Incorrect or poor quality fuel. Old or improper grades of fuel can cause detonation. This causes the piston to rattle, thus the knocking or pinking sound. Drain old fuel and always use the recommended fuel grade.
- [ ] Spark plug heat range incorrect. Uncontrolled detonation indicates the plug heat range is too hot. The plug in effect becomes a glow plug, raising cylinder temperatures. Install the proper heat range plug (Chapter 1).
- [ ] Improper air/fuel mixture. This will cause the cylinders to run hot, which leads to detonation. Clogged jets or an air leak can cause this imbalance. See Chapter 4.

Miscellaneous causes
- [ ] Throttle valve doesn't open fully. Adjust the throttle grip freeplay (Chapter 1).
- [ ] Clutch slipping. May be caused by loose or worn clutch components. Refer to Chapter 2 for clutch overhaul procedures.
- [ ] Timing not advancing. Check ignition timing (Chapter 5).
- [ ] Engine oil viscosity too high. Using a heavier oil than the one recommended in Chapter 1 can damage the oil pump or lubrication system and cause drag on the engine.
- [ ] Brakes dragging. Usually caused by debris which has entered the brake piston seals, or from a warped disc or bent axle. Repair as necessary.

Fault Finding REF•39

4 Overheating

Engine overheats
- [ ] Coolant level low. Check and add coolant (Chapter 1).
- [ ] Leak in cooling system. Check cooling system hoses and radiator for leaks and other damage. Repair or replace parts as necessary (Chapter 3).
- [ ] Thermostat sticking open or closed. Check and replace as described in Chapter 3.
- [ ] Faulty cooling system pressure cap. Remove the cap and have it pressure tested.
- [ ] Coolant passages clogged. Have the entire system drained and flushed, then refill with fresh coolant.
- [ ] Water pump defective. Remove the pump and check the components (Chapter 3).
- [ ] Clogged radiator fins. Clean them by blowing compressed air through the fins from the back.
- [ ] Cooling fan or fan switch fault (Chapter 3).

Firing incorrect
- [ ] Spark plugs fouled, defective or worn out. See Chapter 1 for spark plug maintenance.
- [ ] Incorrect spark plugs.
- [ ] Ignition control unit defective. See Chapter 5.
- [ ] Faulty ignition HT coils (Chapter 5).

Fuel/air mixture incorrect
- [ ] Main jet or fuel injector clogged or wrong size. Dirt, water or other contaminants can clog them. Clean the fuel strainer and/or filter, and the carburettors or injectors (Chapter 4). The standard jetting is for sea level atmospheric pressure and oxygen content.
- [ ] Air filter clogged, poorly sealed, or missing (Chapter 1).
- [ ] Air filter housing poorly sealed. Look for cracks, holes or loose clamps, and replace or repair defective parts.
- [ ] Fuel level too high or too low. Check the float height (Chapter 4).
- [ ] Fuel tank breather hose obstructed.
- [ ] Carburettor intake manifolds loose. Check for cracks, breaks, tears or loose clamps. Replace the rubber intake manifolds if they are split or perished (Chapter 4).

Compression too high
- [ ] Carbon build-up in combustion chamber. Use of a fuel additive that will dissolve the adhesive bonding the carbon particles to the piston crown and chamber is the easiest way to remove the build-up. Otherwise, the cylinder head will have to be removed and decarbonised (Chapter 2).
- [ ] Improperly machined head surface or installation of incorrect gasket during engine assembly.

Engine load excessive
- [ ] Clutch slipping. Can be caused by damaged, loose or worn clutch components. Refer to Chapter 2 for overhaul procedures.
- [ ] Engine oil level too high. The addition of too much oil will cause pressurisation of the crankcase and inefficient engine operation. Check Specifications and drain to proper level (Chapter 1).
- [ ] Engine oil viscosity too high. Using a heavier oil than the one recommended in Chapter 1 can damage the oil pump or lubrication system as well as cause drag on the engine.
- [ ] Brakes dragging. Usually caused by debris which has entered the brake piston seals, or from a warped disc or bent axle. Repair as necessary.

Lubrication inadequate
- [ ] Engine oil level too low. Friction caused by intermittent lack of lubrication or from oil that is overworked can cause overheating. The oil provides a definite cooling function in the engine. Check the oil level (*Daily (pre-ride) checks*).
- [ ] Poor quality engine oil or incorrect viscosity or type. Oil is rated not only according to viscosity but also according to type. Some oils are not rated high enough for use in this engine. Check the Specifications section and change to the correct oil (Chapter 1).

Miscellaneous causes
- [ ] Modification to exhaust system. Most aftermarket exhaust systems cause the engine to run leaner, which make them run hotter. When installing an accessory exhaust system, always rejet the carburettors.

REF•40 Fault Finding

5 Clutch problems

Clutch slipping

- [ ] Insufficient clutch cable freeplay. Check and adjust (Chapter 1).
- [ ] Friction plates worn or warped. Overhaul the clutch assembly (Chapter 2).
- [ ] Plain plates warped (Chapter 2).
- [ ] Clutch springs broken or weak. Old or heat-damaged (from slipping clutch) springs should be replaced with new ones (Chapter 2).
- [ ] Clutch release mechanism defective. Replace any defective parts (Chapter 2).
- [ ] Clutch centre or housing unevenly worn. This causes improper engagement of the plates. Replace the damaged or worn parts (Chapter 2).

Clutch not disengaging completely

- [ ] Excessive clutch cable freeplay. Check and adjust (Chapter 1).
- [ ] Clutch plates warped or damaged. This will cause clutch drag, which in turn will cause the machine to creep. Overhaul the clutch assembly (Chapter 2).
- [ ] Clutch spring tension uneven. Usually caused by a sagged or broken spring. Check and replace the springs as a set (Chapter 2).
- [ ] Engine oil deteriorated. Old, thin, worn out oil will not provide proper lubrication for the plates, causing the clutch to drag. Replace the oil and filter (Chapter 1).
- [ ] Engine oil viscosity too high. Using a heavier oil than recommended in Chapter 1 can cause the plates to stick together, putting a drag on the engine. Change to the correct weight oil (Chapter 1).
- [ ] Clutch housing sleeve seized on shaft. Lack of lubrication, severe wear or damage can cause the sleeve to seize on the shaft. Overhaul of the clutch, and perhaps transmission, may be necessary to repair the damage (Chapter 2).
- [ ] Clutch release mechanism defective. Overhaul the clutch cover components (Chapter 2).
- [ ] Loose clutch centre nut. Causes housing and centre misalignment putting a drag on the engine. Engagement adjustment continually varies. Overhaul the clutch assembly (Chapter 2).

6 Gearchanging problems

Doesn't go into gear or lever doesn't return

- [ ] Clutch not disengaging. See above.
- [ ] Selector fork(s) bent or seized. Often caused by dropping the machine or from lack of lubrication. Overhaul the transmission (Chapter 2).
- [ ] Gear(s) stuck on shaft. Most often caused by a lack of lubrication or excessive wear in transmission bearings and bushings. Overhaul the transmission (Chapter 2).
- [ ] Selector drum binding. Caused by lubrication failure or excessive wear. Replace the drum and bearing (Chapter 2).
- [ ] Gearchange lever return spring weak or broken (Chapter 2).
- [ ] Gearchange lever broken. Splines stripped out of lever or shaft, caused by allowing the lever to get loose or from dropping the machine. Replace necessary parts (Chapter 2).
- [ ] Stopper arm broken or worn. Full engagement and rotary movement of selector drum results. Replace the arm (Chapter 2).
- [ ] Stopper arm spring broken. Allows arm to float, causing sporadic shift operation. Replace spring (Chapter 2).

Jumps out of gear

- [ ] Selector fork(s) worn. Overhaul the transmission (Chapter 2).
- [ ] Gear groove(s) worn. Overhaul the transmission (Chapter 2).
- [ ] Gear dogs or dog slots worn or damaged. The gears should be inspected and replaced. No attempt should be made to service the worn parts.

Overshifts

- [ ] Stopper arm spring weak or broken (Chapter 2).
- [ ] Gearchange shaft return spring post broken or distorted (Chapter 2).

Fault Finding

7 Abnormal engine noise

Knocking or pinking

☐ Carbon build-up in combustion chamber. Use of a fuel additive that will dissolve the adhesive bonding the carbon particles to the piston crown and chamber is the easiest way to remove the build-up. Otherwise, the cylinder head will have to be removed and decarbonised (Chapter 2).
☐ Incorrect or poor quality fuel. Old or improper fuel can cause detonation. This causes the pistons to rattle, thus the knocking or pinking sound. Drain the old fuel and always use the recommended grade fuel (Chapter 4).
☐ Spark plug heat range incorrect. Uncontrolled detonation indicates that the plug heat range is too hot. The plug in effect becomes a glow plug, raising cylinder temperatures. Install the proper heat range plug (Chapter 1).
☐ Improper air/fuel mixture. This will cause the cylinders to run hot and lead to detonation. Clogged jets or an air leak can cause this imbalance. See Chapter 4.

Piston slap or rattling

☐ Cylinder-to-piston clearance excessive. Caused by improper assembly. Inspect and overhaul top-end parts (Chapter 2).
☐ Connecting rod bent. Caused by over-revving, trying to start a badly flooded engine or from ingesting a foreign object into the combustion chamber. Replace the damaged parts (Chapter 2).
☐ Piston pin or piston pin bore worn or seized from wear or lack of lubrication. Replace damaged parts (Chapter 2).
☐ Piston ring(s) worn, broken or sticking. Overhaul the top-end (Chapter 2).
☐ Piston seizure damage. Usually from lack of lubrication or overheating. Replace the pistons and bore the cylinders, as necessary (Chapter 2).
☐ Connecting rod upper or lower end clearance excessive. Caused by excessive wear or lack of lubrication. Replace worn parts.

Valve noise

☐ Incorrect valve clearances. Adjust the clearances by referring to Chapter 1.
☐ Valve spring broken or weak. Check and replace weak valve springs (Chapter 2).
☐ Camshaft or cylinder head worn or damaged. Lack of lubrication at high rpm is usually the cause of damage. Insufficient oil or failure to change the oil at the recommended intervals are the chief causes. Since there are no replaceable bearings in the head, the head itself will have to be replaced if there is excessive wear or damage (Chapter 2).

Other noise

☐ Cylinder head gasket leaking.
☐ Exhaust pipe leaking at cylinder head connection. Caused by improper fit of pipe(s) or loose exhaust flange. All exhaust fasteners should be tightened evenly and carefully. Failure to do this will lead to a leak.
☐ Crankshaft runout excessive. Caused by a bent crankshaft (from over-revving) or damage from an upper cylinder component failure. Can also be attributed to dropping the machine on either of the crankshaft ends.
☐ Engine mounting bolts loose. Tighten all engine mount bolts (Chapter 2).
☐ Crankshaft bearings worn (Chapter 2).
☐ Camshaft drive gear assembly defective. Replace according to the procedure in Chapter 2.

8 Abnormal driveline noise

Clutch noise

☐ Clutch outer drum/friction plate clearance excessive (Chapter 2).
☐ Loose or damaged clutch pressure plate and/or bolts (Chapter 2).

Transmission noise

☐ Bearings worn. Also includes the possibility that the shafts are worn. Overhaul the transmission (Chapter 2).
☐ Gears worn or chipped (Chapter 2).
☐ Metal chips jammed in gear teeth. Probably pieces from a broken clutch, gear or selector mechanism that were picked up by the gears. This will cause early bearing failure (Chapter 2).
☐ Engine oil level too low. Causes a howl from transmission. Also affects engine power and clutch operation (Chapter 1).

Final drive noise

☐ Chain not adjusted properly (Chapter 1).
☐ Front or rear sprocket loose. Tighten fasteners (Chapter 6).
☐ Sprockets worn. Replace sprockets (Chapter 6).
☐ Rear sprocket warped. Replace sprockets (Chapter 6).
☐ Cush drive dampers worn (Chapter 6).

REF•42 Fault Finding

9 Abnormal frame and suspension noise

Front end noise

- ☐ Low fluid level or improper viscosity oil in forks. This can sound like spurting and is usually accompanied by irregular fork action (Chapter 6).
- ☐ Spring weak or broken. Makes a clicking or scraping sound. Fork oil, when drained, will have a lot of metal particles in it (Chapter 6).
- ☐ Steering head bearings loose or damaged. Clicks when braking. Check and adjust or replace as necessary (Chapters 1 and 6).
- ☐ Fork yokes loose. Make sure all clamp pinch bolts are tightened to the specified torque (Chapter 6).
- ☐ Fork tube bent. Good possibility if machine has been dropped. Replace tube with a new one (Chapter 6).
- ☐ Front axle bolt or axle clamp bolts loose. Tighten them to the specified torque (Chapter 7).
- ☐ Loose or worn wheel bearings. Check and replace as needed (Chapter 7).

Shock absorber noise

- ☐ Fluid level incorrect. Indicates a leak caused by defective seal. Shock will be covered with oil. Replace shocks as a pair (Chapter 6) or seek advice from a suspension specialist.
- ☐ Defective shock absorber with internal damage. This is in the body of the shock and can't be remedied. The shocks must be replaced as a pair (Chapter 6).

Brake noise

- ☐ Squeal caused by dust on brake pads (disc brake) or shoes (drum brake). Usually found in combination with glazed pads. Clean using brake cleaning solvent (Chapter 7).
- ☐ Contamination of brake pads (disc brake) or shoes (drum brake), causing brake to chatter or squeal. Clean or replace pads/shoes (Chapter 7).
- ☐ Pads glazed (disc brake). Caused by excessive heat from prolonged use or from contamination. Do not use sandpaper, emery cloth, carborundum cloth or any other abrasive to roughen the pad surfaces as abrasives will stay in the pad material and damage the disc. A very fine flat file can be used, but pad replacement is suggested as a cure (Chapter 7).
- ☐ Disc warped (disc brake). Can cause a chattering, clicking or intermittent squeal. Usually accompanied by a pulsating lever and uneven braking. Replace the disc (Chapter 7).
- ☐ Loose or worn wheel bearings. Check and replace as needed (Chapter 7).

10 Oil pressure warning light comes on

Engine lubrication system

- ☐ Engine oil pump defective, blocked oil strainer gauze or failed relief valve. Carry out oil pressure check (Chapter 1).
- ☐ Engine oil level low. Inspect for leak or other problem causing low oil level and add recommended oil (*Daily (pre-ride) checks*).
- ☐ Engine oil viscosity too low. Very old, thin oil or an improper weight of oil used in the engine. Change to correct oil (Chapter 1).
- ☐ Camshaft or journals worn. Excessive wear causing drop in oil pressure. Replace cam and/or/cylinder head. Abnormal wear could be caused by oil starvation at high rpm from low oil level or improper weight or type of oil (Chapter 1).
- ☐ Crankshaft and/or bearings worn. Same problems as above. Check and replace crankshaft and/or bearings (Chapter 2).

Electrical system

- ☐ Oil pressure switch defective. Check the switch according to the procedure in Chapter 9. Replace it if it is defective.
- ☐ Oil pressure warning light circuit defective. Check for pinched, shorted, disconnected or damaged wiring (Chapter 9).

Fault Finding REF•43

11 Excessive exhaust smoke

White smoke

- [ ] Piston oil ring worn. The ring may be broken or damaged, causing oil from the crankcase to be pulled past the piston into the combustion chamber. Replace the rings with new ones (Chapter 2).
- [ ] Cylinders worn, cracked, or scored. Caused by overheating or oil starvation. The cylinders will have to be rebored and oversize pistons fitted.
- [ ] Valve oil seal damaged or worn. Replace oil seals with new ones (Chapter 2).
- [ ] Valve guide worn. Perform a complete valve job (Chapter 2).
- [ ] Engine oil level too high, which causes the oil to be forced past the rings. Drain oil to the proper level (Chapter 1).
- [ ] Head gasket broken between oil return and cylinder. Causes oil to be pulled into the combustion chamber. Replace the head gasket and check the head for warpage (Chapter 2).
- [ ] Abnormal crankcase pressurisation, which forces oil past the rings. Clogged breather is usually the cause.

Black smoke

- [ ] Air filter clogged. Clean or replace the element (Chapter 1).
- [ ] Main jet too large or loose. Compare the jet size to the Specifications (Chapter 4).
- [ ] Choke cable or linkage shaft stuck, causing fuel to be pulled through choke circuit (Chapter 4).
- [ ] Fuel level too high. Check and adjust the float height(s) as necessary (Chapter 4).
- [ ] Float needle valve held off needle seat. Clean the float chambers and fuel line and replace the needles and seats if necessary (Chapter 4).

Brown smoke

- [ ] Main jet too small or clogged. Lean condition caused by wrong size main jet or by a restricted orifice. Clean float chambers and jets and compare jet size to Specifications (Chapter 4).
- [ ] Fuel flow insufficient. Float needle valve stuck closed due to chemical reaction with old fuel. Float height incorrect. Restricted fuel line. Clean line and float chamber and adjust floats if necessary.
- [ ] Carburettor intake manifold clamps loose (Chapter 4).
- [ ] Air filter poorly sealed or not installed (Chapter 1).

12 Poor handling or stability

Handlebar hard to turn

- [ ] Steering head bearing adjuster nut too tight. Check adjustment as described in Chapter 1.
- [ ] Bearings damaged. Roughness can be felt as the bars are turned from side-to-side. Replace bearings and races (Chapter 6).
- [ ] Races dented or worn. Denting results from wear in only one position (e.g., straight ahead), from a collision or hitting a pothole or from dropping the machine. Replace races and bearings (Chapter 6).
- [ ] Steering stem lubrication inadequate. Causes are grease getting hard from age or being washed out by high pressure car washes. Disassemble steering head and repack bearings (Chapter 6).
- [ ] Steering stem bent. Caused by a collision, hitting a pothole or by dropping the machine. Replace damaged part. Don't try to straighten the steering stem (Chapter 6).
- [ ] Front tyre air pressure too low (Chapter 1).

Handlebar shakes or vibrates excessively

- [ ] Tyres worn or out of balance (Chapter 7).
- [ ] Swingarm bearings worn. Replace worn bearings (Chapter 6).
- [ ] Wheel rim(s) warped or damaged. Inspect wheels for runout (Chapter 7).
- [ ] Wheel bearings worn. Worn front or rear wheel bearings can cause poor tracking. Worn front bearings will cause wobble (Chapter 7).
- [ ] Handlebar clamp bolts loose (Chapter 6).
- [ ] Fork yoke bolts loose. Tighten them to the specified torque (Chapter 6).
- [ ] Engine mounting bolts loose. Will cause excessive vibration with increased engine rpm (Chapter 2).

Handlebar pulls to one side

- [ ] Frame bent. Definitely suspect this if the machine has been dropped. May or may not be accompanied by cracking near the bend. Replace the frame (Chapter 6).
- [ ] Wheels out of alignment. Caused by improper location of axle spacers or from bent steering stem or frame (Chapter 6).
- [ ] Swingarm bent or twisted. Caused by age (metal fatigue) or impact damage. Replace the arm (Chapter 6).
- [ ] Steering stem bent. Caused by impact damage or by dropping the motorcycle. Replace the steering stem (Chapter 6).
- [ ] Fork tube bent. Disassemble the forks and replace the damaged parts (Chapter 6).
- [ ] Fork oil level uneven. Check and add or drain as necessary (Chapter 6).

Poor shock absorbing qualities

- [ ] Too hard:
    a) Fork oil level excessive (Chapter 6).
    b) Fork oil viscosity too high. Use a lighter oil (see the Specifications in Chapter 6).
    c) Fork tube bent. Causes a harsh, sticking feeling (Chapter 6).
    d) Fork internal damage (Chapter 6).
    e) Shock shaft or body bent or damaged (Chapter 6).
    f) Shock internal damage.
    g) Shock pre-load setting incorrect (Chapter 6).
    h) Tyre pressure too high (Chapter 1).
- [ ] Too soft:
    a) Fork or shock oil insufficient and/or leaking (Chapter 6).
    b) Fork oil level too low (Chapter 6).
    c) Fork oil viscosity too light (Chapter 6).
    d) Fork springs weak or broken (Chapter 6).
    e) Shock internal damage or leakage (Chapter 6).
    f) Shock pre-load setting incorrect (Chapter 6).

Fault Finding

13 Braking problems

Brakes are spongy, don't hold (disc brake)
- [ ] Air in brake line. Caused by inattention to master cylinder fluid level or by leakage. Locate problem and bleed brakes (Chapter 7).
- [ ] Pad or disc worn (Chapters 1 and 7).
- [ ] Brake fluid leak. See paragraph 1.
- [ ] Contaminated pads. Caused by contamination with oil, grease, brake fluid, etc. Clean or replace pads. Clean disc thoroughly with brake cleaner (Chapter 7).
- [ ] Brake fluid deteriorated. Fluid is old or contaminated. Drain system, replenish with new fluid and bleed the system (Chapter 7).
- [ ] Master cylinder internal parts worn or damaged causing fluid to bypass (Chapter 7).
- [ ] Master cylinder bore scratched by foreign material or broken spring. Repair or replace master cylinder (Chapter 7).
- [ ] Disc warped. Replace disc (Chapter 7).

Brake lever or pedal pulsates (disc brake)
- [ ] Disc warped. Replace disc (Chapter 7).
- [ ] Axle bent. Replace axle (Chapter 7).
- [ ] Brake caliper bolts loose (Chapter 7).
- [ ] Brake caliper sliders damaged or sticking, causing caliper to bind. Lubricate the sliders or replace them if they are corroded or bent (Chapter 7).
- [ ] Wheel warped or otherwise damaged (Chapter 7).
- [ ] Wheel bearings damaged or worn (Chapter 7).

Brake lever or pedal pulsates (drum brake)
- [ ] Brake drum out of round. Renew wheel or have brake drum skimmed (Chapter 7).
- [ ] Wheel bearings damaged or worn (Chapter 7).

Brakes drag (disc brake)
- [ ] Master cylinder piston seized. Caused by wear or damage to piston or cylinder bore (Chapter 7).
- [ ] Lever balky or stuck. Check pivot and lubricate (Chapter 7).
- [ ] Brake caliper binds. Caused by inadequate lubrication or damage to caliper sliders (Chapter 7).
- [ ] Brake caliper piston seized in bore. Caused by wear or ingestion of dirt past deteriorated seal (Chapter 7).
- [ ] Brake pad damaged. Pad material separated from backing plate. Usually caused by faulty manufacturing process or from contact with chemicals. Replace pads (Chapter 7).
- [ ] Pads improperly installed (Chapter 7).

Brakes drag (drum brake)
- [ ] Brake pedal freeplay insufficient (Chapter 1).
- [ ] Brake springs weak or broken. Renew the springs (Chapter 7).
- [ ] Brake cam sticking due to lack of lubrication (Chapter 1).

ABS indicator light comes on – CBF500A models
- [ ] If the light remains on after start-up or comes on while riding, investigate the fault as described in Chapter 7, Section 20.

14 Electrical problems

Battery dead or weak
- [ ] Battery faulty. Caused by sulphated plates which are shorted through sedimentation. Also, broken battery terminal making only occasional contact (Chapter 9).
- [ ] Battery cables making poor contact (Chapter 9).
- [ ] Load excessive. Caused by addition of high wattage lights or other electrical accessories.
- [ ] Ignition (main) switch defective. Switch either grounds (earths) internally or fails to shut off system. Replace the switch (Chapter 9).
- [ ] Regulator/rectifier defective (Chapter 9).
- [ ] Alternator stator coil open or shorted (Chapter 9).
- [ ] Wiring faulty. Wiring grounded (earthed) or connections loose in ignition, charging or lighting circuits (Chapter 9).

Battery overcharged
- [ ] Regulator/rectifier defective. Overcharging is noticed when battery gets excessively warm (Chapter 9).
- [ ] Battery defective. Replace battery with a new one (Chapter 9).
- [ ] Battery amperage too low, wrong type or size. Install manufacturer's specified amp-hour battery to handle charging load (Chapter 9).

Fault Finding Equipment REF•45

Checking engine compression

● Low compression will result in exhaust smoke, heavy oil consumption, poor starting and poor performance. A compression test will provide useful information about an engine's condition and if performed regularly, can give warning of trouble before any other symptoms become apparent.
● A compression gauge will be required, along with an adapter to suit the spark plug hole thread size. Note that the screw-in type gauge/adapter set up is preferable to the rubber cone type.
● Before carrying out the test, first check the valve clearances as described in Chapter 1.

1 Run the engine until it reaches normal operating temperature, then stop it and remove the spark plug(s), taking care not to scald your hands on the hot components.
2 Install the gauge adapter and compression gauge in No. 1 cylinder spark plug hole **(see illustration 1)**.

Screw the compression gauge adapter into the spark plug hole, then screw the gauge into the adapter

3 On kickstart-equipped motorcycles, make sure the ignition switch is OFF, then open the throttle fully and kick the engine over a couple of times until the gauge reading stabilises.
4 On motorcycles with electric start only, the procedure will differ depending on the nature of the ignition system. Flick the engine kill switch (engine stop switch) to OFF and turn the ignition switch ON; open the throttle fully and crank the engine over on the starter motor for a couple of revolutions until the gauge reading stabilises. If the starter will not operate with the kill switch OFF, turn the ignition switch OFF and refer to the next paragraph.
5 Install the plugs back into their caps and arrange the plug electrodes so that their metal bodies are earthed (grounded) against the cylinder head; this is essential to prevent damage to the ignition system **(see illustration 2)**. Position the plugs well away from the plug holes otherwise there

All spark plugs must be earthed (grounded) against the cylinder head

is a risk of atomised fuel escaping from the plug holes and igniting. As a safety precaution, cover the valve cover with rag. Turn the ignition switch ON and kill switch ON, open the throttle fully and crank the engine over on the starter motor for a couple of revolutions until the gauge reading stabilises.
6 After one or two revolutions the pressure should build up to a maximum figure and then stabilise. Take a note of this reading and on multi-cylinder engines repeat the test on the remaining cylinders.
7 The correct pressures are given in Chapter 1 Specifications. If the results fall within the specified range and on multi-cylinder engines all are relatively equal, the engine is in good condition. If there is a marked difference between the readings, or if the readings are lower than specified, inspection of the top-end components will be required.
8 Low compression pressure may be due to worn cylinder bores, pistons or rings, failure of the cylinder head gasket, worn valve seals, or poor valve seating.
9 To distinguish between cylinder/piston wear and valve leakage, pour a small quantity of oil into the bore to temporarily seal the piston rings, then repeat the compression tests **(see illustration 3)**. If the readings show a

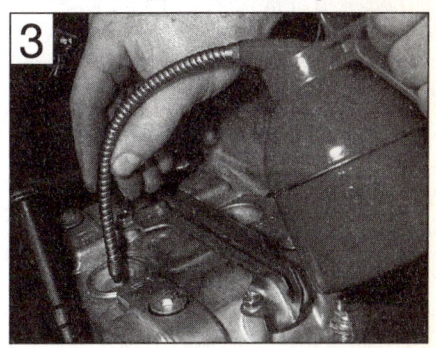

Bores can be temporarily sealed with a squirt of motor oil

noticeable increase in pressure this confirms that the cylinder bore, piston, or rings are worn. If, however, no change is indicated, the cylinder head gasket or valves should be examined.
10 High compression pressure indicates excessive carbon build-up in the combustion chamber and on the piston crown. If this is the case the cylinder head should be removed and the deposits removed. Note that excessive carbon build-up is less likely with the used on modern fuels.

Checking battery open-circuit voltage

⚠️ *Warning: The gases produced by the battery are explosive - never smoke or create any sparks in the vicinity of the battery. Never allow the electrolyte to contact your skin or clothing - if it does, wash it off and seek immediate medical attention.*

Fault Finding Equipment

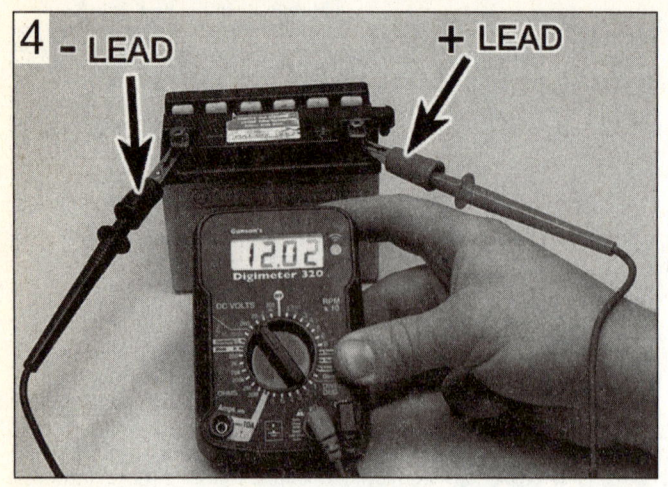

Measuring open-circuit battery voltage

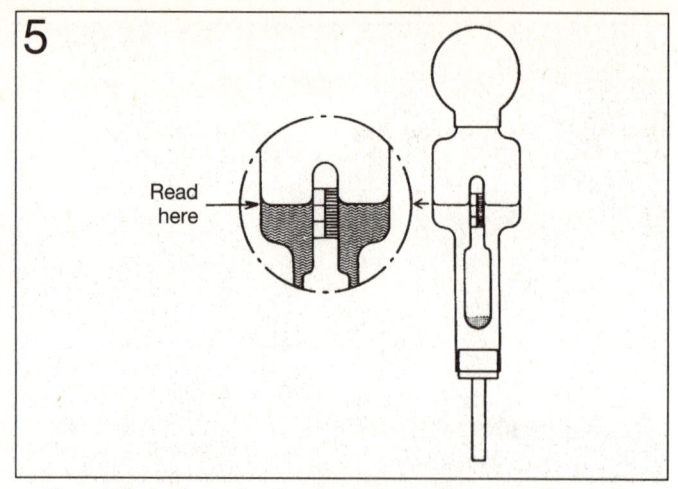

Float-type hydrometer for measuring battery specific gravity

- Before any electrical fault is investigated the battery should be checked.
- You'll need a dc voltmeter or multimeter to check battery voltage. Check that the leads are inserted in the correct terminals on the meter, red lead to positive (+ve), black lead to negative (-ve). Incorrect connections can damage the meter.
- A sound fully-charged 12 volt battery should produce between 12.3 and 12.6 volts across its terminals (12.8 volts for a maintenance-free battery). On machines with a 6 volt battery, voltage should be between 6.1 and 6.3 volts.

1 Set a multimeter to the 0 to 20 volts dc range and connect its probes across the battery terminals. Connect the meter's positive (+ve) probe, usually red, to the battery positive (+ve) terminal, followed by the meter's negative (-ve) probe, usually black, to the battery negative terminal (-ve) **(see illustration 4)**.

2 If battery voltage is low (below 10 volts on a 12 volt battery or below 4 volts on a six volt battery), charge the battery and test the voltage again. If the battery repeatedly goes flat, investigate the motorcycle's charging system.

Checking battery specific gravity (SG)

⚠️ *Warning: The gases produced by the battery are explosive - never smoke or create any sparks in the vicinity of the battery. Never allow the electrolyte to contact your skin or clothing - if it does, wash it off and seek immediate medical attention.*

- The specific gravity check gives an indication of a battery's state of charge.
- A hydrometer is used for measuring specific gravity. Make sure you purchase one which has a small enough hose to insert in the aperture of a motorcycle battery.
- Specific gravity is simply a measure of the electrolyte's density compared with that of water. Water has an SG of 1.000 and fully-charged battery electrolyte is about 26% heavier, at 1.260.
- Specific gravity checks are not possible on maintenance-free batteries. Testing the open-circuit voltage is the only means of determining their state of charge.

1 To measure SG, remove the battery from the motorcycle and remove the first cell cap. Draw some electrolyte into the hydrometer and note the reading **(see illustration 5)**. Return the electrolyte to the cell and install the cap.

2 The reading should be in the region of 1.260 to 1.280. If SG is below 1.200 the battery needs charging. Note that SG will vary with temperature; it should be measured at 20°C (68°F). Add 0.007 to the reading for every 10°C above 20°C, and subtract 0.007 from the reading for every 10°C below 20°C. Add 0.004 to the reading for every 10°F above 68°F, and subtract 0.004 from the reading for every 10°F below 68°F.

3 When the check is complete, rinse the hydrometer thoroughly with clean water.

Checking for continuity

- The term continuity describes the uninterrupted flow of electricity through an electrical circuit. A continuity check will determine whether an **open-circuit** situation exists.
- Continuity can be checked with an ohmmeter, multimeter, continuity tester or battery and bulb test circuit **(see illustrations 6, 7 and 8)**.

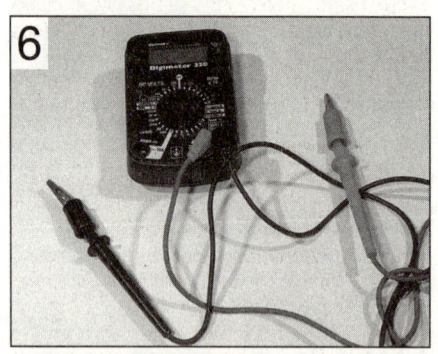

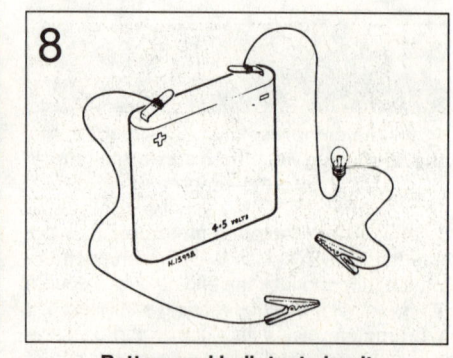

Digital multimeter can be used for all electrical tests

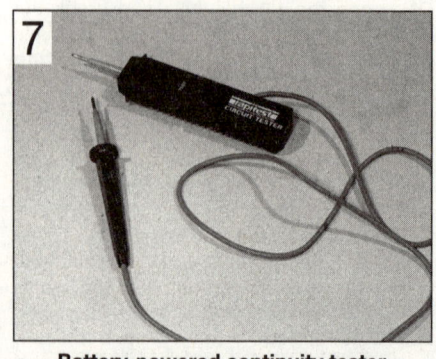

Battery-powered continuity tester

Battery and bulb test circuit

Fault Finding Equipment REF•47

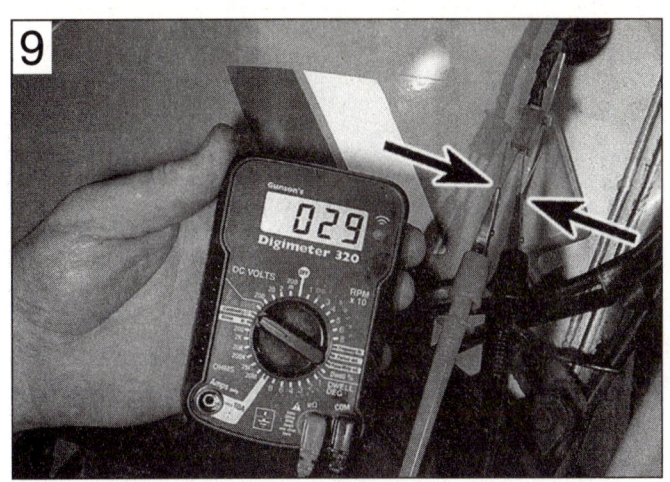

Continuity check of front brake light switch using a meter - note split pins used to access connector terminals

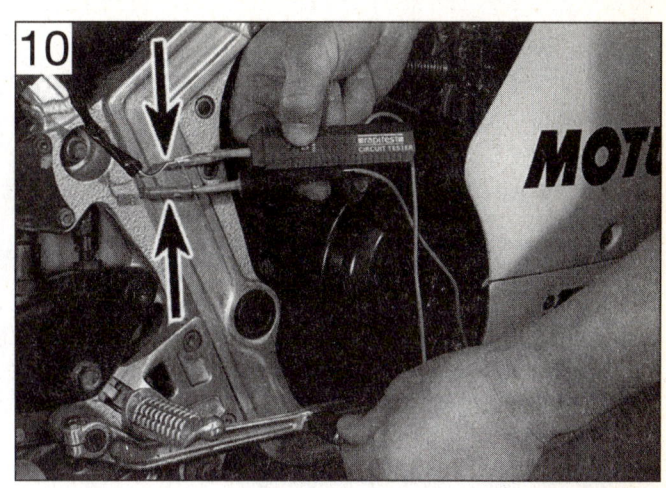

Continuity check of rear brake light switch using a continuity tester

● All of these instruments are self-powered by a battery, therefore the checks are made with the ignition OFF.
● As a safety precaution, always disconnect the battery negative (-ve) lead before making checks, particularly if ignition switch checks are being made.
● If using a meter, select the appropriate ohms scale and check that the meter reads infinity (∞). Touch the meter probes together and check that meter reads zero; where necessary adjust the meter so that it reads zero.
● After using a meter, always switch it OFF to conserve its battery.

Switch checks

1 If a switch is at fault, trace its wiring up to the wiring connectors. Separate the wire connectors and inspect them for security and condition. A build-up of dirt or corrosion here will most likely be the cause of the problem - clean up and apply a water dispersant such as WD40.
2 If using a test meter, set the meter to the ohms x 10 scale and connect its probes across the wires from the switch **(see illustration 9)**. Simple ON/OFF type switches, such as brake light switches, only have two wires whereas combination switches, like the ignition switch, have many internal links. Study the wiring diagram to ensure that you are connecting across the correct pair of wires. Continuity (low or no measurable resistance - 0 ohms) should be indicated with the switch ON and no continuity (high resistance) with it OFF.
3 Note that the polarity of the test probes doesn't matter for continuity checks, although care should be taken to follow specific test procedures if a diode or solid-state component is being checked.
4 A continuity tester or battery and bulb circuit can be used in the same way. Connect its probes as described above **(see illustration 10)**. The light should come on to indicate continuity in the ON switch position, but should extinguish in the OFF position.

Wiring checks

● Many electrical faults are caused by damaged wiring, often due to incorrect routing or chaffing on frame components.
● Loose, wet or corroded wire connectors can also be the cause of electrical problems, especially in exposed locations.

1 A continuity check can be made on a single length of wire by disconnecting it at each end and connecting a meter or continuity tester across both ends of the wire **(see illustration 11)**.
2 Continuity (low or no resistance - 0 ohms) should be indicated if the wire is good. If no continuity (high resistance) is shown, suspect a broken wire.

Checking for voltage

● A voltage check can determine whether current is reaching a component.
● Voltage can be checked with a dc voltmeter, multimeter set on the dc volts scale, test light or buzzer **(see illustrations 12 and 13)**. A meter has the advantage of being able to measure actual voltage.
● When using a meter, check that its leads are inserted in the correct terminals on the meter, red to positive (+ve), black to negative (-ve). Incorrect connections can damage the meter.
● A voltmeter (or multimeter set to the dc volts scale) should always be connected in parallel (across the load). Connecting it in series will not harm the meter, but the reading will not be meaningful.
● Voltage checks are made with the ignition ON.

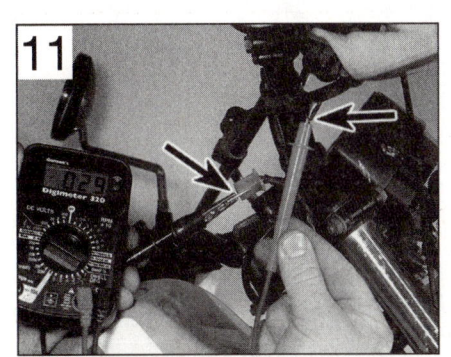

Continuity check of front brake light switch sub-harness

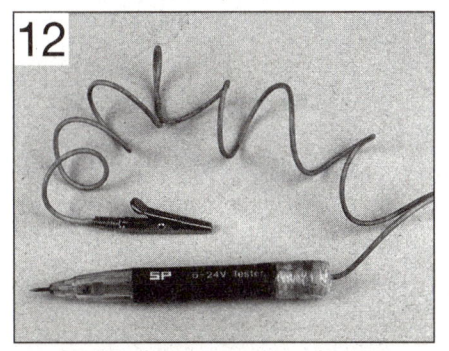

A simple test light can be used for voltage checks

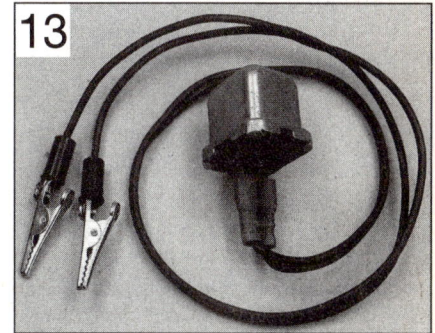

A buzzer is useful for voltage checks

Fault Finding Equipment

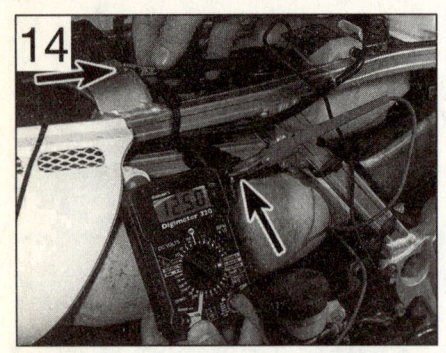

Checking for voltage at the rear brake light power supply wire using a meter . . .

1 First identify the relevant wiring circuit by referring to the wiring diagram at the end of this manual. If other electrical components share the same power supply (ie are fed from the same fuse), take note whether they are working correctly - this is useful information in deciding where to start checking the circuit.

2 If using a meter, check first that the meter leads are plugged into the correct terminals on the meter (see above). Set the meter to the dc volts function, at a range suitable for the battery voltage. Connect the meter red probe (+ve) to the power supply wire and the black probe to a good metal earth (ground) on the motorcycle's frame or directly to the battery negative (-ve) terminal **(see illustration 14)**. Battery voltage should be shown on the meter

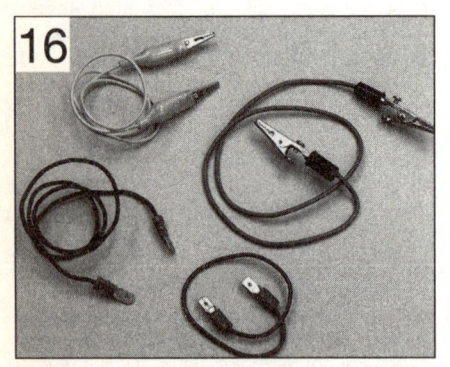

A selection of jumper wires for making earth (ground) checks

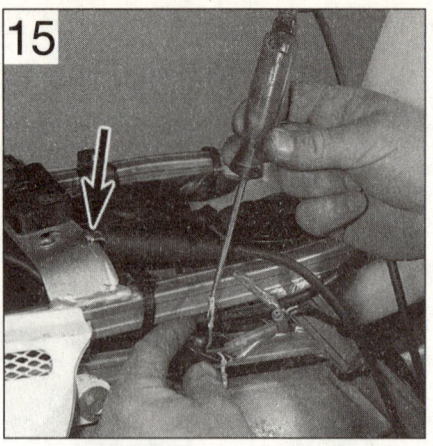

. . . or a test light - note the earth connection to the frame (arrow)

with the ignition switched ON.

3 If using a test light or buzzer, connect its positive (+ve) probe to the power supply terminal and its negative (-ve) probe to a good earth (ground) on the motorcycle's frame or directly to the battery negative (-ve) terminal **(see illustration 15)**. With the ignition ON, the test light should illuminate or the buzzer sound.

4 If no voltage is indicated, work back towards the fuse continuing to check for voltage. When you reach a point where there is voltage, you know the problem lies between that point and your last check point.

Checking the earth (ground)

● Earth connections are made either directly to the engine or frame (such as sensors, neutral switch etc. which only have a positive feed) or by a separate wire into the earth circuit of the wiring harness. Alternatively a short earth wire is sometimes run directly from the component to the motorcycle's frame.
● Corrosion is often the cause of a poor earth connection.
● If total failure is experienced, check the security of the main earth lead from the negative (-ve) terminal of the battery and also the main earth (ground) point on the wiring harness. If corroded, dismantle the connection and clean all surfaces back to bare metal.

1 To check the earth on a component, use an insulated jumper wire to temporarily bypass its earth connection **(see illustration 16)**. Connect one end of the jumper wire between the earth terminal or metal body of the component and the other end to the motorcycle's frame.

2 If the circuit works with the jumper wire installed, the original earth circuit is faulty. Check the wiring for open-circuits or poor connections. Clean up direct earth connections, removing all traces of corrosion and remake the joint. Apply petroleum jelly to the joint to prevent future corrosion.

Tracing a short-circuit

● A short-circuit occurs where current shorts to earth (ground) bypassing the circuit components. This usually results in a blown fuse.

● A short-circuit is most likely to occur where the insulation has worn through due to wiring chafing on a component, allowing a direct path to earth (ground) on the frame.

1 Remove any bodypanels necessary to access the circuit wiring.
2 Check that all electrical switches in the circuit are OFF, then remove the circuit fuse and connect a test light, buzzer or voltmeter (set to the dc scale) across the fuse terminals. No voltage should be shown.
3 Move the wiring from side to side whilst observing the test light or meter. When the test light comes on, buzzer sounds or meter shows voltage, you have found the cause of the short. It will usually shown up as damaged or burned insulation.
4 Note that the same test can be performed on each component in the circuit, even the switch.

Index

Note: *References throughout this index are in the form - "Chapter number" • "Page number"*

A

ABS (Anti-lock Brake System)
  check – 7•24
  operation and fault finding – 7•23
Air filter and sub-air cleaner – 1•20
Air filter housing – 4•15
Air/fuel mixture adjustment – 4•4
Alternator – 9•25

B

Battery
  charging – 9•4
  check – 1•14
  removal, installation and inspection – 9•3
Balancer shaft and bearings – 2•45
Bearings
  main and connecting rod – 2•37
  sprocket coupling – 7•22
  steering head – 1•18, 1•25, 6•12
  swingarm – 1•25
  wheel – 1•19, 7•21, 7•22
Brake
  ABS – 7•23
  bleeding – 7•15
  calipers – 7•5, 7•10
  caliper and master cylinder seals – 1•24
  cam (R and T) models – 1•25
  check – 1•12
  disc – 7•7, 7•12
  drum (R and T models) – 7•16
  fault finding – REF•44
  fluid level check – 0•14, 1•20
  general information – 7•4
  hoses – 1•25, 7•14
  light – 9•6
  light bulb – 9•8
  light switches – 9•10
  master cylinder – 7•8, 7•13
  pads – 1•8, 7•4, 7•10
  pedal (R and T models) – 1•9, 6•3
  pedal (all other models) – 6•4
  specifications – 7•1
  torque settings – 7•3
Bulbs – 9•2

C

Cables
  choke – 1•15, 4•14
  clutch – 1•9, 2•23
  speedometer – 9•13
  throttle – 1•15, 4•13
Calipers – 7•5, 7•10
Cam chain and guides – 2•46
Cam chain tensioner – 2•11
Camshafts – 2•12
Carburettors – 1•16, 4•4 to 4•13
Centre stand – 1•17
Chain (final drive)
  adjustment and lubrication – 1•7
  check – 0•14
  renewal – 6•17
  slider check – 1•20
  sprockets – 6•18
Charging system – 9•24
Clutch
  cable – 1•9, 2•23
  check and adjustment – 1•9
  fault finding – REF•40
  removal, inspection and installation – 2•24
  specifications – 2•2
  switch – 9•18
Component locations – 1•4
Connecting rods and bearings – 2•37
Conversion factors – REF•26
Cooling system – 3•1 *et seq*
  check – 1•12
  draining, flushing and refilling – 1•24
  fan and fan switch – 3•3
  general information – 3•2
  hoses – 3•8
  pressure cap – 3•2
  radiator – 3•5
  reservoir – 3•2
  specifications – 3•1
  temperature gauge and sender – 3•4, 9•15
  thermostat and thermostat housing – 3•4
  water pump – 3•7
Coolant level check – 0•16
Crankcase breather – 1•10
Crankcase halves – 2•34
Crankshaft and main bearings – 2•43
Cylinder compression – 1•25
Cylinder head
  disassembly, inspection and reassembly – 2•18
  removal and installation – 2•16

Index

D

Disc – 7•7, 7•12
Dimensions – 0•10
Diode – 9•18
Drive chain and sprockets – 1•7
Drive chain
 removal, cleaning and installation – 6•17
 slider – 1•20, 6•14

E

Electrical system – 9•1 *et seq*
 alternator – 9•25
 battery – 1•14, 9•3
 brake light – 9•6
 brake light bulb – 9•8
 brake light switches – 9•10
 bulbs – 9•2
 charging system – 9•24
 clutch switch – 9•18
 diode – 9•18
 general information – 9•2
 fault finding – 9•2, REF•44
 fuses – 9•2, 9•4
 handlebar switches – 9•19
 headlight assembly – 9•7
 headlight bulb – 9•6, 9•7
 horn – 9•20
 ignition (main) switch – 9•17
 instrument and warning light bulb – 9•16
 instrument cluster – 9•12
 instruments – 9•13
 licence plate bulb – 9•8
 lighting system – 9•5
 neutral switch – 9•17
 oil pressure switch – 9•17
 regulator/rectifier – 9•26
 sidelight bulb – 9•6, 9•7
 sidestand switch – 9•18
 specifications – 9•1
 speedometer cable/speed sensor – 9•13
 starter motor – 9•21
 starter relay – 9•20
 tail light – 9•6
 tail light assembly – 9•9
 tail light bulb – 9•8
 turn signal assemblies – 9•11
 turn signal bulbs – 9•11
 turn signal circuit – 9•10
 wiring diagrams – 9•27 to 9•33
Engine – 2•1 *et seq*
 camchain and guides – 2•46
 camchain tensioner – 2•11
 camshafts – 2•12
 compression – 1•25
 crankcase halves and cylinder bores – 2•55
 crankcases – 2•34
 crankshaft and main bearings – 2•43
 cylinder head – 2•16, 2•18
 disassembly and reassembly – 2•9
 fault finding – REF•36
 filter and change – 1•11
 general information – 2•5
 initial start-up – 2•56
 main and connecting rod bearings – 2•37
 oil level check – 0•13
 oil pressure – 1•25
 oil pressure relief valve – 2•32
 oil pump – 2•30
 oil sump – 2•32
 pistons – 2•40
 piston rings – 2•42
 removal and installation – 2•6
 running-in – 2•56
 specifications – 2•1
 valve cover – 2•10
 valves/valve seats/valve guides – 2•18
Engine number – 0•9
Exhaust system – 1•26, 4•15

F

Fairing (S models) – 8•4
Final drive chain
 adjustment and lubrication – 1•7
 check – 0•14
 renewal – 6•17
 slider check – 1•20
 sprockets – 6•18
Footrests – 6•3
Forks
 disassembly, inspection and reassembly – 6•7
 oil change – 1•26
 removal and installation – 6•6
 specifications – 6•1
Frame – 6•2
Frame number – 0•9
Frame side panels – 8•1
Front mudguard – 8•5
Front wheel – 7•18
Front wheel speed sensor and pulser ring – 7•25
Fuel hoses – 1•25
Fuel system – 4•1 *et seq*
 air/fuel mixture adjustment – 4•4
 carburettors – 1•16, 4•4 to 4•13
 check – 1•13
 specifications – 4•1
 tank – 4•2
Fuses – 9•2, 9•4

G

Gearbox – 2•46 to 2•53
Gearchange
 fault finding – REF•40
 lever – 6•3
 mechanism – 2•33
 selector drum – 2•53

H

Handlebars and levers – 6•5
Handlebar switches – 9•19
Headlight
 aim – 1•16
 assembly – 9•7
 bulb – 9•6, 9•7
 check – 9•5
Horn – 9•20

Index

Hoses
   brake – 1•25, 7•14
   coolant – 3•8
   fuel – 1•25
HT coils – 5•3

I

Idle speed – 1•8
Ignition system – 5•1 *et seq*
   check – 5•2
   control unit – 5•4
   general information – 5•2
   HT coils – 5•3
   main switch – 9•17
   pulse generator coil – 5•4
   specifications – 5•1
   timing – 5•5
Instruments – 9•13
Instrument and warning light bulbs – 9•16
Instrument cluster and speedometer cable/speed sensor – 9•12
Introduction – 0•4

L

Levers, stand pivots and cable lubrication – 1•10
Licence plate bulb – 9•8
Lighting system – 9•5
Lubricants and fluids – 1•2, REF•23

M

Main and connecting rod bearings – 2•37
Maintenance schedule – 1•3
Master cylinder – 7•8, 7•13
Model development – 0•11
MOT test checks – REF•27

N

Neutral switch – 9•17
Nuts and bolts – 1•20

O

Oil (engine)
   filter and change – 1•11
   level check – 0•13
Oil (front forks) – 1•26, 6•1
Oil pressure check (engine) – 1•25
Oil pressure relief valve – 2•32
Oil pressure switch – 9•17
Oil pump – 2•30
Oil strainer – 2•32
Oil sump – 2•32

P

Pads – 1•8, 7•4, 7•10
Pistons – 2•40
Piston rings – 2•42
Pulse generator coil assembly – 5•4
Pulse secondary air injection (PAIR) system (CBF models) – 4•17
Pump
   oil – 2•30
   water – 3•7

R

Radiator – 3•5
Radiator side panels (R, T, V, W, X, Y and 2 models) – 8•4
Rear brake cam (R and T models) – 1•25
Rear shock absorbers – 6•12
Rear view mirror – 8•4
Rear wheel – 7•20
Rear wheel speed sensor and pulser ring – 7•26
Regulator/rectifier – 9•26
Running-in – 2•56

S

Safety – 0•16
Seat cowling – 8•3
Seats - 8•2
Selector drum and forks – 2•53
Security – REF•20
Sensors
   speed sensor – 9•12
   throttle position sensor (CBF models) – 4•16
Sidestand – 1•17, 6•4
Sidestand switch – 9•18
Sidelight bulb – 9•6, 9•7
Spark plugs
   check and renewal – 1•14
   gap – 1•14
   renewal – 1•23
Speedometer cable – 9•13
Speed sensor – 7•25, 9•13
Specifications – 1•2, 2•1, 3•1, 4•1, 5•1, 6•1, 7•1, 9•1
Sprockets – 6•17
Sprocket coupling/rubber damper – 6•18
Sprocket coupling bearing – 7•22
Starter clutch and idle gear – 2•21
Starter motor – 9•21
Starter relay – 9•20
Steering check – 0•14
Steering head bearings
   check and adjustment – 1•18
   inspection and renewal – 6•12
   lubrication – 1•25
Steering stem – 6•10
Storage – REF•32

Index

Suspension
   adjustments – 6•14
   check – 0•14, 1•17
   front forks – 1•26, 6•6, 6•7
   rear shocks – 6•12
Swingarm and drive chain slider – 1•20, 6•14
Swingarm bearings – 1•25, 6•16
Switches
   brake light – 9•10
   clutch – 9•18
   fan – 3•3
   handlebar – 9•19
   ignition – 9•17
   neutral – 9•17
   oil pressure – 9•17
   sidestand – 9•18

T

Tachometer (CB models) – 9•14
Tail light – 9•6
Tail light assembly – 9•9
Tail light bulb – 9•8
Temperature gauge and sender – 3•4
Timing rotor, primary drive gear and balancer gear – 2•29
Thermostat and thermostat housing – 3•4
Throttle and choke cables – 1•15, 4•13
Throttle position sensor – 4•16
Tools and workshop tips – REF•2 to REF•19
Torque settings – 1•2, 2•5, 3•1, 4•1, 5•1, 6•2, 7•3, 8•1, 9•2
Transmission shafts – 2•48
Transmission shafts and bearings – 2•46
Turn signal
   assemblies – 9•11
   bulbs – 9•11
   circuit – 9•10
Tyres – 0•15, 7•3, 7•23

V

Valve clearances – 1•2, 1•21
Valve cover – 2•10
Valve/valve seats/valve guides – 2•18

W

Water pump – 3•7
Weights – 0•10
Wheel
   bearings – 1•19, 7•21, 7•22
   check – 1•20
   inspection and repair – 7•17
   removal and installation – 7•18, 7•20
Wiring diagrams – 9•27 to 9•33

Haynes Motorcycle Manuals – The Complete List

Title	Book No
APRILIA RS50 (99 - 06) & RS125 (93 - 06)	4298
Aprilia RSV1000 Mille (98 - 03) ♦	4255
Aprilia SR50	4755
BMW 2-valve Twins (70 - 96) ♦	0249
BMW F650 ♦	4761
BMW K100 & 75 2-valve Models (83 - 96) ♦	1373
BMW R850, 1100 & 1150 4-valve Twins (93 - 04) ♦	3466
BMW R1200 (04 - 06) ♦	4598
BSA Bantam (48 - 71)	0117
BSA Unit Singles (58 - 72)	0127
BSA Pre-unit Singles (54 - 61)	0326
BSA A7 & A10 Twins (47 - 62)	0121
BSA A50 & A65 Twins (62 - 73)	0155
Chinese Scooters	4768
DUCATI 600, 620, 750 and 900 2-valve V-Twins (91 - 05) ♦	3290
Ducati MK III & Desmo Singles (69 - 76) ◊	0445
Ducati 748, 916 & 996 4-valve V-Twins (94 - 01) ♦	3756
GILERA Runner, DNA, Ice & SKP/Stalker (97 - 07)	4163
HARLEY-DAVIDSON Sportsters (70 - 08) ♦	2534
Harley-Davidson Shovelhead and Evolution Big Twins (70 - 99) ♦	2536
Harley-Davidson Twin Cam 88 (99 - 03) ♦	2478
HONDA NB, ND, NP & NS50 Melody (81 - 85) ◊	0622
Honda NE/NB50 Vision & SA50 Vision Met-in (85 - 95) ◊	1278
Honda MB, MBX, MT & MTX50 (80 - 93)	0731
Honda C50, C70 & C90 (67 - 03)	0324
Honda XR80/100R & CRF80/100F (85 - 04)	2218
Honda XL/XR 80, 100, 125, 185 & 200 2-valve Models (78 - 87)	0566
Honda H100 & H100S Singles (80 - 92) ◊	0734
Honda CB/CD125T & CM125C Twins (77 - 88) ◊	0571
Honda CG125 (76 - 07) ◊	0433
Honda NS125 (86 - 93) ◊	3056
Honda CBR125R (04 - 07)	4620
Honda MBX/MTX125 & MTX200 (83 - 93) ◊	1132
Honda CD/CM185 200T & CM250C 2-valve Twins (77 - 85)	0572
Honda XL/XR 250 & 500 (78 - 84)	0567
Honda XR250L, XR250R & XR400R (86 - 03)	2219
Honda CB250 & CB400N Super Dreams (78 - 84) ◊	0540
Honda CR Motocross Bikes (86 - 01)	2222
Honda CRF250 & CRF450 (02 - 06)	2630
Honda CBR400RR Fours (88 - 99) ◊ ♦	3552
Honda VFR400 (NC30) & RVF400 (NC35) V-Fours (89 - 98) ◊ ♦	3496
Honda CB500 (93 - 02) & CBF500 03 - 08 ♦	3753
Honda CB400 & CB550 Fours (73 - 77)	0262
Honda CX/GL500 & 650 V-Twins (78 - 86)	0442
Honda CBX550 Four (82 - 86) ◊	0940
Honda XL600R & XR600R (83 - 08) ♦	2183
Honda XL600/650V Transalp & XRV750 Africa Twin (87 to 07)	3919
Honda CBR600F1 & 1000F Fours (87 - 96) ♦	1730
Honda CBR600F2 & F3 Fours (91 - 98) ♦	2070
Honda CBR600F4 (99 - 06) ♦	3911
Honda CB600F Hornet & CBF600 (98 - 06) ◊ ♦	3915
Honda CBR600RR (03 - 06) ♦	4590
Honda CB650 sohc Fours (78 - 84)	0665
Honda NTV600 Revere, NTV650 and NT650V Deauville (88 - 05) ◊ ♦	3243
Honda Shadow VT600 & 750 (USA) (88 - 03)	2312
Honda CB750 sohc Four (69 - 79)	0131
Honda V45/65 Sabre & Magna (82 - 88)	0820
Honda VFR750 & 700 V-Fours (86 - 97) ♦	2101
Honda VFR800 V-Fours (97 - 01) ♦	3703
Honda VFR800 V-Tec V-Fours (02 - 05) ♦	4196
Honda CB750 & CB900 dohc Fours (78 - 84)	0535
Honda VTR1000 (FireStorm, Super Hawk) & XL1000V (Varadero) (97 - 08) ♦	3744
Honda CBR900RR FireBlade (92 - 99) ♦	2161
Honda CBR900RR FireBlade (00 - 03) ♦	4060
Honda CBR1000RR Fireblade (04 - 07) ♦	4604
Honda CBR1100XX Super Blackbird (97 - 07) ♦	3901
Honda ST1100 Pan European V-Fours (90 - 02) ♦	3384
Honda Shadow VT1100 (USA) (85 - 98)	2313
Honda GL1000 Gold Wing (75 - 79)	0309

Title	Book No
Honda GL1100 Gold Wing (79 - 81)	0669
Honda Gold Wing 1000 (USA) (84 - 87)	2199
Honda Gold Wing 1500 (USA) (88 - 00)	2225
KAWASAKI AE/AR 50 & 80 (81 - 95)	1007
Kawasaki KC, KE & KH100 (75 - 99)	1371
Kawasaki KMX125 & 200 (86 - 02) ◊	3046
Kawasaki 250, 350 & 400 Triples (72 - 79)	0134
Kawasaki 400 & 440 Twins (74 - 81)	0281
Kawasaki 400, 500 & 550 Fours (79 - 91)	0910
Kawasaki EN450 & 500 Twins (Ltd/Vulcan) (85 - 07)	2053
Kawasaki EX500 (GPZ500S) & ER500 (ER-5) (87 - 08) ♦	2052
Kawasaki ZX600 (ZZ-R600 & Ninja ZX-6) (90 - 06) ♦	2146
Kawasaki ZX-6R Ninja Fours (95 - 02) ♦	3541
Kawasaki ZX-6R (03 - 06) ♦	4742
Kawasaki ZX600 (GPZ600R, GPX600R, Ninja 600R & RX) & ZX750 (GPX750R, Ninja 750R) ♦	1780
Kawasaki 650 Four (78 - 79)	0373
Kawasaki Vulcan 700/750 & 800 (85 - 04) ♦	2457
Kawasaki 750 Air-cooled Fours (80 - 91)	0574
Kawasaki ZR550 & 750 Zephyr Fours (90 - 97) ♦	3382
Kawasaki Z750 & Z1000 (03 - 08) ♦	4762
Kawasaki ZX750 (Ninja ZX-7 & ZXR750) Fours (89 - 96) ♦	2054
Kawasaki Ninja ZX-7R & ZX-9R (94 - 04) ♦	3721
Kawasaki 900 & 1000 Fours (73 - 77)	0222
Kawasaki ZX900, 1000 & 1100 Liquid-cooled Fours (83 - 97) ♦	1681
KTM EXC Enduro & SX Motocross (00 - 07) ♦	4629
MOTO GUZZI 750, 850 & 1000 V-Twins (74 - 78)	0339
MZ ETZ Models (81 - 95) ◊	1680
NORTON 500, 600, 650 & 750 Twins (57 - 70)	0187
Norton Commando (68 - 77)	0125
PEUGEOT Speedfight, Trekker & Vivacity Scooters (96 - 08) ♦	3920
PIAGGIO (Vespa) Scooters (91 - 06) ♦	3492
SUZUKI GT, ZR & TS50 (77 - 90)	0799
Suzuki TS50X (84 - 00)	1599
Suzuki 100, 125, 185 & 250 Air-cooled Trail bikes (79 - 89)	0797
Suzuki GP100 & 125 Singles (78 - 93) ◊	0576
Suzuki GS, GN, GZ & DR125 Singles (82 - 05) ◊	0888
Suzuki GSX-R600/750 (06 - 09) ♦	4790
Suzuki 250 & 350 Twins (68 - 78)	0120
Suzuki GT250X7, GT200X5 & SB200 Twins (78 - 83) ◊	0469
Suzuki GS/GSX250, 400 & 450 Twins (79 - 85)	0736
Suzuki GS500 Twin (89 - 06) ♦	3238
Suzuki GS550 (77 - 82) & GS750 Fours (76 - 79)	0363
Suzuki GS/GSX550 4-valve Fours (83 - 88)	1133
Suzuki SV650 & SV650S (99 - 08) ♦	3912
Suzuki GSX-R600 & 750 (96 - 00) ♦	3553
Suzuki GSX-R600 (01 - 03), GSX-R750 (00 - 03) & GSX-R1000 (01 - 02) ♦	3986
Suzuki GSX-R600/750 (04 - 05) & GSX-R1000 (03 - 06) ♦	4382
Suzuki GSF600, 650 & 1200 Bandit Fours (95 - 06) ♦	3367
Suzuki Intruder, Marauder, Volusia & Boulevard (85 - 06) ♦	2618
Suzuki GS850 Fours (78 - 88)	0536
Suzuki GS1000 Four (77 - 79)	0484
Suzuki GSX-R750, GSX-R1100 (85 - 92), GSX600F, GSX750F, GSX1100F (Katana) Fours ♦	2055
Suzuki GSX600/750F & GSX750 (98 - 02) ♦	3987
Suzuki GS/GSX1000, 1100 & 1150 4-valve Fours (79 - 88)	0737
Suzuki TL1000S/R & DL1000 V-Strom (97 - 04) ♦	4083
Suzuki GSF650/1250 (05 - 09) ♦	4798
Suzuki GSX1300R Hayabusa (99 - 04) ♦	4184
Suzuki GSX1400 (02 - 07) ♦	4758
TRIUMPH Tiger Cub & Terrier (52 - 68)	0414
Triumph 350 & 500 Unit Twins (58 - 73)	0137
Triumph Pre-Unit Twins (47 - 62)	0251
Triumph 650 & 750 2-valve Unit Twins (63 - 83)	0122
Triumph Trident & BSA Rocket 3 (69 - 75)	0136
Triumph Bonneville (01 - 07)	4364
Triumph Daytona, Speed Triple, Sprint & Tiger (97 - 05) ♦	3755
Triumph Triples and Fours (carburettor engines) (91 - 04) ♦	2162
VESPA P/PX125, 150 & 200 Scooters (78 - 06)	0707
Vespa Scooters (59 - 78)	0126
YAMAHA DT50 & 80 Trail Bikes (78 - 95) ◊	0800
Yamaha T50 & 80 Townmate (83 - 95) ◊	1247

Title	Book No
Yamaha YB100 Singles (73 - 91) ◊	0474
Yamaha RS/RXS100 & 125 Singles (74 - 95) ◊	0331
Yamaha RD & DT125LC (82 - 95) ◊	0887
Yamaha TZR125 (87 - 93) & DT125R (88 - 07) ◊	1655
Yamaha TY50, 80, 125 & 175 (74 - 84) ◊	0464
Yamaha XT & SR125 (82 - 03) ◊	1021
Yamaha YBR125	4797
Yamaha Trail Bikes (81 - 00)	2350
Yamaha 2-stroke Motocross Bikes 1986 - 2006	2662
Yamaha YZ & WR 4-stroke Motocross Bikes (98 - 08)	2689
Yamaha 250 & 350 Twins (70 - 79)	0040
Yamaha XS250, 360 & 400 sohc Twins (75 - 84)	0378
Yamaha RD250 & 350LC Twins (80 - 82)	0803
Yamaha RD350 YPVS Twins (83 - 95)	1158
Yamaha RD400 Twin (75 - 79)	0333
Yamaha XT, TT & SR500 Singles (75 - 83)	0342
Yamaha XV550 Vision V-Twins (82 - 85)	0821
Yamaha FJ, FZ, XJ & YX600 Radian (84 - 92)	2100
Yamaha XJ600S (Diversion, Seca II) & XJ600N Fours (92 - 03) ♦	2145
Yamaha YZF600R Thundercat & FZS600 Fazer (96 - 03) ♦	3702
Yamaha FZ-6 Fazer (04 - 07) ♦	4751
Yamaha YZF-R6 (99 - 02) ♦	3900
Yamaha YZF-R6 (03 - 05) ♦	4601
Yamaha 650 Twins (70 - 83)	0341
Yamaha XJ650 & 750 Fours (80 - 84)	0738
Yamaha XS750 & 850 Triples (76 - 85)	0340
Yamaha TDM850, TRX850 & XTZ750 (89 - 99) ◊ ♦	3540
Yamaha YZF750R & YZF1000R Thunderace (93 - 00) ♦	3720
Yamaha FZR600, 750 & 1000 Fours (87 - 96) ♦	2056
Yamaha XV (Virago) V-Twins (81 - 03) ♦	0802
Yamaha XVS650 & 1100 Drag Star/V-Star (97 - 05) ♦	4195
Yamaha XJ900F Fours (83 - 94) ♦	3239
Yamaha XJ900S Diversion (94 - 01) ♦	3739
Yamaha YZF-R1 (98 - 03) ♦	3754
Yamaha YZF-R1 (04 - 06) ♦	4605
Yamaha FZS1000 Fazer (01 - 05) ♦	4287
Yamaha FJ1100 & 1200 Fours (84 - 96) ♦	2057
Yamaha XJR1200 & 1300 (95 - 06) ♦	3981
Yamaha V-Max (85 - 03) ♦	4072
ATVs	
Honda ATC70, 90, 110, 185 & 200 (71 - 85)	0565
Honda Rancher, Recon & TRX250EX ATVs	2553
Honda TRX300 Shaft Drive ATVs (88 - 00)	2125
Honda Foreman (95 - 07)	2465
Honda TRX300EX, TRX400EX & TRX450R/ER ATVs (93 - 06)	2318
Kawasaki Bayou 220/250/300 & Prairie 300 ATVs (86 - 03)	2351
Polaris ATVs (85 - 97)	2302
Polaris ATVs (98 - 06)	2508
Yamaha YFS200 Blaster ATV (88 - 06)	2317
Yamaha YFB250 Timberwolf ATVs (92 - 00)	2217
Yamaha YFM350 & YFM400 (ER and Big Bear) ATVs (87 - 03)	2126
Yamaha Banshee and Warrior ATVs (87 - 03)	2314
Yamaha Kodiak and Grizzly ATVs (93 - 05)	2567
ATV Basics	10450
TECHBOOK SERIES	
Twist and Go (automatic transmission) Scooters Service and Repair Manual	4082
Motorcycle Basics TechBook (2nd Edition)	3515
Motorcycle Electrical TechBook (3rd Edition)	3471
Motorcycle Fuel Systems TechBook	3514
Motorcycle Maintenance TechBook	4071
Motorcycle Modifying	4272
Motorcycle Workshop Practice TechBook (2nd Edition)	3470

◊ = not available in the USA   ♦ = Superbike

The manuals on this page are available through good motorcycle dealers and accessory shops.
In case of difficulty, contact: **Haynes Publishing**
(UK) +44 1963 442030   (USA) +1 805 498 6703
(SV) +46 18 124016
(Australia/New Zealand) +61 3 9763 8100

Preserving Our Motoring Heritage

The Model J Duesenberg Derham Tourster. Only eight of these magnificent cars were ever built – this is the only example to be found outside the United States of America

Almost every car you've ever loved, loathed or desired is gathered under one roof at the Haynes Motor Museum. Over 300 immaculately presented cars and motorbikes represent every aspect of our motoring heritage, from elegant reminders of bygone days, such as the superb Model J Duesenberg to curiosities like the bug-eyed BMW Isetta. There are also many old friends and flames. Perhaps you remember the 1959 Ford Popular that you did your courting in? The magnificent 'Red Collection' is a spectacle of classic sports cars including AC, Alfa Romeo, Austin Healey, Ferrari, Lamborghini, Maserati, MG, Riley, Porsche and Triumph.

A Perfect Day Out

Each and every vehicle at the Haynes Motor Museum has played its part in the history and culture of Motoring. Today, they make a wonderful spectacle and a great day out for all the family. Bring the kids, bring Mum and Dad, but above all bring your camera to capture those golden memories for ever. You will also find an impressive array of motoring memorabilia, a comfortable 70 seat video cinema and one of the most extensive transport book shops in Britain. The Pit Stop Cafe serves everything from a cup of tea to wholesome, home-made meals or, if you prefer, you can enjoy the large picnic area nestled in the beautiful rural surroundings of Somerset.

John Haynes O.B.E., Founder and Chairman of the museum at the wheel of a Haynes Light 12.

The 1936 490cc sohc-engined International Norton – well known for its racing success

The Museum is situated on the A359 Yeovil to Frome road at Sparkford, just off the A303 in Somerset. It is about 40 miles south of Bristol, and 25 minutes drive from the M5 intersection at Taunton.
Open 9.30am - 5.30pm (10.00am - 4.00pm Winter) 7 days a week, *except Christmas Day, Boxing Day and New Years Day*
Special rates available for schools, coach parties and outings Charitable Trust No. 292048